ELEPHANT
MANAGEMENT

Contributing Authors

Brandon Anthony, Graham Avery, Dave Balfour,
Jon Barnes, Roy Bengis, Henk Bertschinger,
Harry C Biggs, James Blignaut, André Boshoff,
Jane Carruthers, Guy Castley, Tony Conway,
Warwick Davies-Mostert, Yolande de Beer,
Willem F de Boer, Martin de Wit, Audrey Delsink,
Saliem Fakir, Sam Ferreira, Andre Ganswindt,
Marion Garaï, Angela Gaylard, Katie Gough,
C C (Rina) Grant, Douw G Grobler, Rob Guldemond,
Peter Hartley, Michelle Henley, Markus Hofmeyr,
Lisa Hopkinson, Tim Jackson, Jessi Junker,
Graham I H Kerley, Hanno Killian, Jay Kirkpatrick,
Laurence Kruger, Marietjie Landman, Keith Lindsay,
Rob Little, H P P (Hennie) Lötter, Robin L Mackey,
Hector Magome, Johan H Malan, Wayne Matthews,
Kathleen G Mennell, Pieter Olivier, Theresia Ott,
Norman Owen-Smith, Bruce Page,
Mike Peel, Michele Pickover, Mogobe Ramose,
Jeremy Ridl, Robert J Scholes, Rob Slotow, Izak Smit,
Morgan Trimble, Wayne Twine, Rudi van Aarde,
JJ van Altena, Marius van Staden, Ian Whyte

ELEPHANT
MANAGEMENT

A Scientific Assessment for South Africa

Edited by R J Scholes and K G Mennell

WITS UNIVERSITY PRESS

Wits University Press
1 Jan Smuts Avenue
Johannesburg
2001
South Africa
http://witspress.wits.ac.za

ISBN 978 1 86814 479 2

Cover photograph by Donald Cook at stock.xchng
Cover design, layout and design by Acumen Publishing Solutions, Johannesburg
Printed and bound by Creda Communications, Cape Town

FOREWORD

SOUTH AFRICA and its people are blessed with diverse and thriving wildlife. We are also a developing economy with a growing population. From these facts emerges the particular situation of having most of our protected areas surrounded by land that has been transformed, to a greater or lesser extent, by human development. Large mammals, such as elephants, no longer roam the entire landscape, and their populations are no longer completely governed by the laws of nature. Protecting elephants and the ecological systems in which they exist in a practical and sustainable way that balances the needs of humans, elephants and the environment is a challenge to which I am committed.

This Assessment was undertaken to reduce the degree of scientific uncertainty associated with decisions that must be made very soon and in the medium-to-long term. It helps to evaluate the costs and benefits associated with each choice, both in economic and ecological terms, and clarifies the legal framework within which they must be made. Collectively the chapters in this report reveal the many successes our country's experts, in collaboration with their peers in neighbouring countries and abroad, have achieved in understanding elephants and their needs, in fields as diverse as veterinary science, ecology, animal behaviour, population and resource modelling. Importantly, the Assessment exposes important gaps in our understanding and thus outlines necessary future avenues of research. This Assessment represents a key milestone in an ongoing Elephant Research Programme.

Science does not provide all the information required to resolve the difficult issues raised by the management of elephant in a changing and human-dominated world. Many of the required decisions have a strong element of human values implicit in them. How do South Africans wish to treat the other species with which they share our land? Extensive consultation and careful consideration of the values expressed by a wide range of stakeholders is also an essential part of the process of managing elephant in a democratic country. I am grateful to the many experts and interested persons who invested their time, experience and intellect to deliver this Assessment. I look forward to their continued engagement on the issue of elephant management, which is of great interest to many.

Marthinus van Schalkwyk
Minister of Environmental Affairs and Tourism, 2008

CONTENTS

LIST OF FIGURES

LIST OF TABLES

LIST OF BOXES

ABOUT THE AUTHORS AND CONTRIBUTORS

Brandon Anthony is an assistant professor at the Department of Environmental Sciences & Policy, Central European University. His interests are in human-wildlife conflicts and the social impacts of protected areas, particularly in emerging democracies and developing countries.

Graham Avery is archaeozoologist at Iziko South African Museum, with expertise in the study of ancient and modern bones. He has been involved in ground-breaking long-term studies on beached seabirds, as well as the excavation of brown hyaena dens, and jackal and eagle prey. He has been Chairman of the Southern African Association of Archaeologists and President of the Wildlife and Environment Society of South Africa.

Dave Balfour has 15 years' experience in scientific services, working with protected area managers, and has worked on elephant-related research and management in a number of reserves in South Africa. He is a member of the IUICN African Elephant specialist group and currently heads up the Scientific Services of the Eastern Cape Parks.

Jonathan Barnes is based in Namibia, specialising in environmental and resource economics. His professional experience has embraced economic and policy analysis, land use and development planning, and natural resources assessment, mainly concerning wildlife, fisheries, forestry, rangelands, and agriculture.

Roy Gordon Bengis is a wildlife veterinarian and disease epidemiologist. He has been employed as a State Veterinarian in the Kruger National Park for the past 30 years. During this period he has had opportunities to observe and research various indigenous infections and diseases in Kruger's free-ranging wildlife populations. He was also intensively involved in Veterinary Public Health aspects of lethal population management operations in the Park.

Henk Bertschinger is emeritus professor of the University of Pretoria's Faculty of Veterinary Science. He is a veterinary reproductive specialist (European College of Animal Reproduction) and in recent years has devoted his research efforts to control of reproduction and aggressive behaviour in wildlife species such as lion, cheetah, African wild dog, African elephant, and a number of primate species.

Dr Harry Biggs qualified originally as a veterinarian, but has also worked as an epidemiologist, database administrator, and biometrist, later moving into systems ecology as a programme manager. Today his main interests are in adaptive management and governance as it relates to freshwater biodiversity outcomes and ecosystem services. He is actively involved in mentoring. He has spent his career in Namibia and Kruger National Park, but now works across South Africa

James N Blignaut is an ecological economist attached to the University of Pretoria, South Africa, and director of three companies aimed at economic development and the restoration of natural capital. He focuses primarily on environmental fiscal reform, natural resource accounting, payments for ecosystem goods and services, combating invasive alien plant species, restoring natural capital, and local economic development.

André Boshoff is a Research Fellow at the Centre for African Conservation Ecology, Nelson Mandela Metropolitan University. His current research focus is on the historical and present distributions of the larger mammals in southern Africa.

Jane Carruthers, Associate Professor in the Department of History at the University of South Africa, has published widely in the fields of environmental history and the history of science. She is a Fellow of the Royal Society of South Africa and of Clare Hall, Cambridge, and has held visiting Fellowships at the Australian National University and the University of Western Australia. She is President of the South African Historical Society.

Guy Castley is a terrestrial ecologist with an interest in large mammal ecology and conservation. He spent nine years at South African National Parks as the principal animal ecologist responsible for monitoring wildlife populations in all national parks excluding the Kruger National Park. He currently lectures in Ecology, Wildlife Management and Conservation Biology at Griffith University in Australia.

Tony Conway served in Zimbabwe's Department of National Parks and Wildlife Management for 12 years, working in the Zambezi Valley, Hwange National Park and Chirisa (Sebungwe) areas. He has worked for the Natal Parks Board/ Ezemvelo KZN Wildlife since 1983, and is currently the EKZNW Conservation Manager for the iSimangaliso Wetland Park, World Heritage Site.

Warwick Davies-Mostert has been involved in biodiversity conservation for 14 years and the management of large conservation areas within the De Beers Group and Ecology Division since 1995. Recently, he has been involved with the expansion of land under recognised biodiversity conservation management within the Shashe-Limpopo TFCA in order to ensure that keystone species which have large spatial needs can benefit.

Yolandi de Beer completed Honours and Master's degrees at the University of Pretoria. Her research focused on the spatial movements of elephants with regard to environmental variables such as water distribution and landscape heterogeneity in arid environments.

Willem F (Fred) de Boer is a lecturer at Wageningen University, The Netherlands. His main interest is in species diversity and species interactions of large herbivores, with emphasis on plant-herbivore interactions, competition and facilitation. He is coordinator of the Tembo integrated programme, aimed at predicting elephant spatio-temporal distribution.

Audrey Delsink is a freelance elephant ecologist and resident research ecologist at Makalali Private Game Reserve. She is a Ph.D. student of the University of KwaZulu-Natal's Amarula Elephant Research Programme. As an expert on elephant immunocontraception, she has been involved in this ground-breaking study for 10 years. As a consultant, she works with protected area managers in a number of reserves in South Africa.

Martin de Wit is an applied environmental and natural resource economist and specialises in managing the risks of and finding solutions to environmental problems. He is an extraordinary associate professor at the University of Stellenbosch and an associate researcher at the Stellenbosch-based Sustainability Institute.

Saliem Fakir is senior lecturer at the Department of Public Administration and Planning and Associate Director of the Centre for Renewable and Sustainable Energy Studies at the University of Stellenbosch. Director of the World Conservation Union South Africa (IUCN–SA) office for eight years, between 2004 and 2005 he served as a chair of the Board of the National Botanical Institute, and is now a non-executive member.

Sam M Ferreira's research focuses on mammal and bird conservation biology, with an emphasis on temporal dynamics and the factors influencing these. He also has a keen interest in restoration ecology and the application of theoretical ecology to address ecological problems. He currently coordinates aspects of the Elephant Programme at CERU.

Andre Ganswindt, University of Pretoria. He received his Ph.D. in Biology at the University of Muenster for his work on endocrine, physical and behavioural correlates of musth in African elephants. His current interest is reproductive biology in mammals, with a special focus on male elephant reproductive strategies.

Marion E Garaï has a Ph.D. in animal behaviour from the University of Pretoria. She has been a member of the IUCN/African Elephant Specialist Group for the past 12 years. She founded the EMOA and was its Chairperson for 12 years. She is currently Chairperson of the Space for Elephants Foundation.

Angela Gaylard has focused on terrestrial conservation issues, with a decade of work in the Kruger National Park. Her doctorate research advocates the use of a scaled, spatially explicit approach to understanding and managing elephants, focusing on the relationship between surface water distribution and elephant impacts.

Katie Gough is a Ph.D. student at the Centre for African Conservation Ecology, Nelson Mandela Metropolitan University. Her research focuses on the elephants of Addo Elephant National Park and includes aspects of the population's demography, behavioural ecology and conservation management.

C C (Rina) Grant qualified as a veterinarian but is now the Programme Manager: Systems Ecology in SANParks' Scientific Services, based at Skukuza, Kruger National Park. One of her main study areas in systems ecology is the impact of elephants on biodiversity and to determine when management is required, and what type of management would be most appropriate.

Douw Grobler has worked for the past 20 years in the wildlife industry, 13 years in the Kruger National Park, and now as a private wildlife veterinarian and consultant mainly dealing with specialised wildlife capture, especially elephants; introducing new wildlife species into game sanctuaries and parks; managing breeding programmes of sable and roan antelope, disease-free buffalo projects, and research projects such as elephant contraception and the enhancement of anaesthesia on a variety of game species.

Robert Guldemond conducted his Ph.D. at the Conservation Ecology Research Unit at the University of Pretoria. His research focused on the impact elephants have on their environment and selected aspects of elephants' space and habitat use. He currently coordinates aspects of the Biodiversity and Restoration Programme at CERU.

Peter Hartley joined the Natal Parks Board as a ranger and was the officer in charge (Conservator) of Umfolozi Game Reserve. He now heads the

Conservation Compliance Unit at iSimangaliso Wetland Park Authority. He has been involved in elephant management and control since 1990, specialising in anti-poaching.

Michelle Henley is a research ecologist and programme co-ordinator for the Transboundary Elephant Research Programme under the auspices of Save the Elephants. Her research focuses on monitoring elephant movements, their social interactions and habitat selection.

Markus Hofmeyr received his BVSc from the University of Pretoria in 1994. He has a well-established history in game capture, care and holding facilities, and in veterinary conservation programmes throughout South and southern Africa. He is currently the head of the SANParks Veterinary Wildlife Services.

Lisa Hopkinson is Head of the Corporate Legal Services Division of South African National Parks and has a specialist interest in wildlife and conservation law. Most of her time is devoted to the protection and development of government as well as private interests in all matters related to biodiversity conservation, protected area establishment, expansion and development, land use planning and development, as well as tourism development, to the extent that this relates to biodiversity conservation.

Tim P Jackson has 18 years of research experience in mammalogy. His present research with CERU aims at the conservation and management of elephants throughout southern Africa.

Jessica Junker is a postgraduate student at the University of Pretoria and is currently conducting her MSc project under the auspices of CERU. Her research focuses on the temporal population dynamics and conservation of large mammals.

Graham Kerley is Director of the Centre for African Conservation Ecology and Professor of Zoology at Nelson Mandela Metropolitan University, and also serves as a Director of the Eastern Cape Parks Board, and as a member of the Minister's Science Round Table on Elephant Management. His research interests focus on animal-plant interactions and conservation biology, and he has been working on elephant impacts and biology for over a decade.

Hanno Killian is the Research Ecologist at the Welgevonden Private Game Reserve in the Waterberg of Limpopo. As senior member on the reserve's management team he is very much involved in practical elephant management and heads their contraception programme. He also serves on the committee of the Elephant Managers and Owners Association.

Jay Kirkpatrick is Director of the Science and Conservation Center at ZooMontana, in Billings, Montana. For the past 34 years he has carried out research on fertility control for wild horses and other wildlife, for the purpose of developing non-lethal and humane methods of controlling wildlife populations, and on non-capture methods for studying reproduction in free-ranging wildlife species through the use of urinary and fecal steroid hormones.

Laurence Kruger is director of the Organization for Tropical Studies, South Africa, whose 'study abroad' programme is based in the Kruger National Park. His key interest in ecology is how plant species and communities respond to disturbance in the form of fire, herbivores, and the interaction between the two.

Marietjie Landman is a Ph.D. student within the Centre for African Conservation Ecology, Nelson Mandela Metropolitan University. Her research interests focus mainly on the ecology of large herbivores, their interactions with vegetation, and the factors that determine their foraging and habitat use.

Keith Lindsay joined the Amboseli Elephant Research Project in Kenya in 1977. His continuing research focuses on feeding ecology, habitat interactions, demography and ecosystem change.

Dr Rob Little is Conservation Director at WWF South Africa. He is leader of the WWF-SA Species Programme and a member of the global WWF Species Working Group. He is also the WWF-SA representative on the IUCN Species Survival Commission. Since October 2006, he has been Chairman for the Elephant Management & Owners Association (EMOA).

H P P (Hennie) Lötter is a professor of Philosophy at the University of Johannesburg. He specialises in political philosophy and environmental ethics.

Robin L Mackey is a lecturer in the School of Biological and Conservation Sciences at the University of KwaZulu-Natal. While recently she has focused on elephant population growth and management and other conservation studies, she is best known for her theoretical work on species diversity-disturbance relationships.

Johan Malan is an Operations Manager at Veterinary Wildlife Services with South African National Parks, based in the Kruger National Park. He has spent

20 years in the game capture unit and specialises in wildlife capture, boma training and the transportation of various species, as well as the development of capture and transport equipment and the breeding of disease-free buffalo.

Wayne Matthews is the Ezemvelo KwaZulu-Natal Wildlife Regional Ecologist for Maputaland. Based in the Tembe Elephant Park, he is currently involved in planning the Futi-Tembe Transfrontier Area that will straddle the South Africa/Mozambique border. His research has contributed to an understanding of several rare vegetation types and their associated ecological factors in the Maputaland Centre of Plant Endemism.

Kathleen G Mennell obtained her Honours in Ecology and Conservation from the University of the Witwatersrand. She is currently a Masters student within the Council for Scientific and Industrial Research.

Pieter Olivier is currently a master's student at the Conservation Ecology Research Unit in the Department of Zoology and Entomology, University of Pretoria. His research focuses on metapopulation dynamics in large mammals with the emphasis on the restoration of spatial axis to regain spatial-temporal dynamics to withstand extinctions and overcome local impacts on other species.

Theresia Ott studied at the Nelson Mandela Metropolitan University and completed a master's study at the University of Pretoria. She is a Ph.D. student at CERU, studying elephant movements in a landscape ecology context.

Norman Owen-Smith is Research Professor in African Ecology at the University of the Witwatersrand. He obtained his Ph.D. for a study on the behavioural ecology of the white rhinoceros conducted in the Hluhluwe-iMfolozi Park in South Africa. His book, *Megaherbivores*, generalises his findings for other very large mammals including elephants. He has supervised studies on the vegetation impacts of elephants in Kruger Park and northern Botswana.

Bruce Page has worked on the relationship between elephants and their habitats over the past 32 years in many localities in southern Africa. His primary research interest is an understanding of how elephant populations are regulated and how they influence the composition and dynamics of the systems in which they occur.

Mike J S Peel is Senior Researcher at the Agricultural Research Council – Livestock Business Division (Range and Forage). He is project leader of the Savanna Ecosystem Dynamics Project, which investigates the potential of the natural resources of the Savanna Biome to contribute to the economy and development of the southern African sub-region.

Michele Pickover is Curator of Manuscripts: Historical Papers at the library of the University of the Witwatersrand. She is also a Board Member for the South African History Archives and Chairperson of the National Committee of the South African Society of Archivists. She is the author of the book *Animal Rights in South Africa* and a trustee and spokesperson of the organisation Animal Rights Africa.

Mogobe Ramose is Professor of Philosophy in the University of South Africa. He is currently Director of the University of South Africa Regional Learning Center in Ethiopia. His publications include the book, *African Philosophy through Ubuntu.*

Jeremy Ridl is an attorney and environmental law specialist having founded the first specialist environmental law practice in South Africa in 1990. He served as an associate professor and director of the Institute of Environmental Law at the University of KwaZulu-Natal before returning to full-time practice in 2006. Much of his professional time is devoted to environmental activism and the protection of environmental rights in rural communities.

Robert J (Bob) Scholes is a systems ecologist employed by the South African Council for Scientific and Industrial Research. He is an expert on the ecology of African savannas, and has been involved in several international scientific assessments, including the Millennium Ecosystem Assessment. He is a member of the South African National Parks Board.

Rob Slotow is Director of the Amarula Elephant Research Programme at the University of KwaZulu-Natal, Durban, which aims to contribute to conservation of the African elephant through research directed towards management of elephants in wild areas in South Africa and beyond.

Izak Smit is currently employed by the South African National Parks as Research Manager: GIS and Remote Sensing. His main interests revolve around the use of GIS and satellite remote sensing for detecting spatio-temporal patterns which may be of relevance to the effective management of conservation areas.

Morgan Trimble is an MSc student with CERU at the University of Pretoria. Her research interests include large mammal population ecology and conservation.

Wayne Twine is an ecologist who does research at the human-environment interface in rural areas of the former homelands of South Africa. He is based at the Wits Rural Facility in the central lowveld, where he manages a research programme for the School of Animal, Plant and Environmental Sciences, University of the Witwatersrand.

Rudi J van Aarde, Director of the Conservation Ecology Research Unit (CERU), focuses on the restoration of populations and communities as a contribution to conservation. At present, CERU's research on elephants covers populations in Botswana, Malawi, Mozambique, Namibia, South Africa, Zambia and Zimbabwe.

J J van Altena served for seven years in the Kruger National Park during which he was involved with the game capture department and elephant contraception research project. Presently a partner in Catchco Africa, he continues to perform the capture, introduction, and management of wildlife species throughout southern Africa and the immunocontraception of African elephants in many private reserves.

Marius van Staden was admitted as an attorney in 1991. He takes a keen interest in nature conservation and environmental law. His client base includes South African National Parks. One of the environmental law matters he was involved in was decided by the Constitutional Court.

Ian Whyte was involved in the research environment in Kruger National Park for 37 years of which 24 were as co-ordinator of elephant research. He has been a member of the IUCN's Species Survival Commission – African Elephant Specialist Group since 1992.

LIST OF REVIEWERS

Special acknowledgement and thanks to David Cumming, Holly Dublin and Brian Huntley who each undertook the important role of Review Editor for this Assessment. They supervised the extensive review process and saw that it was conducted in a balanced manner by ensuring that all comments generated were responded to in a consistent and justified manner.

Acknowledgement and thanks to the following people who contributed to this Assessment through the peer and stakeholder review processes and the review of the Summary for Policymakers:

Arson, James
Beinart, William
Botha, Pieter
Bradshaw, Gay
Brockett, Bruce
Bell-Leask, Jason
Cameron, Elissa
Clegg, Bruce
Crookes, Douglas
du Toit, Johan
Ebedes, Hym
Eckhardt, Holger
Ferrar, Tony
Freitag-Ronaldson, Stefanie
Friedman, Yolan
Garaï, Marion
Getz, Wanye
Gillson, Lindsey
Glazewski, Jan
Govender, Navashni
Grossman, David
Guldemond, Robert
Hanks, John
Henley, Steve
Joubert, Salomon
Knight, Mike
Kruger, Judith

Lindsey, Keith
Maas, Barbara
MacFayden, Sandra
Mackenzie, Bruce
Moore, Kevin
Nel, Pieter
Neluvhalani, Edgar
Novellie, Peter
Payne, Katy
Pienaar, Danie
Pitts, Neville
Pimm, Stuart
Pretorius, Yolanda
Prins, Herbet
Rogers, Kevin
Roux, Dirk
Shannon, Graeme
Sharp, Robin
Sholto-Douglas, Angus
Shrader, Adrian
Symonds, Alexis
Trollope, Winston
Van Houven, Wouter
Venter, Freek
Von Maltitz, Graham
Voster, Shaun

ACRONYMS AND ABBREVIATIONS

AENP	Addo Elephant National Park
AFESG	African Elephant Specialist Group
AM	Adaptive Management
APNR	Associated Private Nature Reserves
ASF	African swine fever
ASIS	Assateague Island National Seashore
BCFR	Budongo Central Forest Reserve
BNR	Balule Nature Reserve
BP	Years before present. As a convention, 1950 is the year from which BP dates are calculated
BTB	Bovine tuberculosis
BWMA	Botswana Wildlife Management Association
CBNRM	Community-based natural resource management
CITES	The Convention on International Trade in Endangered Species of Wild Fauna and Flora
CONNEPP	Consultative National Environmental Policy Process
CoP	Conference of Parties
CWI	Care for the Wild International
DEAT	Department of Environmental Affairs and Tourism
EE2	Ethinyl oestradiol
EIA	Environmental impact assessment
EKZNW	Ezemvelo KwaZulu Natal Wildlife
EL	Environmental loading
ENSO	El Niño-Southern-Oscillation
ESSA	Ethics Society of South Africa
EZP	elephant zona pellucida
FMA	Freund's Modified Adjuvant
FMD	Foot-and-mouth disease
FSH	Follicle stimulating hormone
GLTP	Great Limpopo Transfrontier Park
GMPGR	Greater Makalali Private Game Reserve
GNI	Gross national income
GnRH	Gonadotropin Releasing Hormone
HEC	Human–elephant conflict
ICI	Intercalving interval
IPCC	Intergovernmental Panel on Climate Change
IUCN	International Union for Conservation of Nature and Natural Resources, or World Conservation Union
KLH	keyhole limpet haemocyanin
KNP	Kruger National Park
KY	Thousands of years ago

LFI	Landscape Functionality Index
LH	Luteinising hormone
LSTCA	Limpopo–Shashe Transfrontier Conservation Area
LTCA	Lubombo Transfrontier Conservation Area
MA	Millennium Ecosystem Assessment
MFPN	Murchison Falls National Park North
MFPS	Murchison Falls National Park South
MK	Mokomasi Game Reserve
MTP	D, L-lactide
MY	Millions of years ago
N&S	Norms and Standards
NDVI	Normalised Difference Vegetation Index
NEMA	National Environmental Management Act of 1998
NEMBA	National Environmental Management: Biodiversity Act 10 of 2004
NEMPAA	National Environmental Management: Protected Areas Act 57 of 2003
NGO	Non-governmental Organisation
NNI	Net national income
NP	National Park
OH&S	Occupational health and safety
OIE	World Organisation for Animal Health
PES	Payments for ecosystem goods and services
PTSD	Post-traumatic Stress Disorder
pZP	Porcine zona pellucida
SANBI	South African National Biodiversity Institute
SANParks	South African National Parks
SRT	Scientific Round Table
SSC	The IUCN Species Survival Commission
TEV	Total Economic Value
TFCA	Transfrontier Conservation Area
TNP	Tsavo National Park
TP	Testosterone propionate
TPC	Thresholds of Potential Concern
TPNR	Timbavati Private Nature Reserve
VOC	Dutch East India Company
WTP	Willingness to pay

PREFACE

Kathleen G Mennell and Robert J Scholes

AS A CONSEQUENCE of the rising number of elephants in protected areas[1] in South Africa, the ecosystems that contain elephants and the people that live adjacent to elephant populations are perceived to be coming under increasing threat. The control of elephant populations by culling has been under a moratorium since the mid-1990s. Attempts to resolve differences of opinion between the authorities responsible for elephant management in the country, private elephant owners, animal rights and biodiversity conservation organisations in South Africa and abroad, and representatives of local communities, have to date not led to a widely agreed future course of action. In 2006, the Minister for Environment Affairs and Tourism convened a Science Round Table to advise on the issue. The Round Table recommended that a Scientific Assessment of Elephant Management be undertaken.

This book is the result of that Assessment, undertaken during 2007, on the authority of the Minister. The Assessment is the first activity in a proposed elephant research programme, which aims to reduce the uncertainties regarding the consequences of various elephant management strategies. The purpose of this Assessment is to:

- document what is known, unknown, and disputed on the topic of elephant–ecosystem–human interactions in South Africa
- synthesise and communicate the information in such a way that decision making and the reaching of social consensus is facilitated.

Note that the Assessment itself does not constitute policy at any level, although it is hoped that it is relevant to the process of policy making at all levels, from the individual protected area through provincial, local, national, regional and international policy.

The Assessment of South African Elephant Management focuses on the interactions between elephants, humans and the ecosystems in which they occur and, in particular, on the possible way elephants could be managed based on their ecology, biology and social significance.

The Assessment addresses more-or-less wild elephants of the species *Loxodonta africana*, in South Africa. Some of these elephant populations are shared with neighbouring countries. Elephants in captive environments, as defined by the Norms and Standards (DEAT, 2008), are not discussed – that is, elephants that require intensive human intervention in the form of food, water, artificial housing and veterinary care, and which are kept in an area of less than 2000 ha designed to prevent escape.

The Assessment is largely based on information in the peer-reviewed scientific literature, along with associated datasets and models. The Assessment has drawn on material from outside of South Africa where it is relevant to the task. Where non-peer reviewed studies were deemed important, this Assessment itself constitutes the peer-review process. Cited documents that are not easily accessible (i.e. in the 'grey literature') have been placed in the public domain by submitting a copy to the IUCN Elephant Specialist Group library in Nairobi. The Assessment did not aim to generate new primary knowledge but instead sought to add value to existing information by collating, summarising, interpreting, and communicating it in a form that would be useful to decision makers. An important feature of an assessment like this one is the explicit use of expert judgement to evaluate the state of existing knowledge.

This volume has four main sections and an overarching Summary for policymakers:

- What background information is necessary to understand the situation, and what are the current trends in elephant-containing ecosystems? (Chapters 1–4)
- What tools have been developed to manage the growth of elephant populations? (Chapters 5-8)
- What are the ethical, economic and legal issues regarding elephants and elephant-containing ecosystems? (Chapters 9-11)
- What management systems can assist in the responsible management of elephant-containing ecosystems? (Chapter 12)

The Assessment is driven by the issues underlying the management of elephants and not by representation of all elephant-containing areas in South Africa. Given the long history of the Kruger National Park, case histories, decisions and actions from the park are often employed as examples to illustrate various principles. These examples are well documented and have similar parallels in other parks.

A 14-member Technical Board comprising the lead authors, the Assessment leader and the Assessment coordinator were responsible for driving and directing the process (figure 1).

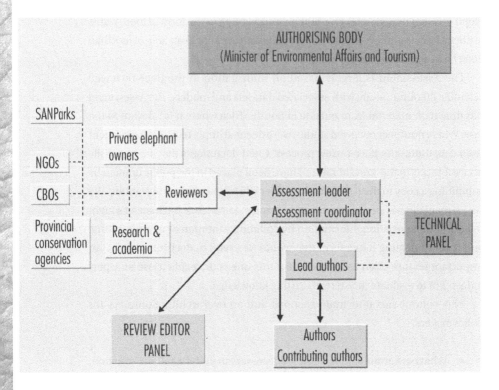

Figure 1: Overview of Assessment role players

Approximately 62 experts were involved as authors and members of the Review Editor Panel. The Assessment underwent two rounds of open review, first by experts who commented on the technical accuracy of the content and secondly by stakeholders who ensured that all issues were addressed adequately and in a balanced fashion (figure 2).

Review comments were received from 73 individuals, of which 21 were submitted by authors of other chapters. Reviewers represented the national and provincial conservation authorities, provincial parks, private managers and owners, conservancies, NGOs, animal welfare groups, academics and individuals involved in the private sector.

By identifying gaps in data and information that prevent policy-relevant questions from being answered, the Assessment can help to guide future elephant research and monitoring that may allow the questions that remain inadequately addressed to be answered in future assessments.

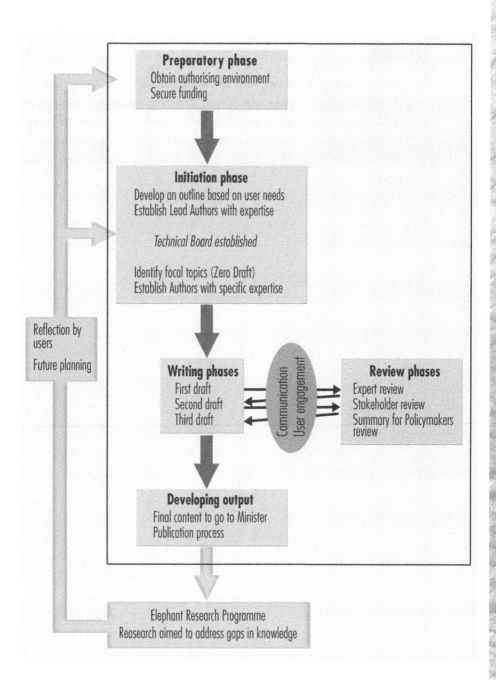

Figure 2: Schematic of the Assessment process and post-Assessment activities. This Assessment is not an isolated process; future research, feedback and suggestions will determine future assessment structure and goals. The iterative review process contributes to the credibility, clarity and balance of the Assessment findings and to the communication with users. The final Assessment findings are communicated to intended users and a wider audience

ENDNOTE

1   The phrase 'protected area' will be used throughout this Assessment as shorthand for areas whose main purpose is the conservation of biodiversity of the legal status or ownership of the land. This includes National Parks, Provincial and Local Government Nature and Game Reserves, and a variety of formal and informal arrangements on private or communally owned land.

SUMMARY FOR POLICYMAKERS

THE NATURE OF THE ISSUE

PRIOR TO European colonisation, elephants occurred virtually everywhere in the area that comprises the modern South Africa, as well as in much of the rest of sub-Saharan Africa. By the beginning of the twentieth century, elephants were in decline over most of their former African range and almost extinct in South Africa. The main causes of the decline were hunting (for ivory, hides, and meat) and loss of habitat, mainly to agriculture. The establishment of protected areas has led to a remarkable recovery in elephant numbers in South Africa, Namibia, Botswana, and Zimbabwe. Elephants remain relatively numerous in Zambia and Mozambique. In most of the rest of Africa, elephant populations are either very low (West Africa), or declined precipitously in the 1970s and 1980s and are now more or less stable (East Africa). The forest-dwelling elephants of Central Africa, almost certainly a different species, continue to decline at an alarming rate. Although the African savanna elephant is not at imminent risk of extinction (figure 1), its population trend has been, and continues to be, of international concern. Actions taken to manage elephant populations in Africa are subject to intense scrutiny and often political pressure. Legal international trade in elephant products is strictly regulated in terms of the Convention on International Trade in Endangered Species of Wild Fauna and Flora (CITES), to which South Africa is a signatory.

This assessment deals exclusively with the management of near-wild populations of the savanna-dwelling African elephant (*Loxodonta africana*) in South Africa. It does not deal with captive elephants. Information on the elephant populations of southern and East Africa is clearly relevant to this Assessment and has been cited, but the social and ecological conditions under which they occur differ significantly from the circumstances in South Africa. The South African situation, where elephant and human distributions are completely spatially separate, is unique.

The elephant population density (i.e. the number of elephants per square kilometre of current elephant range, for a given period of time) has risen in parts of the southern African states listed above to the point where it raises concerns regarding impacts on the environment and people. The key concerns in South Africa are the appearance and ecological functioning of the landscape, the potential impacts on other species of plants and animals, and the livelihoods and safety of people adjacent to the elephant range. There is a vigorous, and often acrimonious, debate as to whether elephant numbers need to be curbed or reduced in South Africa, and if so, how.

Although elephants have been scientifically studied for over half a century, some of the information that could help to guide appropriate decisions is unavailable to the decision-makers, contested by experts, or simply unknown. The Minister for Environmental Affairs and Tourism, who has ultimate responsibility for elephant management within South Africa, convened a Science Round Table to advise him on the issue. One of the recommendations of the Round Table was that an assessment be carried out to gather, evaluate, and present all the relevant information on the topic. This section of the Assessment is a summary of the longer document entitled 'Scientific Assessment of Elephant Management in South Africa'. The assessment process involved 64 experts as chapter authors, and a further 56 persons, including scientists, policymakers, and stakeholders, in the extensive review process.

WHY ELEPHANTS WARRANT SPECIAL MANAGEMENT

There are three main reasons. First, elephants are the largest of the extant land mammals. They are known, along with rhinoceros, hippopotamus and giraffe, as 'megaherbivores' (plant-eaters weighing more than 1000 kg). Elephants are capable of transforming the ecosystems in which they occur in dramatic ways, for instance by debarking or pushing over large trees. Along with large size come the attributes of longevity (up to 60 years) and a relatively slow population growth rate (a long-term rate of up to 7 per cent per annum), which make elephant populations slow to respond to management or changes in resource availability. Because of their large size, low relative metabolic rate and hindgut digestive system, elephants consume a wide range of plant parts, including grass, herbs, tree and shrub foliage, fruit, woody stems, bark, and roots. Further consequences of a large body size are that elephants have few natural predators and a large home range, now substantially constrained.

Second, elephants have a large and complex brain. They are capable of learning and remembering. They experience fear, pain, and (apparently) a sense of loss. They are inferred to be among the more intelligent animals. Third, elephants exhibit complex social behaviour that includes the lifetime persistence of extended family linkages (figure 2).

While none of these attributes are completely unique to elephants, they exhibit them in combination, and to such a degree, that people of many different cultures and backgrounds agree that elephants must be managed with a degree of respect greater than that afforded to most other species of wild animals.

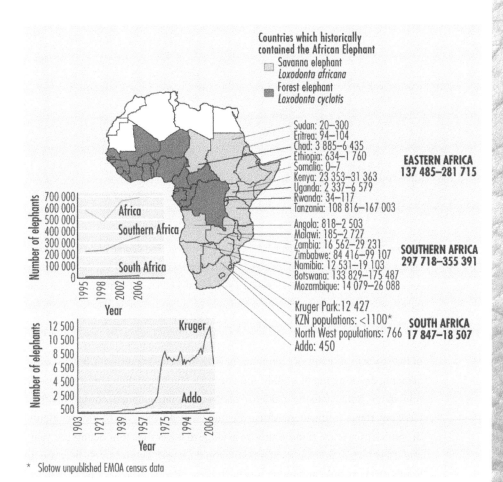

Figure 1: Elephant distribution and population trends in Africa, southern Africa and South Africa. The range in elephant numbers is due to the differences in survey type as defined by Blanc *et al.* (2007), with the first number indicating 'definite' elephant numbers and the second a combination of 'probable', 'possible', and 'speculative'. These categories have decreasing levels of data reliability

HUMAN–ELEPHANT INTERACTIONS

The African elephant and humans both evolved in Africa, where they have a 250 000-year history of cohabitation. For most of that time, humans have been predators of elephants. In the modern period, the interactions between elephants and people take a great variety of forms. Positive interactions include the excitement and awe felt by tourists who look at them. Negative interactions include loss of crops and infrastructure due to elephant damage, infection

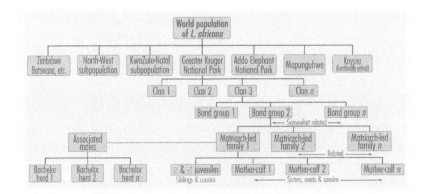

Figure 2: The structure of elephant social organisation. The degree of coherence and importance of the 'bond group' and 'clan' levels remains a matter of disagreement among researchers. The 'subpopulation' level is defined by the ability of the groups to exchange genetic information, and is largely determined by geographical separation. It can be maintained or altered by the translocation of breeding individuals between groups

of livestock as a result of elephants having breached veterinary fences, thus allowing the mingling of wildlife and domestic stock, and direct injury or loss of human life. Relative to the incidence of direct human–elephant conflict in other elephant-range countries in Africa, the frequency and severity of such incidents in South Africa is low (amounting on average to fewer than four deaths per year, and a few tens of thousands of rand of crop damage). This is largely because in South Africa people and elephants have been effectively separated by fences. Most of the incidents involving death or injury of humans take place within the protected areas, or under captive conditions. The levels of conflict may escalate as elephant and human population densities rise further, and if palatable crops are planted and fragile infrastructure is constructed adjacent to elephant-containing areas. Even low levels of human–elephant conflict have a negative effect on people's perception of elephants and conservation, if inappropriately handled.

Human values with respect to elephants cannot be classified into a simple preference for protection or consumptive use. For example, some 'consumptive use' groupings, such as recreational hunters, are highly committed to elephant conservation, and some protection-orientated groups see sustainable use as the key to long-term conservation. Furthermore, attitudes towards elephants are constantly changing, as is the relative power between the various groups of interested parties. There are no definitive surveys in South Africa regarding the size of the stakeholder groups, nor a definitive description of the opinions they hold.

THE 'MORAL STANDING' OF ELEPHANTS

This Assessment considered elephant management from several relevant and documented ethical perspectives, including mainstream 'Western', African traditional and animal rights-centred viewpoints. Human intervention in natural processes in general, and in the lives of elephants in particular, is permissible under defined and restricted conditions in all these ethical frameworks. Under certain circumstances interventions may even be ethically required, when non-intervention has consequences that are ethically unacceptable.

There are scientific reasons to suggest that elephants have a higher degree of sentience than the vast majority of other mammal species. Nevertheless, their capacity for self-consciousness, empathy for other elephants, and problem-solving ability is, on the basis of available information, very much lower than that of humans, and it is humans who define the values framework. Ethically, this suggests that in the tradeoffs between elephants and other species, the needs of elephants might receive a somewhat higher weighting than other species (though not to the point where other species are threatened with extinction), but a lower weighting than the needs of people.

The killing of elephants is defensible in terms of all the above ethical frameworks in the case of imminent danger to human life. A strongly ecologically-orientated ethic would also permit culling where there are strong grounds for believing that the persistence of other species is under threat. A human-centred ethic would permit elephants to be killed if human life or livelihoods were threatened, and in some versions, permits elephants to be killed for human use, including sport. Given the plurality of ethical positions on killing elephants, it is a practical necessity in a participatory democracy such as South Africa for non-lethal options to be seriously considered and found lacking before the lethal option is selected.

The level of self-awareness and empathy exhibited by elephants suggests that they might be considered to have a limited form of a 'right to privacy', in other words, they should be harassed as little as possible. Knowingly causing unnecessary suffering to any sentient organism is unacceptable and forbidden by law. In elephants, there is reasonable cause to suggest that suffering includes emotional stress, for instance through fear based on past experience, or through witnessing harm to other elephants, especially those in the same family group.

CONTROLLING THE DISTRIBUTION OF ELEPHANTS

High levels of elephant impact result from the concentration of animals in specific habitats or areas at particular times of the year, rather than the absolute numbers of elephants. Therefore, methods of altering the distribution of elephants in the landscape are an important way of managing impacts. Fencing is the main current option, though behavioural modification holds some promise. Fences can be used to keep elephants inside protected areas, or keep them out of sensitive locations within the protected area. The effectiveness of elephant fencing varies greatly according to its design and location, and so does its cost. Electrified fences costing R120 000 per kilometre to erect, can almost entirely contain elephants (substantially less than one elephant breakout/ km/y). More expensive mechanical fencing including high impact cable (e.g., Addo's 50-year old 'Armstrong fence', which is estimated to cost R150 000 per kilometre to erect at current prices) can reduce this to virtually no breakouts – one recorded case in 50 years. The minimum legal requirement for electric fencing designed for elephant control costs R34 000 per km to construct, and is anticipated to reduce breakouts to less than 1 per km per year. Ordinary game or livestock fencing has little control value for elephants.

Fences have a maintenance cost over the lifetime of the fence (which is typically several decades, but differs for the type of fence – electric fences have a shorter lifetime and are more expensive to maintain) of 4 to 8 times the initial cost of the fence (expressed in inflation corrected terms). Nevertheless, the cost of constructing and maintaining elephant-restraining fences is lower than the potential damage costs if the fences are not present or ineffective. The damage costs caused by elephant can be direct, in terms of loss of human life, injury, disruption of livelihoods, loss of crops, and damage to property; or indirect, through allowing other damaging or disease-causing animals in or out of the fence breaks caused by elephants. In the South African context, the indirect costs are the main component of damage, and have added up to tens of millions of rand for individual disease epidemics traceable to fence-breaching usually, but not always, caused by elephants. Averaged over the period 2001–2006, the veterinary costs of containing major foot-and-mouth disease outbreaks due to the mixing of wildlife and domestic livestock works out at R28 500 000 per year (in 2007 values).

Research elsewhere in Africa has shown that elephant movements can also be influenced by non-physical barriers (such as chemical repellents, sound or disturbance, referred to as conditional aversion methods), but the control is partial and often temporary.

Elephant distribution can potentially be altered by the manipulation of water availability. Cow herds with calves need to drink every day, and seldom move more than 16 km from surface water. Bull elephants drink less frequently, and range further than cow-and-calf herds. If areas of at least 40 km diameter could be rendered free of surface water for large parts of the year, they would theoretically be only lightly and seasonally used by elephants. The local density in the areas that did have water would be increased as a result, accelerating the transformation of the vegetation there, and hypothetically leading to the onset of elephant density-dependent self-regulation at lower overall densities than would otherwise have been the case. This idea is unproven in practice, and would only be feasible in very large reserves with a sparse natural distribution of water, such as the Mozambican part of the Great Limpopo Transfrontier Conservation Area.

ELEPHANTS AND BIODIVERSITY

African biodiversity has evolved in the presence of elephants for several million years. Elephant are simultaneously an iconic element of biodiversity, and an important agent that shapes the environment, making it more or less suitable for other forms of biodiversity. Therefore it is not simply a question of 'elephants versus biodiversity': elephants are part of biodiversity and biodiversity depends, to some degree, on the presence and abundance of elephants. There are no comprehensive records, at regional scales, of overall biodiversity changes due to presence or absence of elephants. Much of the discussion below is based on reasonable inference from limited studies.

Elephants are said to be a 'keystone species', in other words, a species which is essential for the integrity of the ecosystem. This assertion is difficult to test critically, but is probably true to a degree. While all herbivores have the capacity to change vegetation structure and composition, the effects of feeding and breakage by elephants affect structural components like canopy trees and are greater in magnitude and extent than the effects of most other herbivores, and thus transform landscape features to a greater degree. Recovery time for the woody plant populations affected is longer than it is for grasses.

In certain circumstances high local elephant densities can contribute to the conversion of savanna woodlands into largely treeless grassland or shrubby coppice states (figure 3). Evergreen succulent thicket can be changed to remnant shrub clumps interspersed with grassy patches. The most extreme vegetation transformations have occurred where elephants have attained densities of 2–3 animals/km^2, generally in association with other factors like drought and range compression by humans. Hotter or more frequent fires contribute to the maintenance of open grassy conditions, with a concomitant reduction in shade-loving grasses. The capacity to form a coppice is widespread in African savannas, and is especially common on sandy soils and where mopane trees predominate. The habitat changes resulting from high levels of elephant impact are generally adverse for other plant and animal species, although some species may benefit, especially at low-to-moderate levels of impact. No global species extinctions have yet occurred as a result of the presence of elephants. The local extinction ('extirpation') of some plant species has occurred in succulent thicket. Sensitive and preferred species, such as baobab trees, certain aloes and other species, may be approaching this threshold in savanna protected areas that lack safe refuges, inaccessible to elephants.

Figure 3: These photographs were taken in exactly the same location and direction ten years apart on the clayey soils of the eastern Kruger National Park. Other photo pairs show little change in woody cover, and some show an increase. On average, the cover by tall trees has decreased in the Kruger Park since the 1970s. Not all of this loss can be directly and unequivocally attributed to elephant impacts

Where elephants consume a large fraction of the forage, other herbivores are likely to be reduced in numbers, especially given the ability of elephants to consume both woody plants and grasses. Elephants selectively favour certain woody plant species over others. Since the most severe impacts of elephants on woody vegetation occur during the dry season, the distribution of perennial surface water can restrict the region over which severe vegetation impacts occur. Bull elephants have greater damaging impacts than female elephants on plants. There are reports of elephants killing other animal species, such as rhinoceros. These incidents are usually associated with animals that were translocated when young, and without the presence of adult animals.

Elephants contribute positively to biodiversity by dispersing seeds, opening thickets, making browse more available to smaller herbivores, making water accessible in dry river beds, and promoting nutrient re-cycling. Thus both the absences of elephants from much of their former range, and the overabundance of elephants in the areas to which they are now restricted, have consequences for the appearance and function of the landscapes and the variety and proportions of species found there. The biodiversity consequences depend not only on the local severity of the impacts, but also on their spatial extent and the period over which they are maintained. Theoretically, a patchy mosaic of severe and light impacts could enhance regional biodiversity. Some species benefit from the more open conditions and vegetation regeneration promoted in heavily impacted localities. Reserves much smaller than the typical home range of elephants may not have sufficient space for heterogeneity in the biodiversity impacts of elephants to be expressed, as essentially the entire area becomes heavily impacted. In large protected areas, the major ecological concern is the lags arising from slow plant recovery following damage by elephants. These can potentially lead to decades-long oscillations in the abundance of elephants and the impacts on affected vegetation and other species.

WILL ELEPHANT NUMBERS REGULATE THEMSELVES?

Because of the long time frames inherent in the interaction between elephant and slow-growing trees, there are no observational data to answer this question definitively, nor are there likely to be within the next decade. The following hypotheses are based on model results and the extensive experience with short-lived herbivores.

There is no reason to believe that elephant populations will behave qualitatively differently to those of other herbivores if left to their own devices: the population within a restricted area will grow to a maximum number, and thereafter could follow one of several possible trajectories (figure 4). The interacting time lags between elephants and tree demographics make the smooth rise to a stable equilibrium (the conceptual model on which simplest version of elephant 'carrying capacity' is based) the least likely scenario under the current circumstances in South Africa. Some degree of overshoot and oscillation is more likely, but there is currently no reliable way of predicting the magnitude and duration of the fluctuations.

'Density dependent' mechanisms will eventually reduce the growth rate of elephant populations to zero or below. There is very little direct evidence from any elephant populations in South Africa that such mechanisms are sufficiently effective at current densities to have resulted in an observable depression of the population growth rate. This does not mean that they are not operative – they may simply be masked by natural variability and the long time-lags involved. There is evidence from southern Africa that nutritional stress, and the high metabolic and time cost of foraging when food is sparse, delay the mean age of first conception in elephants and increase the average period between successive births. The mortality rate of calves increases when food and water are scarce, particularly during periods of drought. Where movement is possible, it is inferred (but not demonstrated) that the deteriorating quality of the highly elephant-impacted habitat will encourage the net emigration of elephant to better habitats. When the birth rate falls below the combined death and emigration rate, the population will decline.

It has further been suggested that at some point, the habitat conditions will recover under the new, lower elephant densities, birth rates will rise again, death rates fall, and elephants may immigrate rather than emigrate. This would lead to a periodic oscillation in elephant and habitat within defined upper and lower limits, known as a stable limit cycle. If many interconnected but uncoordinated locations experience such fluctuations, the result could be an approximately steady total elephant population when averaged over a period of many decades and an area of thousands of square kilometres. This idea has not been empirically tested, and at this stage in the development of southern Africa may only be amenable to theoretical modelling since the practical options for such dramatic range expansions no longer exist. There is no evidence that density-dependent population regulation is itself scale-dependent – in other words, that elephants in larger protected areas will self-regulate at a different mean density than elephants in smaller protected areas. Neither is there evidence in support of or against the hypothesis that in very large ranges, the elephant and tree populations form a stable limit cycle (see figure 4) more or less readily than in smaller areas.

The population density at which the density-dependent mechanisms become effective, the maximum density that the population would reach and the magnitude and period of the subsequent fluctuations, can only be guessed at this point. These numbers would certainly differ between ecosystems, depending among other things on the heterogeneity and size of the elephant range, the species of plant present, the amount, quality and availability of food resources produced by the ecosystem, the degree of competition from other herbivores, the availability of other necessary resources (such as water), or other factors affecting elephant mortality, such as disease, hunting, and predation.

It is clear from existing situations in southern Africa that the appearance of the vegetation in the areas favoured by elephants is already highly transformed at densities well below those where self-regulation of elephant numbers occurs. It is less clear how permanent and significant in terms of ecosystem function these changes are. Most of the elephant-induced changes to ecosystem structure and processes are probably reversible in the very long term (the next 100 years). The loss of tall, slow-growing trees in the savanna biome, such as baobabs, would take such a long time to restore that it can be regarded as irreversible with respect to the current generation of stakeholders, and thus becomes an issue of intergenerational equity.

SETTING NUMERICAL LIMITS TO ELEPHANT DENSITIES

The opinion among the experts who were part of this Assessment is that the setting of a nationwide target maximum elephant density ('elephant carrying capacity') is unfeasible, since the ecological circumstances and management objectives vary so greatly across the country. The evidence is as yet inadequate to permit the rigorous setting of such guidelines on a highly situation-specific basis either. A way forward in the absence of such clear guidelines is to manage elephant populations on a case-by-case basis in relation to land use objectives, rather than directly in relation to their numbers. This could be achieved by setting thresholds of acceptable change in key indicators that are sensitive to elephant impact. Such indicators and thresholds should be tailored to the objectives and circumstances of the area under management. Where the thresholds are reached (or there is a reasonable risk that they will be transgressed within the time necessary to manage the elephant population) then appropriate actions would be triggered. In time, the information that arises from such a learning approach may make it possible to determine defensible rules-of-thumb for elephant density under given circumstances.

INCREASING THE SIZE OF THE ELEPHANT RANGE

The effective range of elephant in South Africa has significantly expanded in the past two decades, through three mechanisms: addition of land to existing protected areas (e.g. Addo); translocation of elephant into new areas, particularly private reserves; and by the creation of transfrontier conservation areas, notably the Great Limpopo and Limpopo/Shashe Conservation Areas. These strategies reduce the effective rate of increase of elephant densities in the source areas, and thus delay the onset of elephant impacts, but do not reduce

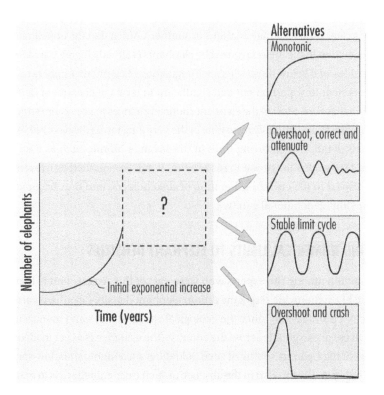

Figure 4: Hypothetical trajectories of elephant numbers. The recovery of elephant populations in South Africa following their near-extirpation in the nineteenth century follows the initial part of these graphs closely, but there is great uncertainty regarding what may occur as elephant numbers rise towards their limit. The number of elephants may (1) continue to increase, with each successive annual increase being slightly smaller than the preceding year until a relatively stable population size is reached; (2) vacillate between high and lower numbers; with diminishing oscillations until relatively stable population size is reached; (3) increase and decrease in a sustained pattern; or (4) increase dramatically and later collapse

the overall elephant population growth rate. By making new resources available, they are likely to allow the population growth rate to remain high for a longer period of time. The elephant density in the new range will, within a few decades, reach similar value to those that have raised concerns in the source regions, and further net migrations or translocations will no longer be possible. The potential within South Africa for further expansion of the elephant range is limited by the density of human settlement and the high degree of transformation for crop agriculture. The scope for future translocation of large numbers of elephant, for the purposes of reducing the elephant density in the source area, is rapidly declining as the recipient areas fill up.

The elephants in South Africa take the form of a few large populations and a large number of small, isolated populations. There are three other large, separated populations in neighbouring countries. Making it possible for elephants to interact between populations (known as metapopulation management), by means of removing fences, connecting populations using migration corridors and translocating elephants between populations, has genetic conservation advantages but no known long-term population control benefits. Simulating a larger, unbounded range by the creation of dispersal sinks through local capture or culling within smaller protected areas may have other benefits (for instance, by creating zones of low elephant impact), but has no population control advantages over non-localised culling or removal.

TRANSLOCATION OF ELEPHANTS

The techniques for capturing, immobilising, transporting and releasing elephants into new environments have been developed in South Africa to the point where elephant mortality is low and the procedure can be done safely (figure 5). It remains expensive and stressful to the animals involved, as well as those within sensory range of the capture operation. The stress can be reduced and the success of the outcome improved by following best-practice guidelines. Proper planning of the translocation operation and the selection of habituated or 'well-behaved' elephants at the capture site are important factors leading to the ultimate success of the operation. Studies of the suitability of the receiving environment for elephants from the particular source population are a necessity. Family groups should be translocated together, and adequately acclimated in a specialised holding pen before release into their new habitat. Translocation does not cure the behaviour of individual elephants with a record of aggression; it simply relocates the problem and should never be attempted.

As most translocations are to fenced protected areas, the dissemination of genetic material is restricted by the small numbers of elephants available to form viable breeding nuclei. It is imperative that management interventions be focused on genetic diversification; if not, a population bottleneck situation, in terms of reduced genetic diversity, will occur on smaller game reserves. Currently, the lack of new receiving areas is the greatest limitation for using translocation as a means of controlling elephant population size. Cost and logistical constraints limit the applicability of translocation as a population control mechanism to relatively small populations.

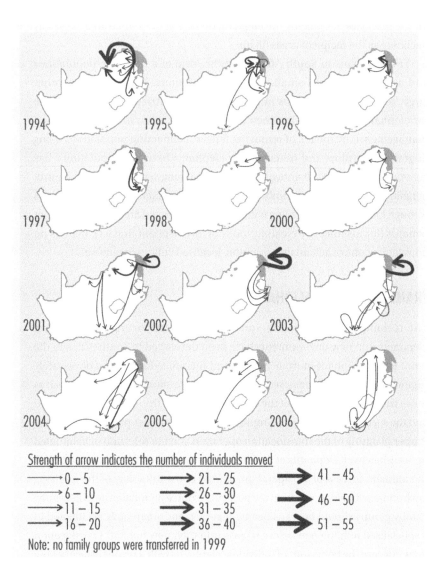

Strength of arrow indicates the number of individuals moved

⟶ 0 – 5	⟹ 21 – 25	⟹ 41 – 45
⟶ 6 – 10	⟹ 26 – 30	⟹ 46 – 50
⟶ 11 – 15	⟹ 31 – 35	⟹ 51 – 55
⟶ 16 – 20	⟹ 36 – 40	

Note: no family groups were transferred in 1999

Figure 5: Translocations of elephant family groups occurring in South Africa over the period 1994–2006. Prior to 1994 individual elephants were translocated

REDUCING THE BIRTH RATE IN ELEPHANT POPULATIONS

Immuno-contraception, particularly of female elephants using Porcine Zona Pellucida (pZP) vaccine, has proven to be an effective and viable way of reducing elephant fertility in many situations. It requires that breeding-age cows be injected with a vaccine several times (two to three times during the first year, followed by an annual booster to sustain contraception). The duration of effect following cessation of vaccination in individual cows is thought to be equivalent to the number of years it has been employed. The injection is administered remotely and does not require that the cow be captured or drugged, and has few known direct side effects. Other technologies (such as one-shot vaccines and Gonadotropin Releasing Hormone (GnRH) vaccine), which could apply to either male or female elephants, are promising but not yet proven in elephant field trials. Hormone-based contraception of female elephants has been shown in South African trials to result in unacceptable levels of aggression, as has castration of bulls. Vasectomies are effective and can be performed without long-term health consequences for the bull elephant, but expense is likely to restrict their use to populations in small, protected areas.

The long-term physiological, behavioural and ecological consequences of widespread contraception of wild elephants are not known, since the trials have been under way for less than eight years. Immuno-contraception is reversible in individuals that have been vaccinated once or twice, but it is not known if it is reversible after three or more treatments. Indications are that reversal after multiple immunisations is slow.

At present, mass contraception has been limited to elephant populations of fewer than about 300 individuals, often under intensely studied conditions where individual cows can be identified and located for re-immunisation. In principle, the vaccination technique could be applied to much larger populations, targeting all mature females rather than specific identified individuals. Accidental vaccination of a pregnant cow has no known negative impacts on the foetus, and multiple-vaccination within a few weeks is not likely to be detrimental to the health of the cow other than its effects on fertility. Successful contraception of about four-fifths of all breeding-age females would lead to a birth rate that approximately matches the inherent mortality rate – that is, population stabilisation.

Because of the longevity of elephants and the 22-month gestation period, contraception is not a technique for reducing elephant numbers in the short term (within a decade or so). It is therefore ineffective for reducing the ecological impacts of elephants once they are already apparent. It is a preventative measure that must commence suitably in advance of the time when unacceptable elephant impacts are anticipated. Primary pZP vaccinations of cows during translocation could be considered as a useful tool to control populations in their destination area. A single vaccination does not cause infertility, but causes the animal to respond quickly to vaccination boosters at a later date, and so reduces the lag period and cost of subsequent contraception.

LETHAL MANAGEMENT

Culling and translocation are the only management options for reducing elephant densities (and thus local impacts) where intervention is urgent – that is, taking effect immediately or within five years. Shooting the animal is usually the only option for the control of individual elephants, in cases where rapid response to threats to human life is required. It is also the most practical option for persistently aggressive or damage-causing animals. Key ecological concerns associated with culling include the partly uncertain (but probably substantial) impacts on the behaviour of the surviving elephants, and the increase in the underlying population growth rate that can result from reducing elephant numbers and disturbing the age and sex ratios. It is possible, but unproven, that an elephant population made artificially younger by age-selective culling could be more prone to overshoot its resource limitations. It is likely that once culling is adopted as an elephant population control method, it must be continued indefinitely or until replaced by another method. Culling, like other high-consequence management options, should be guided by a structured decision-making process. Culling is indicated only once other population management options have been considered, evaluated, and rejected. The preferred method is a single lethal shot to the brain, delivered by a skilled marksman from a helicopter. In the case of females and young animals, current best practice is for the entire family group to be culled at once, and not in the near proximity of other elephants.

THE ECONOMIC VALUE OF ELEPHANTS

Elephants are generators of economic value in the ecosystems in which they occur, and under current circumstances in South Africa only at a minor cost relative to the value of the animals. Having more elephants does not, however, imply that the net economic value will increase proportionately. At some population size the costs associated with destruction to property, threat to human life and degradation of ecosystems by elephants would exceed the benefits.

The total economic value of elephants consists of both direct and indirect values. Some of the former are consumptive (i.e. the elephant must die before the value can be realised) while others are non-consumptive. Currently the total economic value of elephants in southern African countries is overwhelmingly dominated by the so-called 'existence' and 'bequest' values – in other words the hypothetical price that people, mostly not living near to the elephants (or even in southern Africa), are willing to pay to know that wild elephants exist and will continue to do so (table 1). Only a small fraction of this notional value is currently realised through wildlife tourism. Recent advances concerning payments for ecosystem goods and services and the development of markets for these services indicate that through institutional change it is indeed possible to harness more of the existence and option values of elephants in future.

The direct use value that could be realised through the sustainable harvesting of elephant for ivory, hides, and meat in South Africa is limited by the CITES ban on the trade in elephant products to a few millions of rand per year, but is likely to rise again in the future. Nevertheless, even under unrestricted market conditions the direct use value is likely to remain much smaller than either the existence value or the non-consumptive ecotourism value. The degree to which consumptive use (e.g. hunting) might reduce the realisation of the existence value (e.g. by deterring tourism) is unknown.

Component		Value per elephant (ZAR/y)	Total value (ZAR billion/y)
Non-consumptive use	Existence and bequest	29 614	14.7
	Viewing by tourists	10 506	3.9
Defensive expenditures	Protection cost	2 010	1.1
	Damage compensation	1 173	0.6
	Translocation cost	19 095	0.01
Consumptive use*	Ivory	1 291	0.8
	Hunting	290 000–500 000	0.08–0.04
	Live elephant sales	15 000–500 000	0.01–0.3

*This is currently a restricted market and the values are therefore skewed towards the low side.

Table 1: Indicative values in 2007 for the various components of the 'Total Economic Value' of elephants in South Africa. Extensive use has been made of studies in Zimbabwe, Botswana, and Namibia to estimate some of these values, assuming an exchange rate of 6.7 ZAR/US$. They may not be accurate for South Africa, but the relative ranges of the values are likely to be correct. A billion is 109

THE LEGAL ISSUES RELATING TO ELEPHANTS

South African wildlife laws are rooted in Roman-Dutch common law and are expressed in many overlapping (and at times conflicting) statutory enactments at the national, provincial and local level.

Central to the treatment of wild animals in South Africa's law is their definition in the law of property as *res nullius* (belonging to no one) in certain circumstances, which may include large state-owned protected areas. It is submitted that this notion, and the law that is built on it, is inconsistent with South Africa's customary law, the National Environmental Management Act and the Constitution. It is also out of step with the current social perception that wild animals, particularly those occurring in protected areas or escaping from protected areas, are part of the national heritage and should be protected as such. Similarly, the recognition in international law of the concept of a global commons, of which wildlife heritage is an inextricable part, is also not reflected in South African common law.

The application of the *res nullius* principle in many instances results in the national heritage being diminished in circumstances that are not reasonable, justifiable or in the public interest. This increasingly untenable situation could be addressed through a redefinition of the common law by way of judicial intervention and interpretation. This could be a starting point for a revision of relevant legislation and policy regarding elephants, but would also be relevant to securing the status of thousands of other species.

MANAGING COMPLEX SYSTEMS CONTAINING ELEPHANTS

The existence of clear strategies for elephant management, conscious of the social and ecological factors involved, and explicit about the conceptual models on which they are based, would assist in guiding coherent and effective elephant management actions. The approaches that have been applied to elephant management have evolved over time, and will continue to do so. Despite the widely shared respect for elephants, 'moral plurality' about how to manage them (i.e. fully or partly incompatible views) is likely to be a reality for elephant-related issues in South Africa for the foreseeable future, since no single set of values is clearly dominant or ascendant.

For issues such as elephant management, where both the ecological and human systems involved are complex and incompletely understood, the current best practice approach is 'active adaptive management' (figure 6). In adaptive management, actions are accepted as being provisional, and are undertaken as deliberate experiments, with the necessary controls and before-, during-, and after-the-fact data collection. The results of the experiment are then used to refine future management, including the possibility of changing the goals which it seeks to achieve if they prove unattainable or inappropriate (figure 6).

There is consensus among the elephant experts engaged by this assessment that a single set of policies and management rules cannot be applied to all situations where elephants occur in South Africa. Nonetheless, useful guidelines, based on research, can already be provided for the main situations that occur in South Africa (table 2). The appropriate management depends on both ecological factors (such as the type and condition of the habitat, the elephant density and the size of the area, and the presence and status of other species) and human factors (such as the objectives for which the area is managed, the proximity to other land uses, and the economic and technical capacity to undertake certain actions).

The policies appropriate for South Africa are not necessarily applicable in other African countries. Management of elephant populations that straddle international frontiers (such as those in Maputaland, Limpopo, and Mapungubwe) should be at least coordinated, and preferably harmonised, on both sides of the border. Similarly, populations that move between private and public protected areas would benefit from being managed in an integrated and consistent way, but not necessarily identically in both tenures.

WHAT DO WE STILL NEED TO KNOW?

All areas of research into elephants, the ecosystems that contain them, and the societies that care about them, contain residual uncertainty that could be reduced (but not entirely eliminated) by further research. However, there are certain topics on which better understanding is particularly urgent or important from the perspective of elephant management, at all scales from the individual protected area to the subcontinent.

- The trends and societal distribution of human value systems that underlie conflict around the management of elephants, and better ways of managing issues that occur within a context of conflicting value systems.
- The economics of elephants in South Africa, in particular the ways of ensuring that the potential benefits from elephants reach those with the greatest need for them, and the strength of the trade-off between use values and non-use values.
- The long-term physiological and behavioural consequences of contraception, and the practical implications of contraception in large elephant populations.
- The importance and persistence of stress in elephants induced by exposure to culling or hunting, capture, translocation, and separation from clan members.
- Examining stress, behaviour and demographic vital statistics in elephant populations at differing densities – what are the effects of being subjected

to, or being maintained at, high densities, and what are the biodiversity consequences?

- The effects on various elements of biodiversity – including composition, structure, and function – of increasing levels of elephant pressure, in all major ecosystems in which they co-occur in South Africa.
- The potential to control elephant distribution by behavioural modification.
- The feasibility and consequences of achieving elephant population self-regulation by concentrating elephant densities in a portion of the potential habitat, through for instance manipulation of water availability.

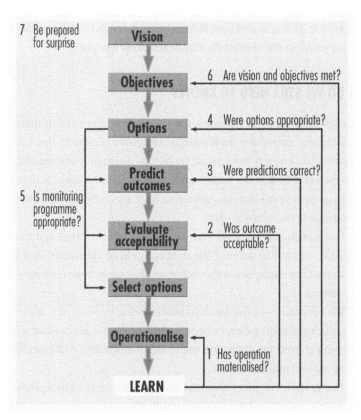

Figure 6: Diagram showing the process of adaptive management, illustrating the sequence of actions and analyses aimed at deriving and implementing the objectives, and enhancing management over time. This version of the adaptive management approach was developed by S Pollard, K Rogers, and H Biggs

Ecosystem type	Primary management objective	
	Biodiversity conservation (mainly state protected areas)	Tourism income (mainly private or communal areas)
Semi-arid savannas and their included riparian forests	*Large areas (>5 000 km²):* Laissez-faire may work in certain areas or particular circumstances. If not, and if sufficiently arid, attempt limiting elephant range by controlling perennial water supply. Internal translocation and localised mass contraception to protect areas of high sensitivity. Appropriate fencing on all boundaries adjacent to inhabited areas. Impact indicators relate to the maintenance of landscape-scale biodiversity and thresholds linked to degree of reversibility in 25- to 50-year timeframe. *Medium sized (50–5 000 km²):* Laissez-faire unlikely to work. Long-term population control can be considered by individual contraception. Short-term control, if unavoidable, by translocating or culling. Elephant-proof fencing on all crop, agricultural or human settlement boundaries. Impact indicators tied to sustainability and the preservation of unique features of the area, such as patches of specialised habitat. *Small areas (<50 km²):* Do not introduce elephant, translocate out if already present.	*Medium and small areas:* Long-term population control by individual contraception, short term by translocation to other private areas, or culling if no recipients are available. Elephant-proof fencing of any boundary adjacent to crop agriculture or human settlement. Key indicator of elephant overpopulation is effect on the overall economic viability of the land use.
Species-rich restricted-range ecosystems*	Elephant-resistant exclusion fences around the most threatened plant communities. Long-term population control by individual or mass contraception. Short-term control, if unavoidable, by culling. Impact on rare species not adequately represented outside of management area is key indicator of the need to limit elephant densities, threshold is minimum viable population (plus safety margin) of these or indicators of significant landscape degradation (e.g. Landscape Functionality Index [LFI]).	
Arid shrubland (Karoo)	Stocking with elephant not recommended. Historical evidence for the necessity of continuous presence of elephant is weak and seasonal stocking is unfeasible.	Should be contemplated in medium to large areas only. Restriction of access by limiting distribution of perennial water should control elephant impact.

* e.g., Thembe dune forest and Addo succulent thicket

Table 2: An example of differentiated guidelines for the management of elephants in South Africa

THE ELEPHANT IN SOUTH AFRICA: HISTORY AND DISTRIBUTION

Lead author: Jane Carruthers
Author: André Boshoff
Contributing authors: Rob Slotow, Harry C Biggs, Graham Avery, and Wayne Matthews

INTRODUCTION

So geographers, in Afric maps,
With savage pictures fill their gaps;
And o'er unhabitable downs
Place elephants for want of towns.
On Poetry: A Rhapsody

THESE LINES by Jonathan Swift (1667-1745) are often quoted as a satire on the cartography of the age. However, they also contain three observations about the elephant populations of Africa that illuminate aspects of elephant distribution and human–elephant contact and that continue to influence elephant management. The first is that elephants are the iconic and most charismatic mammals of Africa – indeed, its very symbol. In a continent renowned for its megafauna and wealth of raw materials, elephants and their ivory hold premier positions. The second observation is that elephants were once very widely distributed on the African continent, occurring wherever there was suitable habitat, while the third is that where large settled concentrations of humans occur, one will find either no elephants or very few.

This chapter considers the shifting economic and political dynamics, value systems and technologies that have impacted on Africa's elephant populations, with detailed attention being given to South Africa. It explains how the current (2006) presence of the African elephant *Loxodonta africana* indicates that it was once abundant throughout the continent in suitable habitat. While the process of the dramatic decline in elephant range and numbers did not play out in the same way throughout Africa, as far as South Africa is concerned it was accelerated in the nineteenth century by a growing market for ivory and by significant habitat transformation within a modern state. By the early twentieth century the once large elephant population in the region had been virtually exterminated except for a few small relict populations in remote localities. In the later twentieth century, however, owing to a combination of factors that are outlined below, an elephant population that is highly restricted to limited areas (relative to pre-colonial distribution) in South Africa has undergone a period of sustained growth. Since its near-extinction in the region owing to hunting and dense human settlement and rural land exploitation, elephant population growth is rebounding in strictly protected preserves and being manipulated through intensive management and translocations.

ELEPHANT SPECIES IN AFRICA

Linnaeus placed African and Asian elephants together in a single genus, *Elephas*, but they were separated in 1797 by Johann Blumenbach into the Asian *Elephas* and the African *Loxodonta* (Skinner & Chimimba, 2005). There is evidence from DNA-based studies that two species of African elephant exist, namely, *L. africana* (savanna elephant) and *L. cyclotis* (forest elephant) (e.g. Roca *et al.*, 2001). In addition to those at the genetic level, the two taxa exhibit certain other more obvious differences; most notably, in habitat – *cyclotis* occupies mainly the forested parts of Central Africa, whereas *africana* occupies mainly the savannas of eastern and southern Africa – and in morphology – *cyclotis* is smaller than *africana*, and there are differences in ear structure (Skinner & Chimimba, 2005). In West Africa, elephants live in both forest and savanna habitats (Blanc *et al.*, 2007) and here a new species has been postulated (Eggert *et al.*, 2002). However, there is no consensus in the scientific community as to the current number of species of elephant in Africa (Debruyne, 2005). Consequently, the International Union for Conservation of Nature and Natural Resources (IUCN) Species Survival Commission (SSC) recognises only one species on this continent, namely *Loxodonta africana*.

THE AFRICAN ELEPHANT FOSSIL AND PALAEO-ANTHROPOLOGICAL RECORD

The fossil history of elephants in Africa is relatively patchy, in spite of the fact that they may suffer die-offs involving large numbers, particularly of sub-adults, during droughts (Haynes, 1992). Haynes compared data from Zimbabwe drought mortality with Pleistocene *Mammuthus* and concluded that severe drought conditions in parts of the southern United States may have led to the Pleistocene extinction of *Mammuthus*. It follows that wet–dry climatic patterns may have provided a similar mechanism in Africa, where droughts have been a regular part of the climatic cycles that arose during the Late Pliocene and Lower

Pleistocene. Conditions for periodic droughts and concomitant high elephant mortality may have placed populations under sufficient stress to have led to adaptive evolution or even extinctions.

Enormous probocidean diversity is evident from the fossil record (Coppens *et al.*, 1978), and the group forms a major part of the Tertiary and Quaternary faunas of the world. The fossil history of the Proboscidea was punctuated by a number of adaptive shifts that resulted in varied types such as mastodonts, gomphotheres, stegodonts and elephants, which are subdivided into three families: the Gomphotheriidae (very diverse Late Eocene to Middle Pleistocene, which includes *Anancus* species); Elephantidae (Late Miocene to Recent, which includes the genera *Elephas* and *Loxodonta*); and Mammutidae (Early Miocene to sub-Recent ancestral mammoth).

Two of the most important fossil localities that have yielded proboscideans are Langebaanweg and Elandsfontein in the Western Cape. Elephants are extremely rare in the South African hominin cave breccias, which presents a problem, since elephants, like pigs, were evolving and diversifying rapidly and are thus of value in correlation and relative dating. Chronological and taxonomic inferences have been made in the light of what has been found in East African sequences (Cooke, 1993). It appears, however, that no *Loxodonta* were present in the Australopithecine-bearing deposits of Gauteng, probably because of the nature of the traps or because the habitat was better suited to *Elephas*. *Loxodonta* are rare in sites with *Elephas* and vice versa, and it seems that this pattern follows that in North Africa.

The earliest elephants in southern Africa belong to the genus *Elephas,* of which the earliest is *E. ekorensis* (Maglio, 1973). They date to between 4.5 and 3.0 my ago and are ancestral to *E. recki*, a species sub-divided into several units of which *E. recki brumpti* and *E. recki iolensis* are known to have occurred in South African hominin-bearing deposits (Cooke, 1993; Coppens *et al.*, 1978), although a number of specimens have been re-assigned from *Elephas ekorensis* to *Elephas recki recki* or *E. r. brumpti* (Cooke, 1993). A series of time-successive *Elephas* species has been used for biochrononological determinations. However, a review of *Elephas recki* indicates significant chronological and morphological variation between the currently recognised taxa. *E. recki* may therefore comprise more as yet undescribed taxa (Todd, 2005). The upshot is that the use of *Elephas* sub-species as biochronological indicators may be compromised. It is not known for sure, but the disappearance of *E. recki* from Africa by the Late Pleistocene was probably related to vegetation changes which became more suited to *Loxodonta*.

Kalb *et al.* (1996) give Middle Pliocene dates for what were considered the earliest loxodont elephants, *Loxodonta exoptata* and *L. adaurora*. Coppens *et al.*

(1978) thought that *E. exoptata* was probably *L. adaurora*) in Eastern Africa (Middle Pliocene to Early Pleistocene). It is now thought, however, that the earliest loxodont species comes from the earlier Varswater Formation of the Muishond Fontein Peletal Phosphorite Member (Hendey, 1970; Roberts, in press) at Langebaanweg, where Sanders (2007) has recently erected a new loxodont species *Loxodonta cookei* sp. nov. (Late Miocene–Early Pliocene between 5.4 and 4.0 my ago), which pre-dates *Elephas*. *L. cookei* therefore extends the *Loxodonta exoptata*–*Loxodonta africana* lineage further back into the Late Miocene, with *L. cookei* being more primitive than *L. exoptata*. It is thus probably ancestral to the *L. exoptata*–*L. africana* lineage, and *L. cookei* sp. nov. therefore represents the earliest record for the genus so far. Sanders also recognised a second elephant, referable to *Mammuthus subplanifrons*, which is probably the most archaic stage of mammoth evolution. The *Loxodonta* group appears to have remained conservative in its dental specialisations, however, as other groups became extinct during the Pliocene, or changed rapidly during the Pleistocene. *Loxodonta* changed very little from *Loxodonta adaurora*. Indeed, the southern African *Loxodonta adaurora* is very similar to *Loxodonta africana*.

From the Langebaanweg faunal list and isotopic analyses of congeners from other African sites, it may be inferred that woodland and open conditions were present locally with abundant grazing in the ecosystem of Langebaanweg (Sanders, 2007). Other evidence suggests that the vegetation was more lush than now, with woodland browsing to mixed feeding dominated by open C3 grassland. The climate was one of wet winters and dry summers; dry periods may have cycled into periods of drought and increasing aridity (Franz-Odendaal *et al.*, 2002; Franz-Odendaal, 2006).

Loxodonta africana and *L. cyclotis* have been treated conservatively as a single species although the savanna elephant lives in savanna, bush and lightly forested regions, while the forest elephant, normally found in tropical forests, is more of a browser and frugivore. Recent cranial-morphological and molecular studies have shown, however, that they are distinct savanna and forest species respectively and diverged from a common late Pliocene ancestor about 3 my ago. The differences found were predominantly genetic rather than intra-group genetic variation or hybridisation (Roca & O'Brien, 2005).

Loxodonta atlantica, primarily from Middle Pleistocene Elandsfontein (400–700 ky ago) (Klein *et al.*, 2007), was a later development contemporary in Africa with *Elephas iolensis*. But although found in the same regions they have rarely been recorded from the same localities. The reason for their survival was possibly due to their being ecologically separated. During the Middle

Pleistocene *L. africana* occurs together with *L. atlantica* at Elandsfontein. By the Late Pleistocene, however, *Elephas* was extinct in Africa, leaving only the extant *Loxodonta africana* (Coppens *et al.*, 1978) and *L. cyclotis* (Roca & O'Brien, 2005). *L. atlantica* was previously thought to be in North Africa only, but it has since been identified in East and southern Africa. It includes Middle Pleistocene fossils from Elandsfontein and Late Pliocene fossils from the Omo. *L. atlantica* probably derives from *L. adaurora* and by Plio-Pleistocene times was already very distinct (Coppens *et al.*, 1978).

Scott (1907) established a new species *Elephas zulu* from KwaZulu-Natal although Coppens *et al.* (1978) synonymise *Elephas (Loxodonta) zulu* with the North African *Loxodonta atlantica* from Ternifine. Cooke (1960) noted the similarities, but the issue could not be resolved as long as Scott's types were the only material from South Africa. The distinctions were later confirmed by additions to the Elandsfontein material, which is attributable to Scott's taxon. The two populations vary, however, and do not warrant specific separation and are best given sub-specific status *L. atlantica atlantica* (northern Africa) and *L. atlantica zulu* (southern Africa) (Coppens *et al.*, 1978).

The Middle Pleistocene Elandsfontein habitat was undoubtedly mixed fynbos with C3 scrub or woodland and a significant grass component, given the fauna (Kaiser & Franz-Odendaal, 2004; Klein *et al.*, 2007; Klein & Cruz-Uribe, 1991; Luyt *et al.*, 2000) and evidence of pollens from coprolites and sediments (Singer & Wymer, 1968).

ELEPHANTS IN THE ARCHAEOLOGICAL RECORD

A survey of bone material from archaeological sites in southern Africa suggests that elephants were widespread in the subregion. Outside of the present South Africa, elephant material has been identified from sites in the present countries of Zimbabwe for the periods 30 000–25 000 and 1 500–500 BP, Botswana for the period 1 500–500 BP, and Mozambique for the period 1 000 BP to the present. Within the borders of the present South Africa, elephant material has been identified from sites in the present provinces of Western Cape (18 000–1 500 BP, 500–recent BP), Gauteng (2 000–1 500 BP, 500–recent BP), North West (500–recent BP), Limpopo (1 500–recent BP), Mpumalanga (1 500–1 000 BP) and KwaZulu-Natal (1 500–1 000 BP, 500–recent BP) (Plug & Badenhorst, 2001).

One needs to be cautious about using archaeological material to determine the early distribution of the elephant, owing to the paucity of information on the nature of the finds, and the possibility that the material at some sites may have been transported by humans from elsewhere (Plug & Badenhorst, 2001).

This potential problem may be particularly applicable to the 500–recent BP period, as it includes the colonial period, when large-scale hunting of elephants with firearms occurred, along with the transport of ivory to faraway markets (e.g. Skead, 2007). Nevertheless, unearthed bone material has provided a useful indicator of the early incidence of elephants, for example in the Eastern Cape Province of South Africa (figure 1).

ELEPHANTS, IVORY AND AFRICAN HISTORY

While it is difficult to generalise for a continent such as Africa with the wide ecological, climatic, cultural and economic diversity of its entire rich indigenous fauna, elephants have played a crucial role in Africa's history.

Ivory, 'white gold' (Kunkel, 1982), is by far the most significant by-product of Africa's elephants and it has played the most important role, in addition to human occupation of elephant habitat, in shaping the status of elephant populations. Unlike numerous other natural resources, ivory does not deteriorate quickly, people can transport it, and over the centuries it has retained its high commercial value. White, opaque, flexible, smooth, and fine-grained throughout the tusk, the African variety of ivory is softer and easier to work than the Asian by cutting, sawing, painting, staining, slicing, and carving (Luxmoore, 1991; Alpers, 1992; Meredith, 2001). Its aesthetic beauty has been internationally appreciated for millennia and, together with slaves and gold, it has at times been Africa's major export. Ivory has linked the people of Africa with the outside world and shaped perceptions of the continent. Long before Africans were colonised, the elephant herds were being exploited and many regions became enmeshed in international trade through ivory.

For more than 10 000 years the 'subtle glowing colour and sensual surface' of ivory (Luxmoore, 1991) has ensured its prominent position among the luxury goods of the world, but in the later years of the twentieth century the assault on the elephants of Africa to procure it has been unprecedented. This market is particularly sensitive to taste and fashion and demand has at times led to the virtual extinction of local elephant herds. For example, in Roman times ivory was so coveted and the market so insatiable that in AD 77 Pliny complained about an ivory shortage, commenting that the North African elephant herds had been wiped out (Meredith, 2001; Wickens, 1981; Alpers, 1992; Luxmoore, 1991).

Another of the major peaks in supply and demand occurred during the nineteenth century with the industrialisation of Europe and the United States, when ivory was in general use for trivial manufactured products. In the early

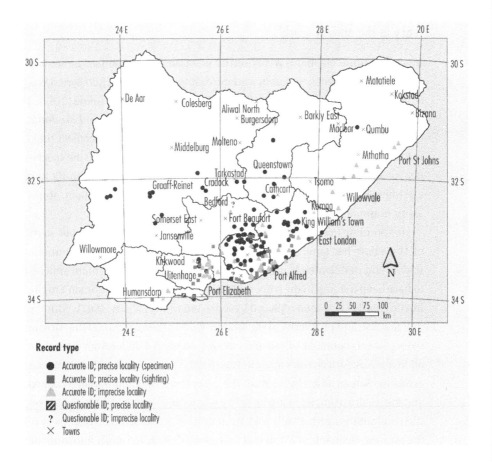

Figure 1: Historical distribution of the African elephant in the broader Eastern Cape (Skead, 2007)

decades of the twentieth century, ivory was used less frequently in the West, but in the 1970s demand from Asia – for ornaments and for popular Chinese and Japanese signature seals, *hankos* – peaked and this took its toll particularly on the herds of East Africa, leading to a ban on ivory exports from Africa (Parker & Amin, 1983; Parker, 2004). Between 1979 and 1988 the value of ivory exported from Africa increased more than three-fold (Barbier *et al.*, 1990: 36–37).

Importantly too, savanna elephants occupied those niches most suitable for human agriculture, leading to a reduction in what was available for elephants – as much as 10–25 per cent of the total range over Africa in general (Milner-Gulland & Beddington, 1993; Oliver & Atmore, 1967).

Ivory in the ancient world

In past centuries elephants were used for entertainment in Roman games and circuses, and played their part in warfare; examples being the Third Syrian War (c.240 BC) and the Carthaginian Hannibal's campaigns against Rome (218 BC) (Alpers, 1992; Meredith, 2001). There are also isolated records of *Loxodonta* being tamed as beasts of burden, for example at Garanga in the Belgian Congo in the early twentieth century (African Parks Foundation, 2007), but the species has never been domesticated to the extent that *Elephas* has in India and elsewhere in South Asia. It has also not become a religious or symbolic object as, for example, Ganesha has in India (Sukumar, 1989; Jhala, 2006).

Responding to changing regional and international politics, various parts of Africa have had strong ties to ivory importing destinations, usually through traders and middlemen (Curtin *et al.,* 1995; Shillington, 1995). There appears to have been a slump in the ivory trade with the collapse of the Roman Empire but this market was soon replaced by India and China (Alpers, 1992). With the political rise and prosperity of the Islamic world around AD 1000 the African east coast was commercialised and city states emerged under Arab control. In addition, West African ivory made its way to India, China and the Mediterranean across the Sahara desert by caravan. The 'Ivory Coast' was appropriately named and the wealth of ancient kingdoms like Ghana and Asante was predicated on this commodity together with gold. In contrast to the softer and whiter ivory of the savanna elephants of North and East Africa that carved easily into intricate shapes, the harder variety from West Africa was used mainly for knife handles (Curtin *et al.,* 1995). The role that elephants have played in the southern African cultural record is not well documented (Hammond-Tooke, 1974), but for the rest of Africa it has been determined that elephants – either ivory or elephants carved or painted on other materials – featured in art, in initiation and other rituals and in myth, folklore, oral traditions, song and dance (Ross, 1992).

Over millennia, therefore, ivory determined and altered Africa's relationship with the rest of the world. In the course of so doing, it also transformed many African communities and their mutual interactions. Control of the trade in ivory as well as in ostrich feathers, slaves and gold brought power and wealth to many African leaders, enabled Africans to control and exploit each other, and laid the foundations of strong states with the emergence of hierarchies based on wealth and class (Gordon & Gordon, 1996). Whole tusks were brought in as tribute to chiefs from vassals and clients, and ivory was used for personal adornment in an ostentatious display of wealth. More importantly for economic prosperity and political authority, ivory was exchanged for iron and other useful metals that contributed to improved methods of cultivation, as well as for cloth, beads and other goods – in later centuries, firearms and liquor. Thorbahn (1979) has argued that sources of ivory far into the African interior were already exhausted by 1500 and that the Indian Ocean ivory trade not only significantly affected elephant populations but was responsible for the emergence of a distinct East African culture that extended far inland and was not confined to the coastal regions alone.

Ivory extraction never promoted a strong and sustainable economy, but instead a fragile one based on a single product (Curtin *et al.*, 1995). Moreover, the quest for ever increasing amounts of ivory led to a traders' frontier moving further into the interior taking with it warfare and slavery, often leaving economic collapse in its wake (Curtin *et al.*, 1995; Gordon & Gordon, 1996).

Mapungubwe and Great Zimbabwe

Southern Africa first became locked into the international ivory trade around AD 900 with the rise of the Limpopo valley states – Schroda, K2 and Mapungubwe. The Lydenburg area supported a community that thrived earlier, from c. AD 300 to 1000, but although this settlement engaged in mixed farming, pottery and the smelting of iron and copper, ivory and other trade commodities are absent from the archaeological record (Hall, 1987; Esterhuysen & Smith, 2007).

By contrast, there is abundant evidence of gold, beads, ivory, ceramics, and other trading items at Schroda, K2 and Mapungubwe. These people were both agro-pastoralists and hunters and the Limpopo valley and its hinterland appears to have been prime habitat for elephant at that time (Mitchell, 2002). According to Huffman (2005) the Schroda settlements were located away from rivers and floodplains, possibly to avoid elephants destroying their fields, and it is highly likely that humans would have competed with elephants for the fertile banks of Africa's rivers. Slivers of ivory have been excavated at Schroda and examples of ivory that has been sawn, trimmed and polished have been found at Mapungubwe, suggesting that ivory supported a local industry as well as being an export product (Hall, 1987; Voigt, 1983). The trading activities of Mapungubwe created a complex society of considerable power and wealth, from which emerged southern Africa's first state and political hierarchy, possibly with ivory as one of its primary economic bases. After AD 1300, consequent upon a variety of factors, Zimbabwe eclipsed Mapungubwe as the regional power (Carruthers, 2006a). One reason for Mapungubwe's demise may be related to a reduction in elephant numbers, because at the end of the period of settlement there is less evidence of ivory and more of gold, although it is possible also that the trade of Africa's east coast had then shifted because gold superseded ivory in value throughout the Islamic world between the ninth and the twelfth centuries (Hall, 1987).

Ivory and other products from Mapungubwe were traded from coastal ports along the African east coast, the so-called 'Zanj', the littoral stretching from Zanzibar and Kilwa in the north to Sofala in the south (Axelson, 1969; Axelson, 1973; Curtin *et al.*, 1995; Leslie & Maggs, 2000). The great stone towns of the Swahili coast were built between 1200 and 1500; the major trading commodity from the interior was gold (which Zimbabwe could supply) but ivory and slaves were also exported to the East (Curtin *et al.*, 1995; Axelson, 1973; Shillington, 1995).

Thula Mela hill is a stone-walled site in the Pafuri area of the northern Kruger National Park. It was occupied from the fifteenth to the seventeenth century (Küsel, 1992) and it is somewhat similar to Great Zimbabwe in terms of the archaeological evidence and spatial geography uncovered so far. Meskell (2007) explains that Thula Mela was also tied into international trade and that this settlement of some 3 000 people lay close to a known trade route from the interior to the coast. Fragments of Chinese porcelain similar to those from Zimbabwe and Khami have been found. The engagement in trade impacted on social relations in terms of creating an economic and political hierarchy as had been the case at Mapungubwe. Both Meskell (2007) and Küsel (1992) describe mining and metallurgy as being critical to Thula Mela's prosperity, gold apparently being smelted and worked on site (Küsel, 1992). Reports on the excavated fauna at Thula Mela indicate little evidence of ivory, for although there is material from game and domestic animals, sea shells, glass and ostrich egg beads, only one 'carved ivory bangle' has been found (Küsel, 1992) and, it seems, no slivers and off-cuts of ivory as might have been expected if ivory had been a major export product.

By the early 1500s Europe was expanding and the Christian Portuguese, who were developing the spice trade with the East Indies, had started to tap into the Islamic traffic on the African east coast. It is also probable that southern Africa's northern Nguni people in present-day southern Mozambique and northern KwaZulu-Natal were trading with the cities of the Zanj and also with Europeans, first with ivory and then with cattle in exchange for iron and copper (Etherington, 2001).

The ceaseless demand for ivory as the African export staple in exchange for cloth, beads and metal, meant that the elephant resources of the deep interior of southern and eastern Africa were increasingly exploited even in pre-colonial times. Axelson quotes a figure of 9 656 kg of ivory being used to buy cloth for the Sofala trade in 1517 and mentions the great herds of elephants seen by Lourenço Marques at the Rio da Lagoa (Axelson, 1973), which seems to indicate that extensive trade had not yet reached that far south. The Indian traders and Hindu merchants captured the ivory market (for betrothal bangles) along the Zanj in the seventeenth century and became far more prosperous in this trade than did the Portuguese, whose overseas empire soon faded (Hall, 1987).

PRE-COLONIAL SOUTHERN AFRICA

South Africa is situated at the southern tip of the continent and, apart from Mapungubwe in the Limpopo valley and possibly some northern Nguni, until the Dutch East India Company's (VOC) settlement at the Cape and the expansion of a settler community into the interior, much of the present Republic of South Africa was generally isolated both from international trade and many cultural currents connected with luxury goods such as ivory (Mitchell, 2002). In consequence of this isolation, together with its distinctive pattern of colonial settlement that radiated from Cape Town northwards and eastwards, South Africa's elephant history played out somewhat differently from other parts of Africa that were earlier and more effectively integrated into global patterns of trade.

It has become customary to consider the South African human past in terms of the economies in which its inhabitants have engaged. Although there have been times and places in which it is impossible to distinguish communities on the basis of their socio-political and economic organisation because of assimilation, acculturation and fluid socio-economic relationships, the broad periodisation into hunter-gatherers and foragers (San Bushmen); herders (Khoekhoen) and mixed farmers (Early and Late 'Iron Age' societies), and colonisers and settlers (Dutch East India Company, British control of the Cape and Natal colonies and the Boer republics) will be followed here. It is important also to recognise that a lack of homogeneity – for entirely different reasons – characterises South African society still and influences the worldviews and value systems of various groups and communities in diverse ways.

Hunter-gatherers and herders: San and Khoekhoen

Generally speaking, the economy of the semi-nomadic San communities was based on hunting and gathering. Food had to be acquired every day, technology was rudimentary, band size was small and wealth was not accumulated. Political and economic hierarchy was flexible or absent and an ethic of sharing was strong (Mitchell, 2002). Current scholarship, particularly by Lewis-Williams, has led to an improved understanding of the spiritual foundations of San society, and it is now indisputable that beliefs and worldviews find expression in the abundant rock paintings and engravings located throughout southern Africa (e.g. Lewis-Williams & Dowson, 1999). It has been shown that these paintings and engravings are generally the work of shamans, people with special powers in the community, able through trance and painting to link the temporal with the spiritual and to ensure the health and well-being of the community through the intervention of the spirit world. In this respect, eland *Taurotragus oryx* had a pre-eminent role, being a major intermediary between the two domains, and this animal dominates the rock art. How elephant were integrated into this belief system is not as clear. There are, however, numerous examples of painted and engraved elephant figures sometimes shown being hunted by a large party of men. Fragments remain of a painted frieze of elephants at Grootkraal near Wodehouse (Lee & Woodhouse, 1970), while in the Western Cape there are paintings of elephants surrounded by zigzags and crenellated lines. There are also a few therianthropes with elephant heads and trunks. Deacon believes that elephant may be linked symbolically with water and some scholars think that the !Kung consider elephant to have remarkable potency because its meat is of all three types: red, black, and white (Hollmann 2004; Lewis-Williams & Dowson, 1999). Dowson has observed that images of elephants become more common in the later, eighteenth and nineteenth century, San rock art (e.g. at Taung, Wepener, and in the Drakensberg), and postulates that because shamans depicted certain animals in order to establish control over the hunt, elephants would have increasingly appeared in order to bolster or reinforce San involvement in the ivory trade (Dowson, 1995).

The transhumant herding communities of the western and south-western Cape, the Khoekhoen, were familiar with elephants and the value of ivory – they traded ivory bracelets with the Portuguese navigator and explorer Vasco da Gama in Mossel Bay on 25 July 1497 – but there is no direct evidence that they engaged in extensive operations (Axelson, 1973).

Pre-colonial and early colonial African trade

As mentioned earlier, South Africa's autochthonous people at Mapungubwe (between AD 900 and 1300) were the first to export large quantities of ivory. They were followed by the northern Nguni in southern Mozambique/northern KwaZulu-Natal (in the 1500s) who traded ivory, horns, and cattle in exchange for beads, cloth, iron, and copper with the Portuguese and others along the south-east coast (Delius, 1983). By the early 1600s it appears that ivory also left south-east Africa from what is now the bay of Port Natal (Durban) – the Portuguese discovered elephant pits at the bay in 1626 and 1643 – but it is unlikely that local hunting methods depleted the local elephant population to any great extent (Ellis, 1998).

By 1750 the Portuguese had established a permanent settlement at present-day Maputo and soon this became an extremely significant port which encouraged economic growth and political consolidation (Eldredge, 1995). The traffic in ivory boomed, allowing the Tembe and then the Madubu to become formidable traders. For local Africans, ivory quickly changed from being a by-product of the hunt to an objective of it, and age regiments were organised to kill elephants. New commodities were imported into southern Africa and, by controlling this exchange, chiefs were able to build up relatively strong polities during this period. However, by the 1790s it appears that the demand for ivory had somewhat fallen off, but by then parts of southern Africa were firmly integrated into a market economy (Bonner, 1983). Because elephants were not a sustainable resource that quickly reproduced itself, when they became scarce and the stocks of ivory depleted, a political crisis arose in a number of southern African polities. Ivory had not promoted productive land use or an agricultural surplus. Extraction could not be indefinitely supported and chiefs competed with each other for declining supplies (Eldredge, 1995).

How elephant by-products other than ivory were utilised is not well documented. Fitzsimons (1920) records that nineteenth century African societies ate the flesh, converted the skin into whips and bartered these and other by-products to traders along with the ivory (see figure 2). In addition, the skin of the stomach was made into a blanket and the bones were broken up and boiled for their marrow.

Although detailed analyses are lacking, a number of historians of South Africa's pre-colonial Bantu-speaking communities have commented on the role that elephants and ivory played in these societies in the eighteenth and early nineteenth centuries, i.e. the period before whites intruded with firearms and wagons, and prior to the enormous demand for ivory from Europe and the United States. In respect of the Xhosa, for example, Peires (1981) has explained how elephant hunting was a co-operative enterprise. Groups of men followed the targeted animal for days, finally encircling it with fire before killing it. Ivory was an item of barter and San exchanged ivory with the Xhosa for cattle and 'dagga'. Like other valuable products of the hunt, tusks were reserved for the chief who awarded them to followers as a mark of distinction. Ivory was used for armlets, worn on the left arm. Some of these armlets survive, but none show evidence of intricate carved patterning or design (Peires, 1981). It is possible that the Xhosa participated in the Delagoa Bay trade, because they were aware of ivory's commercial potential and traded it among themselves (Hall, 1987). In the late 1700s the first trekboers who entered Xhosaland with firearms shot elephant in order to barter ivory with the local Xhosa. By 1752 there was a substantial trade in ivory, but it was probably still linked to Delagoa Bay (and thus destined for the market in India and the Arab world) rather than to Cape Town (Peires, 1981). Mpondo people appear to have become involved later in the ivory trade. In 1850, the explorer-hunter Gordon Cumming, for example, commented that the Mpondo killed elephants for meat rather than for ivory, although he observed that they wore ivory rings and other ornaments on their fingers and arms (Beinart, 1982).

In his work on the Pedi, Delius (1983) also suggests an ivory boom for this community in the mid-eighteenth century coinciding with the Portuguese settlement of Delagoa Bay but unrelated to white settlement at the Cape. Receiving ivory as tribute, which they then passed on into the trade, the power of Pedi chiefs increased (Delius, 1983). The Tswana clans were also involved

Figure 2: 'Choice bits of an elephant. The feet and trunk' (1862) by Thomas Baines (RGS X229/021960. In M. Stevenson (ed.) 1999. *Thomas Baines: An artist in the service of science*. Christies, London, p.23)

in elephant hunting and conducted a grand winter hunt, *letsholo*, from which every male benefited, but the bulk of the by-products (ivory and feathers) was reserved for chiefs (Shillington, 1985). By 1822 the Tswana in what is now North West Province, Northern Cape and Botswana were trading ivory with the Tsonga in the east. It is likely that the expansion of Tswana communities was related to their proximity to the rich ivory grounds of the western part of southern Africa, an area that was penetrated not long afterwards by white commercial and sport hunters via the 'Hunters' Road' (or 'Missionary Road') that snaked from Durban and Grahamstown westwards to Shoshong, thus avoiding the regions most heavily infested by tsetse fly (Mitchell, 2002; Carruthers, 2007).

Cape Colony and the interior of South Africa

It appears that, almost inexorably, the extensive exploitation of elephants was spreading southwards to the very tip of Africa. This was hastened by the Dutch East India Company's settlement at the Cape of Good Hope which opened up new ivory routes from south to north and introduced firearms to South Africa. As time passed, company employees, free burghers, trekboers and visitors alike spread into the subcontinent seeking a lifestyle based on hunting and extensive pastoralism rather than on labour and capital intensive agriculture (Guelke, 1989; Van der Merwe, 1938; 1945; Pollock & Agnew, 1963). In the eighteenth century, formal expeditions set off from Stellenbosch, travelling for a number of months and returning with wagons loaded with ivory (Wilson, 1969b). Two professional trading expeditions left the Cape in 1736 under Hermanus Heupenaer and returned with profitable stocks of ivory bartered from the Thembu and Xhosa (MacKenzie, 1988).

While there was a dynamic and vibrant ivory trade in pre-colonial southern Africa, the advent of firearms combined with the specific industrial economy of nineteenth century Europe and the United States led to increasing extraction that decimated the elephant herds of South Africa. Radiating outwards from Cape Town, Grahamstown, Durban and Potchefstroom, the frontier of settlement and ivory worked its way northwards and eastwards exterminating elephant as it expanded. As wildlife of all kinds diminished and African groups lost their independence through colonial conquest, elephant habitat was transformed by agricultural development and the establishment of towns (Carruthers, 1995a; 1995b). It is likely that this transformed landscape was the major factor that prevented elephants from recolonising the areas in which they had been hunted.

THE NINETEENTH-CENTURY SLAUGHTER IN THE INTERIOR

The mid-nineteenth century witnessed the era of greatest wildlife slaughter in southern Africa. Britain took control of the Cape Colony in 1806 at the end of the Napoleonic Wars. Colonial settlement densified, particularly in the Eastern Cape, which had been losing its elephants fairly slowly, but incrementally, since the early 1700s to the extent that in 1804, according to Barrow's account at the time, there were very few left (Roche, 1996). Similarly, in the Northern Cape, new powerful, fully armed, raiding and hunting communities such as the Korana, Griqua, Afrikaners, and Basters emerged with deleterious consequences for the elephants in that region (Guelke, 1989; Morris, 1992; Penn, 2005).

With permanent, if very small, white settlement at Port Natal in 1824, ivory exploitation began in earnest in present-day KwaZulu-Natal. Traders conducted negotiations with leading chiefs, like Dingane, for permission to hunt in Zululand. Ivory extraction gave employment to many Africans as well as the opportunity to acquire firearms and thereby to intensify their own hunting and trading activities (Ellis, 1998). Because KwaZulu-Natal's ivory was exported through both Delagoa Bay and Cape Town, it is difficult to determine exactly how much ivory was extracted. In 1824, ivory exports from Cape Town were approximately 9 072 kg and by 1836 had risen to 48 080 kg (Ellis, 1998). The value of elephant hides also increased from £2 324 in 1820 to £23 544 in 1825 (MacKenzie, 1988). In 1836 a certain B. Norden traded 2 500 kg ivory from Dingane, and it is likely that the amount of ivory exported from KwaZulu-Natal increased even further with the arrival in 1837 of about 4 000 Boer settlers who established the short-lived Republic of Natalia. Very little of this hunting was recreational, but rather for commercial purposes (Ellis, 1998). Furthermore, despite these significant estimates of large quantities of ivory, Ross (1989) has shown that until 1835 the value of ivory exports was actually very small in comparison with agriculture and pastoralism and contributed very little to the economy of the Cape Colony as a whole, even after Port Elizabeth was established. Van Sittert (2005) reminds us that ivory is a 'high-bulk, low-value commodity' of far less worth than ostrich feathers and that, moreover, hunting required substantial capital to transport ivory to the coast and to support and provision hunting parties.

As far as the rest of southern Africa is concerned, from the late 1700s and early 1800s small parties of trekboers and traders penetrated the northern interior in search of ivory and other commercial products of the hunt, either opening up new routes or using traditional pathways. Soon others followed in their wake, including for example the expeditions of Andrew Smith and recreational sport hunters such as William Cornwallis Harris (figure 3) and Roualeyn Gordon Cumming, who funded their adventures by selling the ivory they collected, but who were not hunting specifically for ivory (Carruthers, 1995a). More significant were the large Voortrekker parties who were not visitors and itinerants, but potential settlers cementing partnerships with African mercenary hunters and seeking to establish independent polities in the interior that bypassed the British colonies and linked up with Delagoa Bay in present day Mozambique (Lye, 1975; Harris, 1840; Harris, 1852; Carruthers, 1995a; Carruthers, 1995b). With this ferment of activity and the intrusions of the highly mobile military state established by the Ndebele leader Mzilikazi, the South African interior was in political and social turmoil. To this must be added the Mfecane, a complex and little understood social and political upheaval which aggravated the general insecurity and fluidity of the time (Hamilton, 1995).

Figure 3: 'Hunting the wild elephant', by William Cornwallis Harris (1852, opp. p. 70)

Ivory and other products of hunting enabled mercantile and subsistence polities to survive before more complex forms of social, political and economic organisation came into being. When agriculture, towns and settlements were ultimately established, the elephant habitat was significantly transformed in areas developed for crop production and there was no question of the animals returning. Elephant hunting in particular was a major factor accelerating colonial expansion into the southern African interior and the available abundant ivory was possibly the major product that sustained it. In the 1830s

Harris recorded large herds of elephant around the Magaliesberg, the range that stretches from Tshwane to Rustenburg and straddles the North West Province and Gauteng (Harris, 1840), and he had so much ivory that at one stage he was forced to abandon it (MacKenzie, 1988). By the time that Thomas Baines visited the Transvaal in the early 1850s, however, these great herds had already disappeared (Carruthers & Arnold, 1995) and later in the century wildlife had been virtually exterminated by hunting and disease (the rinderpest of the mid-1890s played a particular role in this regard although elephants were not affected by the disease). However, until minerals were discovered and settled agriculture and private property established, the wealth of the region was predicated on a variety of products of the hunt and sales were regularly held from the 1840s onwards in the Eastern Cape, at Fort Willshire and Grahamstown (Peires, 1981) (figure 4).

Figure 4: 'Wagons on Market Square, Grahamstown' (1850) by Thomas Baines (Albany Museum), (Carruthers & Arnold, 1995, p. 120)

While ivory hunting led, at times, to conflict among groups (the destruction of Schoemansdal in Limpopo Province at the hands of the Venda in 1867 is an example), it also created areas of co-operation. Africans, skilled ivory hunters in their own right, were quick to adopt firearms. Emerging African leaders demanded tribute in ivory and partnerships were formed between whites and

Africans. At times *zwarteskutters* (local African hunters) were employed in their hundreds, and they were able to hunt in the summer months, well beyond the tsetse fly and malaria belts and on foot, unlike the whites. In 1855 it was estimated that 90 000 kg of ivory was exported from the Transvaal, together with vast quantities of hide and horn. Apart from the sport-hunters who were relatively few in number, all the communities in the interior lived by hunting and raiding, and without their exports they would not have been able to sustain themselves (Carruthers, 1995a; Wagner, 1980; Delius, 1983). However, more than ivory, the most significant trade items of that period were firearms and ammunition, which allowed groups to dominate one another but also to offer protection (and ivory) to followers. Shillington's figures for the official firearms trade into southern Bechuanaland (Botswana) are given as rising from 55 in 1858 to 7 902 in 1872 (with a concomitant rise in powder, lead and percussion caps), but the illegal trade was, of course, everywhere very much higher (Shillington, 1985). Prowess at elephant hunting brought acclaim and fame, and names such as Jan Viljoen (210 elephants killed on a single expedition), Henry Hartley (1 000–1 200 killed in the course of his career), William Finaughty (95 elephants yielding 2 200 kg of ivory), Jakob Makoetle, and the Venda chief Makhado, are leading personalities in this regard (Delius, 1983; Wagner, 1980; MacKenzie, 1988).

As elephant and other species declined in certain areas in southern Africa, they had to be followed ever further into the interior towards the Zambezi River in northern Botswana and Zimbabwe (figure 5). Wagons were required for transporting hides and ivory and a transport industry developed. An outbreak of lung-sickness among Tswana cattle in what is now North West Province and Botswana in 1856–1857 put even more pressure on wildlife (Shillington, 1985). Africans controlled the ivory market in the more arid parts of southern Africa ('the Great Thirst Land') and while hunters such as Frederick Selous and George Westbeech derived income from ivory, their trade in firearms was far more lucrative (Shillington, 1985). There are also accounts of travellers who wanted to barter or buy cattle and corn having ivory pressed on them by African communities. Missionary-explorer David Livingstone noted ivory rotting in piles near Lake Ngami and used for fencing cattle kraals because there was so much of it with transport unavailable to take it to the coast. David Hume and other traders with wagons were able to maximise stockpiles of ivory as well as to obtain new supplies (Wilson, 1969a).

Figure 5: 'The elephant killed by Transvaal hunter Henry Hartley in the Zimbabwe area for its ivory lies on a seam of gold', by Thomas Baines (National Archives of Zimbabwe), (Carruthers & Arnold, p. 122)

As can be seen in figure 6, the boom years for ivory exported from the Cape were from 1860 to the early 1880s, whereafter there was a dramatic decrease. However, these numbers must be treated with caution because the origin of this ivory is not recorded. Nevertheless, historical statistics from Delagoa Bay also suggest a crash around that time and there is supporting evidence also that the great boom years of ivory and ostrich feathers were in the 1860s and 1870s (Shillington, 1985).

During the nineteenth century, while the price of ivory generally remained stable (Beachey, 1967), its value increased in comparison with the lower prices that were required to purchase manufactured goods (cloth, beads) that were produced in greater quantities and more cheaply by the rapidly industrialising West. Practical as well as luxury objects made of ivory became very popular with the growing middle class in Europe and the market expanded (Oliver & Atmore, 1967). Because of its abundance, ivory became the 'plastic of the age', being turned into knife handles, piano-keys, billiard balls, games, scientific instruments, tool handles and ornaments (Oliver & Atmore, 1967).

There seems to have been little or no appreciation that the herds of elephant were being negatively affected by increased hunting, even though ivory had to be sought by venturing further and further into the interior. Ivory at that time was not considered a 'rare' or 'precious' commodity, merely a useful one. Hunters were driven by a commercial motive, but also by the belief that there was an inexhaustible supply of ivory and that declining numbers of elephants merely involved their moving out of reach (Carruthers, 1995a & b).

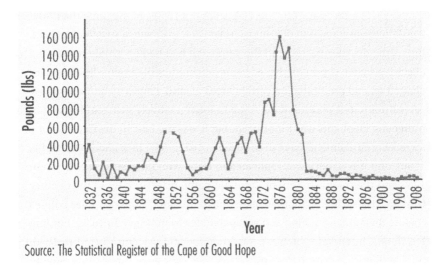

Source: The Statistical Register of the Cape of Good Hope

Figure 6: Ivory exports from the Cape Colony, 1832–1909. Figures for 1851 not obtained (Roche, 1996)

CHANGING CONSERVATION THINKING AND PROTECTIONIST PARADIGMS IN SOUTHERN AFRICA

There was unbridled exploitation of elephant and other wildlife in the nineteenth century. Hunting was not a recreational pastime that would respond to voluntary or imposed restrictive measures, but a means of economic survival in a harsh environment and in unsettled political and social circumstances. Moreover, a wide variety of people and groups was involved. Hunting elephant was encouraged by settler society, through the belief that ridding the countryside of wildlife was a pioneering and patriotic necessity in order to create a 'civilised' state (Kruger, 1902; Carruthers, 1995a & b). In this regard, the ivory trade can be argued to have been a by-product of the competition between humans and elephants for land (Luxmoore, 1991).

Elephant hunting was also considered to be the prerogative of visiting sportsmen and an indication of high social status (Carruthers, 1995a & b). Africans also maximised their involvement in elephant hunting and the ivory trade because this enabled them to acquire firearms that could be used to resist colonial expansion, that provided useful products in exchange, and that enabled stronger chiefs to attract large followings (Delius, 1983; Shillington, 1985).

Laws against exterminating wildlife species – based on the European model – were introduced by the VOC from the 1650s. In the Cape the initial legislation was framed to discourage waste and to ensure a sustainable yield. But as the decades passed and this approach had had no visible restraining effect, legislation was strengthened to prevent hunting of certain wildlife species altogether, and prohibited on certain areas of land, known as 'game reserves'. After Britain took control of the Cape Colony, relatively stringent legislation protecting elephants was introduced, but it was too late. In the Cape Colony increasing curtailments were introduced in 1822, 1857, 1886 and 1908 (Roche, 1996), and this pattern was replicated in Natal with fairly strict legislation in 1866 and the establishment of Zululand game reserves later in the century (Cubbin, 1992). In the Soutpansberg, one of the four Transvaal Boer polities (before their amalgamation and independence from Britain in 1852), hunting legislation intended to achieve sustainable yield was introduced in 1846, but elephants were excluded from it because they were simply too valuable to settler society (Carruthers, 1995a). Thereafter more and more legislation was introduced in the Transvaal (1858, 1870, 1880 and 1891), which, on each occasion, tightened up the rules and generally increasingly excluded Africans from hunting. Even when European sporting ethics became more prevalent with an increased number of immigrants coming to southern Africa after the mineral discoveries and with the increasing number of visiting British sportsmen and 'scientific collectors', the attitude of imperial 'possession' and aesthetic beauty that came to be applied to wildlife was never sufficiently strong to stem the tide of extermination (Carruthers, 1995a & b). First, the animals seemed so abundant; second, the commercial advantage of hunting was too great, and, third, in Roman-Dutch law wild animals are *res nullius* – they belong to no one (see Glazewski, 2005). So even if perpetrators were apprehended, killing wild animals was not regarded as a serious offence and it was not regarded as a crime in the same category as theft of livestock.

Legislation was simply ineffective in the face of a pioneering mentality, ideas of inexhaustibility and a desire to accumulate capital. Studying debates on the game laws proves useful in determining the motives for hunting.

The commercial hunters could not understand the pleasure of sport hunting, which they regarded as wasteful, while the sport-hunters vilified those who thought so little of trophies, using horns and hides for domestic articles and turning meat from wildlife into biltong. Africans were often accused of having caused the most destruction of wildlife, but when this belief was tested, it proved not to be the case, particularly when it is considered that ownership of firearms was withdrawn from them (Carruthers, 1995a & b). Preservation measures in South Africa grew almost in proportion to the decline in wildlife. By the end of the 1800s there were game reserves in many parts of South Africa – in the Cape, the Transvaal and Zululand – but in all of these areas, except possibly for a few individuals between the Olifants and Letaba rivers (Stevenson-Hamilton, 1934) and in northern Zululand (see case study below), there were no elephant left to protect (Carruthers, 1995a; Cubbin, 1992; Roche, 1996). During the rinderpest epidemic that saw massive loss of southern African livestock (1896) and during the South African (Anglo-Boer) War (1899–1902), game legislation was rescinded or merely ignored as people sought alternative sources of food (Carruthers, 1995a & b).

Effective administration – then as now – is required to control unbridled exploitation of natural resources, whether elephant, forests, whales, ostriches or sharks. In 1910, when the four British colonies united as the Union of South Africa, wildlife protection devolved to the provinces. By the First World War the South African agricultural economy was modernising (tsetse fly had not reappeared in the Transvaal) and there were fears that the game reserves in the Transvaal and Zululand would be deproclaimed by the stroke of an administrator's pen. For this and other reasons (Carruthers, 1995a & b) the Kruger National Park was established in 1926 (*National Parks Act* No. 56 of 1926) by the amalgamation of the Sabi and Singwitsi Reserves and the excision, expropriation and exchange of private land. The fate of the Zululand game reserves was quite different. In that area, a government intent on encouraging commercial (settler) agriculture waged *de facto* war on the wildlife as a consequence of research on the control of tsetse fly, killing very large numbers of wildebeest and other wildlife, by which time elephants had long disappeared (Brooks, 2001).

It was around this time that the Addo elephants also came under pressure for their depredations on farmland (see below). In 1931 the Addo Elephant National Park (located in the Eastern Cape) was included in the prevailing template of establishing National Parks to preserve specific species, such as Kalahari Gemsbok (1931), Bontebok (1931) and Mountain Zebra (1937). The Dongola Wild Life Sanctuary was established as a national park in 1947 and it might have provided some protection for the elephants in that area (Shashe-Limpopo confluence and Limpopo River valley), but pressure from the electorate caused its abolition in 1949 (Carruthers, 1992).

THE INTERNATIONAL DIMENSION

In tandem with internal developments around South Africa's protected area estate and its growing body of wildlife and environmental legislation, the beginning of the twentieth century saw the start of an international wildlife protection movement in which elephant conservation played a large part. In April 1900 a conference was held in London, instigated by Hermann von Wissman (governor of German Tanganyika) and hosted by the British government and the Society for the Preservation of the Wild Fauna of the Empire (presently the Fauna and Flora Preservation Society). A month later a Convention was signed in which elephants (Clause 11) and other wildlife species were afforded differing degrees of protection. While the South African colonies were prepared to ratify the Convention, many other colonial powers and their territories were not – hides, ivory and other hunting products being too economically important – and the Convention was unsuccessful from the start although negotiations continued sporadically until the outbreak of the First World War (Carruthers, 1995a).

A major difficulty in protecting elephants throughout Africa, historically and currently, is the different prevailing legal systems and the contrasting histories of ivory extraction and export. For example, in East Africa, Sudan and other areas, payment in tusks was made to government officials in lieu of salaries and the settler populations did not develop from the same colonial imperatives or experience the same historical trajectory as had been the case in South Africa (Carruthers, 1997; MacKenzie 1988).

From 1948 South Africa has been integrated – and at times, has played a prominent part – in many of the bodies that manage international nature conservation. These have included the World Conservation Union (IUCN) and its subsidiary and specialist organisational arms (Hall-Martin & Carruthers, 2003).

POST-SECOND WORLD WAR

By the end of the Second World War, apart from the government-attempted extermination of the Addo elephant herd, there had been a hiatus of nearly 50 years in elephant hunting in South Africa. The intervening period had witnessed the evolution of a tradition of wildlife viewing in the country's national parks, in particular the Kruger National Park, the only one that contained elephants in substantial numbers (Carruthers, 1995b). The provincial nature conservation authorities became better organised as an arm of local government in the 1940s. This was initiated by an alteration in the national financing legislation (*Financial Relations Consolidation and Amendment Act* No. 38 of 1945). In 1947 the parastatal Natal Parks, Game and Fish Preservation Board came into being. The Transvaal Nature Conservation Division and the Cape Nature Conservation Department were established in 1952, these being incorporated directly into the civil service (Carruthers, 2006b). By 1950 the South African public, both black and white, had therefore long ceased ivory extraction or elephant hunting. Values had changed and ivory was no longer a product from which individuals might make a living.

THE DISTRIBUTION AND ABUNDANCE OF THE AFRICAN ELEPHANT IN AFRICA

Distribution

The widespread distribution of elephants throughout Africa can be inferred from the centuries-old trade in ivory and by making allowances for transformations of the landscape – by way of habitat loss and fragmentation – that have taken place over the last century. Currently (2006) elephants occur in 37 range states in Africa (figure 7), with their distribution varying considerably across four regions – 'from small, fragmented populations in West Africa to vast, virtually undisturbed tracts of elephant range in Central and Southern Africa. Southern Africa has the largest extent of elephant range of any region, and accounts for 39 per cent of the species' total range area. Central and East Africa follow with 29 per cent and 26 per cent of the continental total, respectively, while West Africa accounts for only 5 per cent' (Blanc *et al.*, 2007, 21).

Numbers

In an assessment of numerical status made in 2006 (Blanc *et al.*, 2007), the quality of the data did not permit a definitive overall statement on trends in the numbers of elephants in West and Central Africa. This was, however, possible for the East Africa region where, overall, there was an increase in the 'definite' category, and for the southern Africa region, where, overall, there was a 19 per cent increase (over the previous Assessment) in this category. In terms of elephant numbers in the four African regions, southern Africa carries the highest total (table 1).

Region	Elephant numbers			
	Definite	Probable	Possible	Speculative
Central Africa	10 383	48 936	43 098	34 129
Eastern Africa	137 485	29 043	35 124	3 543
Southern Africa	297 718	23 186	24 734	9 753
West Africa	7 487	735	1 129	2 939
Total	472 269	82 704	84 334	50 364

Table 1: Regional comparison of elephant numbers in Africa in 2006. The four categories under 'Elephant numbers' refer to population estimates classified according to survey type (from Blanc *et al.*, 2007)

THE AFRICAN ELEPHANT IN SOUTHERN AFRICA, EXCLUDING SOUTH AFRICA

Historically, the African elephant probably occurred throughout the present country of Namibia (De Villiers & Kok, 1984). By the 1930s the species was limited mainly to the region known as the Kaokoveld, and to northern and north-eastern Namibia, including the Caprivi Strip (Shortridge, 1934); it still occurs over much of this area today, and also in the Etosha National Park and the Kaudom Game Park (Blanc *et al.*, 2007). In 2006, it was estimated that Namibia held around 12 500 elephants, but the species' full range and numbers there still need to be accurately surveyed. In present-day Botswana, elephants occurred historically mainly in the northern and eastern parts of the country (e.g. see Skinner & Chimimba, 2005). Currently, the growing Botswana population, estimated to number between 134 000 and 174 000 individuals, is restricted to the far northern part of the country (north of the Nxai and Makgadikgadi pans and including the Okavango Delta and the Chobe area), with a small population of about 1 000 animals in the Tuli area in the east (Blanc *et al.*, 2007).

The African elephant is considered to have been widespread in the past in the area encompassed by present-day Zimbabwe. For example, it was reported that 'vast numbers' of elephants occurred in Matabeleland and Mashonaland in the 1871–1875 period (Bryden, 1889). Owing to persecution and displacement by humans, the total number of elephants dropped to below 4 000 in 1900, before recovering during the twentieth century (Skinner & Chimimba, 2005) to a population of over 80 000 in 2006 (Blanc *et al.*, 2007). The majority of Zimbabwe's elephants occur in and around protected areas along the borders

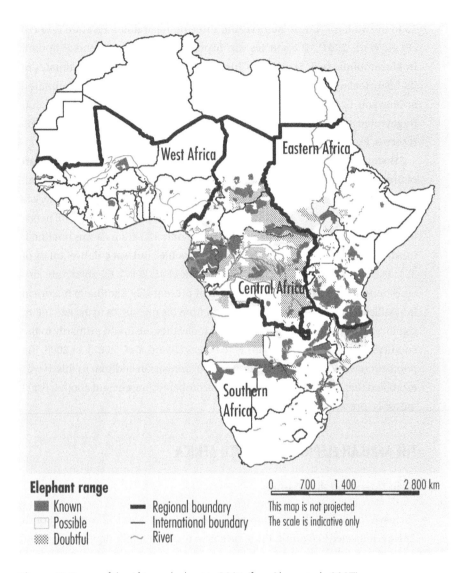

Figure 7: Range of the African elephant in 2006 (from Blanc *et al.*, 2007)

with neighbouring countries, and four main populations exist – namely North-West Matabeleland, Sebungwe, Zambezi Valley and Gonarezhou (Blanc *et al.*, 2007). There is evidence of increased elephant mortality in Zimbabwe through poaching (Dunham *et al.*, 2006).

The elephant occurred historically in Swaziland but its population there succumbed to human pressure – persecution and displacement – and it became locally extinct (see also Skinner & Chimimba, 2005). In 2006, there were only 31 elephants in the country – 13 in the Hlane Royal National Park,

15 in the Mkhaya Nature Reserve and three in the Malolotja Nature Reserve (Blanc *et al.*, 2007). The species was historically widespread and abundant in Mozambique (e.g. Skinner & Chimimba, 2005); however, no estimates of its historical abundance exist. The population in this country, estimated at between 14 000 and 19 000 individuals in 2006, has become somewhat fragmented and is now restricted to 16 areas, made up of national parks, game reserves, forest reserves and state land (Blanc *et al.*, 2007).

Historically, the elephant is considered to have occurred over a large part of present-day Malawi. Today, elephant populations in this country are small and fragmented, being confined almost entirely to eight parks and reserves (Blanc *et al.*, 2007). The actual size of Malawi's elephant population still needs to be determined. In 2006 a 'definite' total of only 185 animals was obtained; in addition to this, there were 'probable', 'possible' and 'speculative' totals of 323, 632 and 1 587 animals, respectively (Blanc *et al.*, 2007). Elephants are also considered to have occurred over much of present-day Zambia in historical times. Illegal hunting for ivory, particularly from the mid-1970s to the late 1980s, significantly reduced the population and today they are found primarily in the country's extensive system of protected areas (Blanc *et al.*, 2007). In 2006, the population was estimated at around 16 500 animals; in addition to this, it was estimated that around 6 000 animals were 'probably' present and another 6 000 'possibly' present (Blanc *et al.*, 2007).

THE AFRICAN ELEPHANT IN SOUTH AFRICA

Early/historical distribution

The early/historical distribution of elephants in the area today called South Africa is indicated in figure 8. The information used to compile figure 8 represents a combination of records from skeletal material, indigenous art and historical records (e.g. sightings); a radius (= travelling distance) of 50 km from each spot locality was used to estimate the past distribution. Notwithstanding the fact that some of the records used to create figure 8 may be related to human movement (e.g. translocation of bone material by humans), it would appear that at some or other time in the past, elephants could or did occur over much of what is now South Africa, including the arid north-western parts. However, the available information is not systematic and distribution gaps may be the result of a lack of information and/or reliable historical records. In addition, the distribution pattern 'telescopes' time and should not be interpreted to show that elephants occurred at all the given localities at the same time. For example, it is likely that by the end of the eighteenth century there were no longer elephants in the arid parts of the central and north-western parts of the country.

It is noteworthy that there is evidence that elephants were present in the dry interior of the country, especially in the north-west (figure 8). African elephants are known, throughout their range, to undertake long-distance movements, especially in arid areas, such as the Kaokoveld in Namibia (Viljoen, 1989). Based on archaeological finds from the Karoo and on historical accounts it has been surmised that as recently as 1750 elephants moved freely between the coastal and sub-coastal parts of the present Eastern Cape to the Orange River in the north (Vernon, 1990).

A more recent analysis and appraisal of available information from the broader Eastern Cape (Boshoff *et al.*, 2002) has led to the creation of three likely zones of historical elephant occurrence in that region (table 2). Given the dependence of elephants on surface water, along with suitable forage (Skinner

Figure 8: Early/historical distribution of elephants in the area covered by the present South Africa, based on skeletal material, indigenous art and historical records (adapted from Ebedes *et al.*, 1995)

& Chimimba, 2005), it is considered that these distribution patterns were determined by the availability of these key resources.

	Coastal zone	Sub-coastal zone (S of the Great Escarpment)	Inland zone (N of the Great Escarpment)
Density of elephants	Relatively high	Relatively low	Largely absent, or at a very low density
Status of elephants	Mainly resident, but local movements undertaken	Some may have been resident but most were local migrants or nomads	Present only as occasional migrants or nomads, mainly as travellers between the coastal and sub-coastal zones and the Orange River
Habitats occupied by elephants	Present throughout most of the mosaics of forest, thicket and savanna	Present mainly in the wide river valleys, vegetated with riverine forest and thicket. Interfluves also used	In transit through karroid vegetation. The riparian and kloof vegetation was most likely also utilised

Table 2: The density, status and habitats of elephants in three likely zones of occurrence in the broader Eastern Cape (after Boshoff *et al.*, 2002)

The map of the early/historical distribution of elephants in the broader Eastern Cape (figure 1) is devoid of sight records of elephants north of the Great Escarpment; this is considered to be the result of a combination of the highly ephemeral nature of the species' occurrence there and the paucity of observers there during the eighteenth century, when elephants may still occasionally have passed through that area. It is surmised that, by the end of the eighteenth century, elephants no longer moved between the coastal areas and the hinterland north of the Great Escarpment.

The size of the population

The number of elephants that occurred in pre-colonial South Africa cannot be determined, but it has been suggested that there may have been around 100 000 before 1652 (Hall-Martin, 1992). Recent, substantiated estimates have the historical (pre-colonial) population of the Cape Floristic Region (mainly in the Western Cape) at close to 3 000 elephants (Kerley *et al.*, 2003), and that of the core of the subtropical thicket biome (mainly in the Eastern Cape, west of the Kei River) at almost 6 000 elephants (Kerley & Landman, 2006). Allowing for partial overlap between these two study areas, a reasonable estimate of around 8 000 elephants is reached for the area stretching from south of the Orange River to the Kei River in the mid-nineteenth century. The density of elephants in the arid central, northern and north-western parts of this region is likely to have been very low.

The decline of the population

By the 1890s, almost all the elephants in South Africa had been exterminated (Whyte, 2001; Hall-Martin, 1992; Skead, 1980, 2007). At the time, there were three relict populations: in the Knysna area of today's Western Cape (30–50 individuals), in the Addo area of today's Eastern Cape (130–140), and an unknown number in the Sihangwane (Tembe) area of Maputaland, in today's KwaZulu-Natal. Hall-Martin (1992) reports a fourth relict population, in the Olifants Gorge area in the east of the former Transvaal Province, which was proclaimed a game reserve in 1898 (but see the account for the Lowveld below). According to Hall-Martin (1992), by 1920 there were only 120 elephants in these four populations (Lowveld, Knysna, Addo, and Tembe). This is considered to be a slight underestimate, since, at this time, the Knysna population comprised 13 individuals, the Addo population 16 individuals and the Lowveld population 100 individuals (Hall-Martin, 1992), and there was a small population of unknown size in the Tembe area.

The recovery of the population

From these small relict populations, elephants in South Africa have undergone a period of sustained growth, this as a result of a combination of the proclamation and fencing of national parks that contained elephants, natural population growth, the establishment of new national parks and provincial nature reserves and their stocking with elephants, of some immigration from Mozambique, and, more recently, the establishment of small herds in private nature reserves and on game ranches. The Kruger National Park has been the source of almost all of the translocated elephants in South Africa, although Addo has had a small role in this regard.

Between 1979 and 2001 over 800 elephants were transferred to 58 reserves in South Africa, these being mainly in Limpopo and KwaZulu-Natal provinces (Garaï et al., 2004). An example of the continuation in this trend beyond 2001 comes from the Eastern Cape (figure 9), where the number of individual populations increased from one (in the Addo Elephant National Park, in 1931) to 14 by 2006; one of the 14 new populations is in a provincial reserve (Great Fish River Reserve) while the remainder are in private reserves (Kerley & Landman, 2006).

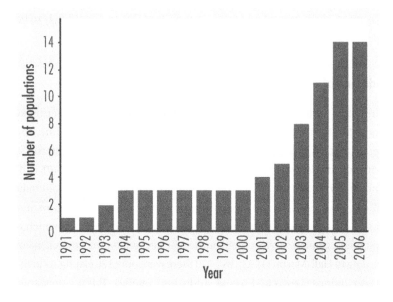

Figure 9: Increase in the number of populations of elephants in the Eastern Cape; the year of establishment is shown (from Kerley, 2006)

Elephants have also been translocated beyond the borders of South Africa. In 2000, a group of 16 animals was moved from the Madikwe Game Reserve in North West Province to the Quiçama National Park in Angola (Van Hoven & Du Toit, 2001). In the early 2000s, three groups (25 in 2001; 48 in 2002; 38 in 2003) were translocated from the Kruger National Park to neighbouring Mozambique, within the area that comprises the Kruger National Park-Gonarezhou-Limpopo Transfrontier Conservation Area (Johan Malan, South African National Parks, pers. comm. 2007).

According to Blanc *et al.* (2007), the species' current (2006) range in South Africa (figure 10) comprises at least 34 individual populations, and a 35th, collectively termed 'Private reserves', in the Western Cape, Eastern Cape, KwaZulu-Natal, Mpumalanga, Limpopo and North West provinces. The increase in the number of elephant populations in private reserves in these provinces is also reflected in the 2006 data presented by Blanc *et al.* (2007); here, of 35 populations listed, 12 are in national parks or provincial nature reserves, or other state land (Knysna Forest Reserve), with the remainder being located within private reserves (the latter includes a category 'Private reserves', most of them located along the western edge of the Kruger National Park). By 1990, however, the largest elephant populations on private land were those in the Klaserie (395 animals) and Timbavati (167) private nature reserves, to the west of Kruger (Hall-Martin, 1992). By 2006, these totals had increased to 569 and 712, respectively (Blanc *et al.*, 2007), probably relating to immigration from the Kruger Park as the intervening fences were removed.

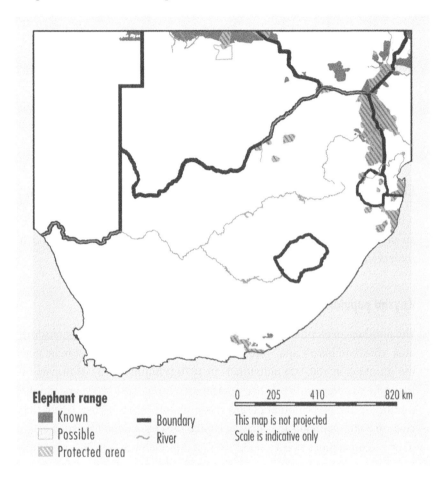

Elephant range

▮ Known   ▬ Boundary
☐ Possible   ～ River
▨ Protected area

0   205   410   820 km

This map is not projected
Scale is indicative only

Figure 10: The distribution of the African elephant in South Africa in 2006 (from Blanc *et al.*, 2007)

From the information presented above, it is clear that the numbers of individual elephant populations in South Africa, while limited to small reserves for the most part, have increased substantially, with the increase on private land being particularly noteworthy. The summary totals for elephants in the country for 2006 are provided in table 3; there is an increase of 26.8 per cent in the 'Definite' category between 2002 (14 071 individuals) and 2006 (17 847 individuals). The individual populations in South Africa vary from tiny to large; in 2006 the smallest was in the Great Fish River Reserve in the Eastern Cape (2 individuals) and the largest, by far, was in the Kruger National Park and surrounding areas (12 427 individuals).

	Definite	Probable	Possible	Speculative
Totals (2002)	14 071	0	855	0
Totals (2006)	17 847	0	638	22

Table 3: Summary totals of African elephants in South Africa. The four categories under 'Elephant numbers' refer to population estimates classified according to survey type (from Blanc *et al.*, 2007)

CASE STUDIES OF FOUR ELEPHANT POPULATIONS IN SOUTH AFRICA

A brief history of each of the relict populations of Knysna, Addo, and Tembe is given below, together with the history of distribution in the Lowveld, in which there were no resident elephants at the turn of the century.

Knysna population

The numbers of elephants that occurred in the Swellendam to Humansdorp area – the 'Southern Cape' – are not known. A 'well-informed' estimate puts the numbers at 400–500 individuals in 1876 (Phillips, 1925). However, by that time the elephants had been persecuted for at least 100 years and the number in the pre-colonial period was probably much higher. Given that close to 3 000 elephants may have occupied the entire Cape Floristic Region in pre-colonial times (Kerley *et al.*, 2003), it is probably reasonable to assume that approximately 1 000, or more, elephants once occupied the Outeniqua-Tsitsikamma area, where Knysna is situated (Boshoff *et al.*, 2002). The demise of the elephants accelerated during the early part of the eighteenth century, when trekboers and other individuals began to expand the boundaries of the Dutch East India Company settlement at the Cape (Skead, 1980, 2007); during the nineteenth century, 'residents of the Outeniqua and Tsitsikamma region ... regularly hunted elephants, some for sport and entertainment, and others for meat and ivory' (Roche, 1996). The Southern Cape elephants were also hunted to protect crops, property and human lives (Roche, 1996). According to Roche (1996), the rising population and economic growth of Knysna between 1856 and 1886 was 'crucial to the decrease in elephant numbers' in the district and by 1900, only 30–50 animals remained (Hall-Martin, 1992). A century later, in

2000, fewer than five individuals were left and only an estimated four in 2006 (Blanc *et al.*, 2007). The decline of the elephant in the George–Tsitsikamma area is documented in table 4. In 1994, three sub-adult females from the Kruger National Park were translocated to the Knysna forest area to supplement the declining population (Seydack *et al.*, 2000), but within four years of the translocation one animal had died and the remaining two were moved to a private nature reserve in the Eastern Cape. A recent study (Eggert *et al.*, 2007) has provided evidence for the presence of five elephants, their levels of genetic diversity being similar to those in the Kruger National Park, thereby suggesting that the Knysna elephants represent a remnant of the once widespread population of elephants in South Africa.

Year/period	Estimated number of elephants
1650	1 000?
1876	400–500
1879	not less than 200
1884	200
1900	40–50
1905	20
1910	15
1920	13
1925	12
1969/70	10–13
1990	4
2001	3

Table 4: The decline of the George–Tsitsikamma elephant population, 1650–2001, based on historical evidence. Data from Boshoff *et al.* (2002) and Hall-Martin (1992)

The forests of the Knysna district are not typical elephant habitat, but elephants moved into the forests from adjacent open areas to find refuge from persecution and displacement by humans and their activities (Skinner & Chimimba, 2005). The decline of this population is attributed to the poor quality of their atypical habitat (Seydack *et al.*, 2000).

Addo population

In pre-colonial and pre-Xhosa southern Africa, the elephant was a prominent herbivore on the landscapes of the Eastern Cape that were vegetated with dense, spiny and succulent subtropical thicket, including those areas where the thicket formed mosaic communities with other major vegetation types, e.g. savanna (Skead, 2007). This is considered to be prime habitat for elephants (Cowling *et al.*, 2003; Kerley *et al.*, 2003). FitzSimons (1920) reported often finding elephant remains – tusks, teeth and skeletons – in the vicinity of Port Elizabeth and said that he believed that they had been hunted mostly for their meat by 'Boer hunters of old'. An indication of the extent of the former distribution of elephants in this region is provided in figure 1.

Following the large-scale destruction of elephants, and the transformation or degradation of their habitat, the region's population underwent a progressive decrease, until, by the 1880s, the remaining herds were confined to the Uitenhage, Addo, and Alexandria districts, and eventually a remnant herd of 130–140 individuals survived in the dense thicket of the Addo area (Skead, 2007). In the early 1900s agricultural development led to conflicts between the elephants and local farmers in the district (Hoffman, 1993), and this resulted in demands for the extermination of the remaining elephants in the area. Both national and provincial political authorities favoured agricultural development and protection for the farming community, and a professional hunter – Major P.J. Pretorius – was contracted to eliminate the elephant population. Between July 1919 and August 1920 he shot approximately 120 of the creatures, leaving 16 survivors when the programme was halted, apparently by a change of heart towards protection on the part of Pretorius and rising public opinion to save the remaining elephants (Hoffman, 1993).

The numbers continued to dwindle until, in 1931, only 11 individuals remained (Hall-Martin, 1992; Whitehouse & Hall-Martin, 2000). The proclamation of the Addo Elephant National Park (AENP) in 1931 finally provided some protection for the Addo herd, but there was no strategy to manage them except plans for the erection of a boundary fence around the park. But even before the erection of the fence, there was an overall trend of a progressive increase in the population. By 1954, when Addo was fenced, the original herd of 11 individuals had increased to 22 elephants (Hall-Martin, 1992; Hall-Martin & Carruthers, 2003) and since then the population has shown an exponential increase in numbers (figure 11). The history of the Addo population has been well researched (Whitehouse & Hall-Martin, 2000; Whitehouse & Kerley, 2002), and by 2006, it comprised 459 individuals (Blanc *et al.*, 2007), with an average annual growth rate of 5.8 per cent (Kerley & Landman, 2006).

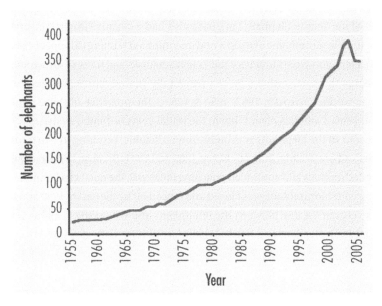

Figure 11: Overall increase in elephant population size in the Addo Elephant National Park since the park was fenced in 1954. The decline from 2003 was caused by the translocation of a large group of animals to a concession area (adapted from Kerley & Landman, 2006)

Maputaland (Tembe) population

In southern Mozambique and northern KwaZulu-Natal, the Maputo Elephant Reserve, Futi Corridor and Tembe Elephant Park presently have elephant populations separated from each other either by electric fences or human-made barriers. These sub-populations represent the remaining fragments of the coastal plain population that, until 1855, roamed as far south as the White Umfolozi River (Klingelhoeffer, 1987). Zululand's last elephant – allegedly the last of Cetshwayo's 'herd' – was found dead on a Mr van Rooyen's farm on the northern bank of the Umfolozi River in January 1918, having been killed in order to provide medicine for the local human inhabitants (Fitzsimons, 1920).

In 1983 the Tembe Elephant Park was established in response to increasing levels of human-elephant conflict, particularly the raiding of crops, mainly by bull elephants at night. The southern, western, and eastern borders were fenced with an electric game-proof fence, while initially the northern boundary was left open. Since then, both the Maputo Elephant Reserve and Tembe Elephant Park elephant populations have suffered heavy poaching (Smithers & Tello,

1976; Hatton *et al.*, 2001). Political and military circumstances in Mozambique north of the Tembe Elephant Park have also had a negative impact, involving considerable movements of people and uncontrolled hunting (Kloppers, 2001 & 2004). Elephants were hunted for their meat and tusks and some of the wounded elephants that were either shot or incidentally snared returned to Tembe to recover or die (Ostrosky, 1987, 1988 & 1989). The presence of these injured and unsettled animals meant that in the initial years of Tembe's existence, the behaviour of the elephants was highly unpredictable, resulting in two human deaths and one serious injury between 1985 and 1994. Because of the danger of wounded animals and continued poaching incidents, the northern boundary of Tembe with Mozambique was closed and fenced off by the end of 1989 and this effectively cut the link between the floodplains and the northern Maputaland Coastal Plain elephant population (Hall-Martin, 1988; Ostrosky, 1989).

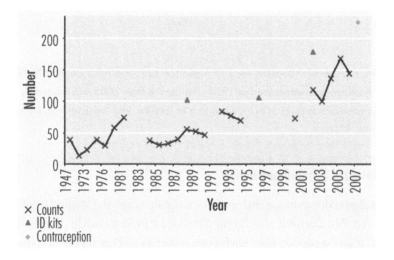

Figure 12: Summary of Tembe Elephant Park elephant counts from 1947 to 2006. Based on the data of Ferraz & Lugg, in Bruton & Cooper (1980); Dutton, in Ostrosky (1988); Thomson (1974); Hall-Martin (1976, 1988); Klingelhoeffer (1987); Ostrosky (1988); Ward (1986–90); Matthews (1992–1994; 2002–2006); Ridgeway & Jenkins (1996); Morley & van Aarde (2002); Matthews *et al.* (2007)

In 1940 around 40 elephants were thought to be resident in this region, with estimates varying between 30 and 85 animals. It has subsequently been difficult to count the elephants in this area although ground (1971 and 1973) and aerial counts (1974-2000) have been done. Although transect sampling was done as described by Norton-Griffiths (1978), this was not standardised and therefore not reliable even though waterhole counts were used as supporting evidence. Currently the rate of increase has been estimated as being between 5.1 per cent pre-fencing and 8.3 per cent post-fencing of the park (Morley & van Aarde, 2002).

The growing elephant population in Tembe Elephant Park (figure 12) comprises 200–226 animals (Matthews, 2002–2006), and that of the Maputo Elephant Reserve approximately 329–350 (Matthews & Momade, 2006), which are essentially free roaming. An unknown number of elephants are resident in the Futi Corridor but they are under constant human threat (Ostrosky & Matthews, 1995). Local opinion maintains that a more or less stable breeding group has probably been resident in the region of Tembe Elephant Park for a long time. The limits to its home range are thought to have been the Muzi swamps in the south-east and the Rio Maputo floodplains, some 28 km to the north-west in Mozambique. Previously, this area would have included all of the northern and most of the central sections of the current extent of the Tembe Elephant Park.

There are ongoing initiatives to form the Futi–Tembe Transfrontier Conservation Area straddling this South Africa–Mozambique border (Hanks, 2000; Hall-Martin & Carruthers, 2003; Peace Parks Foundation, 2006; Porter *et al.*, 2004) and once this has been accomplished, the former range and historical roaming patterns of these elephants will be restored.

Lowveld population

It is clear from the thriving ivory trade out of Delagoa Bay from the mid-1700s that elephants occurred in the savanna areas of the Lowveld. However, there is a general paucity of precise records for the region and those that exist are open to different interpretations. It has been suggested that if elephants occurred in this region, they would have been at relatively low densities. Nevertheless, at the end of the nineteenth century, despite the prevalence of diseases, such as malaria, that inhibited human occupation, the elephant population in the Lowveld was heavily exploited and between 1880 and 1896 the last remaining elephants in that region were exterminated. Thus, when the Sabi Game Reserve was proclaimed in 1898, it was devoid of a resident population of elephants although a few may have survived in the Olifants Gorge area outside the game reserve to the north (Hall-Martin, 1992).

However, a population of elephants remained in Portuguese territory (Mozambique, where the species was once abundant); by 1905 a few individuals originating from this territory had been observed in the Sabi Game Reserve, and by 1910 a small population had established itself there.

Increasingly, elephants moved across into the reserve and eventually, by the early 1930s, a breeding population had established itself. The Kruger National Park was proclaimed in 1926 after consolidating various portions of state and private land. Warden Stevenson-Hamilton (1947) stated that the 'favourite' country for elephants lay between the Letaba and Olifants rivers and that in 1926 there were probably about 100 elephants in this area having immigrated from Mozambique (Stevenson-Hamilton, 1934), making a total of 400 by 1938 in the park (Stevenson-Hamilton, 1947).

The number of individuals and breeding herds in Kruger steadily increased, through natural growth and immigration (Hall-Martin, 1992), until the park supported 6 586 elephants in 1967 (table 5). The rate of increase was initially slow (between 1905 and 1925), then relatively rapid between 1925 and 1945, but subsequently slowed again (Whyte, 2001), prior to a significant increase through immigration in the 1960s from Mozambique and Zimbabwe (Hall-Martin, 1992). Active management of the population, in the form of culling, was initiated in 1967 with the aim of maintaining the population at around 7 000 elephants (see below).

Year	Number	Nature of estimate
1903	0	Estimate
1905	10	Estimate
1908	25	Estimate
1925	100	Estimate
1931	135	Estimate
1932	170	Estimate
1933	200	Estimate
1936	250	Estimate
1937	400	Estimate
1946	450	Estimate
1947	560	Estimate
1954	740	Estimate
1957	1 000	Estimate
1960	1 186	Aerial survey
1962	1 750	Fixed-wing survey
1964	2 374	Helicopter count
1967	6 586	Helicopter count

Table 5: Estimates of numbers of elephants in the Kruger National Park from 1903 to 1967, when population management through culling was initiated (from Whyte, 2001)

Before the erection of a fence along the South African–Mozambique border (1974–1976) elephants could move freely between the two territories. In 1994, the western boundary fence dividing the Kruger Park from the private nature reserves between the Sabie and Olifants rivers was removed in order to allow the free movement of elephants and other wildlife (Whyte, 2001). Elephant numbers increased from 7 806 in 1994 to 10 459 in 2002. An additional development was the incorporation of the Kruger National Park into the Greater Limpopo Transfrontier Conservation Area, which opened up new areas of habitat to elephants and other species in Mozambique, where animal numbers are currently very low. Social and economic changes have profoundly affected the habitat for elephants and other wildlife in this region. For example, large areas outside of parks have been closed off to most species through intensified settlement and agricultural activity. In addition, access to water, migration routes and vegetation has diminished since the eighteenth century.

HISTORY OF ELEPHANT MANAGEMENT IN SOUTH AFRICA

Changing philosophies of elephant management

Elephants have been managed in South Africa for many decades but this has not occurred in isolation from the management of other forms of wildlife in the region. In this context, it is useful to briefly review the change in philosophies of elephant management within the context of those pertaining to the management of wildlife in general in South Africa.

Until after the Second World War there was no formal scientific basis for wildlife management in South Africa's protected areas (Carruthers, in press). Game reserves of the early twentieth century were based on very old ideas of custodianship, founded on managing for 'sport hunting', which involved eradicating predators on 'game species' (generally antelope) and preventing poaching. In the 1930s and 1940s it seemed that the discipline that would co-opt wildlife conservation and management would be veterinary science. However, there were contrary underlying trends in these decades that spawned a suite of ecological and environmental sciences that have more recently come to enjoy high-ranking academic status and to influence wildlife management.

Modern wildlife management principles emerged principally from the rise of ecology in the second half of the twentieth century. Plant ecology, the first to gain credibility, was predicated on its possible benefits for increasing the production capacity of healthy grassland. Then, in 1927, Charles Elton published his famous book, *Animal Ecology*, which conceptualised nature literally as an economy, with energy flowing in terms of supply and demand. This idea was named an 'ecosystem' by Arthur Tansley (in response to John Phillips, a renowned South African botanist who was influenced by the 'holism' of Jan Smuts) in 1935 (Anker, 2001; Kingsland, 2005; Phillips, 1934 & 1935). Later, together with the ideas of G. Evelyn Hutchinson (who taught zoology at the University of the Witwatersrand for two years before moving to the United States), a leading imaginative and innovative theorist in developing cybernetics and modelling, ecology moved away from its 'natural history' base and evolved into systems theory through the work of Hutchinson's students including R. Lindeman and H.T. Odum (Golley, 1993; Hagen, 1992; Worster, 1993; Kingsland, 2005). In the South African wildlife management arena of the 1930s, Elton corresponded with Stevenson-Hamilton in 1937 on the possibilities of doing controlled census counts of animals along accessible roads at fixed points and even inventing some kind of 'marking bullet' so that animals might be tagged and recognised (Carruthers, 2001). But these were ideas ahead of their technical possibility and nothing emanated in terms of management.

The point to note is that these early ecologists were investigating the notion of ecological 'climax' and how to construct and maintain a stable environmental state (Clements, 1916; Phillips, 1959). There was a long time-lag between these ideas reaching scientific acceptance and their application in the management practices of national parks in Africa. After a period of fluctuating ideas in the 1950s, these emerging ecological notions developed fully as what has subsequently come to be referred to as a 'command-and-control' methodology (Holling & Meffe, 1996) in the 1960s when some protected area scientists attempted to stabilise, maintain, and engineer the ecosystems they managed. There were flaws in this thinking. For example, we have subsequently learnt that selection for resilience and long-term survival prevails over selection for maximum current production (Denison *et al.*, 2003) or stability. Wildlife management strategies following command-and-control (or management by intervention) were aimed at producing vast landscapes of perennial grass with the 'correct' and 'best' balance of herbivores. In addition, there was scant attention to scale: if an elephant in a 1ha paddock trial exhausted its food supply within a month, then it 'followed' that 1 000 elephants would denude an area 1 000 times as large in the same period. Cause and effect were usually seen as straightforward and a system's complexity was regarded as resolvable if it were reduced to a series of simple cause-effect chains.

By the 1970s, however, some scientists had begun to question why these complicated models of the natural world seemed to match it so poorly in practice. They adopted ideas from 'complexity' (Cilliers, 1998), a field which had earlier theoretical origins. Complexity differs from complicatedness in that a complex system may have few or many components but it has feedbacks (desirable or undesirable 'vicious circles' which cause non-linear reactions, i.e. the system heads off in a completely different direction from that expected). This newer thinking – a dramatic shift conceptually and philosophically (Gunderson & Holling, 2002; Bradshaw & Bekoff, 2001; Bradshaw & Borchers, 2000) – provided for a series of alternate system states that change over time and this better resonated with what scientists experienced in practice, namely ever-changing systems, the dynamics of which were difficult to predict, but which, when analysed retrospectively, made perfect sense. In other words, it reflects the realities facing on-the-ground researchers and management. For instance, instead of seeking one ultimate final state towards which an ecosystem would evolve, complexity accommodates 'what happens' more often than not, and raises the possibility of different outcomes, thus undermining the paradigm of science and management with certainty. In addition, scale became explicit, and spatial variation (called heterogeneity) became critical in the management of a system (Walker & Salt, 2006). Although all these fundamental ways in which the world appeared to work led to different management strategies, each has its particular use and the important art is to recognise when and where to use them and when to change.

While 'command-and-control' still has validity in specific circumstances, it is increasingly being replaced by 'adaptive management', which accords with the scientific paradigm described above. This style of decision-making for natural resource management deals with complexity and uncertainty by processes of probing and testing (so-called 'active' adaptive management), and supports ongoing learning (Rogers, 2006). It is characterised by feedbacks continuously influencing action. 'Strategic' adaptive management deals with future action by setting objectives, but it nevertheless anticipates surprises and the structured re-setting of objectives as people learn.

There are other important dimensions to take into account when analysing changing ideas around wildlife management. For instance, there was a fairly clear continuum of development from individual species' concerns (before 1950), followed by a period during which populations and communities of organisms were a central focus of management, leading to full ecosystem management in recent years (McNeely, 1993). A focus on biodiversity (see Chapter 3) has also had an influence, widening its scope to embrace structure, composition and, finally, function (= process). Not all African countries have managed their elephant populations in the same way over the last 100 years, nor have the same principles underlain their scientific and conservation practice. This is not the appropriate place to provide a comparison of the varying geographical scales, economic or other policy drivers, rural and urban population dynamics and tourism profiles of differing African states, but South Africa may be exceptional in Africa in terms of some of its management strategies (Leakey, 2001; Parker, 2004).

Understanding relative input of role-players, and the underlying drivers in elephant management decision-making: A case study from the Kruger National Park

The Kruger National Park has a long and well documented history of elephant management and for this reason is presented as a case study. (It must be noted that some smaller provincial parks, e.g. North West, were not managed to the same extent of command-and-control as Kruger, although elephant management did not differ markedly.) A number of documents have been scrutinised for the purpose: KNP Masterplan (Vol. I through V, Joubert 1986; KNP Masterplan Vol. VI, Joubert, 2007; KNP Masterplan Vol. VII and VIII, Braack, 1997; KNP Park Management Plan (draft), SANParks 2006; Minister's SRT process, Owen-Smith *et al.*, 2006; Du Toit *et al.*, 2003). What follows is not a critique, because past decisions were taken with the best available information available and within the ecological framework of the time.

The first key change in elephant management in the Kruger National Park was the shift from preservation to culling in the 1960s, which was motivated by a number of factors: (1) intra-specific competition for space by elephants precluding population growth, (2) inter-specific competition with other large herbivores, (3) by that time the western fence had created an artificial system and elephants could therefore not be allowed to continue to increase in number (precautionary principle as defined by the Rio Declaration on Environment and Development in 1992), (4) disturbance culling along the Crocodile River in order to prevent excursions. The ecological basis of culling was to optimise production of the elephant and larger herbivore populations within perceived fodder constraints, while the trigger was the number of elephants, based on a potential stocking density concept. The number to be culled was based on observed densities at which elephant automatically dispersed, and on predicted damage to vegetation around rivers and waterholes (Van Wyk & Fairall, 1969). The decision to cull was taken internally by the National Parks Board, but it was appreciated that the general public might be alarmed at this action and an awareness campaign was advocated (Pienaar, 1960 unpublished report cited in Joubert, 1986). When working through the records one can observe that there were dichotomies even at that time, e.g. ecological considerations versus tourism development with difficult trade-offs being made and often motivated as being conservation-friendly.

This command-and-control policy (the system being interpreted as essentially simple, linearly predictable and manipulable) with an optimisation trigger continued without fundamental review for almost 30 years, this despite developments in scientific theory (ecosystem theory, heterogeneity and biodiversity), the emergence of a strong animal rights movement (changing societal values), and global concern about the decline of elephants (IUCN classification as Endangered). Eventually, there was a major shift in South African values, as well as in science, the former owing to democratisation, animal rights, CITES and the people and parks movement (table 6). In view of all this uncertainty, in 1994 Dr G.A. Robinson, then head of SANParks, placed a moratorium on elephant culling through the offices of his Director of Research, Dr A.J. Hall-Martin. Robinson challenged scientists to produce an adaptive management plan that would lead to putting together sufficient evidence to control elephants.

At the time, the IUCN African Elephant Specialist Group (AfESG) observed that the Kruger Park had poorly defined management objectives which isolated elephants from other ecosystem components. This, combined with the moratorium, prompted a complete revision of the KNP Masterplan (Vols VII & VIII; Braack, 1997a, 1997b), including a revised mission statement, an objectives hierarchy deriving from it, and the entrenchment of an explicit adaptive management approach. There was a broad public consultation process. Two key points were the continued explicit inclusion of the precautionary principle, and adoption of Thresholds of Potential Concern (TPCs) as triggers for decision-making. The new plan was significant in shifting from command-and-control to active adaptive management, and from using numbers to environmental indicators. Elephant management, however, would still be guided by 'controlled fluctuations of elephant numbers', but these were not in fact triggered by the TPC having been exceeded. The elephant policy would still use numbers as triggers (to create variation, and see what would happen at these differing densities), with the TPCs only as endpoints for judging when resultant high or low impacts were becoming unacceptable. In the sense that the primary action was not driven by TPCs, this policy differed from other 1997 policies. The changes that had been made were apparently not sufficient to warrant its acceptance at that time.

	1960s	1970s	1980s	Early 1990s	1994 moratorium	1997/99 plan	2004 indaba	2005 debates	2006 debates	2006 plan	2007 post-SRT
Management	•	•	•	•	□	•	□	□	□	•	□
Internal scientists	•	□	□	□	□	•	•	•	□	•	□
Board	•	□	□	□	•	□	•	□	□	□	□
External scientists	□	–	–	–	•	□	□	□	•	□	•
General public	□	□	□	□	□	□	□	•	•	□	□
Claimants/neighbours	–	–	–	–	–	□	•	□	□	□	□
Animal rights	–	–	–	□	•	□	•	•	•	□	□
Wilderness lobby	–	–	–	□	□	•	□	□	□	□	□

Relative contribution to the decision: – = Zero; □ = Medium; • = High

Table 6: Extent to which different interest groups influenced KNP elephant management policy

Because SANParks was not able to gain approval for the elephant policy (despite a strongly held belief that numbers of elephants urgently required management), there was a broad consultative process in 2004 (Indaba) and in 2005 (SANParks, undated), with both community stakeholders and scientists (Luiperdskloof). These developments were both a reaction by broader society to the inability of Kruger management to make progress – either to get on with culling or to get away from it, depending on their particular viewpoints – and by a concerned SANParks management that was prepared to engage even more widely. All this caused political and public pressure to mount.

Based on uncertainty and the apparent impasse, the Minister of Environmental Affairs and Tourism convened a 'Round Table' of scientific advice (SRT), the key outcome of which was: 'There is no compelling evidence for the need for immediate, large-scale reduction of elephant numbers in the Kruger National Park', although the next statement reads: 'Nevertheless, in some parks, including the KNP, elephant density, distribution and population structure may need to be managed locally to meet biodiversity and other objectives' (Owen-Smith *et al.*, 2006). This independent outside review and advice (but with SANParks represented) enabled a reflection of current knowledge that was as unbiased as possible and reflected an effort to obtain a broader expertise that was assumed to be less contingent on localised national park agendas.

A draft Kruger Management Plan (SANParks 2006) has made several important shifts in comparison with past frameworks. First, the precautionary principle has been explicitly removed and, second, all terrestrial ecosystem concerns, including elephant management, have been integrated into a unified objectives hierarchy.

Drivers and role players have played their respective parts in changing ideas around elephant management. For example, the 1965 culling decision was primarily internal to Kruger Park management, with little input either from the National Parks Board itself, or from the public. In more recent developments around the culling moratorium, the Board has taken the major role, with contributions from 'external' scientists and the animal welfare movement. Between 1997 and 2006 plans were driven mainly by a joint management/internal scientist team, with low input from the Board itself, and a moderate one both from external scientists and neighbouring communities. As the general public became more involved in the issue, there has been a recent key shift as external scientists at the Scientific Round Table have taken on an instrumental role and there appears to have been a relative decline in the role of animal welfare groups (see table 6.) Two important initiatives which have assisted SANParks in dealing with this changing situation effectively are an explicit articulation of its own management and conservation values, and a concerted thrust within Kruger to engage outside collaborative scientists, including an annual science networking meeting (see figure 13).

Observations from other parks

Very few other national or provincial parks have yet confronted the question of controlling elephant numbers. Elephant introductions to new reserves were based on a stocking density approach (Slotow, pers. obs.), and early decisions followed a command-and-control philosophy. Examples include removal of elephants from Madikwe (Chapter 8), first fencing Addo and then removing the fence to allow elephants to roam (see above and Chapter 7), and the contraception of the Makalali elephants (Chapter 6). Managers are revising their management planning to adaptive management, and recent examples of this shift in planning are the removal of one-third of the elephants (Chapter 8), fencing out sand forest patches (Chapter 7) and a comprehensive contraception programme in Phinda (Chapter 6) and Tembe in 2007 (Chapter 6 and see above). In these cases an objectives driven management plan exists, key indicators of unsustainable impact (on sand forest) have been measured and assessed, and interventions are attempting to move the current state to a predetermined target. In addition, there is a comprehensive research programme in both reserves following up on the consequences of the interventions so that learning takes place.

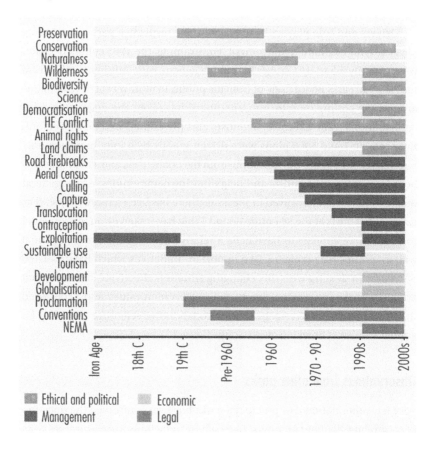

Figure 13: Diagrammatic indication of key values in elephant management in South Africa

CONCLUSION

Archaeological and historical records indicate that the African elephant once occurred, or potentially occurred, permanently or ephemerally, over most of present-day South Africa and indeed southern Africa. However, relative regional densities cannot be now determined. Through a combination of direct persecution by humans (hunting for sport, meat and ivory, and to protect human lives and crops) and habitat loss (transformation or degradation of habitat, and displacement by humans), the species was almost exterminated within South Africa by the early 1900s. Concomitant declines in range and numbers occurred in other southern African countries. The decline in the ivory market at that time broke the cycle of elephant hunting and ivory sale which had been the primary factors shaping elephant populations and distributions. As the twentieth century progressed, effective conservation measures in South Africa, and in the Kruger National Park in particular, and also in other protected areas, along with the more recent practice of establishing small populations on private land, have resulted in a burgeoning elephant population in this country. This is in contrast to some other parts of the species' range in Africa, where the population continues to decline. The history of elephants in South Africa has also illuminated how values that have been placed on elephants, as well as their exploitation, management strategies and conservation, have shifted over time to mirror the priorities and values of different societies in the region as well as global and international pressures.

REFERENCES

African Parks Foundation. http://www.africanparks-conservation.com/what_ parks_congo.html (accessed 17 September 2007).

Alpers, E. 1992. The ivory trade in Africa: An historical review. In: D. Ross (ed.) *Elephant: The animal and its ivory in African culture*. University of California, Los Angeles, 349–360.

Anker, P. 2001. *Imperial ecology: Environmental order in the British Empire, 1895–1945*. Cambridge University Press, Cambridge.

Axelson, E. 1969. *Portuguese in South-East Africa 1600–1700*. Witwatersrand University Press, Johannesburg.

Axelson, E. 1973. *Portuguese in South-East Africa 1488–1600*. Struik, Cape Town.

Barbier, E.B., J.C. Burgess, T.M. Swanson & D.W. Pearce 1990. *Elephants, economics and ivory*. Earthscan, London.

Beachey, R.W. 1967. The East African ivory trade in the nineteenth century. *The Journal of African History* 8(2), 269–290.

Beinart, W. 1982. *The political economy of Pondoland 1860 to 1930*. Ravan Press, Johannesburg.

Blanc, J.J., R.F.W. Barnes, G.C. Craig, H.T. Dublin, C.R. Thouless, I. Douglas-Hamilton & J.A. Hart 2007. *African elephant status report: An update from the African Elephant Database*. IUCN, Gland.

Bonner, P. 1983. *Kings, commoners and concessionaires: The evolution and dissolution of the nineteenth-century Swazi State*. Cambridge University Press, Cambridge.

Boshoff, A., J. Skead & G. Kerley 2002. Elephants in the broader Eastern Cape – an historical overview. In: G. Kerley, S. Wilson & A. Massey (eds) *Elephant conservation and management in the Eastern Cape.* Proceedings of a workshop held at the University of Port Elizabeth, 5 February 2002. Terrestrial Ecology Research Unit, University of Port Elizabeth. Report No. 35, 3–15.

Braack, L. 1997a. *A revision of parts of the management plan for the Kruger National Park.* Volume VII: *An objectives hierarchy for the Kruger National Park.* South African National Parks, Skukuza.

Braack, L. 1997b. *A revision of parts of the management plan for the Kruger National Park.* Volume VIII: *Policy proposals regarding issues relating to biodiversity maintenance, maintenance of wilderness qualities, and provision of human benefits.* South African National Parks, Skukuza.

Bradshaw, G.A. & M. Bekoff 2001. Ecology and social responsibility: The re-embodiment of science. *Trends in Ecology and Evolution*, 16(8), 460–465.

Bradshaw, G.A. & J.G. Borchers 2000. Narrowing the science-policy gap: Uncertainty as information. *Conservation Ecology* 4(1), 7.

Brooks, S.J. 2001. Changing nature: A critical historical geography of the Umfolozi and Hluhluwe Game Reserves, Zululand, 1887–1947. Ph.D. thesis, Queen's University, Kingston, Ontario.

Bruton, M.N. & K.H. Cooper (eds) 1980. *Studies of the ecology of Maputaland.* Rhodes University and Wildlife Society of South Africa, Grahamstown and Durban.

Bryden, H.A. 1889. *Kloof and karroo: Sport, legend and natural history in the Cape Colony, with a notice of the game birds and of the present distribution of the antelopes and the larger game.* Longmans Green, London.

Carruthers, J. 1992. The Dongola Wild Life Sanctuary: 'psychological blunder, economic folly and political monstrosity' or 'more valuable than rubies and gold'?' *Kleio* 24, 82–100.

Carruthers, E.J. 1995a. *Game protection in the Transvaal, 1846 to 1926.* Government Printer, Pretoria.

Carruthers, J. 1995b. *The Kruger National Park: A social and political history.* Natal University Press, Pietermaritzburg.

Carruthers, J. 1997. Lessons from South Africa: War and wildlife in the Southern Sudan. *Environment and History* 3(3), 299–321.

Carruthers, J. 2001. *Wildlife and warfare: The life of James Stevenson-Hamilton.* University of Natal Press, Pietermaritzburg.

Carruthers, J. 2005. Changing perspectives on wildlife in southern Africa, c.1840 to c.1914. *Society and Animals* 13(3), 183–199.

Carruthers, J. 2006a. Mapungubwe: An historical and contemporary analysis of a World Heritage cultural landscape. *Koedoe* 41(1), 1–14.

Carruthers, J. 2006b. Science, conservation and apartheid: South Africa's national parks c.1948–c.1960. Paper presented to 'Scientists and social commitment' Conference of the British Society for the History of Science, London 15–17, September 2006.

Carruthers, J. 2007. Early Boer republics: Changing political forces in the Cradle of Humankind, 1830s to 1890s. In: P. Bonner, A. Esterhuysen, & T. Jenkins (eds). *The search for origins: Science, history and South Africa's 'Cradle of Humankind'*. Wits University Press, Johannesburg, 180–199.

Carruthers, J. in press. Influences on wildlife management and conservation biology in South Africa c.1900 to c.1940. *South African Historical Journal* 58, 65–90.

Carruthers, J. & M. Arnold. 1995. *The life and work of Thomas Baines*. Fernwood, Cape Town.

Cilliers, P. 1998. *Complexity and postmodernism: Understanding complex systems*. Routledge, London.

Clements, F.E. 1916. *Plant succession: An analysis of the development of vegetation*. Publication No. 242. Carnegie Institution, Washington.

Cooke, H.B.S. 1960. Further revision of the fossil Elephantidae of southern Africa. *Palaeontologia Africana* 7, 59–63.

Cooke, H.B.S. 1993. Fossil proboscidean remains from Bolt's Farm and other Transvaal Cave breccias. *Palaeontologia Africana* 30, 25–34.

Coppens Y., V.J. Maglio, C.T. Madden & M. Beden. 1978. Proboscidea. In: V.J. Maglio & H.B.S. Cooke (eds) *Evolution of African mammals*. Harvard University Press, Cambridge MA, 336–367.

Cowling, R.M., A.T. Lombard, M. Rouget, G.I.H. Kerley, T. Wolf, R. Sims-Castley, A.T. Knight, J.H.J. Vlok, S.M. Pierce, A.F. Boshoff, A.F. & S.L. Wilson 2003. *A conservation assessment for the Subtropical Thicket Biome*. Terrestrial Ecology Research Unit, University of Port Elizabeth. Report No. 43.

Cubbin, A.E. 1992. An outline of game legislation in Natal, 1866–1912 (i.e. until the promulgation of the Mkhuze Game Reserve). *Journal of Natal and Zulu History* 14, 37–47.

Curtin, P., S. Feierman, L. Thompson & J. Vansina 1995. *African history: From earliest times to independence*. Second edition. Longmans, London.

Debruyne, R. 2005. A case study of apparent conflict between molecular phylogenetics: the interrelationships of African elephants. *Cladistics* 21, 31–50.

Delius, P. 1983. *The land belongs to us*. Ravan Press, Johannesburg.

Denison R.F., E.T. Kiers & S.A. West 2003. Darwinian agriculture: when can humans find solutions beyond the reach of natural selection? *The Quarterly Review of Biology* 78, 145–168.

De Villers, P.A. & O.B. Kok 1984. Verspreidingspatrone van olifante (*Loxodonta africana*) in Suidwes-Afrika met spesiale verwysing na die Nasionale Etoshawildtuin. *Madoqua* 13, 281–296.

Dowson, T.A. 1995. Hunter-gatherers, traders and slaves: The 'Mfecane' impact on Bushmen, their ritual and their art. In: C.A. Hamilton (ed) *The Mfecane aftermath: Reconstructive debates in South African history*. Witwatersrand University Press and University of Natal Press, Johannesburg and Pietermaritzburg, 51–70.

Dunham, K.M., C.S.Mackie, O.C. Musemburi, D.M. Chipesi, N.C. Chiwese, R.D. Taylor, T. Chimuti, C. Zhuwau & M.A.H. Brightman 2006. Aerial survey of elephants and other large herbivores in the Sebungwe region, Zimbabwe. Unpublished Report. Harare: WWF SARPO.

Dunlap, T.R. 1988. *Saving America's wildlife: Ecology and the American mind, 1850–1990*. Princeton University Press, Princeton.

Du Toit, J.T., K.H. Rogers & H.C. Biggs (eds) 2003. *The Kruger experience: Ecology and management of savanna heterogeneity*. Island Press, Washington.

Ebedes, H., C. Vernon & I. Grundling 1995. *Past, present and future distribution of elephants in southern Africa*. Department of Agricultural Development, Transvaal Region. Paper presented at an Elephant Symposium, Berg-en-Dal Camp, Kruger National Park, 29–30 April.

Eggert, L.S., G. Patterson & J.E. Maldonado 2007. The Knysna elephants: A population study conducted using faecal DNA. *African Journal of Ecology* (OnlineEarly Articles) doi:10.1111/j.1365-2028.2007.00794.x.

Eggert, L.S., C.A. Rasner & C.S. Woodruff 2002. The evolution and phylogeography of the African elephant inferred from mitochondrial DNA sequence and nuclear satellite markers. *Proceedings of the Royal Society, London* (B) 269 (1504), 1993–2006.

Eldredge, E.A. 1995. Sources of conflict in southern Africa c.1800–1830: The 'Mfecane' reconsidered. In: C.A. Hamilton (ed) *The Mfecane aftermath: Reconstructive debates in South African history*. Witwatersrand University Press and University of Natal Press, Johannesburg and Pietermaritzburg, 123–162.

Ellis, B. 1998. The impact of the white settlers on the natural environment of Natal, 1845–1870. MA thesis (History), University of Natal, Pietermaritzburg.

Esterhuysen, A. & J. Smith 2007. Stories in stone. In: P. Delius (ed) *Mpumalanga: History and Heritage*. University of KwaZulu-Natal Press, Pietermaritzburg.

Etherington, N. 2001. *The Great Treks: The transformation of southern Africa, 1815–1854*. Pearson, Harlow.

Fitzsimons, F.W. 1920. The African elephants. In *Mammals*. Vol. 3. *The natural history of South Africa*. Longmans Green, London, 242–276.

Franz-Odendaal, T., J.A. Lee-Thorp & A. Chinsamy 2002. New evidence for the lack of C4 grassland expansions during the early Pliocene at Langebaanweg, South Africa. *Paleobiology* 28, 378–388.

Franz-Odendaal, T.A. 2006. Analysis of the dental pathologies in the Pliocene herbivores of Langebaanweg and their palaeoenvironmental implications. *African Natural History* 2, 184–185.

Garaï, M.E., R. Slotow, R.D. Carr & B. Reilly 2004. Elephant reintroductions to small fenced reserves in South Africa. *Pachyderm* 37, 28–36.

Glazewski, J. 2005. *Environmental law in South Africa*. Second edition. LexisNexis Butterworths, Durban.

Golley, F.B. 1993. *A history of the ecosystem concept in ecology: More than a sum of the parts*. Yale University Press, New Haven.

Gordon, A.A. & D.L. Gordon 1996. *Understanding contemporary Africa*. Second edition. Lynne Reinner, Boulder.

Guelke, L. 1989. Freehold farmers and frontier settlers, 1657–1780. In: R. Elphick & H. Giliomee (eds) *The shaping of South African society, 1652–1840*. Maskew Miller Longman, Cape Town, 66–108.

Gunderson, L.H. & C.S. Holling 2002. *Panarchy: Understanding transformations in human and natural systems*. Island Press, Washington.

Hagen, J.B. 1992. *An entangled bank: The origins of ecosystem ecology*. Rutgers University Press, New Brunswick.

Hall, M. 1987. *The changing past: Farmers, kings and traders in southern Africa, 200–1860*. David Philip, Cape Town.

Hall-Martin, A.J. 1976. Tongaland, IUCN Pan African Elephant Survey Questionnaire, Naivasha, Kenya.

Hall-Martin, A.J. 1988. Comments on the proposed fencing of the northern boundary of the Tembe Elephant Park. Report to the KwaZulu Bureau of Natural Resources, Ulundi.

Hall-Martin, A.J. 1992. Distribution and status of the African elephant *Loxodonta africana* in South Africa, 1652–1992. *Koedoe* 35, 65–88.

Hall-Martin, A.J. & J. Carruthers (eds) 2003. *South African National Parks: A celebration*. Horst Klemm, Johannesburg.

Hamilton, C.A. (ed.) 1995. *The Mfecane aftermath: Reconstructive debates in South African history*. Witwatersrand University Press and University of Natal Press, Johannesburg and Pietermaritzburg.

Hammond-Tooke, W.D. (ed.) 1974. *The Bantu-speaking peoples of southern Africa.* Second edition. Routledge & Kegan Paul, London.

Hanks, J. 2000. The role of Transfrontier Conservation Areas in southern Africa in the conservation of mammalian biodiversity. In: A. Entwistle & N. Dunstone (eds) *Priorities for the Conservation of Mammalian Diversity. Has the panda had its day? Conservation Biology* 3. Cambridge University Press, Cambridge, 239–256.

Harris, W.C. 1840. *Portraits of the game and wild animals of southern Africa.* n.p., London.

Harris, W.C. 1852. *The wild sports of southern Africa.* Henry Bohm, London.

Hatton, J., M. Couto & J. Ogelthorpe 2001. *Biodiversity and war: A case study of Moçambique.* WWF Biodiversity Support Program, Washington D.C.

Haynes, G. 1992. *Mammoths, mastodonts and elephants: Biology, behaviour and the fossil record.* Cambridge University Press, Cambridge.

Hendey, Q.B. 1970. The age of the fossiliferous deposits at Langebaanweg, Cape Province. *Annals of the South African Museum* 69, 215–247.

Hoffman, M.T. 1993. Major P.J. Pretorius and the decimation of the Addo elephant herd in 1919–1920: Important reassessments. *Koedoe* 36(2), 23–44.

Holling. C.S. & G.K. Meffe 1996. Command and control and the pathology of natural resource management. *Conservation Biology* 10, 328–337.

Hollmann, J. (ed.) 2004. *Customs and beliefs of the /Xam Bushmen.* Wits University Press, Johannesburg.

Huffman, T. 2005. *Mapungubwe: Ancient African civilisation on the Limpopo.* Wits University Press, Johannesburg.

Jhala, J. 2006. Journey with Ganesh. *South Asian Popular Culture* 4(1), 35–47.

Joubert, S.C.J. 1986. *Masterplan for the management of the Kruger National Park.* Volume I–VI. Skukuza archives, Kruger National Park, SANParks, Skukuza.

Joubert, S.C.J. 2007. *The Kruger National Park: A History.* Volume III. High Branching, Johannesburg.

Kaiser, T.M. & T.A. Franz-Odendaal 2004. A mixed-feeding *Equus* species from the Middle Pleistocene of South Africa. *Quaternary Research* 62, 316–323.

Kalb, J.E., D.J. Froehlich & G.L. Bell. 1996. Phylogeny of African and Eurasian Elephantoidea of the late Neogene. In: J. Shoshani & P. Tassy (eds) *The Proboscidea: Evolution and palaeoecology of elephants and their relatives.* Oxford University Press, Oxford.

Kerley, G. 2006. Status of elephant populations in the eastern Cape. In: G. Kerley & J. Lessing (eds) *Proceedings of the Second Eastern Cape Elephant and Conservation Management Workshop.* Centre for African Conservation Ecology, Nelson Mandela Metropolitan University. Report No. 57, 2–4.

Kerley, G.I.H. & M. Landman 2006. The impacts of elephants on biodiversity in the Eastern Cape subtropical thickets. *South African Journal of Science* 102, 395–402.

Kerley, G.I.H., R.L.Pressey, R.M. Cowling, A.F. Boshoff & R. Sims-Castley 2003. Options for the conservation of large and medium-sized mammals in the Cape Floristic Region hotspot, South Africa. *Biological Conservation* 112, 169–190.

Kingsland, S.E. 2005. *The evolution of American ecology, 1890–2000.* Johns Hopkins University Press, Baltimore.

Klein, R.G., G. Avery, K. Cruz-Uribe & T.E. Steele 2007. The mammalian fauna associated with an archaic hominin skullcap and later Acheulean artifacts at Elandsfontein, Western Cape Province, South Africa. *Journal of Human Evolution* 52, 164–186.

Klein R.G. & K. Cruz-Uribe 1991. The bovids from Elandsfontein, South Africa, and their implications for the age, palaeoenvironment, and origins of the site. *The African Archaeological Review* 9, 21–79.

Klingelhoeffer, E.W. 1987. Aspects of the ecology of the elephant *Loxodonta africana* and a management plan for the Tembe Elephant Reserve in Tongaland, KwaZulu. M.Sc. dissertation, University of Pretoria.

Kloppers, R.J. 2001. The utilisation of the natural resources in the Matutuine District of southern Moçambique: Implications for Transfrontier Conservation. M.Sc. dissertation, University of Pretoria.

Kloppers, R.J. 2004. Border crossings: Life in the Moçambique/South Africa borderland. Ph.D. thesis, University of Pretoria.

Kruger, P. 1902. *The Memoirs of Paul Kruger.* 2 vols. T. Fisher Unwin, London.

Kunkel, R. 1982. *Elephants.* Harry N. Adams, New York.

Küsel, M.M. 1992. A preliminary report on settlement layout and gold melting at Thula Mela, a Late Iron Age site in the Kruger National Park. *Koedoe* 35(1), 55–64.

Leakey, R.E. 2001. *Wildlife wars: My fight to save Africa's natural resources.* St Martin's Press, New York.

Lee, D.N. & H.C. Woodhouse. 1970. *Art on the rocks of southern Africa.* Purnell, Cape Town.

Leslie, M. & T. Maggs (eds) 2000. *African Naissance: The Limpopo Valley 1000 years ago.* The South African Archaeological Society, Goodwin Series Vol. 8.

Lewis-Williams, D. & T. Dowson 1999. *Images of power: Understanding San rock art.* Second edition. Struik, Cape Town.

Luxmoore, R. 1991. The ivory trade. In: S.K. Eltringham (consultant). *The illustrated encyclopedia of elephants.* Salamander, London, 148–157.

Luyt, J., J.A. Lee-Thorp & G. Avery 2000. New light on Middle Pleistocene environments from Elandsfontein, Western Cape Province, South Africa. *South African Journal of Science* 96, 399–404.

Lye, W.F. (ed.) 1975. *Andrew Smith's journal of his expedition into the interior of South Africa, 1834–1836.* Balkema, Cape Town.

MacKenzie, J.M. 1988. *The empire of nature: Hunting, conservation and British imperialism.* Manchester University Press, Manchester.

Maglio, V.J. 1973. Origin and evolution of the Elephantidae. *Transactions of the American Philosophical Society* 6, 1–149.

Matthews, W.S. 1992-1994. Aerial game survey for Tembe Elephant Park. Kosi Bay Nature Reserve unpublished internal reports.

Matthews, W.S. 2000. Large herbivore population estimates for Tembe Elephant Park: October 2000. KwaZulu-Natal Nature Conservation Services unpublished report, 1–15.

Matthews, W.S. 2002–2006. Large herbivore population estimates for Tembe Elephant Park: KwaZulu-Natal Nature Conservation Services unpublished reports.

Matthews, W.S., D. Cooper & H. Bertschinger 2007. Contraception of Tembe elephant, May 2007 KwaZulu-Natal Wildlife unpublished report, 1–5.

Matthews, W.S. & N. Momade 2006. Aerial survey report for Maputo Special Reserve. KwaZulu-Natal Wildlife & Direcção Nacional de Florestas e Fauna Bravia unpublished report, 1–21.

McNeely, J.A. (ed.) 1993. *Parks for life: Report of the fourth World Congress on National Parks and Protected Areas, 10–21 February 1992.* IUCN Protected Areas Programme; WWF, Gland.

Meredith, M. 2001. *Elephant destiny: Biography of an endangered species in Africa.* Public Affairs, New York.

Meskell, L. 2007. Falling walls and mending fences: Archaeological ethnography in the Limpopo. *Journal of Southern African Studies* 33(2), 383–400.

Milner-Gulland, E.J. & J.R. Beddington 1993. The exploitation of elephants for the ivory trade: An historical perspective. *Proceedings: Biological Sciences* 252(1333), 29–37.

Mitchell, P. 2002. *The archaeology of southern Africa.* Cambridge University Press, Cambridge.

Morley, R.C. & R.J. van Aarde 2002. A historical assessment of elephant numbers in Maputaland. In: R. van Aarde & T. Jackson (eds) Restoration of the Tembe-Futi-Maputo Coastal Plains elephant population. Unpublished, Appendices to final report submitted to the Peace Parks Foundation.

Morris, A.G. 1992. *Skeletons of contact: A study of prehistoric burials from the lower Orange River Valley, South Africa*. Witwatersrand University Press, Johannesburg.

Norton-Griffiths, M. 1978. *Counting animals*. Serengeti Ecological Monitoring Programme, Nairobi.

Oliver, R. & A. Atmore 1967. *Africa since 1800*. Cambridge University Press, Cambridge.

Ostrosky, E.W. 1987. Monitoring of elephant movements across the international border between South Africa and Moçambique in the Tembe Elephant Park. KwaZulu Bureau of Natural Resources. Unpublished report, Ulundi.

Ostrosky, E.W. 1988. Monitoring of elephant movements across the international border between South Africa and Moçambique in the Tembe Elephant Park, 2nd annual report, 1988. Internal report, KwaZulu Bureau of Natural Resources, Ulundi.

Ostrosky, E.W. 1989. Monitoring of elephant movements across the international border between South Africa and Moçambique in the Tembe Elephant Park, 3rd annual report, 1989. Internal report, KwaZulu Bureau of Natural Resources, Ulundi.

Ostrosky, E. & W.S. Matthews 1995. The Transfrontier conservation initiatives in southern Maputo Province, Moçambique, comments on feasibility of the Futi Corridor. Prepared for Direcção Nacional de Florestas e Fauna Bravia (DNFFB), Moçambique.

Owen-Smith, N., G.I. Kerley, B. Page, R. Slotow & R. van Aarde 2006. A scientific perspective on the management of elephants in the Kruger National Park and elsewhere. *South African Journal of Science* 102, 389–394.

Parker, I. 2004. *What I tell you three times is true: Conservation, ivory, history and politics*. Librario, Moray.

Parker, I. & M. Amin 1983. *Ivory crisis*. Chatto & Windus, London.

Peace Parks Foundation. 2006. Project profiles. http://www.peaceparks.org/profiles/index.html. (accessed: 6 October 2006).

Peires, J.B. 1981. *The House of Phalo: A history of the Xhosa people in the days of their independence*. Ravan Press, Johannesburg.

Penn, N. 2005. *The forgotten frontier: Colonist and Khoisan on the Cape's Northern Frontier in the eighteenth century*. Ohio University Press and Double Storey, Athens, Ohio and Cape Town.

Phillips, J.F.V. 1925. The Knysna elephant: A brief note on their history and habits. *South African Journal of Science* 22, 287–293.

Phillips, J.F.V. 1934 & 1935. Succession, development, the climax, and the complex organism: An analysis of concepts, Part 1. *Journal of Ecology* 22, 2 (1934), 554–71; Part 2, *Journal of Ecology*, 23, 1 (1935), 210–46; Part 3, *Journal of Ecology*, 23, 2 (1935), 488–588.

Phillips, J.F.V. 1959. *Agriculture and ecology in Africa: A study of actual and potential development south of the Sahara.* Faber & Faber, London.

Pienaar, U. de V. 1983. Management by intervention: The pragmatic/economic option. In: N. Owen-Smith (ed.) *Management of large mammals in African conservation areas.* HAUM, Pretoria.

Plug, I. & S. Badenhorst 2001. The distribution of macromammals in southern Africa over the past 30 000 years, as reflected in animal remains from archaeological sites. *Transvaal Museum Monograph* No. 12. Pretoria.

Pollock, N.C. & S. Agnew 1963. *An historical geography of South Africa.* Longmans, London.

Porter, R., D. Potter & B. Poole 2004. Revised draft of the NTF TFCA Concept Development Plan. Unpublished Ezemvelo KZN Wildlife report.

Ridgeway, E. & L. Jenkins 1996. Elephant monitoring at Tembe Elephant Park. Unpublished internal report, KwaZulu-Natal Wildlife.

Roberts, D.L. (in press) *Lithostratigraphy of the Sandveld Group.* Council for Geoscience, Pretoria.

Roca, A.L. & S.J. O'Brien 2005. Genomic inferences from Afrotheria and the evolution of elephants. *Current Opinion in Genetics and Development* 15(6): 652-659.

Roca, A.L., N. Georgiadis, J. Pecon-Slattery & S.J. O'Brien 2001. Genetic evidence for two species of elephant in Africa. *Science* 293, 1473–1477.

Roche, C. 1996. 'The elephants at Knysna' and 'The Knysna elephants'. From exploitation to conservation: Man and elephants at Knysna 1856–1920. B.A. (Hons) thesis, Department of History, University of Cape Town.

Rogers, K.H. 2006. *Biodiversity custodianship in SANParks: A protected area management planning framework.* Report to South African National Parks, Pretoria.

Ross, D.H. (ed). 1992. *Elephant: The animal and its ivory in African culture.* Fowler Museum of Cultural History, University of California, Los Angeles.

Ross, R. 1989. The Cape and the world economy, 1652–1835. In: R. Elphick & H. Giliomee (eds) *The shaping of South African society, 1652–1840.* Maskew Miller Longman, Cape Town, 243–280.

Sanders, W.J. 2007. Taxonomic review of fossil Proboscidea (Mammalia) from Langebaanweg, South Africa. *Transactions of the Royal Society of South Africa* 62, 1–16.

Scott, W.B. 1907. *A collection of fossil mammals from the coast of Zululand*. Third and Final Report: Geological Survey of Natal and Zululand, 253–263.

Seydack, A.H.W., C. Vermeulen & J. Huisamen 2000. Habitat quality and the decline of an African elephant population: Implications for conservation. *South African Journal of Wildlife Research* 30, 34–42.

Shillington, K. 1985. *The colonisation of the Southern Tswana 1870–1900*. Ravan Press, Johannesburg.

Shillington, K. 1995. *History of Africa*. Revised edition. St Martin's Press, New York.

Shortridge, G.C. 1934. *The mammals of South West Africa*. Vol. 1. Heinemann, London.

Singer, R. & J. Wymer 1968. Archaeological investigations at the Saldanha skull site in South Africa. *South African Archaeological Bulletin* 23, 63–74.

Skead, C.J. 1980. *Historical mammal incidence in the Cape Province: Vol. 1 – The Western and Northern Cape*. Department of Nature and Environmental Conservation, Provincial Administration of the Cape of Good Hope, Cape Town.

Skead, C.J. 2007. *Historical incidence of the larger land mammals in the broader Eastern Cape*. Second edition. Boshoff, A.F., G.I.H. Kerley, & P.H. Lloyd (eds). Port Elizabeth, Centre for African Conservation Ecology, Nelson Mandela Metropolitan University.

Skinner, J.D. & C.T. Chimimba 2005. *The mammals of the southern African subregion*. Cambridge University Press, Cape Town.

Smithers, R.H.N. & J.L.P.L. Tello 1976. *Check list and atlas of the mammals of Moçambique*. National Museums and Monuments of Rhodesia, Salisbury.

South African National Parks n.d. The Great Elephant Indaba. Finding an African Solution to an African Problem. Minutes of a meeting held at Berg-en-Dal 19–21 October 2004. SANParks, Pretoria.

South African National Parks 2006. Kruger National Park Management Plan. (Draft as submitted to Department of Environmental Affairs and Tourism). South African National Parks, Pretoria. Available at: http://www.sanparks.org.

Stevenson-Hamilton, J. 1934. *The low-veld: Its wild life and its people*. Second edition. Cassell, London.

Stevenson-Hamilton, J. 1947. *Wild life in South Africa*. Cassell, London.

Sukumar, R. 1989. *The Asian elephant: Ecology and management*. Cambridge University Press, New York.

Thomson, G. 1974. The Muzi-Sihangwana elephants. Report to the Natal Parks Board, Pietermaritzburg.

Thorbahn, P.F. 1979. The precolonial ivory trade of East Africa: Reconstruction of a human-elephant ecosystem. Ph.D. thesis, University of Massachusetts.

Todd, N.E. 2005. Reanalysis of African *Elephas recki*: Implications for time, space and taxonomy. *Quaternary International* 126-128, 65-72.

Van der Merwe, P.J. 1938. *Die trekboer in die geskiedenis van die Kaapkolonie*. Nasionale Pers, Cape Town.

Van der Merwe, P.J. 1945. *Trek*. Nasionale Pers, Cape Town.

Van Hoven, W. & J.G. du Toit 2001. Airlifting elephant families to Angola. In: Penzhorn, B.L. (ed.) *Proceedings of a symposium on the relocation of large African mammals*. Faculty of Veterinary Science, University of Pretoria, Pretoria.

Van Sittert, L. 2005. Bringing in the wild. *Journal of African History* 46, 269-291.

Van Wyk, P. & N. Fairall 1969. The influence of the African elephant on the vegetation of the Kruger National Park. *Koedoe* 12, 57-89.

Vernon, C.J. 1990. Famous hunters of the past: 1 – The Krugers and the possibility of elephant in the Karoo. *Pelea* 9, 115-117.

Viljoen, P.J. 1989. The ecology of the desert dwelling elephants *Loxodonta africana* of western Damaraland and Kaokoland. Ph.D thesis, University of Pretoria.

Vincent, J. 1970. The history of Umfolozi Game Reserve, Zululand, as it relates to management. *Lammergeyer* 11, 7-49.

Voigt, E.A. 1983. *Mapungubwe: An achaeozoological interpretation of an Iron Age community*. Transvaal Museum, Pretoria.

Wagner, R. 1980. Zoutpansberg: The dynamics of a hunting frontier, 1848-1867. In: S. Marks & A. Atmore (eds) 1980. *Economy and society in pre-industrial South Africa*. Longman, London.

Walker, B. & D. Salt 2006. *Resilience thinking: Sustaining ecosystems and people in a changing world*. Island Press, Covelo.

Ward, M.C. 1986-1990. Aerial survey of Tembe Elephant Park. Kosi Bay Nature Reserve, unpublished internal reports.

Whitehouse, A.M. & A.J. Hall-Martin 2000. Elephants in the Addo Elephant National Park, South Africa: Reconstruction of the population's history. *Oryx* 34, 46-55.

Whitehouse, A.M. & G.I.H. Kerley 2002. Retrospective assessment of long-term conservation management of elephants in Addo Elephant National Park, South Africa. *Oryx* 36(3), 243–248.

Whyte, I.J. 2001. Conservation management of the Kruger National Park elephant population. Ph.D. thesis, University of Pretoria.

Whyte, I.J., H.C. Biggs, A. Gaylard & L.E.O. Braack 1999. A new policy for the management of the Kruger National Park's elephant population. *Koedoe* 42, 111–131.

Whyte, I.J., R. van Aarde & S.L. Pimm 2003. Kruger's elephant population: Its size and consequences for ecosystem heterogeneity. In: J.T. du Toit, K.H. Rogers & H.C. Biggs (eds) *The Kruger experience: Ecology and management of savanna heterogeneity*. Island Press, Washington, 332–348.

Wickens, P.L. 1981. *An economic history of Africa from the earliest times to partition*. Oxford University Press, Cape Town.

Wilson, M. 1969a. The Sotho, Venda and Tsonga. In: M. Wilson & L.M. Thompson (eds) *The Oxford history of South Africa*. Vol. 1. *South Africa to 1870*. Clarendon Press, Oxford, 131–186.

Wilson, M. 1969b. Co-operation and conflict: The Eastern Cape frontier. In: M. Wilson & L.M. Thompson (eds) *The Oxford history of South Africa*. Vol. 1. *South Africa to 1870*. Clarendon Press, Oxford, 187–232.

Worster, D. 1977. *Nature's economy: The roots of ecology*. Cambridge University Press, Cambridge.

Worster, D. 1993. *The wealth of nature: Environmental history and the ecological imagination*. Oxford University Press, New York.

2

ELEPHANT POPULATION BIOLOGY AND ECOLOGY

Lead author: Rudi van Aarde
Authors: Sam Ferreira, Tim Jackson, and Bruce Page
Contributing authors: Yolandi de Beer, Katie Gough, Rob Guldemond,
Jessi Junker, Pieter Olivier, Theresia Ott, and Morgan Trimble

INTRODUCTION

THE ELEPHANT debate deals largely with population size, how elephant numbers change over time, how they may affect other species (e.g. Owen-Smith *et al.*, 2006; Van Aarde *et al.*, 2006), and how elephants should be managed (e.g. Whyte *et al.*, 2003; Van Aarde & Jackson, 2007). Changes in elephant numbers are the basis of many management plans and policies. For instance, the Convention on International Trade in Endangered Species of Wild Fauna and Flora (CITES) utilises trends in numbers and poaching data to inform ivory trade decisions (Hunter & Milliken, 2004). Past decisions to cull elephants in several parks across the southern African subcontinent have also been motivated by numbers and trends in numbers over time (Cumming & Jones, 2005).

The focus of past management on numbers, rather than impact, may have detracted from the ultimate goal of controlling or reducing the effect elephants had on vegetation, other species, and people. The limited options available when managing numbers (see chapters on contraception, translocation and culling) and the emotive issues that surround this may also detract from its popularity and effectiveness. However, a multitude of options exists and can be developed to manage impact (see Chapter 12). Ultimately, the effectiveness of management hinges on monitoring the outcomes for impact, which include the response of affected species, ecological processes, elephant range utilisation, and elephant numbers. This monitoring may be done on a local scale (e.g. around waterholes), at the park level (e.g. to monitor the effectiveness of contraception and culling), or on the regional scale (e.g. to monitor the effectiveness of restoring seasonal and large-scale movement patterns). Therefore it is important to unravel and understand the mechanisms that determine spatial utilisation patterns and how numbers vary across space and time. This chapter focuses on assessing our understanding of the factors that determine these variables.

In this chapter we compare the social, spatial, and demographic profiles of South Africa's elephant populations to those of elephant populations elsewhere in Africa. We also make a concerted effort to explain similarities and differences, and we use these to evaluate the response of elephant populations to their living conditions in South Africa's conservation areas. For the spatial aspects, we compare South Africa's elephants with those living across the environmental gradient typical of southern Africa. For the demographic component, we compare data on South Africa's populations to all other information available from elephant populations in Africa. We also provide brief summaries of elephant sociology (box 1) and intelligence (box 2) that may modify our understanding of the spatial and dynamic responses of elephants to the environment, people, and management. Additionally, we discuss the effects of various management actions on population biology. We conclude this chapter with recommendations on how to accommodate elephant population responses to management in South Africa. We consider all of this as relevant to the assessment of South Africa's elephant populations.

SPATIAL UTILISATION

Distribution

Elephants need to drink regularly and therefore occur where surface water is available (e.g. Smit *et al.,* 2007a; Harris *et al.,* 2008). Through southern Africa, 70–80 per cent of elephant range occurs outside protected areas (Blanc *et al.,* 2007; Van Aarde & Jackson, 2007). Fencing partly restricts regional distributions in southern Africa (Van Aarde *et al.,* 2005; Mbaiwa & Mbaiwa, 2006). In unfenced areas, human population density and agriculture influence elephant distribution (Hoare & Du Toit, 1999), but elephants and humans continue to coexist across most of the southern African distributional range of elephants (Jackson *et al.,* 2008). This is not the case in South Africa, where elephants are fenced off to live on land set aside for conservation and where people do not inhabit the land.

Historically, elephants ranged through much of South Africa (Hall-Martin, 1992). However, by 1920, human population growth, expanding settlement, the ivory trade, and crop protection decimated elephant numbers in the country to an estimated 120 (Hall-Martin, 1992). These few elephants were restricted to areas around Knysna, Addo, Tembe, and the Olifants Gorge (later proclaimed as part of Kruger) (Hall-Martin, 1992).

Box 1: Social aspects of African savanna elephants

Elephants live in a well structured and complex society. Their so-called fission-fusion social structure influences the way they interact with each other and with their natural environment (McComb *et al.*, 2001; Wittemyer *et al.*, 2005a; Archie *et al.*, 2006; Wittemyer & Getz, 2007). It is therefore critical that conservation management efforts consider the consequences for elephant society (Couzin, 2006). For instance, destroying part of a social unit may have consequences for the surviving members of that unit. Furthermore, by keeping elephants in relatively small areas their social structuring may not provide for behavioural inbreeding avoidance (see Archie *et al.*, 2006), or for the spatial segregation of herds based on dominance (Wittemyer *et al.*, 2007a).

Cows form the foundation of the social structure – they generally spend their entire lives in tightly knit social groups and live in a specific area (Moss, 1988). Adult bulls, on the other hand, are generally solitary though they associate with female groups (breeding herds) for brief periods of travel and to mate (Moss & Poole, 1983; Poole & Moss, 1989). The female social structure has been described as comprising six hierarchical tiers of organisation. From lower to higher levels of organisation, these tiers include mother-calf units, families, bond groups, clans, subpopulations, and populations (Wittemyer *et al.*, 2005a). The basic unit of social structure, however, is the matriarch-led family unit, typically consisting of 1–20 adult cows, their daughters, and immature male offspring (Archie *et al.*, 2006).

Families are highly stable across time and season (Wittemyer *et al.*, 2005a). Because most female elephants remain with the group into which they were born, relatedness within families is high (Archie *et al.*, 2006). Though permanent fissions are rare, families may break up into smaller subgroups for short periods, or fuse with other families to form larger groups (Wittemyer *et al.*, 2005a; Archie *et al.*, 2006). Families that consistently fuse to form larger groups are known as bond groups. Similarly, coalitions of bond groups are known as clans. Sub-populations and populations are higher-order tiers that group lower-order tiers together based on geography.

This multi-tiered structure probably evolved to balance the costs and benefits of sociality (Wittemyer *et al.*, 2005a). Potential benefits include the defence of resources and territories, joint protection from predators, shared parenting duties, collective social and ecological knowledge, and increased inclusive fitness (Archie *et al.*, 2006). Higher tiered structures such as bond groups and clans might also enable the exchange of ecological information over relatively long distances (Foley, 2002). However, social living also has costs; it may intensify competition. The balance of the costs and benefits of associating at various tiers in the hierarchy differs temporally and seasonally in response to resource variability, the number of individuals in each group, and the spatial distribution of groups (Wittemyer *et al.*, 2005a).

Therefore, size and composition of social units may be influenced by human manipulation of resources such as the availability of water or the reduction of habitable space. Other interventions, such as culling, hunting, poaching, contraception, and translocation may also alter size and composition of groups (Ferreira *et al.*, 2008). The implications of these influences at the population level are poorly understood and require more research. However, McComb *et al.* (2001) show that families with older matriarchs have greater reproductive success, potentially due to the superior ability of older matriarchs to distinguish between the calls of known and unknown elephants. Therefore, hunters or poachers focusing their efforts on large tusked individuals may disproportionately affect the population through the removal of a few key individuals (McComb *et al.*, 2001). Furthermore, kinship is a primary driver of social relationships, and bond groups consist largely of related families (Archie *et al.*, 2006). Therefore, when population control measures remove a family's close relatives in other family units, the bond group and any associated fitness benefit may dissolve. Archie *et al.* (2006) recommend that elephant conservation measures strive to maintain patterns of maternal kinship.

Additionally, group dominance, primarily determined by the age of the matriarch (Wittemyer & Getz, 2007), plays an important role in spatial structuring (Wittemyer *et al.*, 2007a). Dominant groups enjoy disproportionate access to preferred habitats during the dry season, thereby minimising exposure to predation and conflict with humans and expending less energy than subordinate groups. Conversely, subdominant groups are relegated to marginal areas often outside protected reserves (Wittemyer *et al.*, 2007a). This research highlights the importance of social mechanisms and open ecosystems to population control and to the mitigation of the impacts elephants may have on ecosystems (Van Aarde & Jackson, 2007).

Within the southern African region, the local distribution of elephants varies seasonally. This can be ascribed to variation in resource availability across space and time (O'Connor *et al.,* 2007). For example, towards the end of the dry season when surface water is scarce, elephant density increases near rivers (Stokke & Du Toit, 2002; Jackson *et al.,* 2008). Similar effects occur around artificial waterholes (De Beer *et al.,* 2006), where dry season elephant densities are related to the density of added water points (Chamaillé-Jammes *et al.,* 2007). Thus, elephant distribution varies in space and time and is modified by water provision.

Factors determining the distribution of elephants

Within regions where elephants occur, several factors influence their local distribution. These factors include landscape type, food and water availability, rainfall-related changes in food quality and water availability, elephant density, social structures, management, and people.

Landscape type affects distribution because elephants do not move randomly through the terrain. Some landscape types, such as riparian environments and wetlands, support more elephants than others (e.g. Ntumi *et al.,* 2005; Kinahan *et al.,* 2007; Smit *et al.,* 2007a; Harris *et al.,* 2008), whereas steep hills tend to be avoided by elephants (Nellemann *et al.,* 2002; Wall *et al.,* 2006), despite their ability to negotiate such terrain under exceptional conditions.

Food and water are key requirements of elephants and affect their distribution. The water requirements of elephants are central to understanding patterns of their spatial use. For instance, in Kruger elephants drink on average every two days during the dry season (Young, 1970). In drier environments, bull elephants probably drink every 3–5 days and breeding herds every 2–4 days (Viljoen, 1988; Leggett, 2006b). Elephants, especially breeding herds, therefore seldom roam far away from drinking water.

Across southern Africa, we generally distinguish between dry and wet seasons. During the wet season, food resources are more abundant and higher in quality (Owen-Smith, 1988). Water is also distributed widely during the wet season and may not therefore restrict elephant spatial use and roaming distances (Leuthold, 1977; Western & Lindsay, 1984; Verlinden & Gavor, 1998; Gaylard *et al.,* 2003; De Beer *et al.,* 2006). In the dry season, however, the quality of food resources deteriorates, and seasonal water sources dry up. Therefore, elephants may use different habitats in a different part of their range

Box 2: Elephant intelligence

Once corrected for body size, the African elephant has a brain comparable in size and complexity to those of humans and other primates (Cozzi *et al.*, 2001). This certainly contributes to the popular belief that elephants have exceptional brainpower. Observations of elephants helping others and their apparent grief when facing dead conspecifics may strengthen the belief that elephants possess almost human-like awareness and intelligence. However, recent literature suggests that elephants are not extraordinarily intelligent but are, like many species, well adapted to cope with the natural spatial and temporal variability they face (Hart *et al.*, 2007).

Hart *et al.* (2008) also suggest that elephants perform poorly when compared to chimpanzees and humans in cognitive feats such as the use of tools, visual discrimination learning, and tests of 'insight behaviour' such as solving puzzles to reap rewards. However, elephants do have long-term, extensive spatial and temporal memory (Foley, 2002; Hakeem *et al.*, 2005; Leggett, 2006a). For herds to survive it is critical that there should be individuals within the herd that can successfully find isolated water holes and new foraging grounds over vast distances. Thus, long-term memory may enhance the ability to find scarce resources. Additionally, elephants, like many other species, can discriminate between different sounds. They can recognise individual calls from 1–1.5 km away (McComb *et al.*, 2003), and know the individual calls of about 100 other elephants (McComb *et al.*, 2000). Such auditory recognition may enable social associations between groups (McComb *et al.*, 2000). African elephants use olfaction and vision to identify different types of people in their local area and to vary their reactions appropriately to probable danger (Bates *et al.*, 2007). This may also be the case for other species that have not yet been studied.

Another aspect of elephant behaviour is their reaction to other elephants that are disabled or dead (Hart *et al.*, 2007). They can distinguish between elephant remains and those of other species, and often spend time investigating elephant corpses (Moss, 1988; McComb *et al.*, 2006). Responses to the death of an elephant calf include exploratory behaviour, fear and alarm behaviour, support efforts to lift the dying calf, body-guarding reactions and even aggression towards the body (Payne, 2003). There are many anecdotes of elephants trying to help others disabled by immobilisation drugs or bullets (see Douglas-Hamilton *et al.*, 2005). Behaviour consistent with Post-Traumatic Stress Syndrome in humans has been observed in elephants (Bradshaw *et al.*, 2005). Inferences that such instances represent higher-order emotional expression or intelligence are subjective.

at the height of the wet season compared to the dry season (Western & Lindsay, 1984; Verlinden & Gavor, 1998; CERU, unpublished data).

In theory, if home range size is dependent on habitat productivity (see Harestad & Bunnel, 1979), elephants should range further during the dry season to include food resources otherwise available within smaller areas during the wet season. Contradictory to this expectation, elephants tend to concentrate their foraging activities in relatively small ranges close to water during the dry season (Gaylard *et al.*, 2003; Osborn & Parker, 2003; Redfern *et al.*, 2003; De Beer *et al.*, 2006; Leggett, 2006a; Smit *et al.*, 2007a). This suggests that elephants seek key resources such as water (see Scoone, 1995; Illius, 2006), regardless of the spatial distribution of other resources. Thus, in the dry season, water availability is a determinant of elephant spatial use (De Beer *et al.*, 2006) while selection for vegetation is often secondary (Harris *et al.*, 2008; Chamaillé-Jammes *et al.*, 2007).

Elephants also seek vegetation that is available near water; consequently, they may avoid water sources that are not associated with suitable vegetation (Harris *et al.*, 2008). In the arid Etosha National Park (Namibia), vegetation is sparsely distributed, and elephants select areas near water with high vegetation cover. However, here they will move greater distances during the dry season to obtain food (Harris *et al.*, 2008). In the evergreen savannas of Maputo Elephant Reserve (Mozambique), high vegetation cover is often associated with the distribution of water, and during the dry season, elephants do not have to move far from water to obtain food (Harris *et al.*, 2008). Thus, elephants meet their nutritional requirements within the constraints set by the location of water sources (Redfern *et al.*, 2003).

In savannas, there is a relationship between rainfall and primary productivity (e.g. Coe *et al.*, 1976). More recently, the remotely sensed Normalised Difference Vegetation Index (NDVI) has been used as a surrogate for primary productivity (e.g. Pettorelli *et al.*, 2005). Primary productivity (measured by NDVI) does apparently influence elephant spatial use, and during the dry season elephant densities tend to be higher in more productive areas, though the relationship is weak (Chamaillé-Jammes *et al.*, 2007; Young *et al.*, 2008).

Temporal (time) scales determine our interpretation of the way that elephants utilise the landscapes where they live. For instance, on a short time scale (hourly), the relative position of food resources (the distribution of individual forage and non-forage plants), water and shade can explain elephant movements. To study these movements, the distribution of path lengths and turn angles might be related to these resources (Dai *et al.*, 2007). Scaling up to a daily interval, movements usually consist of elephants foraging and travelling to and away from water and shade (De Villiers & Kok, 1988; Kinahan *et al.*, 2007). On a seasonal scale, within the same locality, elephants travel daily over longer distances during the wet than the dry season (Wittemyer *et al.*, 2007a; CERU, unpublished data).

Distribution across the landscape is also affected by the density of elephants (their number per unit area). As elephant numbers increase, distribution may change in two ways. First, local densities may remain relatively constant while the population extends its range. This may be the case in northern Botswana, where Junker *et al.* (2008) show that increased elephant numbers were associated with expansion of their range, whereas elephant densities did not increase. Here, space was not limiting, and elephants were able to extend their distribution outwards into unoccupied areas. Alternatively, if fencing, human populations, or other factors limit the area elephants can occupy, density may increase within specific areas. Young *et al.* (2008) studied elephant populations in Kruger and observed that as numbers increased after culling stopped, at a time when increases in land area were limited (the study period was prior to the removal of parts of the fence between Kruger and Limpopo National Park in Mozambique), the number of patches occupied by elephants increased. Thus, as densities increased, elephants became more evenly distributed across Kruger.

Furthermore, the social hierarchy of elephants may underlie spatial use, with dominant herds in Kenya having a greater proportion of their range within protected areas compared to subordinate herds (Wittemyer *et al.*, 2007a; see box 1). Here, dominant herds also spend more time near water and move shorter distances when measured at hourly, daily, or seasonal time intervals (Wittemyer *et al.*, 2007a). We are not aware of similar studies in any South African parks. In South Africa, fences limit temporal patterns of spatial use – all 63 populations in the country live in fully or partially fenced areas (see later). Consequently seasonal changes in the location and sizes of ranging areas (home ranges) in fenced-in populations were less pronounced than in free-ranging populations elsewhere in Africa (CERU, unpublished data).

As most of the elephant range in Africa occurs outside protected areas (Blanc *et al.*, 2007), human and elephant ranges overlap in many places. Inevitably, this leads to interactions between elephants, people, and their livelihoods (Van Aarde & Jackson, 2007). Elephants come into greater contact with people where their ranges increase. In northern Botswana, for instance, an increase in the distributional range of elephants led to a substantial increase in conflict between people and elephants (Alexander *et al.*, 2006).

Elephants appear to use space in a manner that reduces contact with people. On a daily basis, they achieve this by altering their drinking behaviour. For instance, along the Okavango River in north-western Botswana, people are active in fields during the day, while elephants visit areas close to the river at night only (Jackson *et al.*, 2008), thereby limiting overlap in times that elephants and people are in the same area. Spatially, elephants may avoid areas close to human settlements and leave areas entirely when human densities reach a particular threshold (Hoare & Du Toit, 1999). Hoare (1999) suggests that breeding herds are more likely than bulls to avoid people. When the distributions of people and elephants do overlap, conflict is often reported. Incidences of conflict, therefore, appear to be correlated with spatial factors such as human density, land transformation, agriculture, roads, and proximity to protected areas (Hoare & Du Toit, 1999; Parker & Osborn, 2001; Sitati *et al.*, 2003).

Elephant home ranges

The home range of an elephant represents the area it traverses in its normal activities of food gathering, mating, and caring for young. Home ranges can be measured on various time scales (e.g. monthly, seasonally, annually), and provide a measure of elephant spatial use in relation to various biotic and abiotic factors. Rainfall apparently plays an important role in determining home range size and location (Thouless, 1995; Osborn, 2004). Furthermore, across southern Africa, rainfall generally increases from southwest to northeast, creating a gradient of vegetation types (e.g. Sankaran *et al.*, 2005). In dry areas towards the west of the subcontinent where rainfall is relatively low, elephants tend to have larger home ranges than in wetter areas to the east (Van Aarde *et al.*, 2005).

Resources such as water, food, and shelter are unevenly distributed across the landscape, which gives rise to a mosaic of different land type patches (habitats or vegetation classes) (Forman & Godron, 1986). Heterogeneity refers to the complexity and variability of the spatial pattern contained by these patches within this landscape mosaic (Li & Reynolds, 1994). At the landscape scale, some aspects of heterogeneity influence the location and/or size of elephant home ranges (Grainger *et al.,* 2005; Murwira & Skidmore, 2005; De Beer, 2007; Ott, 2007). In general, elephants favour areas where vegetation patches are more complex and diverse (Ott, 2007). Relatively high levels of heterogeneity, due to an increase in the length of habitat edges (Tufto *et al.,* 1996; Saïd & Servanty, 2005), may further benefit elephants by providing better opportunities to obtain resources (De Beer, 2007; Ott, 2007). In relatively wet (mesic) savannas (see Sankaran *et al.,* 2005), cows tend to occur in areas with higher levels of heterogeneity than where bulls occur, and for both sexes, heterogeneity levels are higher within their wet season ranges than within dry season ranges (Ott, 2007).

In Kruger, only one measure of heterogeneity that Grainger *et al.* (2005) examined explains variability in elephant home range sizes, possibly because the distribution of artificial water resources (e.g. dams, drinking troughs and waterholes maintained by water from boreholes) masks patterns in landscape use. Here, the areas of elephant home ranges tend to decrease as the density of waterholes increases (Grainger *et al.,* 2005), as is also the case in the Etosha National Park and the Khaudum Game Reserve in northern Namibia (De Beer, 2007). This once again points to water and the distribution thereof being an important determinant of the manner in which elephants utilise landscapes. Tampering with the distribution of water through the construction of dams and waterholes therefore will alter the ranging behaviour of elephants.

In South Africa fences that separate conservation areas where elephants live from the surrounding landscape influence the home range. Consequently, elephants in South Africa have relatively small home ranges (breeding herds mean = 595 km², range: 21 km²–2 766 km², n = 51; bulls mean = 153 km² range: 32 km²–1 707 km², n = 43; figure 1), compared to those of elephants throughout the rest of the region (breeding herds mean = 1 678 km² , range 4 km²–10 738 km², n = 73; bulls mean = 2 095 km², range 3 km²–12 800 km², n = 23; figure 1). Home range sizes of both bulls and breeding herds are smaller in South Africa compared to those of elephants in other areas of southern Africa with similar rainfall (figure 2). Significantly, all South Africa's elephants (at least for the time these data were available) occur in fenced areas, while the movements of those in the rest of the region, except for Etosha in Namibia, are not restricted in the same way.

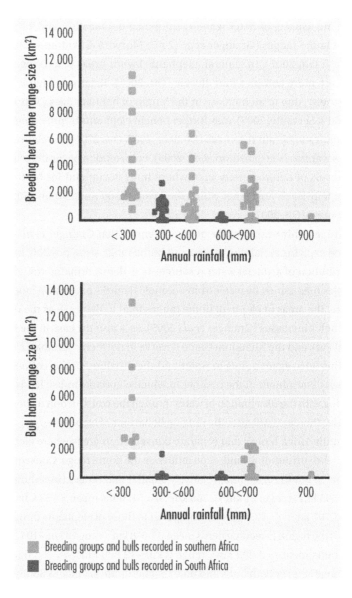

Figure 1: The home range sizes of elephant bulls and breeding herds in South Africa (dark symbols) compared to those recorded elsewhere in southern Africa (light symbols), within different annual rainfall classes. We recognise that this comparison may be confounded by factors such as season. Even so, for both bulls and cows in South Africa, home range sizes appear to be smaller and to vary less in area than those of elephants elsewhere in the region. This leads to concerns about management practices in South Africa, such as fencing, that restrict elephant range use, with consequences for the intensity at which they will use the landscape (figure adapted from Guldemond, 2006)

This raises three principal concerns regarding the home ranges of elephants in South Africa. First, home ranges here are relatively small compared to those of elephants throughouty the rest of the region. Second, given the relatively small sizes of most protected areas in South Africa, the home ranges of individual elephants here may cover a greater proportion of these protected areas than elsewhere. Third, unlike some other areas, there may be little spatial segregation in land use between the dry and wet seasons (see Western & Lindsay, 1984; Verlinden & Gavor, 1988).

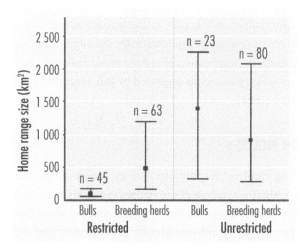

Figure 2: The home range sizes (range [min, max] with mean) across southern Africa of elephant bulls and breeding herds whose movements are restricted by fencing, compared to those whose movements are unrestricted. The comparison is limited to areas within the annual rainfall ranges similar to that in South African study sites (376–748 mm per year). All elephants in South Africa occur in areas where fences restrict movements, while those in the rest of the region do not. Thus, grouping elephants into areas where their movements are compromised by fencing, also groups them into South African and non-South African populations and underlies a fundamental reason for the small ranges characteristic of elephants in South Africa (figure adapted from Guldemond, 2006)

Together, these factors suggest that elephants in South Africa make more intensive use of the land available to them than elsewhere. In turn, the impact they have on vegetation is likely to be more severe, giving vegetation little chance to recover from elephant damage (see Van Aarde *et al.*, 2006). A decrease in home range area induced by fencing thus will enhance the impact that elephants can have on the landscapes where they live.

Our present understanding of the distribution and spatial use patterns of elephants in South Africa are incomplete. However, technological improvements have enhanced our ability to track the movements of elephants over vast areas and for extended periods, thus expanding our capacity to address important research questions. Such research, especially when conducted as parts of an adaptive management strategy that manipulates landscape variables such as the distribution of water and fences, should allow us to assess why spatial use patterns of elephants in South Africa differ from patterns throughout the rest of southern Africa. This could also address the impact elephant spatial use may have on the landscape, vegetation, and other species. However, preventing elephants from moving outside small fenced reserves precludes the application of management options that restore their large-scale spatial use patterns, as suggested by Van Aarde *et al.* (2006) and Van Aarde & Jackson (2007).

POPULATION BIOLOGY

Understanding elephant population biology can empower conservation managers to predict the response of populations to various management actions. As part of population biology, studies of the dynamics of populations focus on factors that change their attributes over time and explain how such changes determine population numbers. These population attributes include the size, density (numbers per square kilometre or per square mile), distribution, birth rates, death rates, and dispersal rates of a collection of individuals that share space. For research purposes, a population must comprise enough individuals from which to collect data to estimate these vital rates and provide for statistical limitations of analytical procedures (Akçakaya, 2002). Populations that comprise only a few breeding herds and bulls therefore do not lend themselves to estimates of vital rates. This certainly holds for most of the newly established populations confined to relatively small areas in South Africa. The factors that influence births, deaths, immigration, and emigration determine population size and change in numbers over time (population growth). In this section we compare the attributes of elephant populations and discuss the factors that may limit population sizes. We also compare the dynamics of South Africa's elephant populations to populations elsewhere in Africa.

Box 3: Assigning ages to elephants

Monitoring population changes is important for implementing appropriate management actions and evaluating their effectiveness (Gibbs, 2000). Authorities could use demographic parameters, such as age at first calving, calving interval, and survival rates to predict population changes over time. They seldom do so. One reason is that estimates of these population parameters require accurate determination of the ages of individuals within a population.

Methods to determine the chronological ages of elephants include measuring molar tooth wear and progression (e.g. Laws, 1966; Sikes, 1966; Fatti *et al.*, 1980; Jachmann, 1988), elephant tusk dimensions (Hanks, 1972; Sukumar *et al.*, 1988), back lengths (Croze, 1972), shoulder heights (e.g. Laws, 1966; Douglas-Hamilton, 1972; Jachmann, 1988; Lee & Moss, 1995; Shrader *et al.*, 2006a), hind foot lengths (Western *et al.*, 1983; Lee & Moss, 1995), and dung boli diameters (e.g. Reilly, 2002; Morrison *et al.*, 2005). All these methods rely on the relationship between a particular morphological feature and age to determine the age of an individual elephant. Only three body size measures have formally been related to known age. Lee & Moss (1995) provided a relationship between footprint diameters and known age while Morrison *et al.* (2005) did that for dung boli in Amboseli National Park. Shrader *et al.* (2006a) showed that the Addo Elephant National Park and Amboseli elephants had the same relationship between shoulder height and known age elephants. These relationships are the best available to assign ages for cows up to age 15 and for bulls up to age 25.

Several factors may impede the success of age determination techniques. Dense vegetation may hamper direct measurements of free-ranging elephants, and many earlier measurements could only be taken from captive or immobilised animals (Lee & Moss, 1995). Measuring tusk dimensions requires close access to elephants (Hanks, 1972). Studies examining the rates of tooth eruption have yet to be carried out on living, free-ranging elephants, though studies of the lower and upper jaw tooth rows of shot elephants in Uganda (Laws, 1966; Laws *et al.*, 1975) saw the development of age determination techniques based on eruption and wear patterns. Measurements of footprints are subject to terrain, substrate, incline and other environmental factors (Western *et al.*, 1983; Reilly, 2002; Morrison *et al.*, 2005). Measuring the back length or shoulder height of elephants in the field is only practical where visibility is good and animals can be photographed (Morrison *et al.*, 2005). Furthermore, this technique requires expensive equipment such as digital range finders and cameras, may be time-consuming, and may be prone to measurement error (Jachmann, 1980; Morrison *et al.*, 2005; Shrader *et al.*, 2006a). However, digital photogrammetry, a recently developed method to measure shoulder heights of elephants (Shrader *et al.*, 2006b), requires less time and produces more accurate and precise results than other measuring techniques.

Births

The number of calves that an average cow will have in her lifetime is determined by the ages at which cows have their first and last calves, and the years that elapse between births. The number of calves produced by each cow influences the rate at which a population grows. Generally populations will grow faster when cows have their first calves when relatively young, when the time that elapses between births (calving intervals) is short, and when they continue to breed to old age. The age at first calving, calving interval, and age at last calving, are therefore key traits of a population. Quantifying these traits and understanding how they vary across space, time, between elephants of different ages, and between populations, enables us to decipher the dynamics of a population.

Scientists use different methods to estimate age at first calving. Some of them study elephants over a long time to follow individual life histories (e.g. Whitehouse & Hall-Martin, 2000; Moss, 2001; Wittemyer *et al.*, 2005b; Gough & Kerley, 2006); others observe family units and identify cow-calf associations (e.g. Jachmann, 1980; Jachmann, 1986); others examine breast development in cows (e.g. McKnight, 2000), or note the reproductive activity of killed cows by assessing whether a cow is pregnant and counting how many placental scars (i.e. pigmented scars on the uterus that represent the number of times a cow has been pregnant) she carries (e.g. Hanks, 1971; Lewis, 1984; Lindeque 1991; Whyte 2001). All these methods rely on assigning ages accurately to individual elephants (see box 3).

Long-term observations and cow-calf associations return the age at which a cow had her first calf, while the other methods give the age at which she conceived or is likely to conceive. Age at first conception can be converted to age at first calving by adding 22 months, the gestation period in elephants (Hodges *et al.,* 1994). Estimates of the length of calving intervals can be influenced by the deaths of calves, incorrect assignment of the ages of elephants, and allomothering – when cows look after calves that are not their offspring (Lee, 1987). Comparisons are limited when considering that different techniques were used to estimate age (see box 3) and age at first calving.

Published estimates (table 1) show a wide range of ages at which cows have their first calf. For instance, in Addo some cows can conceive when seven years old, thus giving birth at nine (Gough & Kerley, 2006). The mean ages of first calving tend to be lower for South African populations compared to elsewhere in Africa (figure 3A). Cows in South Africa tend to have their first calves at an average age of 11.3 years (median = 11.9, SD = 1.8, n = 8 estimates). Those elsewhere have their first calves at an age of 14.1 years (median = 13.5, SD = 3.0, n = 16 estimates). In addition, the range and confidence limits of estimates of age at first calving tend to be wider for populations elsewhere compared to populations in South Africa (figure 3B). This suggests that most cows in South African parks may have their first calf at younger ages than those living elsewhere. Thus, if all the other traits are the same, populations in South Africa will increase faster than elsewhere.

Why would cows in South Africa mature earlier than elsewhere? We know that, for mammals, resource quality affects the age at sexual maturity and therefore the age when they may have their first calves (e.g. Owen-Smith, 1990). This suggests that elephants in South African parks have better resources available than elephants living elsewhere. This could be due to dams and waterholes that are constructed in these parks enabling access to additional resources by allowing elephants the opportunity to forage in otherwise inaccessible areas. Elephants living here may therefore not be constrained by resources and this could be one of the reasons why elephant cows in South African populations may have their first calves at a relatively young age.

Elephant cows across Africa give birth at intervals of 1.8–13.5 years (table 1). The calving intervals of 10.3, 11.0, 11.5, and 13.5 years for elephants in the Tsavo National Park, Kenya (McKnight, 2000), the Amboseli National Park, Kenya (Moss, 2001), the Murchison Falls National Park, Uganda (Buss & Smith, 1966), and the Budongo Forest Reserve, Uganda (Laws *et al.,* 1975), respectively, are exceptionally long when compared to values from elsewhere. Additionally, the 1.8 years noted for an elephant in Amboseli (Moss, 2001) is exceptionally short (table 1). The 22-month gestation period combined with apparent infertility induced by suckling places a lower limit on the length of the calving interval (Hodges *et al.,* 1994). Thus, the extremely short calving interval noted in Amboseli may be due to the early death of the previous calf. However, infertility during suckling has not been confirmed in free-ranging elephants, but indirectly inferred from observation in Amboseli where birth intervals of 3.2 years for cows whose calves died before 2 years of age is shorter than the median of 4.5 years (Moss, 2001). No single factor yet has been identified that can explain the variability in calving intervals in elephants, but Laws *et al.* (1975) suggest that calving intervals tend to increase with density. This observation needs further study but is supported by some South African data (CERU unpublished records).

Calving intervals for elephants varied considerably across Africa (table 1). Mean values for South African populations tend to be similar to the lower end of mean values recorded elsewhere in Africa (figure 3C). Elephants living in South African populations have calves on average every 3.6 years (median = 3.8, SD = 0.7, n = 10 estimates), while those elsewhere have calves every 4.2 years (median = 3.8, SD = 1.8, n = 22 estimates). The length of calving intervals tends to vary less in South African populations than elsewhere in Africa (figure 3D). This may be related to regional rainfall differences. Even so, the confidence intervals and ranges of values of calving intervals suggest that most cows in South Africa tend to have calves more often than those living elsewhere in Africa. The reasons for this are not known, but may be related to the relatively low calf mortalities noted in South Africa (see later), or by resources not being limited as a result of management interventions such as water provision, as we discussed earlier. Compared to age at first calving and calving interval, age at last calving is less well known. We found three estimates in the published literature: (1) 60 years in Kruger, based on ovarian activity noted for killed elephants aged using tooth eruption criteria (Smuts, 1975), (2) 48–55 years in Addo, based on individual life histories with guessed ages (Whitehouse & Hall-Martin, 2000), and (3) guesstimates of 52–56 years for elephants in Amboseli (Moss, 2001). This suggests fertility may begin to decrease in a cow's late forties. Too little information is available to compare elephants from different regions.

The onset and end of breeding are not abrupt in a population. Typically, the age at which cows have their first calf differs from population to population, but the age-specific birth rate remains relatively constant for adult cows within a population, and then declines around the age when elephants stop breeding (Whitehouse & Hall-Martin, 2000; Moss, 2001).

Various measures serve as indices of age-specific reproductive output, which usually is expressed as fecundity, defined as yearly production of female calves per cow of a given age group. In table 2 we present data for different populations on the percentages of cows that were pregnant and/or lactating among culled specimens of a specific age, or the percentage of cows that gave birth. Values for Kruger and Etosha seem similar, but much higher than those for other populations, probably due to the different information being recorded by different workers. We have no comparable statistics on this aspect of reproductive output for different populations.

Locality	Method	Age at first calving (years)					Range	Reference
		Mean	Median	SD	SE	95% CI		
South African populations								
Addo	Individual histories	13.0	–	–	–	–	–	Woodd, 1999
	Individual histories	13.0	–	2.03	0.3	12.5–13.5	10–16	Whitehouse & Hall-Martin, 2000
	Individual histories	12.3	–	1.73	0.2	11.7–12.7	–	Gough & Kerley, 2006
	Cow-calf associations	13.8	–	–	0.8	12.1–15.4	–	Ferreira & Van Aarde, 2008
Kruger	Culled samples	–	–	–	–	–	11.0–17.0	Smuts, 1975
	Culled samples	–	–	–	–	–	9.0–14.0	Whyte, 2001
Mabula	Cow-calf associations	12.3	12.0	–	0.6	11.2–13.4	–	Mackey et al., 2006
Phinda	Cow-calf associations	10.3	10	–	0.6	9.2–11.4	–	Mackey et al., 2006
Pilanesberg	Cow-calf associations	9.2	9	–	0.2	8.8–9.6	–	Mackey et al., 2006
Pongola	Cow-calf associations	8.4	8	–	0.5	7.3–9.5	–	Mackey et al., 2006
Tembe	Cow-calf associations	11.5	–	–	0.5	10.4–12.5	–	Morley, 2005
Other populations								
Amboseli	Individual histories	13.7	14.1	–	–	–	8.9–21.6	Moss, 2001
	Cow-calf associations	13.6	–	–	0.5	12.5–14.6	–	Ferreira & Van Aarde, 2008
Bugongo	Placental scars	22.4	–	–	–	19.9–24.9	–	Laws et al., 1975
Etosha	Placental scars	12.5	–	–	–	–	10.8–12.8	Lindeque, 1988
	Placental scars	13.7	–	–	–	–	12.8–13.8	Lindeque, 1988
	Puberty	13.8	–	–	1.2	11.5–16.2	–	Lindeque, 1988
	Puberty	12.6	–	–	1.5	9.7–15.6	–	Lindeque, 1988
	Culled samples	15.3	–	–	–	–	13.8–17.8	Lindeque, 1988
	Culled samples	13.3	–	–	–	–	9.8–17.8	Lindeque, 1988
Kasungu	Cow-calf associations	12.8	–	2.6	–	–	–	Jachmann, 1986
Kidepo	Cow-calf associations	–	–	–	–	–	8.8–13.8	Croze, 1972
Luangwa	Placental scars	15.8	–	–	–	–	13.0–19.0	Hanks, 1972

Locality	Method	Age at first calving (years)						Reference
		Mean	Median	SD	SE	95% CI	Range	
Maputo	Cow-calf associations	9.8	–	–	0.5	9.3–10.3	–	Morley, 2005
Mkomazi	Placental scars	12.2	–	–	–	11.3–13.1	–	Laws *et al.*, 1975
Mkomazi East	Placental scars	12.2	–	–	–	11.0–13.4	–	Laws *et al.*, 1975
Murchison North	Culled samples	–	–	–	–	–	8.8–12.8	Buss & Smith, 1966
	Placental scars	16.3	–	–	–	15.5–17.1	–	Laws *et al.*, 1975
Murchison South	Placental scars	17.8	–	–	–	16.9–18.6	–	Laws *et al.*, 1975
Tsavo	Cow-calf associations	–	–	–	–	–	12.8–16.8	McKnight, 2000
	Cow-calf associations	–	–	–	–	–	12.8–16.8	McKnight, 2000
	Placental scars	11.7	–	–	–	10.8–12.6	–	Laws *et al.*, 1975
Zambezi	Culled samples	–	–	–	–	–	15.8–16.8	Dunham, 1988
	Culled samples	–	–			–	12.8–14.8	Dunham, 1988

Table 1A: The ages at first calving for elephant populations across Africa. We present published statistics and the method that yielded estimates of these values. Counts of placental scars are for cows culled for either research or management purposes

Locality	Method	Calving interval (years)						Reference
		Mean	Median	SD	SE	95% CI	Range	
South African populations								
Addo	Individual histories	3.8	–	–	–	–	–	Woodd, 1999
	Individual histories	3.8	–	1.29	0.1	3.6–4.0	–	Whitehouse & Hall-Martin, 2000
	Individual histories	3.3	–	0.77	–	–	–	Gough & Kerley, 2006
	Cow-calf associations	4.0	–	–	0.3	3.3–4.6	–	Ferreira & Van Aarde, 2008
Kruger	Placental scars	4.5	–	–	–	4.0–5.0	–	Smuts, 1975
	Culled samples	3.7	–	–	–	–	–	Whyte, 2001
Mabula	Cow-calf associations	2.4	–	–	0.1	2.3–2.5	–	Mackey *et al.*, 2006
Phinda	Cow-calf associations	3.9	–	–	0.2	3.5–4.3	–	Mackey *et al.*, 2006
Pilanesberg	Cow-calf associations	3.3	–	–	0.1	3.1–3.5	–	Mackey *et al.*, 2006
Pongola	Cow-calf associations	3.1	–	–	0.2	2.7–3.5	–	Mackey *et al.*, 2006

Locality	Method	Calving interval (years)						Reference
		Mean	Median	SD	SE	95% CI	Range	
Tembe	Cow-calf associations	4.6	–	–	0.6	3.4–5.8	–	Morley, 2005
Other populations								
Amboseli	Individual histories	4.5	4.2	–	–	–	1.8–11.7	Moss, 2001
	Cow-calf associations	4.6	–	–	0.2	4.1–5.1	–	Ferreira & Van Aarde, 2008
Bugongo	Culled samples	7.7	–	–	–	5.4–13.5	–	Laws et al., 1975
Etosha	Culled samples	3.8	–	–	–	–	–	Lindeque, 1988
	Placental scars	2.1	–	–	–	–	–	Lindeque, 1988
	Placental scars	2.5	–	–	–	–	–	Lindeque, 1988
Kasungu	Cow-calf associations	3.9	–	1.1	–	2.2–5.3	–	Jachmann, 1986
	Cow-calf associations	3.3	–	1.3	–	–	–	Jachmann, 1986
Kidepo	Culled samples	2.2	–	–	–	–	–	Croze, 1972
	Culled samples	3.2	–	–	–	–	–	Croze, 1972
Luangwa	Culled samples	3.0	–	–	–	–	–	Hanks, 1972
	Placental scars	4.0	–	–	–	–	–	Hanks, 1972
Maputo	Cow-calf associations	3.1	–	–	1.1	3.0–4.2	–	Morley, 2005
Mkomazi	Culled samples	2.9	–	–	–	2.6–3.4	–	Laws et al., 1975
Mkomazi East	Culled samples	4.2	–	–	–	3.1–5.0	–	Laws et al., 1975
Murchison North	Culled samples	–	–	–	–	–	2.6–5.8	Buss & Smith, 1966
	Culled samples	9.1	–	–	–	7.5–11.5	–	Laws et al., 1975
Murchison South	Culled samples	5.6	–	–	–	4.8–6.8	–	Laws et al., 1975
Tsavo	Cow-calf associations	4.6	–	–	–	–	–	McKnight, 2000
	Cow-calf associations	5.0	–	1.8	0.9	3.2–6.8	–	McKnight, 2000
	Culled samples	6.8	–	–	–	5.1–10.3	–	Laws et al., 1975
Zambezi	Culled samples	2.8	–	–	–	–	–	Dunham, 1988
	Culled samples	3.4	–	–	–	–	–	Dunham, 1988
	Placental scars	3.8	–	–	0.4	3.0–4.6	–	Dunham, 1988

Table 1B: Lengths of calving intervals (B) for elephant populations across Africa. We present published statistics and the method that yielded estimates of these values. Counts of placental scars are for cows culled for either research or management purposes

When combining reproductive output with the survival likelihood of a cow of a specific age (see table 3), a reproductive value can be assigned to each age group. This so-called reproductive value gives the relative contribution that each age group makes to the increase in population size. Our analyses suggest that the overall pattern is the same for all populations for which we have information (figure 4). Furthermore, in all these populations, elephants that are 15–25 years old contribute most to future growth of populations.

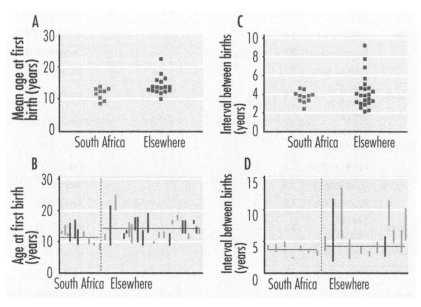

Figure 3: A comparison of reproductive variables of elephant populations living in South Africa with those for elephants living elsewhere in Africa. A) The mean age at first calving recorded for each population. B) The lower and upper confidence limits (lighter lines) or range between minimum and maximum values (darker lines) of age at first calving for each population, depending on published information. The horizontal black lines are the mean values calculated from estimates. South African elephant populations (those at the left of the dotted line) tend to give birth when younger than elephants elsewhere in Africa. C) The mean calving interval for each population. D) The lower and upper confidence limits (lighter lines) or range between minimum and maximum values (darker lines) of birth intervals recorded for each population. The horizontal black lines are the mean values calculated from estimates available for populations. The ranges for South African elephant populations tend to be at the lower end of those elsewhere and suggest that cows living in South Africa have calves more often than cows elsewhere in Africa

Deaths

Under natural conditions elephant populations typically have relatively low yearly death rates. These are usually expressed as high survival rates (Laws, 1969; Hanks, 1979; Whitehouse & Hall-Martin, 2000; Dudley *et al.*, 2001; Moss, 2001; Whyte, 2001; Wittemyer *et al.*, 2005b; Gough & Kerley, 2006). Age- and sex-specific survival values have been published for several populations (table 3). These are often calculated from age distributions of culled samples, but long-term studies of individuals of known age provide the most reliable information (e.g. Whitehouse & Hall-Martin, 2000; Moss, 2001). More recently Ferreira & Van Aarde (2008) developed survey and calculation protocols that are not invasive and that yield estimates comparable to those from long-term studies.

| | South Africa | | Populations elsewhere in Africa | | |
| | Kruger | Amboseli | Etosha | Luangwa | Murchison |
Age (yrs)	Pregnant or lactating	Giving birth	Pregnant or lactating	Pregnant	Lactating
0–4	0	0	0	0	0
5–9	5.5	0	3.6	0	2.0
10–14	52.0	14.0	32.2	5.2	3.0
15–19	91.0	21.0	76.7	56.6	20.0
20–24	80.5	23.0	94.1	50.6	50.0
25–29	93.0	23.0	98.8	50.6	65.0
20–34	86.5	23.0	89.6	50.0	66.0
35–39	93.7	23.0	93.3	50.0	76.0
40–44	92.9	20.0	100.0	42.1	60.0
45–49	94.7	18.0	93.3	42.1	57.0
50–54	89.3	14.0	86.7	33.3	37.0
55–59	85.7	10.0	56.7	33.3	0
60–64	–	0	–	–	–
Reference	Smuts, 1975	Moss, 2001	Lindeque, 1988	Hanks, 1979	Laws *et al.*, 1975

Table 2: Age-specific reproductive rates (given as percentages) as indices of age-specific fecundity for selected elephant populations across southern Africa

Lee & Moss (1995) suggest that in Amboseli many elephants die during the first two years of life, fewer during the next one to two years, and more after they are weaned when about four years old. This is supported by studies on elephants in Addo where survival rates for young elephants tend to be lower than for adults, particularly for juveniles in the first few years of life (Whitehouse & Hall-Martin, 2000; Moss, 2001).

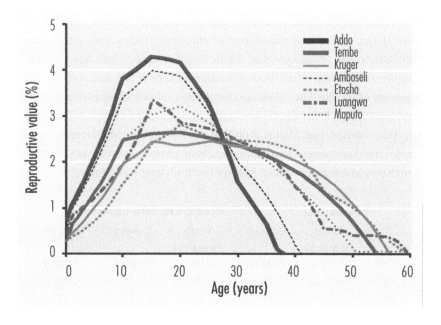

Figure 4: Reproductive values (the percentage contribution of different age groups to future population growth) as a function of age for South African (solid lines) and other (broken lines) populations. We extracted data from the literature and standardised the value for each age class as a fraction of the maximum value across all age classes for each population. We then used survival estimates (table 3) to calculate survival likelihoods (the probability at birth that an individual will survive to a specific age). Combining fecundity and survival likelihood with an independent estimate of population growth yielded the reproductive values following the equations of Case (2000)

Survival rates are relatively high across all ages (table 3). Here a comparison of values we have for South African populations with those for populations elsewhere in Africa yields valuable insights. For instance, for the first age class we note that the lowest survival value for South African populations (0.90) is higher than the lowest value of 0.59 noted for elephants elsewhere in Africa. Survival rates for elephants in older age classes are slightly less variable for South African populations than for populations elsewhere. Some may deem these comparisons invalid because different methods were employed to obtain data for the different populations. Nonetheless, our Assessment suggests that survival is relatively high in South African populations, compared to some populations elsewhere in Africa. If so, and if other population traits remain constant or higher, as has been shown earlier in this chapter, then population sizes should also increase faster here than elsewhere in Africa.

| | Population | \multicolumn{7}{c}{Age (years)} |
		0	1–9	10–19	20–29	30–44	45–60	60+
South African populations	Addo[1]	0.94	0.99	0.99	0.99	0.99	0.98	0.00
	Kruger[2]	0.97	0.97	0.97	0.97	0.97	0.97	–
	Tembe[3]	0.90	0.99	0.99	0.99	0.99	0.99	–
Populations elsewhere in Africa	Amboseli[4]	0.94	0.99	0.98	0.99	0.97	0.95	–
	Buganga[5]	0.97	0.97	0.97	0.97	0.97	0.97	–
	Etosha[6]	0.84	0.87	0.9	0.93	0.92	0.88	0.84
	Kasungu[7]	0.94	0.96	0.96	0.94	0.91	0.86	0.67
	Luangwa[8]	0.59	0.92	0.89	0.87	0.86	0.50	0.00
	Maputo[3]	0.82	0.94	0.95	0.97	0.97	0.97	–
	Mkomazi[5]	0.95	0.95	0.95	0.95	0.95	0.95	–
	Murchison[5]	0.97	0.97	0.97	0.97	0.97	0.97	–
	Sambura[9]	0.98	0.99	0.98	0.98	0.97	0.97	–
	Tsavo[5]	0.95	0.95	0.95	0.95	0.95	0.95	–

[1] Calculated from individual histories. Extracted from Whitehouse & Hall-Martin (2000)

[2] Calculated from the difference between observed and expected population growth rates. Extracted from Whyte (2001)

[3] Calculated from age distributions and fecundity estimates. Extracted from Morley & van Aarde (2006)

[4] Calculated from individual histories. Extracted from Moss (2001)

[5] Calculated from age distributions assuming that exponential growth is zero. Extracted from Laws *et al.* (1975)

[6] Calculated from age distributions assuming that exponential growth is 0.1. Extracted from Lindeque (1988)

[7] Calculated from age distributions assuming that exponential growth is zero. Extracted from Jachmann (1980, 1984)

[8] Calculated from age distributions assuming that exponential growth is zero. Extracted from Hanks (1979)

[9] Calculated from individual histories. Extracted from Wittemyer *et al.* (2005b)

Table 3: Annual survival rates for elephants in different age classes and populations. To compare estimates we grouped estimates into age classes and calculated mean annual survival rates for each group from the published information. Some studies assumed constant survival across all ages

Ivory poaching (e.g. Gillson & Lindsay, 2003; Stiles, 2004; Reeve, 2006; Wasser *et al.*, 2007) and formal culling programmes (e.g. Lindeque, 1991; Cumming *et al.*, 1997; Butler, 1998; Van Aarde *et al.*, 1999) will lower individual survival. At the population level the influence of poaching on age-specific survival rates may be more profound when poachers target older individuals (see Milner-Gulland & Mace, 1991; Ferreira *et al.,* 2008). Alternatively, providing water (e.g. Gaylard *et al.,* 2003) may lower death rates, even during droughts (Walker *et al.,* 1987). Culling of entire breeding herds plus their associated males, such as was the practice in Kruger (Whyte, 2001), may have had no or little influence on the age distribution and hence on estimates of age-specific survival rates for the population.

Droughts (e.g. Corfield, 1973; Walker *et al.,* 1987; Dudley *et al.,* 2001), disease (Berry, 1993; Lindeque, 1988; Turnbull *et al.,* 1991), and predation also affect survival. Lions target unweaned calves in the Hwange National Park, Zimbabwe (Loveridge *et al.,* 2006), and 4- to 15-year-old elephants in the Savuti Region of Botswana (Joubert, 2006). However, in most cases the incidence of predation seems low and may be relatively unimportant for survival rates at the population level.

Elephants seem sensitive to droughts, and several authors reported die-offs during dry spells (Corfield, 1973; Walker *et al.,* 1987; Dudley *et al.,* 2001). When considering that 4–6 dry spells may occur in a 50-year period (e.g. Ogutu & Owen-Smith, 2003), most elephants would be exposed to drought as a mortality agent to which they may be most sensitive when relatively young. Considering the apparent importance of rainfall for survival, the projected climate change across southern Africa, which may result in more frequent and severe droughts across much of the distributional range of elephants (IPCC, 2007) could increase elephant mortality in the coming century.

Immigration and emigration

Immigration (movement into an area) and emigration (movement out of an area) affect population growth and population size. We know that elephants do immigrate to colonise new areas or re-colonise areas they previously occupied. For instance, elephants from Mozambique colonised all of the area of Kruger within 50 years (Whyte *et al.,* 2003), at rates of 7–10 kilometres per year (Whyte, 2001). Elephants also re-colonised the Serengeti National Park in Tanzania after an absence of 40 years (Lamprey *et al.,* 1967). In some cases human actions can spur elephant movements. The provision of water certainly enabled elephants to colonise and permanently occupy areas that were relatively inhospitable, especially during the dry seasons, such as Hwange in Zimbabwe (Chamaillé-Jammes *et al.,* 2007), the Etosha National Park in Namibia (Lindeque & Lindeque, 1991) and the Khaudum Game Reserve in northern Namibia. In the case of Khaudum, civil unrest in southern Angola may have accelerated immigration (see Van Aarde & Jackson, 2007).

Based on count data, elephants apparently immigrated and emigrated in response to management in Kruger and moved into areas where densities were reduced through culling (Van Aarde *et al.*, 1999). When western park fences were removed, emigration from Kruger also gave rise to the rapid increase in elephant numbers on adjacent private land where elephant numbers were previously low (Whyte, 2001; D. Varty, Conservation Corporation, pers. comm.). Furthermore, recent movements across Kruger's eastern boundary into the Limpopo National Park in Mozambique seem to co-occur with a recent decline in elephant numbers in Kruger (H. Magome, SANParks, pers. comm.).

Published information on immigration and emigration rates for elephants is scarce, probably due to the difficulty and costs of monitoring the movements of many elephants for extended periods over vast areas. Study of the breeding herd of elephants that was observed to have colonised the Amboseli ecosystem by gradually shifting its annual home range (Moss, 1988) suggests that dispersal, immigration, and emigration events are relatively rare and hard to detect using conventional survey techniques. Genetic approaches (e.g. Spong & Creel, 2001) may facilitate the study of elephant immigration and emigration. It is likely that density, environmental factors, and physical barriers, both man-made and natural, may affect these rates. This may enhance population growth locally. For instance, preventing movements out of an area through fencing may be followed by population increase despite the limitation of resources. This happened in Kruger where elephant numbers increased at 10.4 per cent per annum prior to its complete fencing in 1976. During the period when Kruger was completely fenced, elephant numbers increased at 6.6 per cent, while numbers increased at only 1.5 per cent per year after some of the fences were removed along the western boundary in 1994 (Whyte, 2001). This may be due to elephants emigrating out of Kruger and to the surrounding areas.

Water provisioning may also influence emigration. For instance, the placing of 10–15 waterholes in Khaudum in Namibia led to the elephant population increasing from 80 in 1976 to 3 400 in 2004 (Van Aarde & Jackson, 2007). For many elephant populations in South Africa, fences that isolate conservation areas from the surrounding landscapes block dispersal, immigration and emigration. This hampers limitation of population growth through dispersal, a scenario very different to that experienced by several populations elsewhere in southern Africa. These aspects need further investigation because immigration and emigration can clearly influence population growth.

Recent literature (e.g. Bulte *et al.*, 2004; Van Aarde *et al.*, 2006; Van Aarde & Jackson, 2007) considers the stimulation and maintenance of dispersal movements of special importance to the maintenance of metapopulation dynamics and the mitigation of impact. Such movements certainly occur, even within conservation areas (Van Aarde *et al.*, 1999). For instance, our recent analysis of landscape-specific yearly counts in Kruger suggests that population growth rates on different landscapes ranged between –20 per cent and 30 per cent annually (CERU, unpublished data). Compared to the mean annual growth rate of 4.0 per cent between 1998 and 2004 for the entire park (Young *et al.*, 2008), such extremes can be ascribed only to large-scale movements within the park. The forces responsible for these apparent large-scale movements need further investigation and are probably associated with changes in habitat conditions in response to heterogeneity in yearly rainfall across the Park.

Numbers and densities

It is difficult to count elephants. Total counts of elephants are usually based on direct censuses of all individuals that live in a study area, but usually include errors, which can be quite large, due to missed or double-counted individuals. Sample counts use statistical sampling techniques such as ground- or aerial-based line-transect surveys to get an estimate of the number of elephants in sub-areas, which are then extrapolated to the whole area (Norton-Griffiths, 1978). The sample methods and intensity of surveys affect the precision of estimates, which are statistically expressed as confidence limits of estimates. This has major implications for the validity of year-to-year comparisons of estimates to deduce trends in population growth. As a statistic, the confidence limits reflect on the precision of a population estimate – when confidence limits are high, estimates are imprecise.

When consulting the 2007 report on the status of African elephants (Blanc *et al.*, 2007) one notes 384 counts and estimates; 19 per cent of these are total counts, 34 per cent are estimates based on sample counts, and 41 per cent are estimates based on guesses. What is more, the 75 confidence limits for estimates calculated from aerial sample counts in this report (Blanc *et al.*, 2007) ranged from 10 to 376 per cent of the value of the estimate (median = 65.3 per cent). These high levels of imprecision clearly limit the value of such estimates for management and assessment of population growth rates. Wide confidence limits also may hamper the analysis of elephant population trends in South Africa where registration counts (e.g. Gough & Kerley, 2006), recapture modelling (e.g. Morley & van Aarde, 2006), and total counts (e.g. Garaï *et al.*, 2004; Whyte, 2001) may yield wide confidence limits or lack indications of the precision of estimates of the sizes of populations.

Population name	Area size (km²)	Elephant numbers	Elephant density (number per km²)	Year of estimate	Exponential growth rate (%±SE)	Time period (number of estimates)
Addo Elephant Park	1 250	459	2.90*	2005	1.7±0.2	1931–2005 (*n*=70)
Andover Game Reserve	71	11	0.15	1994	–	–
Atherstone Nature Reserve	136	60	0.44	2005	12.5±2.2	1994–2005 (*n*=4)
Balule Nature Reserve	400	457	1.14	2006	–	–
Borakalalo National Park	120	2	0.02	1994	–	–
Great Fish River Reserve Complex	440	2	0.01	2005	–	–
Greater Kuduland Safaris	120	6	0.05	1995	–	–
Greater St. Lucia Wetland Park	539	45	0.08	2005	3.4±4.7	2002–2005 (*n*=3)
Hluhluwe-Umfolozi Game Reserve	965	346	0.36	2004	19.8±3.4	1981–2001 (*n*=12)
Ithala Game Reserve	297	84	0.28	2005	7.5±1.6	1990–2005 (*n*=5)
Kaia Ingwe	45	5	0.11	1994	–	–
Kapama Game Farm	246	36	0.15	2005	–	–
Kariega Private Game Reserve	190	11	0.06	2005	–	–
Karkloof Falls Safari Park	14	2	0.14	1990	–	–
Klaserie Private Game Reserve	628	569	0.91	2006	5.6±1.2	1978–2006 (*n*=10)
Kruger National Park**	19 624	12 427	0.63	2006	1.1±0.3 / 4.1±0.6	1964–2006 (*n*=38) / 1996–2006 (*n*=11)[#]
Kwalata Game Ranch	90	22	0.24	1994	–	–
Kwandwe Private Game Reserve	160	27	0.17	2005	–	–
Lalibela Private Game Reserve	75	11	0.15	2005	–	–
Lowhills Game Reserve	30	8	0.27	1994	–	–
Mabula Game Lodge	120	9	0.08	2004	–4.2±2.6	1989–2004 (*n*=4)
Madikwe Nature Reserve	700	455	0.65	2005	5.6±0.9	1995–2005 (*n*=4)
Mahlatini Game Reserve	15	5	0.33	1994	–	–
Makalali Private Game Reserve	140	72	0.51	2005	16.0±5.0	1994–2005 (*n*=4)
Makuya National Park	165	54	0.33	2006	3.3±8.1	1990–2006 (*n*=4)
Manyeleti Game Reserve	228	71	0.31	2006	–0.6±7.4	1990–2006 (*n*=3)
Marakele National Park	380	110	0.29	2005	11.5±0.6	1996–2005 (*n*=4)
Mkuzi Falls Safaris	22	3	0.14	1994	–	–
Mkuzi Game Reserve	380	37	0.1	2005	9.3±3.1	1994–2005 (*n*=4)
Mokolo River Nature Reserve	45	6	0.13	994	–	–
Mpongo Park	25	8	0.32	1990	–	–
Mthethomusha Game Reserve	80	30	0.38	2005	7.8±3.6	1990–2005 (*n*=3)
Mtibi Game Farm	25	6	0.24	1994	–	–
Ndzalama Game Reserve	79	8	0.1	1994	–	–
Pamula Game Lodge	21	5	0.24	1994	–	–
Paradise Game Farm	30	6	0.2	1994	–	–
Phalaborwa Mining Company	41	77	1.88	2006	8.0±6.5	1990–2006 (*n*=6)
Phinda Resource Reserve	150	78	0.52	2004	7.3±3.1	1990–2004 (*n*=4)

Population name	Area size (km²)	Elephant numbers	Elephant density (number per km²)	Year of estimate	Exponential growth rate (%±SE)	Time period (number of estimates)
Pilanesberg National Park	553	140	0.25	2005	11.5±1.3	1980–2005 (n=9)
Pongola Game Reserve	119	55	0.46	2005	10.3±5.1	1997–2005 (n=4)
Pongolapoort Game	80	48	0.6	2005	12.3±2.4	1997–2005 (n=5)
Reserve Pumulanga Game Reserve	27	3	0.11	1994	–	–
Rhinoland Safaris	70	5	0.07	1994	–	–
Rietboklaagte Game Farm	25	3	0.12	1990	–	–
Riverside Lodge	40	6	0.15	1995	–	–
Sabi Sand Game Reserve	572	857	1.5	2006	19.7±3.8	1990–2006 (n=6)
Selati Game Reserve	300	85	0.28	2005	–	–
Shamwari Game Reserve	150	61	0.41	2005	3.7±5.0	1994–2005 (n=4)
Songimvelo Game Reserve	490	60	0.12	2005	10.5±3.0	1992–2002 (n=4)
Sutton Game Ranch	20	4	0.2	1994	–	–
Tembe Elephant Park	300	167	0.56	2005	6.0±0.8	1974–2005 (n=18)
Thaba Tholo	250	17	0.07	1994	–	–
Thornybush Game Lodge	80	18	0.23	1995	–	–
Thukela Biosphere Reserve	240	9	0.04	1994	–	–
Timbavati Game Reserve	494	712	1.44	2006	12.7±1.4	1985–2006 (n=10)
Touchstone Game Farm	75	10	0.13	1994	–	–
Tshukudu Game Lodge	45	2	0.04	1994	–	–
Umbabat Game Reserve	144	163	1.13	2006	6.1±8.1	1994–2006 (n=5)
Venetia Limpopo Nature Reserve	91	61	0.67	2005	25.5±3.9	1990–2005 (n=6)
Vosdal Game Farm	64	3	0.05	1994	–	–
Welcome Game Reserve	21	5	0.24	1990	–	–
Welgevonden Private Game Reserve	330	100	0.3	2005	3.8±1.9	1995–2005 (n=4)
Zulu Nyala Safaris	7	4	0.57	194	–	–

*   Addo's population is in three separate areas each 120 km² in size. The majority of the elephants (348) lived in one of these in 2005. We present density calculated for this area.

** Estimates for Kruger do not include adjacent areas.

\#   Estimated growth for Kruger represents the period after culling stopped.

Table 4: (previous page) A summary of the numerical status of elephant populations in South Africa. Here we provide the property sizes, population sizes, and densities for the year in which the most recent estimate was reported (data extracted from the CERU database). We also estimated exponential growth rate where the data were suitable for calculation. We used densities because in several cases areas surveyed varied from year to year for a particular locality. Exponential growth was the slope of the natural logarithm of density regressed against time ($N_t = N_0 e^{rt}$)(Caughley, 1977). We provide the time period on which the calculation of growth was based as well as the number of population estimates available in a time series for the calculation

Elephants in South Africa make up only 3.8 per cent of Africa's elephants (17 847 in South Africa and 472 269 across the continent as a whole, based on definite estimates as classified by Blanc *et al.*, 2007). Population sizes in South Africa vary considerably, with the largest in Kruger, which had 12 427 elephants in 2006 (table 4). Kruger is also the largest area in South Africa that holds elephants. Of the remaining 62 places that hold elephants, only Addo is larger than 1 000 km², of which only 360 km² is available to elephants.

Comparison of South African population sizes with those elsewhere is troublesome because the areas surveyed at a site often vary from year to year. In such cases, it is useful to calculate density to compare one locality to another or one year to another. However, this standardisation is challenging.

The ecological meaning of density may vary considerably depending on how it is calculated (Gaston *et al.*, 1999), e.g. annual ecological density = numbers per area of each vegetation type per 365 days; seasonal ecological density = numbers per area of each vegetation type per season; decadal limiting density = maximum numbers per area of each vegetation type in limiting year. Interpretation of densities may be most appropriate when measured at times when the population is limited by resources, e.g. for the dry season, when density effects may be strongest because resources then are scarce.

Based on the recent African Elephant Status Report (Blanc *et al.*, 2007) dry season elephant densities vary considerably across Africa (figure 5), probably in response to local resource availability determined by biome and rainfall; management actions such as fencing, water provisioning, and culling; natural predation; and hunting or poaching. The reality is that elephant densities, and hence numbers, vary greatly in both space and time. Densities deduced from Blanc *et al. (*2007) for South Africa ranged from 0.04 to 2.90 n.km^{-2} (table 4). In addition, South Africa tends to have relatively more populations with high densities than elsewhere in Africa (figure 5). This outcome may be explained by the patterns we have noted above for birth and survival rates – in South Africa, cows have their first calf at younger ages, have subsequent calves more often, but have similar survival rates. These factors and the limitations placed on dispersal by fences could lead to higher population growth rates (see later) and result in higher densities.

Population growth

Population growth is usually expressed as a percentage value per annum. It reflects on the contribution that the individual makes to changes in population numbers. It is a summary statistic that can be compared between populations in research on factors that limit population size. Growth rates will vary from year to year because of year-to-year variations in environmental conditions that limit population processes. In spite of this and for ease of interpretation, population ecologists often calculate growth rate from population estimates and assume that rates remain relatively constant from year to year.

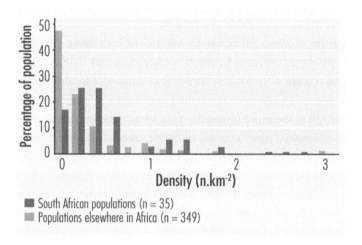

Figure 5: The distribution of elephant densities extracted from the most recent African Elephant Status Report (Blanc *et al.*, 2007). We separated estimates for South African populations from those for populations elsewhere in Africa. We counted the number of estimates falling into density classes that were 0.2 n.km^{-2} wide. The distribution for South African populations has a median (the most central value across the range of densities) of 0.31 n.km^{-2} while that for populations elsewhere in Africa was 0.11 n.km^{-2} even though 12 parks elsewhere in Africa support densities greater than those for the parks in South Africa

Population growth rates vary geographically. In eastern Africa populations are generally stable, while those in southern Africa are increasing (Blanc *et al.*, 2005). However, within southern Africa, numbers in Hwange in Zimbabwe (Chamaillé-Jammes *et al.*, 2007) and northern Botswana appear to be stabilising (Junker *et al.*, 2008). Those in some areas in Zimbabwe (Cumming *et al.*, 1997), Namibia (Lindeque, 1991), and South Africa (Van Aarde *et al.*, 1999; Gough & Kerley, 2006) are increasing, while in places in Zambia, such as the Kafue National Park (Guldemond *et al.*, 2005), the Lower Zambezi National Park and parks in the Luangwa valley, numbers are decreasing or stabilising (Ferreira *et al.*, 2008). In some instances, the estimated annual population growth exceeds the maximum theoretical growth rate. This is particularly the case for small populations in South Africa (see table 4), where synchronised breeding and skewed age structures can cause high, short-term spurts in annual population growth rates which will not persist in the longer term.

South African populations have annual growth rates that range from –0.6 to 25.5 per cent per year (table 4). Of the 29 estimates of annual population growth rate in South Africa, only two were negative while 16 were higher than 7 per cent per annum (table 4). None of these populations were stable. Based on census data, populations elsewhere in Africa grow at annual rates ranging from –87.7 to 148.8 per cent per year (see figure 6). High apparent positive or negative population growth rates result from large-scale movements, particularly when a few elephants comprise the initial population size. Elsewhere in Africa, 70 (46 per cent) of the annual growth rates that we could estimate from population estimates were negative.

All South African populations have been exposed to some form of management that includes fencing, population control through translocations, culling or contraception, and water provision. Contrastingly, most other populations in Africa have relatively little management and are not fenced, allowing large-scale movements. The response of populations to management can best be measured by their growth rates. South Africa's intensely managed populations increased at rates that were both faster and less variable than populations elsewhere in Africa (figure 6), suggesting that conditions created by management stimulate growth. This is not surprising, since elephant populations, like those of all other species, should respond to resource supply and the protection afforded by conservation management. On the other hand, the inhibition of dispersal may also be largely responsible for higher population growth rates in fenced South African populations than for the open populations elsewhere in Africa where immigration and emigration do occur.

As indicated earlier, immigration and emigration rates are hard to determine. However, in the near future, our understanding of the influence of immigration and emigration on populations may be enhanced by comparing growth rates derived from estimated birth and death rates (see Ferreira & Van Aarde 2008) with those calculated from census data. We are aware of few field studies (e.g. Van Aarde et al., 1999; Gough & Kerley, 2006; Ferreira & Van Aarde, 2008) that modelled population growth rates from birth and death rates. A few studies estimated theoretical growth rates (e.g. Hanks & McIntosh, 1973; Calef, 1988), while others used demographic predictors to evaluate population responses to contraception (e.g. Dobson, 1993; Van Aarde et al., 1999) and trophy hunting (e.g. Owen, 2005).

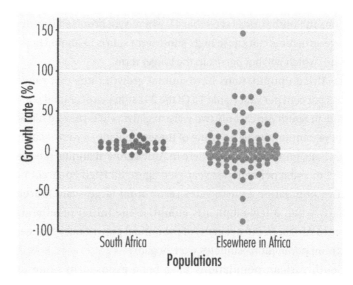

Figure 6: Exponential growth rates (Caughley, 1977) estimated from at least three population estimates in a time series for 28 South African elephant populations and 152 populations elsewhere in Africa. South African populations have a narrow distribution of growth rates (−4.2 per cent to 25.5 per cent) compared to populations elsewhere (−87.7 per cent to 148.2 per cent) and appear to centre above zero (South Africa: median = 7.7 per cent; elsewhere: median = 0.95 per cent; median refers to the most central growth rate across the range of rates)

Studies of density-dependent population growth in elephants are rare (e.g. Van Aarde *et al.*, 1999; Sinclair, 2003; Gough & Kerley, 2006; Junker *et al.*, 2008), yet they are needed to evaluate the consequences of any of the management regimes that elephant populations may be exposed to in the future.

Population limitation

Elephants are generalists and therefore utilise a variety of food resources. Even so, food availability influences vital rates. For instance, the distances that elephants need to travel between water and habitats of high nutritional value may affect energetic expenditure and influence conception and mortality of young animals. Indeed, the very low calf mortality rates found at Addo were attributed to a constant supply of food and water in comparison to other populations (Gough & Kerley, 2006). From studies elsewhere in Africa we know that conception rate varies with primary productivity as proxied by NDVI

Box 4: Individual responses to management

Elephant management techniques including culling, translocation, and contraception may have important consequences for individual elephants. Depending on the scale of action, individual responses may generate an effect at the population level.

The consequences of culling for individual elephants, especially selective culling, are poorly understood and need to be assessed (Slotow et al., 2005). The trauma endured by culled orphans and those raised by inexperienced mothers puts calves at risk for developing symptoms similar to post-traumatic stress disorder in humans – abnormal startle response, depression, unpredictable asocial behaviour and hyper-aggression (Bradshaw et al., 2005).

Translocation can also affect elephant behaviour. The introduction of bulls to new and strange environments occasionally results in 'breakouts' as bulls potentially try to return to their previous home ranges or attempt to gain access to different vegetation or reproductively active females (Garai & Carr, 2001). Additionally, adolescent males require socialisation with older bulls for normal social development (Slotow et al., 2000; Bradshaw et al., 2005), a requirement often neglected by translocation endeavours (e.g. Pilanesberg and Phinda; Slotow & Van Dyk, 2001; Genis et al., 2004). This problem probably holds for all South African populations that were founded through reintroduction prior to 1998 when bulls older than 25 years were not included in founder groups (Slotow & Van Dyk, 2001).

Additionally, despite early optimism that contraception was effective, safe, and reversible (Fayrer-Hosken et al., 2000), it may have side-effects that influence the health and behaviour of cows (Whyte et al., 1998; Pimm & Van Aarde, 2001; Van Aarde & Jackson, 2007). Hormonal treatments may cause cows to remain in sexual heat and be harassed by bulls and evicted from their social groups (Whyte & Grobler, 1997). Furthermore, as elephant society is kin-based (Archie et al., 2006), artificial control of reproduction may have consequences for social hierarchies and, in turn, individual well-being (see McComb et al., 2001).

(Wittemyer et al., 2007b, 2007c). Conception and birth rates therefore should also vary with spatial variation in rainfall and NDVI.

Periodic droughts may also induce variation in vital rates. For instance, severe droughts synchronised both births and the length of calving intervals in Amboseli (Moss, 2001). Droughts also increase death rates in elephants and may occasionally lead to large-scale die offs (Walker *et al.,* 1987; Dudley *et al.,* 2001), as was the case in the Tsavo ecosystem in Kenya between 1975 and 1980, when many elephants died during an extended drought (Corfield, 1973; Ottichilo, 1987).

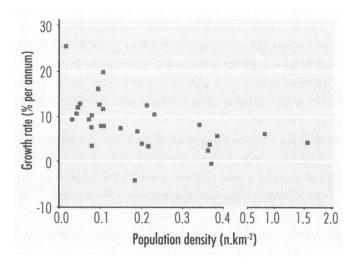

Figure 7: Exponential population growth of South African elephant populations since 1985 as a function of density. We calculated annual growth rate (expressed as percentage) from time series of density extracted for 29 places using $N_t = N_0 e^{rt}$ (Caughley, 1977) and plotted these against the density at the onset of each of the time series. Populations had higher growth rates when the starting density was low

The effect of poaching on populations can be severe (e.g. Douglas-Hamilton, 1972) and may leave demographic signals. For instance, populations in Zambia, an ivory poaching hotspot (Wasser *et al.*, 2007), continued to decline (Ferreira *et al.,* 2008) despite the ivory ban of 1989 (Stiles, 2004). Here, populations had few large and thus old elephants, herds were small (Ferreira *et al.,* 2008) and many elephants had no tusks (Steenkamp *et al.,* 2007).

Although no conclusive analysis of density dependence in African elephant populations has been carried out to date, in at least three studies equilibrium models that include density dependence, best described trends in elephant population numbers over time (Sinclair, 2003; Junker *et al.,* 2008; Chamaillé-Jammes *et al.,* 2007), while one study (Addo) found no evidence of density dependence in population growth (Gough & Kerley, 2006). This lack of evidence for density dependence in Addo is not surprising considering the enlargement of space available to elephants there, and the high relative abundance of resources. In Kruger, Van Aarde *et al.* (1999) inferred density dependence in population growth from changes in densities after culling operations, while Van Jaarsveld *et al.* (1999) also found evidence for density dependence for Kruger and the declining Knysna population. With the exception of one population in the Timbavati, Van Jaarsveld *et al.* (1999) reported density independence in population growth for the recovering South African populations that they studied.

Exponential annual population growth rates that we calculated for South African populations since 1985 tended to be higher when densities were low at the onset of the time series on which we based calculations (figure 7). Although this is not evidence for density dependence, these observations suggest that density may explain between-population variability in population growth rates. Density therefore may be important to explain changes in population growth once densities are high enough to reduce food availability and hence reduce reproductive and survival rates as well as enhance dispersal rates, all of which will inhibit growth. The role of density dependence for the population, as well as for the impact elephants may have on other species, needs further investigation. For instance, reduced population growth at high densities may be negated if populations are artificially reduced through culling (this topic is discussed in Chapter 8). On the other hand, the numbers of elephants at levels where density reduces reproduction and survival may have unacceptable impacts on other species.

THE RESPONSES OF ELEPHANT POPULATIONS TO MANAGEMENT

Inferences on how individual elephants (see box 4) or populations of elephants will respond to management are often based on hear-say. Few measures of such responses have been published (e.g. Van Aarde *et al.*, 1999), and in general, these suffer from poor experimental design, improper replication, and *ad hoc* interpretations (Van Aarde & Jackson, 2007; Guldemond & Van Aarde 2008). For elephants in southern Africa, as for several other species elsewhere in the world, it seems that most past conservation management actions had their origins in experiential rather than experimental evidence (e.g. Pullin & Knight, 2005). For instance, the original decisions to cull elephants in several conservation areas across Africa were motivated by the apparent impact elephants may have had or were having on vegetation (e.g. Pienaar *et al.,* 1966; Laws *et al.,* 1975; Bell, 1983). However, there was little scientific evidence of such impacts, and in the case of Kruger, supporting evidence to motivate the cull was collected after the decision to cull had been taken (e.g. Van Wyk & Fairall, 1969). We also know of no published information to illustrate that a management action such as culling had the desired outcome of reducing the impact that elephants apparently had on vegetation and other species; however, it is hard to know what might have happened in the absence of culling. Proposals to reinstate culling are founded in the so-called precautionary principle (e.g. Whyte, 2004; Mabunda, 2005). Elephant management clearly continues to be a debatable topic (Cumming & Jones, 2005; Mabunda, 2005; SANParks, 2005; Owen-Smith *et al.,* 2006; Van Aarde *et al.,* 2006). More often than not the debate seems to be founded on staunch opinion backed by advocacy, rather than scientific evidence. This is not surprising, because scientists often focus on defining and describing problems rather than on finding solutions for problems.

The response of elephant populations to both direct and indirect management actions may depend on the intensity of the actions applied. For elephant populations, direct management typically aims to reduce numbers by decreasing birth rates (e.g. through contraception), increasing death rates (e.g. through culling), or mimicking dispersal (e.g. translocation). Populations are protected and managed indirectly by erecting fences around conservation areas and by providing additional water. The underlying assumption of direct management actions is that a reduction in elephant numbers will lower the intensity of resource use and will ultimately reduce elephant impact on other species, usually vegetation. This assumption may not be valid (see Van Aarde & Jackson, 2007) because, rather than numbers alone, impact can also depend on the intensity of resource utilisation reflected by spatial use patterns (see Gordon *et al.,* 2004) and dictated by the distribution of key resources. In addition, the ultimate success of management actions to reduce impact has yet to be assessed. We therefore cannot elaborate on the effectiveness of management to reduce impact. However, we can evaluate and speculate on the responses of elephant populations to management actions such as contraception, culling, translocation, and the manipulation of resources such as water and space (e.g. restrictions through fences or providing space through transfrontier conservation areas). Here we focus on a broader comparative evaluation while later chapters focus on specific case studies.

Contraception

This topic is dealt with in detail in Chapter 6. Here we address only aspects relating to population dynamics.

The application of contraceptives to reduce fertility in wildlife is well beyond the research phase (Kirkpatrick, 2007; Perdok *et al.*, 2007). Birth rates may be reduced by treating cows with hormones and their derivates, or with immuno-contraceptives to reduce or control fertility (e.g. Fayrer-Hosken *et al.*, 2000; Pimm & Van Aarde, 2001; see Chapter 6 for detailed methodology).

Reducing reproductive rates may also alter the age and social structures of breeding herds and possibly influence the well-being of cows and their calves (McComb *et al.*, 2001; Pimm & Van Aarde, 2001). Contraceptives may lengthen inter-calving intervals or increase the age of first calving (Perdok *et al.*, 2007). Unlike culling, contraception does not reduce numbers – instead it relies on natural mortality and reduced reproductive output to reduce population size over time.

The efforts needed to stabilise elephant numbers in large populations through birth control are both laborious and costly (Pimm & Van Aarde, 2001). At the population level, birth control is constrained by the number of females needing treatment (Whyte *et al.*, 1998). Age at first calving will only increase effectively if almost 50 per cent of pregnant cows less than 15 years old are on birth control or forced to abort (Mackey *et al.*, 2006). In Kruger, elephant population growth will only stabilise if managers treat nearly 75 per cent of adult cows continuously for 11 years (Van Aarde *et al.*, 1999). We agree with others (Bertschinger *et al.*, 2003; Delsink *et al.*, 2006; Perdok *et al.*, 2007) that immunocontraception can currently only be regarded as a proven and realistic option for reducing population growth in small, confined populations. As for the ultimate goal of management, the ability of contraception to reduce elephant impacts on vegetation still needs to be determined.

Culling

Culling is discussed in detail in Chapter 8. Culling can be directed at reducing the sizes of local populations, stabilising populations, manipulating the number of animals in distinct social groups within a population, or removing elephants from specific parts of their distributional range (e.g. from obvious zones of conflict).

Controversy aside, the 30-year elephant culling regime in Kruger provided a valuable case study. Much has been written on the topic of culling, also for species other than elephant (see Walker *et al.*, 1987; Cumming *et al.*, 1997; Proaktor *et al.,* 2007). In general, it seems that the reduction in density through culling inflates population growth rate, by releasing vital rates (age at first calving and inter-calving interval) from limitations set by density dependence (for elephants see Whyte *et al.*, 1998; Van Aarde *et al.*, 1999). Therefore, elephant culling with the intention of maintaining populations at a level below which resources are limited is a self-perpetuating practice because populations are pushed to densities where reproductive potential and survival may be optimised. Put simply, culling can only be effective to reduce numbers in the medium term if it is maintained indefinitely and at a rate above the population's growth rate.

An interesting issue to consider is whether Kruger's elephants would have stopped increasing through density dependence should culling not have taken place. An analysis presented by Van Aarde *et al.* (1999) provides support that density dependence becomes apparent at 0.37 elephants per km^{-2}, and they suggested that culling was probably unnecessary unless populations remained at densities higher than that value for two or more years. However, this appeared not to be the case, and elephant density in Kruger is approaching much higher values (Blanc *et al.*, 2007).

There are two possible explanations for this discrepancy. First, perhaps the mode of density limitation during the culling era was via migration from non-culled regions at densities greater than 0.37 km^{-2} to other regions in the park where the cull reduced density to relatively low levels (see Van Aarde *et al.*, 1999). In this case, it is unlikely that vital rates would change in response to reduced resource availability because elephants simply migrated to resource-rich areas rather than experiencing the limitations imposed by resource scarcity. The second explanation is that resource limitation truly limited elephant density at densities greater than 0.37 km^{-2}. In this case, tell-tale changes in vital rates would be expected. Unfortunately such information is not available.

The fact that the Kruger elephant population is not currently limited at the density proposed by Van Aarde *et al.* (1999) probably reflects on changes in resource availability. The assessment of Van Aarde *et al.* (1999) was based on data from a dry cycle lasting several years and including a severe drought in 1992 (see Mills *et al.*, 1995; Ogutu & Owen-Smith, 2003). Since then conditions have changed, and drought conditions may no longer limit resources, therefore explaining the lack of immediate density-dependent responses. Additionally, the relatively high densities at which elephants presently occur in Kruger could be a delayed response of reproductive output in response to culling (eruptive growth, discussed in detail in Chapter 8).

Culling apparently can effectively limit population growth only when applied continuously. For instance, following the cessation of culling in the Kruger, growth rates increased dramatically (see Whyte *et al.*, 2003). Furthermore, after the cessation of culling in 1995 in Hwange National Park

(Zimbabwe), elephant numbers almost doubled in just six years, while elsewhere in Zimbabwe, numbers grew about 28 per cent over the same period (Foggin, 2003). Even so, culling does reduce numbers, albeit temporarily.

Where selective culling may target bulls or animals of certain age classes, distorted age structures may enhance, rather than suppress growth rates (see Gordon *et al.*, 2004) and so negate the intention of culling. In addition, at lower densities population growth rate may increase due to the release of density-dependent limitations of reproductive rate (see Sinclair, 2003). Thus, inappropriate culling may effectively increase growth rate.

A major shortcoming of past elephant culling programmes is that none of them employed an evaluation approach to assess efficiency in reducing the apparent impact that motivated the undertaking of the programmes.

Translocation

This topic is discussed in detail in Chapter 5. Initially, policy regarding the translocation of elephants was formulated to establish more elephant populations across southern Africa (Pienaar *et al.*, 1966). This was done on the premise that genetic variability of elephants could be enhanced or maintained through this process. A secondary outcome of elephant translocation developed as a more ethical solution than culling to control and/or reduce elephant numbers in a particular region. The translocation of elephants is, however, not unique to South Africa. Other African countries, such as Kenya, also have experience in shifting elephants, albeit for different reasons. There, elephants were moved from small reserves to larger parks such as the Tsavo National Park to mitigate human-elephant conflict (Njumbi *et al.*, 1996). The efficiency of these translocations still has to be assessed.

Since 1979, elephants from the Kruger have been captured, translocated, and released in other parks and reserves (Garaï *et al.*, 2004), some of them privately owned (Garaï & Carr, 2001). In some of the earlier translocation efforts, only elephant calves were moved, but due to aberrant social behaviour of young bulls (Slotow *et al.*, 2000), intact family units and adult bulls have been included in recent efforts to establish new populations or during re-introductions. Some 58 elephant populations were established in South Africa alone between 1979 and 2001 (Garaï *et al.*, 2004), with the numbers in newly founded elephant populations expected to increase (Slotow *et al.*, 2005). All of these newly established populations live in fenced reserves that are relatively small, ranging in area from 15 to 900 km^2 (Slotow *et al.*, 2005). One particular aspect that stands out is the high growth rates reported for these populations, some as high as

25.5 per cent (table 4). This is well beyond the maximum rate of increase that maximum birth and survival rates predict for elephants living in closed areas (Calef, 1988; Van Aarde *et al.*, 1999). This abnormally high growth rate can most likely be ascribed to synchronised calving and/or unstable population structures typical of small groups. Additionally, most of the recently established elephant populations comprise few individuals (see table 4), and estimates of their vital rates thus may suffer from statistical limitations (Akçakaya, 2002). Theoretically, the conversion of unstable age structures to stable structures will be associated with a reduction in average population growth rate to values around 5 per cent per year when populations are enclosed.

Despite aberrant population growth rates, translocations of elephants are regarded as successful to establish populations (Garaï *et al.*, 2004; Slotow *et al.*, 2005). However, its contribution to conservation needs to be questioned since many researchers warn against the effects on other species of continual increase in elephant numbers in these newly established reserves. In most of these reserves, elephants are confined to relatively small areas where space is so limited that it does not allow natural seasonal roaming. Dispersal also is impossible due to surrounding land use options. Fences that surround these areas and artificial pans and waterholes may lead to small home ranges that are intensely utilised and to high growth rates. This will intensify the impact that elephants will have on the landscape surrounding these artificial sources of water. Thus, the establishment of new populations through translocations may create more population control issues than it solved as many of these populations may soon require management to reduce impact.

More than 800 elephants were moved from the Kruger between 1979 and 2001 (Garaï *et al.*, 2004), with the main translocation efforts between 1990 and 2001 (Slotow *et al.*, 2005). On average, in those years when translocation took place, about 1 per cent of the population was removed from Kruger. Based on the trends in population numbers given by Whyte *et al.* (1998), these translocations clearly had little effect on Kruger's elephant numbers and certainly did not reduce the population's rate of increase during the 1990–2001 period.

Other aspects that may relate to the translocation of elephants, such as the demand for and availability of suitable elephant habitat, management constraints (e.g. costs of capture, care, translocation, and release of elephants), and possible effects (post-traumatic stress) on individual elephants are dealt with in Chapter 5.

Translocation may also have undesirable genetic and conservation consequences. Recent advances in genetic profiling of sub-populations as separable entities provides conservation managers with a powerful tool to locate the sources of illegal ivory and thus to strengthen conservation efforts (see Wasser *et al.*, 2007). Mixing elephants from different regions will destroy these unique genetic signals and therefore detract rather than enhance conservation initiatives.

Translocations that mix elephants of different genetic stocks also interfere with conservation ideologies that centre on the maintenance of biodiversity, for biodiversity conservation also emphasises the maintenance of ecological processes. Of these processes, natural selection is probably the one process that gives rise to sub-population differences as an adaptation to local conditions. Interfering with this detracts from the conservation paradigm to which South Africa and several of its neighbouring countries are signatories.

The translocation of elephants is relatively easy and can give rise to the establishment of new populations, thereby recovering key ecological processes that may have been lost through earlier local exterminations of elephants. This, however, only holds when environmental conditions in areas where new populations are established meet the requirements for the development of an elephant population. This apparently is not the case for most populations established through translocations in South Africa and the conservation management benefits of translocations therefore must be questioned. Low rates of translocations may have little benefits for the donor populations, because the removal of elephants may merely re-distribute elephants in the donor populations, as has been the case when elephants were removed through culling from specific management areas in Kruger (see Van Aarde *et al.*, 1999). In conservation terms the genetic consequences of translocations when mixing individuals of different sub-populations is also not desirable. On the other hand, genetic enrichment in artificially isolated populations such as Addo may be advantageous.

Manipulation of water

Water is a primary determinant of the distribution of elephants (De Beer *et al.*, 2006; Chamaillé-Jammes *et al.*, 2007; Harris *et al.*, 2008). Elephant breeding herds are especially water dependent as the young calves and lactating cows need to drink frequently (e.g. Stokke & Du Toit, 2002). It is therefore not surprising that the manipulation of surface water distribution has major consequences for the way elephants roam and forage across the land they occupy. Such water may alter seasonal movements and enable elephants to inhabit sensitive landscapes for longer periods of the year than they would have under natural conditions. This could intensify impact, especially for plants that are not predisposed to intensive utilisation. The vegetation in such areas therefore does not have the opportunity to recover seasonally.

Water provisioning is a standard procedure many wildlife managers practise across the southern African range of elephants (see Chapter 7 for more details). Such provisioning affects movement patterns (Harris *et al.*, 2008), home range utilisation and size (Grainger *et al.*, 2005; De Beer *et al.*, 2006; De Beer, 2007) and the impact that elephants have on local vegetation (Gaylard *et al.*, 2003; De Beer *et al.*, 2006; O'Connor *et al.*, 2007). For instance, water made available in man-made waterholes could attract elephants to occupy land that they would not otherwise have occupied – habitats avoided under natural conditions may now be utilised, thus resulting in the redistribution of elephants and negating the potential for density related forces to inhibit survival and reproductive output of elephants in preferred habitats. Water provisioning therefore may boost the so-called elephant problem.

Water manipulation may also influence the demography of populations. Recent work in the Hwange National Park in Zimbabwe suggests that density tends to increase with the increase in artificial waterhole densities (Chamaillé-Jammes *et al.*, 2007). Distance to water is also a primary determinant of the densities at which elephants occur (Western, 1975; Stokke & Du Toit, 2002; Redfern *et al.*, 2003; Grainger *et al.*, 2005). Owen-Smith (1996) and Chamaillé-Jammes *et al.* (2007) suggest that the manipulation of artificial surface water can be an important tool through which to manage elephant populations. The effectiveness of water manipulation as a management tool, however, may differ between areas and between populations (Smit *et al.*, 2007b).

Water provision influences populations by enhancing survival, especially of juveniles, during droughts and/or in arid regions. Water provisioning also enhances immigration, as illustrated by our recent and ongoing assessment of population time series from several areas in northern Namibia. In northern Namibia, without exception, water provisioning in both formal and informal conservation areas was followed by an increase in population numbers locally (CERU, unpublished data). This may also explain the trends in numbers in Kruger during the 1960s and 1970s when water availability was increased artificially (Pienaar, 2005) and before a fence isolated elephants in Mozambique from those in Kruger. Therefore, the water provided in human-made structures either attracts elephants from elsewhere (as has been the case in Hwange in Zimbabwe following the establishment of additional water points (Chamaillé-Jammes *et al.*, 2007), or enhances local survival. Presently, elephants appear to be moving out of Kruger, where water sources are apparently being closed, into areas west of Kruger with an extremely high density of artificial waterholes (J. Swart, Sabi Sands Game Reserve, pers. comm.).

Surface water distribution and manipulation may cause population size to increase to artificially high numbers (Van Aarde & Jackson, 2007). Water provided in human-made structures, therefore, may be at the root of the so-called elephant problem. We are not aware of published accounts of the influence of surface water manipulation on reproductive output and survival, both of which may be implicated in the relatively high numbers at which elephants occur when water is artificially provided. This clearly needs further investigation.

Within protected areas, efforts to stabilise the availability and spread of drinking water to regions that were inaccessible during the dry season probably affected elephant survival, as young are particularly susceptible to drought conditions (Dudley *et al.*, 2001; Loveridge *et al.*, 2006). Improved survival may increase population size because survival of young is an important determinant of population growth (e.g. Gaillard *et al.*, 1998).

Surface water distribution may also determine dispersal, which influences population numbers through immigration and emigration. Artificial waterholes attract elephants and result in populations being established in areas where elephants otherwise would not occur, particularly during the dry season (Chamaillé-Jammes *et al.*, 2007; De Beer, 2007; Smit *et al.*, 2007a). This is especially true for the arid savannas where elephant populations became resident in response to water provisioning in Etosha (Lindeque, 1988), Khaudum (De Beer, 2007) and Hwange (Chamaillé-Jammes *et al.*, 2007).

Fencing

The fencing of conservation areas and the establishment of veterinary fences to control the spread of contagious diseases inhibits both seasonal movements and dispersal and thereby has consequences for the size of elephant populations (Mbaiwa & Mbaiwa, 2006; Van Aarde *et al.*, 2006; Van Aarde & Jackson, 2007; see Chapter 7 for details on fencing as a management tool).

Fences have an edge effect on the utilisation intensities of home ranges and, consequently, on the impact that elephants may have on vegetation (CERU, unpublished data). More importantly, however, at the population level, the lack of dispersal opportunities may enhance local population growth (Owen-Smith, 1988). The advent of the dropping of some of the fences surrounding Kruger is too recent for formal literature to have noted emigration events that could have resulted in a decrease in population size. Recent observations suggest a marked increase in elephant numbers in the Limpopo National Park (Mozambique) that adjoins the eastern boundary of Kruger, while at the same time, numbers in Kruger have stabilised (H. Magome, SANParks, pers. comm.). This supports our earlier speculation that dispersal is an important determinant of local population size (Van Aarde *et al.*, 2006). This clearly needs further investigation.

Manipulation of space

The manipulation of space potentially involves the development of linkages, corridors, and/or so-called stepping stones to link sub-populations into a metapopulation structure of some kind (see Van Aarde & Jackson, 2007). The recent literature on elephant social dynamics (Archie *et al.*, 2006) and spatial use patterns of groups of elephants of differing social status (Wittemyer *et al.*, 2007a) also calls for the enhancement of space to ensure social structuring and out-breeding. Population level responses to spatial manipulation have not been recorded, except for incidences where the recent extension of the range of elephants resulted from the lifting of some of the fences of Kruger (De Villiers & Kok, 1997). This gave rise to elephants establishing themselves on vacant land in neighbouring conservation areas.

The present distributional range of elephants is patchy and extends beyond conservation areas in countries other than South Africa, though most elephants do occur in formally protected areas. Elephants do disperse readily into vacant habitats. For instance, historical records show that elephants moved from Mozambique into South Africa's Kruger, which in the early 1900s supported fewer than 10 elephants. Dispersal at annual rates of 7–10 km meant that the Park's approximate 20 000 km^2 was colonised within 50 years (Whyte *et al.*, 2003). Similarly, in 1955 elephants were recorded in the Serengeti after an absence of at least 40 years. Here numbers increased over a 10-year period, mainly through immigration, to some 2 000 individuals (Lamprey *et al.*, 1967).

In areas where managers manipulated water availability, elephant populations expanded rapidly and at rates that exceeded their reproductive capacity. For instance, Etosha's population comprised approximately 50 individuals in 1950 and increased to some 2 000 by 1980 (Lindeque & Lindeque, 1991). Following water supplementation in Khaudum, the population increased from around 80 in 1976 to some 3 400 in 2004 (Ben Beytell, Ministry of the Environment and Tourism, Windhoek, pers. comm.). Civil unrest in southern Angola may have contributed to this increase in Khaudum, which at ~13 per cent per year is almost triple the value that is typical for populations that increase in response to natural values of birth and deaths.

In Kruger, culling induced dispersal of elephants into areas where densities were reduced (Van Aarde *et al.*, 1999). It therefore follows that elephants do disperse when given the opportunity or when circumstances allow or force them. This is critical to the application of landscape conservation models such as the metapopulation model to the conservation management of elephants, since the metapopulation in its true sense can only operate with dispersal (Van Aarde & Jackson, 2007).

We do not know much of the consequences that the effective increase of space would have for elephant demography. Recent arguments favour the restoration of elephant spatial dynamics, which could influence population responses and restore spatial-temporal dynamics (Van Aarde *et al.*, 2006; Van Aarde & Jackson, 2007). This may lead to local instability in elephant numbers that reduces local impact and conflict while inducing a regional stabilisation of numbers that reduces the threat to the long-term persistence of elephants. These predictions need to be evaluated and tested, but are supported by our recent analyses of differences in population growth rates for different landscape types in the Kruger.

CONCLUSIONS

The Assessment allows us to put forward a conceptual framework that can serve as a guideline for management as well as research (figure 8). The framework explicitly recognises the nature of the dilemma that pervades elephant management in South Africa where most elephants live as a single population in a large conservation area (e.g. Kruger) while the remainder live in many highly artificial and distinct populations in small and isolated reserves (see table 4).

The diverse elephant management challenges can be visualised as falling along a continuum of management intensity. Small and isolated areas invariably require intensive management and consequently will be the least natural. Such areas will contribute relatively little to elephant conservation, but they may be critical for other forms of biological diversity. In contrast, large areas require progressively less management as the integrity of natural processes increases. As a result, areas managed for elephants exist along a continuum of artificial to nearly natural, from populations as reproductive isolates to populations as connected spatial entities, and from relatively costly to relatively cost effective. Most importantly, spatial constraints of elephant-containing areas could define management responses ranging from those that focus on the symptoms, i.e. high elephant numbers (in small areas with intensive management), to those that focus on the forces that cause the symptoms, i.e. why elephant numbers are high in the first place (in large connected areas with low intensive management).

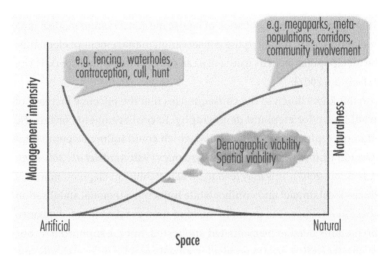

Figure 8: A conceptual model for the management of elephant populations in South Africa. Elephant populations occupy a continuum of size of habitat. Where the available area is small, intensive management is required and the level of 'naturalness' is low. At the other extreme, little management is needed, and the degree of naturalness is high. The aim throughout is to achieve demographic and ecological viability, given the spatial constraint. To the left of the intersection of the curves is the region of demographic and spatial limitations where populations will have to be managed. Populations to the right of the intersection increasingly may need less and less management. The point of intersection represents an approximation rather than a given point

We therefore foresee a scenario where elephants confined to small parks are managed as individuals rather than populations. In this case, the emphasis will be on limiting population size through contraception and/or translocation and protecting species sensitive to elephant impact by manipulating local range use by fencing off selected sensitive areas or trees, perhaps on a long-term rotational basis. Management methods may also include the periodic displacement of elephants from areas of these parks, either through the rotational occupation of landscapes or rotational removal of elephants themselves. Elephants here will most likely live as a breeding herd that will include only the lower tiers of social structuring known for the species (see box 1).

At the other end of the spectrum, where areas have the capacity to provide for all tiers of social organisation up to the population as a unit (see box 1), management can be more relaxed and occasional. In these more natural situations, management no longer centres on elephants, but focuses on the landscape as a spatially and temporally dynamic arena in which all forms of biodiversity, including structural and functional diversity, have an opportunity to persist. Here, management can focus on maintaining spatial linkages for dispersal while allowing for extreme local fluctuations in elephant numbers. Creating larger areas for more effective conservation may require the internationalisation of conservation management, as foreseen in the development of transfrontier conservation initiatives presently driven by several NGOs and supported by several southern African governments.

We also need to be pragmatic. We concede that most elephant-containing areas in South Africa are likely to fall in the region of our conceptual model that proposes intense management. These areas often do not provide for seasonal movements, let alone spatial variability in demography. The managers of such areas cannot aim to achieve demographic viability through natural limiting mechanisms such as density-dependent birth reductions, drought-related mortalities and local dispersal. They will have to resort to active intervention to reduce impact, probably by manipulating population sizes in sensitive places and varying spatial occupation to ameliorate impacts on other species.

REFERENCES

Akçakaya, H.R. 2002. Estimating the variance of survival rates and fecundities. *Animal Conservation* 5, 333–336.

Alexander, K.A., C. Gibbons, M. Ramotadima, M.E.J. Vandewalle, J. Mucheka, I.M. Cattadori & N. Drake 2006. Long-term changes in elephant distribution and seasonal factors: Influence on conflict incidence and impacts on gender specific rural livelihoods. In: L.E.O. Braak & R. Smuts (eds) Towards rationalizing transboundary elephant management and human needs in the Kavango/mid-Zambezi region. Unpublished proceedings of a workshop presented on 23 and 24 May 2006 in Gabarone, Botswana, by Conservation International (Southern Africa Wilderness and Transfrontier Conservation Programme), Cape Town, 36–44.

Archie, E.A., C.J. Moss & S.C. Alberts 2006. The ties that bind: genetic relatedness predicts the fission and fusion of social groups in wild African elephants. *Proceedings of the Royal Society B* 273, 513–522.

Bates, L.A., K.N. Sayaliel, N.W. Njiraini, C.J. Moss, J.H. Poole & R.W. Byrne 2007. Elephants classify human ethnic groups by odor and garment color. *Current Biology* 17, 1–6.

Bell, R.H.V. 1983. Decision making in wildlife management with reference to problems of overpopulation. In: R. Owen-Smith (ed.) *Management of large mammals in African conservation areas.* HAUM Educational Publishers, Pretoria, 145–171.

Bengis, R.G. 1996. Elephant population control in African national parks. *Pachyderm* 22, 83–86.

Berry, H.H. 1993. Surveillance and control of anthrax and rabies in wild herbivores and carnivores in Namibia. *Revue Scientifique et Technique – Office International Des Epizooties*, 137–146.

Bertschinger, H.J., J.F. Kirkpatrick, R.A. Fayrer-Hosken, R.A. Grobler & J.J. van Altena 2003. Immunocontraception of African elephants using porcine zona pellucida vaccine. In: B. Colenbrander, J. de Gooijer, R. Paling, S.S. Stout; T. Stout & W.R. Allen (eds) *Managing African elephant populations: act or let die?* Proceedings of an expert consultation on the Control of Wild Elephant Populations, Faculty of Veterinary Medicine, Utrecht University, Netherlands, 45–47, http://elephantpopulationcontrol. library.uu.nl/paginas/frames.html (accessed 5 April 2006).

Blanc, J.J., R.F.W. Barnes, C.G. Craig, I. Douglas-Hamilton, H.T. Dublin, J.A. Hart & C.R. Thouless 2005. Changes in elephant numbers in major savanna populations in eastern and southern Africa. *Pachyderm* 38, 19–28.

Blanc, J.J., R.F.W. Barnes, G.C. Craig, H.T. Dublin, C.R. Thouless, I. Douglas-Hamilton & J.A. Hart 2007. *African elephant status report: An update from the African Elephant Database.* IUCN, Gland.

Bradshaw, G.A., A.N. Schore, L.B. Brown, J.H. Poole & C.J. Moss 2005. Elephant breakdown. *Nature* 433, 807.

Bulte, E., R. Damania, L. Gillson & K. Lindsay 2004. Space – the final frontier for economists and elephants. *Science* 306, 420–421.

Buss, I.O. & N.S. Smith 1966. Observations on reproduction and breeding behaviour of the African elephant. *Journal of Wildlife Management* 30, 375–388.

Butler, V. 1998. Elephants: trimming the herd. *BioScience* 48, 76–81.

Calef, G.W. 1988. Maximum rate of increase in the African elephant. *African Journal of Ecology* 26, 323–327.

Case, T.J. 2000. *An illustrated guide to theoretical ecology.* Oxford University Press, Oxford.

Caughley, G. 1977. *Analysis of vertebrate populations.* John Wiley & Sons, New York.

Chamaillé-Jammes, S., M. Valeix, & H. Fritz 2007. Managing heterogeneity in elephant distribution: between elephant population density and surface-water availability. *Journal of Applied Ecology* 44, 625–633.

Coe, M.J., D.H. Cumming & J. Phillipson 1976. Biomass and production of large herbivores in relation to rainfall and primary productivity. *Oecologia* 22, 341–354.

Corfield, T.F. 1973. Elephant mortality in Tsavo National Park, Kenya. *African Journal of Ecology* 11, 339–368.

Coulson, T., F. Guinness, J. Pemberton & T. Clutton-Brock 2004. The demographic consequences of releasing a population of red deer. *Ecology* 85, 411–422.

Couzin, I.D. 2006. Behavioral ecology: Social organization in fission-fusion societies. *Current Biology* 16, 169–171.

Cozzi, B., S. Spagnoli & L. Bruno 2001. An overview of the central nervous system of the elephant through a critical appraisal of the literature published in the XIX and XX centuries. *Brain Research Bulletin* 54, 219–227.

Croze, H. 1972. A modified photogrammetric technique for assessing age structure of elephant populations and its use in Kidepo National Park. *East African Wildlife Journal* 10, 91–115.

Cumming, D.H.M., M.B. Fenton, I.L. Rautenbach, R.D. Taylor, G.S. Cumming, M.S. Cumming, J.M. Dunlop, A.G. Ford, M.D. Hovorka, D.S. Johnston, M. Kalcounis, Z. Mahlangu & C.V.R. Portfors 1997. Elephants, woodlands and biodiversity in southern Africa. *South African Journal of Science* 93, 231–236.

Cumming, D.H.M. & B. Jones 2005. *Elephants in southern Africa: Management issues and options.* WWF SARPO, Harare.

Dai, X., G. Shannon, R. Slotow, B. Page & J. Duffy 2007. Short-duration daytime movements of a cow herd of African elephants. *Journal of Mammalogy* 88, 151–157.

De Beer, Y. 2007. Determinants and consequences of elephant spatial use in Southern Africa's arid savannas. MSc thesis, University of Pretoria, Pretoria.

De Beer, Y., W. Kilian, W. Versveld & R.J. van Aarde 2006. Elephants and low rainfall alter woody vegetation in Etosha National Park, Namibia. *Journal of Arid Environments* 64, 412–421.

Delsink, A.K., J.J. van Altena, D. Grobler, H. Bertschinger, J. Kirkpatrick & R. Slotow 2006. Regulation of a small, discrete African elephant population through immunocontraception in the Makalali Conservancy, Limpopo, South Africa. *South African Journal of Science* 102, 403–405.

De Villiers, P.A. & O.B. Kok 1988. Eto-ekologiese aspekte van olifante in die Nasionale Etoshawildtuin. *Madoqua* 15, 319–338.

De Villiers, P.A. & O.B. Kok 1997. Home range, association and related aspects of elephants in the eastern Transvaal Lowveld. *African Journal of Ecology* 35, 224–236.

Dobson, A.P. 1993. Effect of fertility control on elephant population dynamics. *Journal of Reproduction and Fertility Supplement* 90, 293–298.

Douglas-Hamilton, I. 1972. On the ecology of the African elephant. Ph.D. thesis, Oxford University, Oxford.

Douglas-Hamilton, I., T. Krink & F. Vollrath 2005. Movements and corridors of African elephants in relation to protected areas. *Naturwissenschaften* 92, 158–163.

Dudley, J.P., G.C. Craig, D.S. Gibson, G. Haynes & J. Klimowicz 2001. Drought mortality of bush elephants in Hwange National Park, Zimbabwe. *African Journal of Ecology* 39, 187–194.

Dunham, K.M. 1988. Demographic changes in the Zambezi Valley elephants (*Loxodonta africana*). *Journal of Zoology* 56, 382–388.

Fatti, L.P., G.L. Smuts, A.M. Starfield & A.A. Spurdle 1980. Age determination in African elephants. *Journal of Mammalogy* 61, 547–551.

Fayrer-Hosken, R.A., D. Grobler, J.J. van Altena, H.J. Bertschinger & J.F. Kirkpatrick 2000. Immunocontraception of African elephants. *Nature* 407, 149.

Ferreira, S.M. & R.J. van Aarde 2008. A rapid method to estimate some of the population variables for African elephants. *Journal of Wildlife Management* 72, 822–899.

Ferreira, S.M., R.J. van Aarde & J. Junker 2008. Ivory poaching disrupts Zambian savanna elephant population structures (in review).

Foggin, C. 2003. The elephant problem in Zimbabwe: can there be an alternative to culling? In: B. Colenbrander, J. de Gooijer, R. Paling, S.S. Stout; T. Stout & W.R. Allen (eds) *Managing African elephant populations: act or let die?* Proceedings of an expert consultation on the Control of Wild Elephant Populations, Faculty of Veterinary Medicine, Utrecht University, Netherlands, 17–21, http://elephantpopulationcontrol.library.uu.nl/pagi nas/frames.html (accessed 5 April 2006).

Foley, C.A.H. 2002. *The effects of poaching on elephant social systems.* Princeton University, Princeton.

Forman, R.T.T. & M. Godron. 1986. *Landscape Ecology.* John Wiley & Sons, New York.

Gaillard, J.-M., M. Festa-Bianchet & N.G. Yoccoz 1998. Population dynamics of large herbivores: Variable recruitment with constant adult survival. *Trends in Ecology and Evolution* 13, 58–63.

Garaï, M.E. & R.D. Carr 2001. Unsuccessful introductions of adult elephant bulls to confined areas in South Africa. *Pachyderm* 31, 52–57.

Garaï, M.E., R. Slotow, R.D. Carr & B. Reilly 2004. Elephant reintroductions to small fenced reserves in South Africa. *Pachyderm* 37, 28–36.

Gaston, K.J., T.M. Blackburn & R.D. Gregory 1999. Does variation in census area confound density comparisons? *Journal of Applied Ecology* 36, 191–204.

Gaylard, A., N. Owen-Smith & J.V. Redfearn. 2003. Surface water availability: Implications for heterogeneity and ecosystem processes. In: J. du Toit, K.H. Rogers & H.C. Biggs (eds) *The Kruger experience: Ecology and management of savanna heterogeneity.* Island Press, Washington, 171–188.

Genis, H., R. Slotow & K. Pretorius 2004. The effect of mature elephant bull introduction on resident bull population ranging patterns and musth periods: Phinda Private Game Reserve. *Ecological Journal* 6, 14–19.

Gibbs, J.P. 2000. Monitoring populations. In: L. Boitani & T.K. Fuller (eds) *Research techniques in animal ecology – controversies and consequences.* Columbia University Press, New York.

Gillson, L. & K. Lindsay 2003. Ivory and ecology – changing perspectives on elephant management and the international trade in ivory. *Environmental Science & Policy* 6, 411–419.

Gordon, I.J., A.J. Hester & M. Festa-Bianchet 2004. The management of wild large herbivores to meet economic, conservation and environmental objectives. *Journal of Applied Ecology* 41, 1021–1031.

Gough, K. & G.I.H. Kerley 2006. Demography and population dynamics in the elephants *Loxodonta africana* of Addo Elephant National Park, South Africa: is there evidence of density dependent regulation? *Oryx* 40, 434–441.

Grainger, M., R. van Aarde & I. Whyte 2005. Landscape heterogeneity and the use of space by elephants in Kruger National Park, South Africa. *African Journal of Ecology* 43, 369–375.

Guldemond, R.A.R. 2006. The influence of savannah elephants on vegetation: A case study in the Tembe Elephant Park, South Africa. Ph.D. thesis, University of Pretoria, Pretoria.

Guldemond, R. & R. van Aarde 2008. A meta-analysis of the impact of African elephants on savanna vegetation. *Journal of Wildlife Management* 72 (4), 892–899.

Guldemond, R., E. Lehman, S. Ferreira & R. van Aarde 2005. Elephant numbers in Kafue National Park, Zambia. *Pachyderm* 39, 50–56.

Hakeem, A.Y., P.R. Hof, C.C. Sherwood, R.C. Switzer, L.E.L. Rasmussen & J.N. Ellman 2005. Brain of the African elephant (*Loxodonta africana*):

Neuroanatomy from magnetic resonance images. *Anatomical Record Part A* 287A, 1117–1127.

Hall-Martin, A.J. 1992. Distribution and status of the African elephant *Loxodonta africana* in South Africa, 1652-1992. *Koedoe* 35, 65–80.

Hanks, J. 1971. Reproduction of the elephant (*Loxodonta africana*) in the Luangwa Valley, Zambia. *Journal of Reproduction and Fertility* 30, 13–26.

Hanks, J. 1972. Growth of the African elephant (*Loxodonta africana*). *East African Wildlife Journal* 10, 251–272.

Hanks, J. 1979. *A Struggle for Survival: The Elephant Problem*. C. Struik, Cape Town.

Hanks, J. & J.E.A. McIntosh 1973. Population dynamics of the African elephant (*Loxodonta africana*). *Journal of Zoology* 169, 29–38.

Harestad, A.S. & F.L. Bunnel 1979. Home range and body weight: a re-evaluation. *Ecology* 60, 389–402.

Harris, G.M., G.J. Russel, R.J. van Aarde & S.L. Pimm 2008. Habitat use of savanna elephants in southern Africa. *Oryx* 42, 66–75.

Hart, B.L., L.A. Hart & N. Pinter-Wollman 2007. Large brains and cognition: where do elephants fit in? *Neuroscience and Biobehavioural Reviews* doi:10.1016/j.neubiorev.2007.05.012.

Hoare, R.E. 1999. Determinants of human-elephant conflict in a land-use mosaic. *Journal of Applied Ecology* 36, 689–700.

Hoare, R.E. & J. du Toit 1999. Coexistence between people and elephants in African savannas. *Conservation Biology* 13, 633-639.

Hodges, J.K., R.J. van Aarde, M. Heistermann and H.O. Hoppen 1994. Progestin content and biosynthetic potential of the corpus luteum of the African elephant (*Loxodonta africana*). *Journal of Reproduction and Fertility* 102, 163–168.

Hunter, N. & T. Milliken 2004. Clarifying MIKE and ETIS. *Pachyderm* 36, 129–132.

Illius, A.W. 2006. Foraging and population dynamics. In: K. Danell, R. Bergström, P. Duncan & J. Pastor (eds) *Large herbivore ecology, ecosystem dynamics and conservation*. Cambridge University Press, Cambridge.

IPCC 2007. *Climate change 2007*. The IPCC Fourth Assessment Report AR4. Available at http://www.ipcc.ch/

Jachmann, H. 1980. Population dynamics of the elephants in the Kasungu National Park, Malawi. *Netherlands Journal of Zoology* 30, 622–634.

Jachmann, H. 1984. The ecology of the elephants in Kasungu National Park, Malawi, with specific reference to management of elephant populations in the Brachystegia biome of South Central Africa. Ph.D. thesis, University of Groningen, Groningen.

Jachmann, H. 1986. Notes on the population dynamics of the Kasungu elephants. *African Journal of Ecology* 24, 215–226.

Jachmann, H. 1988. Estimating age in African elephants: A revision of Laws' molar evaluation technique. *African Journal of Ecology* 26, 51–56.

Jackson, T.P., S. Mosojane, S. Ferreira & R.J. van Aarde 2008. Solutions for elephant crop raiding in northern Botswana: moving away from symptomatic approaches. *Oryx* 42, 83–91.

Joubert, D. 2006. Hunting behaviour of lions (*Panthera leo*) on elephants (*Loxodonta africana*) in the Chobe National Park, Botswana. *African Journal of Ecology* 44, 279–281.

Junker, J., R.J. van Aarde & S.M. Ferreira 2008. Temporal trends in elephant *Loxodonta africana* numbers and densities in northern Botswana: is the population really increasing? *Oryx* 42, 58–65.

Kinahan, A.A., S.L. Pimm & R.J. van Aarde 2007. Ambient temperature and landscape use in the savanna elephant (*Loxodonta africana*). *Journal of Thermal Biology* 32, 47–58.

Kirkpatrick, J.F. 2007. Measuring the effects of wildlife contraception: The argument for comparing apples with oranges. *Reproduction, Fertility and Development* 19, 548–552.

Lamprey, H.F., P.E. Glover, H.I.M. Turner & R.H.V. Bell 1967. Invasion of the Serengeti National Park by elephants. *African Journal of Ecology* 5, 151–166.

Laws, R.M. 1966. Age criteria for the African elephant *Loxodonta a. africana*. *East African Wildlife Journal* 4, 1–37.

Laws, R.M. 1969. The Tsavo research project. *Journal of Reproduction and Fertility Supplements* 6, 495–531.

Laws, R.M., I.S.C. Parker & R.C.B. Johnstone 1975. *Elephants and their habitats: The ecology of elephants in North Bunyoro, Uganda*. Clarendon Press, Oxford.

Lee, P.C. 1987. Allomothering among African elephants. *Animal Behaviour* 35, 278–291.

Lee, P.C. & C.J. Moss. 1995. Statural growth in the African elephant (*Loxodonta africana*). *Journal of Zoology London* 236, 29–41.

Leggett, K. 2006a. Home range and seasonal movement of elephants in the Kunene Region, northwestern Namibia. *African Zoology* 41, 17–36.

Leggett, K. 2006b. Effect of artificial water points on the movement and behaviour of desert-dwelling elephants of north-western Namibia. *Pachyderm* 40, 40–51.

Leuthold, W. 1977. Spatial organisation and strategy of habitat utilization of elephants in Tsavo National Park, Kenya. *Zeitschrift fur Saugetierkunde* 42, 358–379.

Lewis, D.M. 1984. Demographic changes in the Luangwa Valley elephants. *Biological Conservation* 29, 7–14.

Li, H. & J.F. Reynolds 1994. A simulation experiment to quantify spatial heterogeneity in categorical maps. *Ecology* 75, 2446–2455.

Lindeque, M. 1988. Population dynamics of elephants in Etosha National Park, S.W.A./Namibia. Ph.D. thesis, University of Stellenbosch, Stellenbosch.

Lindeque, M. 1991. Age structure of the elephant population in the Etosha National Park, Namibia. *Madoqua* 18, 27–32.

Lindeque, M. & P.M. Lindeque 1991. Satellite tracking of elephants in northern Namibia. *African Journal of Ecology* 29, 196–206.

Loveridge, A.J., J.E. Hunt, F. Murindagomo & D.W. Macdonald 2006. Influence of drought on predation of elephant (*Loxodonta africana*) calves by lions (*Panthera leo*) in an African wooded savannah. *Journal of Zoology* 270, 523–530.

Mabunda, D. 2005. *Report to the Minister: Environmental Affairs and Tourism on developing elephant management plans for national parks with recommendations on the process to be followed.* SANParks, Pretoria. Available online at http://www.sanparks.org/events/elephants/.

Mackey, R.L., B.R. Page, D. Duffy & R. Slotow 2006. Modelling elephant population growth in small, fenced, South African reserves. *South African Journal of Wildlife Research* 36, 33–43.

Mbaiwa, J.E. & O.I. Mbaiwa 2006. The effects of veterinary fences on wildlife populations in Okavango Delta, Botswana. *International Journal of Wilderness* 13, 17–41.

McComb, K., C. Moss, S. Sayailel & L. Baker 2000. Unusually extensive networks of vocal recognition in African elephants. *Animal Behaviour* 59, 1103–1109.

McComb, K., C.J. Moss, S.M. Durant, L. Baker & S. Sayialel 2001. Matriarchs as repositories of social knowledge in African elephants. *Science* 292, 491–494.

McComb, K., D. Reby, L. Baker, C. Moss & S. Sayailel 2003. Long distance communication of acoustic cues to social identity in African elephants. *Animal Behaviour* 65, 317–329.

McComb, K., L. Baker & C. Moss 2006. African elephants show high level of interest in the skulls and ivory of their own species. *Biology Letters* 2, 26–28.

McKnight, B. 2000. Changes in elephant demography, reproduction and group structure in Tsavo East National Park (1966-1994). *Pachyderm* 29, 15–24.

Mills, M.G.L., H.C. Biggs & I.J. Whyte 1995. The relation between rainfall, lion predation and population trends in African herbivores. *Wildlife Research* 22, 75–88.

Milner-Gulland, E.J. & R. Mace 1991. The impact of the ivory trade on the African elephant *Loxodonta africana* population as assessed by data from the trade. Biological Conservation 55, 215–229.

Morley, R.C. 2005. The demography of a fragmented population of the savanna elephant (*Loxodonta africana* Blumenbach) in Maputaland. Ph.D. thesis, University of Pretoria, Pretoria.

Morley, R.C. & R.J. van Aarde 2006. Estimating abundance for a savanna elephant population using mark-resight methods: A case study for the Tembe Elephant Park, South Africa. *Journal of Zoology, London* 271, 418–427.

Morrison, T.A., P.I. Chiyo, C.J. Moss & S.C. Alberts 2005. Measures of dung bolus size for known-age African elephants (*Loxodonta africana*): Implications for age estimation. *Journal of Zoology, London* 266, 89–94.

Moss, C.J. 1988. *Elephant memories: Thirteen years in the life of an elephant family*. William Morrow, New York.

Moss, C.J. 2001. The demography of an African elephant (*Loxodonta africana*) population in Amboseli, Kenya. *Journal of Zoology, London* 255, 145–156.

Moss, C.J. & J.H. Poole. 1983. Relationships and social structure of African elephants. In: R.A. Hinde (ed) *Primate Social Relations: An integrated approach*. Sinauer, New York, 315–325.

Murwira, A. & A.K. Skidmore 2005. The response of elephants to the spatial heterogeneity of vegetation in a Southern African agricultural landscape. *Landscape Ecology* 20, 217–234.

Nellemann, C., S.R. Moe & L.P. Rutina 2002. Links between terrain characteristics and forage patterns of elephants (*Loxodonta africana*) in northern Botswana. *Journal of Tropical Ecology* 18, 835–844.

Njumbi, S., J. Waithaka, S. Gachago, J. Sakwa, K. Mwathe, P. Mungai, M. Mulama, H. Mutinda, P. Omondi & M. Litoroh 1996. Translocation of elephants: The Kenyan experience. *Pachyderm* 22, 61–65.

Norton-Griffiths, M. 1978. *Counting animals*. AWLF, London.

Ntumi, C.P., R.J. van Aarde, N. Fairall & W.F. de Boer 2005. Use of space and habitat use by elephants (*Loxodonta africana*) in the Maputo Elephant

Reserve, Mozambique. *South African Journal of Wildlife Research* 35, 139–146.

O'Connor, T.G., P.S. Goodman & B. Clegg 2007. A functional hypothesis of the threat of local extirpation of woody plant species by elephant in Africa. *Biological Conservation* 136, 329–345.

Ogutu, J.O. & N. Owen-Smith 2003. ENSO, rainfall and temperature influences on extreme population declines among African savanna ungulates. *Ecology Letters* 6, 412–419.

Osborn, F.V. 2004. The concept of home range in relation to elephants in Africa. *Pachyderm* 37, 37–44.

Osborn, F.V. & G.E. Parker 2003. Towards an integrated approach for reducing the conflict between elephants and people: A review of current research. *Oryx* 37, 80–84.

Ott, T. 2007. Landscape heterogeneity is a determinant of range utilization by African elephant (*Loxodonta africana*) in mesic savannas. MSc thesis, University of Pretoria, Pretoria.

Ottichilo, W.K. 1987. The causes of the recent heavy elephant mortality in the Tsavo ecosystem, Kenya, 1975-80. *Biological Conservation* 41, 279–289.

Owen, C-S. 2005. Is the supply of trophy elephants to the Botswana hunting market sustainable? MSc thesis, University of Cape Town, Cape Town.

Owen-Smith, N. 1983. *Management of large mammals in African conservation areas*. Proceedings of a symposium held in Pretoria, South Africa, April 1982. Haum, Pretoria.

Owen-Smith, R.N. 1988. *Megaherbivores: the influence of very large body size on ecology*. Cambridge University Press, Cambridge.

Owen-Smith, N. 1990. Demography of a large herbivore, the Greater Kudu *Tragelaphus strepsiceros*, in relation to rainfall. *Journal of Animal Ecology* 59, 893–913.

Owen-Smith, N. 1996. Ecological guidelines for waterpoints in extensive protected areas. *South African Journal of Wildlife Research* 26, 107–112.

Owen-Smith, N., G.I.H. Kerley, B. Page, R. Slotow & R.J. van Aarde 2006. A scientific perspective on the management of elephants in the Kruger National Park and elsewhere. *South African Journal of Science* 102, 389–394.

Parker, G.E. & F.V. Osborne 2001. Dual-season crop damage by elephants in eastern Zambezi Valley, Zimbabwe. *Pachyderm* 30, 49–56.

Payne, K. 2003. Sources of social complexity in the three elephant species. In: F.B.M. de Waal & P.L. Tyack (eds) *Animal social complexity*. Harvard University Press, Cambridge, 57–85.

Perdok, A.A., W.F. de Boer & T.A.E. Stout 2007. Prospects for managing African elephant population growth by immuno-contraception: a review. *Pachyderm* 42, 1–11.

Pettorelli, N., J.O. Vik, A. Mysterud, J. Gaillard, C.J. Tucker & N.C. Stenseth 2005. Using the satellite-derived NDVI to assess ecological responses to environmental change. *Trends in Ecology and Evolution* 20, 503–510.

Pimm, S.L. & R.J. van Aarde. 2001. African elephants and contraception. *Nature* 411, 766.

Pienaar, D. 2005. Water provision in the Kruger National Park. In: K.H. Rogers (ed.) *Elephants and biodiversity – a synthesis of current understanding of the role and management of elephants in savanna ecosystems.* SANParks, Skukuza, 227–234.

Pienaar, U. de V., P. van Wyk & N. Fairall 1966. An aerial census of elephant and buffalo in Kruger National Park, and the implications thereof on intended management schemes. *Koedoe* 9, 40–107.

Poole, J.H. & C.J. Moss 1989. Elephant mate searching: group dynamics and vocal and olfactory communication. In: P. Jewell & G. Maloiy (eds) *The biology of large African mammals in their environment: Symposia of the Zoological Society of London* Vol. 61, Clarendon, 111–125.

Proaktor, G., T. Coulson & E.J. Milner-Gulland 2007. Evolutionary responses to harvesting in ungulates. *Journal of Animal Ecology* 76, 669–678.

Pullin, A.S. & T.M. Knight 2005. Assessing conservation management evidence base: A survey of management-plan compilers in the United Kingdom and Australia. *Conservation Biology* 19, 1989–1996.

Redfern, J.V., R. Grant, H. Biggs & W.M. Getz 2003. Surface-water constraints on herbivore foraging in the Kruger National Park, South Africa. *Ecology* 84, 2092–2107.

Reeve, R. 2006. Wildlife trade, sanctions and compliance: Lessons from the CITES regime. *International Affairs* 82, 881–897.

Reilly, J. 2002. Growth in the Sumatran elephant (*Elephas maximus sumatranus*) and age estimation based on dung diameter. *Journal of Zoology, London* 258, 205–213.

Saïd, S. & S. Servanty 2005. The influence of landscape structure on female roe deer home-range size. *Landscape Ecology* 20, 1003–1012.

SANParks. 2005. *The great elephant indaba: finding an African solution to an African problem.* South African National Parks, Pretoria.

Sankaran, M., N.P. Hanan, R.J. Scholes, J. Ratnam, D.J. Augustine, B.S. Cade, J. Gignoux, S.I. Higgins, X. le Roux, F. Ludwig, J. Ardo, F. Banyikwa, A. Bronn, G. Bucini, K.K. Caylor, M.B. Coughenour, A. Diouf, W. Ekaya,

C.J. Feral, E.C. February, P.G.H. Frost, P. Hiernaux, H. Hrabar, K.L. Metzger, H.H.T. Prins, S. Ringrose, W. Seas, J. Tews, J. Worden & N. Zambatis 2005. Determinants of woody cover in African savannas. *Nature* 438, 846–849.

Scoone, I. 1995. Exploiting heterogeneity: Habitat use by cattle in dryland Zimbabwe. *Journal of Arid Environments* 29, 221–237.

Shrader, A.M., S.M. Ferreira, M.E. McElveen, P.C. Lee, C.J. Moss & R.J. van Aarde 2006a. Growth and age determination of African savanna elephants. *Journal of Zoology, London* 270, 40–48.

Shrader, A.M. S.M. Ferreira & R.J. van Aarde 2006b. Digital photogrammetry and laser rangefinder techniques to measure African elephants. *South African Journal of Wildlife Research* 36, 1–7.

Sikes, S.K. 1966. The African elephant, *Loxodonta africana*: A field method for the estimation of age. *Journal of Zoology, London* 150, 279–295.

Sinclair, A.R.E. 2003. Mammal population regulation, keystone processes and ecosystem dynamics. *Philosophical Transactions of the Royal Society London* 358, 1729–1740.

Sitati, N.W., M.J. Walpole, R.J. Smith & N. Leader-Williams 2003. Predicting spatial aspects of human-elephant conflict. *Journal of Applied Ecology* 40, 667–677.

Slotow, R. & G. van Dyk 2001. Role of delinquent young 'orphan' male elephants in high mortality of white rhinos in Pilanesberg National Park, South Africa. *Koedoe* 44, 85–94.

Slotow R. & G. van Dyk 2004. Ranging of older male elephants introduced to an existing small population without older males: Pilanesberg National Park. *Koedoe* 47, 91–104.

Slotow, R., G. van Dyk, J. Poole, B. Page & A. Klocke 2000. Older bull elephants control young males. *Nature* 408, 425–426.

Slotow, R., M.E. Garaï, B. Reilly, B. Page & R.D. Carr 2005. Population dynamics of elephants re-introduced to small fenced reserves in South Africa. *South African Journal of Wildlife Research* 35, 23–32.

Smit, I.P.J., C.C. Grant & B.J. Devereux 2007a. Do artificial waterholes influence the way herbivores use the landscape? Herbivore distribution patterns around rivers and artificial surface water sources in a large African savanna park. *Biological Conservation* 136, 85–99.

Smit, I.P.J., C.C. Grant & I.J. Whyte 2007b. Elephants and water provision: What are the management links? *Diversity and Distributions* 13, 666–669.

Smuts, G.L. 1975. Reproduction and population characteristics of elephants in the Kruger National Park South Africa. *Journal of the Southern African Wildlife Management Association* 5, 1–10.

Spong, G. & S. Creel. 2001. Deriving dispersal distances from genetic data. *Proceedings of the Royal Society London B* 268, 2571–2574.

Steenkamp, G., S.M. Ferreira & M.N. Bester 2007. Tusklessness and tusk fractures in free-ranging African savanna elephants (*Loxodonta africana*). *Journal of the South African Veterinary Association* 78, 75–80.

Stiles, D. 2004. The ivory trade and elephant conservation. *Environmental Conservation* 31, 309–321.

Stokke, S. & J.T. du Toit 2002. Sexual segregation in habitat use by elephants in Chobe National Park, Botswana. *African Journal of Ecology* 40, 360–371.

Sukumar, R., N.V. Joshi & V. Krisnamurthy 1988. Growth in the Asian elephant. *Proceedings of the Indian Academy of Sciences* 97, 561–571.

Thouless, C.R. 1995. Long distance movements of elephants in northern Kenya. *African Journal of Ecology* 33, 321–334.

Tufto, J., R. Anderson & J. Linnell 1996. Habitat use and ecological correlates of home range size in a small cervid: the roe deer. *Journal of Animal Ecology* 65, 715–724.

Turnbull, P.C., R.H. Bell, K. Saigawa, F.E. Munyenyembe, C.K. Mulenga & L.H. Makala 1991. Anthrax in wildlife in the Luangwa Valley, Zambia. *The Veterinary Record* 128, 399–403.

Van Aarde, R., I. Whyte & S. Pimm 1999. Culling and dynamics of the Kruger National Park African elephant population. *Animal Conservation* 2, 287–294.

Van Aarde, R.J., T. Jackson & D.G. Erasmus 2005. Assessment of seasonal home-range use by elephants across southern Africa's seven elephant clusters. Unpublished Report. Conservation Ecology Research Unit, Peace Parks Foundation, University of Pretoria. Available on http://www.up.ac.za/academic/zoology/ceru/Home.htm.

Van Aarde, R.J., T.P. Jackson & S.M. Ferreira 2006. Conservation science and elephant management in southern Africa. *South African Journal of Science* 102, 385–388.

Van Aarde, R.J. & T. Jackson 2007. Megaparks for metapopulations: Addressing the causes of locally high elephant numbers in South Africa. *Biological Conservation* 134, 289–297.

Van Jaarsveld, A.S., A.O. Nicholls & M.H. Knight 1999. Modelling and assessment of South African elephant *Loxodonta africana* population persistence. *Environmental Modelling and Assessment* 4, 155–163.

Van Wyk, P. & N. Fairall 1969. The influence of the African elephant on the vegetation of the Kruger National Park. *Koedoe* 12, 57–89.

Verlinden, A. & I.K.N. Gavor 1998. Satellite tracking of elephants in northern Botswana. *African Journal of Ecology* 36, 105–116.

Viljoen, 1988. The ecology of the desert-dwelling elephants *Loxodonta africana* (Blumenbach, 1797) of Western Damaraland and Kaokoland. Ph.D. thesis (Wildlife Management), University of Pretoria, Pretoria.

Walker, B.H., R.H. Emslie, R.N. Owen-Smith & R.J. Scholes 1987. To cull or not cull: Lessons from a southern African drought. *Journal of Applied Ecology* 24, 381–401.

Wall, J., I. Douglas-Hamilton, & F. Vollrath 2006. Elephants avoid costly mountaineering. *Current Biology* 16, 527–529.

Wasser, S.K.C., R. Mailand, B. Booth, E. Mutayoba, B. Kisamo, B. Clark & M. Stephens 2007. Using DNA to track the origin of the largest ivory seizure since the 1989 trade ban. *Proceedings of the National Academy of Science, U.S.A.* 104, 4228–4233.

Western, D. 1975. Water availability and its influence on the structure and dynamics of a large mammal community. *East African Wildlife Journal* 13, 265–286.

Western, D., C. Moss & N. Georgiadis 1983. Age estimation and population age structure of elephants from footprint dimensions. *Journal of Wildlife Management* 47, 1192–1197.

Western, D. & W.K. Lindsay 1984. Seasonal herd dynamics of a savanna elephant population. *African Journal of Ecology* 22, 229–244.

Whitehouse, A.M. & A.J. Hall-Martin 2000. Elephants in Addo Elephant National Park, South Africa: Reconstruction of the population's history. *Oryx* 34, 46–55.

Whyte, I.J. 2001. Conservation management of the Kruger National Park elephant population. Ph.D. thesis, University of Pretoria, Pretoria.

Whyte, I.J. 2004. Ecological basis of the new elephant management policy for Kruger National Park and expected outcomes. *Pachyderm* 36, 99–108.

Whyte, I.J. & D. Grobler 1997. *The current status of contraception research in the Kruger National Park.* Scientific report 13/97. National Parks Board, Skukuza.

Whyte, I. J., R.J. van Aarde, R.J. & S.L. Pimm 1998. Managing the elephants of Kruger National Park. *Animal Conservation* 1, 77–83.

Whyte, I.J., R.J. van Aarde & S.L. Pimm 2003. Kruger elephant population: its size and consequences for ecosystem heterogeneity. In: J.T. du Toit, K.H. Rogers & H.C. Biggs (eds) *The Kruger experience: Ecology and management of savanna heterogeneity.* Island Press, Washington, 332–348.

Wittemyer, G., I. Douglas-Hamilton & W.M. Getz 2005a. The socioecology of elephants: Analysis of the processes creating multitiered social structures. *Animal Behaviour* 69, 1357–1371.

Wittemyer, G., D. Daballen, H. Rasmussen, O. Kahindi & I. Douglas-Hamilton 2005b. Demographic status of elephants in the Samburu and Buffalo Springs National Reserves, Kenya. *African Journal of Ecology* 43, 1365–2028.

Wittemyer, G. & W.M. Getz 2007. Hierarchical dominance structure and social organization in African elephants, *Loxodonta africana. Animal Behaviour* 73, 671–681.

Wittemyer, G., W.M. Getz, F. Vollrath & I. Douglas-Hamilton 2007a. Social dominance, seasonal movements, and spatial segregation in African elephants: A contribution to conservation behavior. *Behavioral Ecology and Sociobiology* 61, 1919–1931.

Wittemyer, G., A. Ganswindt & K. Hodges 2007b. The impact of ecological variability on the reproductive endocrinology of wild female African elephants. *Hormones and Behaviour* 51, 346–354.

Wittemyer, G., H.B. Rasmussen & I. Douglas-Hamilton 2007c. Breeding phenology in relation to NDVI variability in free-ranging African elephant. *Ecography* 30, 42–50.

Woodd, A.M. 1999. A demographic model to predict future growth of the Addo elephant population. *Koedoe* 42, 97–100.

Young, E. 1970. Water as faktor in die ekologie van wild in die Nasionale Krugerwildtuin. Ph.D. thesis, University of Pretoria, Pretoria.

Young, K.D., S.M. Ferreira & R.J. van Aarde 2008. The influence of increasing population size and vegetation productivity on elephant distribution in the Kruger National Park. *Austral Ecology* (accepted).

EFFECTS OF ELEPHANTS ON ECOSYSTEMS AND BIODIVERSITY

Lead author: Graham IH Kerley
Authors: Marietjie Landman, Laurence Kruger, and Norman Owen-Smith
Contributing authors: Dave Balfour, Willem F de Boer, Angela Gaylard,
Keith Lindsay, and Rob Slotow

> On the following morning we were up before the sun, and, travelling in a
> northerly direction, soon became aware that we were in a district frequented
> by elephants, for wherever we looked, trees were broken down, large branches
> snapped off, and bark and leaves strewn about in all directions, whilst the
> impress of their huge feet was to be seen in every piece of sandy ground.
> *F C Selous (1881, 39), north of Gweru, Zimbabwe, in 1872*

INTRODUCTION

THE ISSUE of the effects of elephants within ecosystems has emerged
strongly since the formulation of the concept of the 'elephant problem and
the concerns that elephants may irrevocably alter the remaining areas which
are available to them' (Caughley, 1976a). Two perspectives need to be kept
in mind when these concerns are raised. Firstly, the order of Proboscideans
(including the modern elephants) evolved in Africa as part of a unique group
of mammals, the Afrotheria (Robinson & Seiffert, 2003), with their roots going
back 80 million years. Proboscideans of various forms subsequently colonised
all continents except for Australia and Antarctica; mammoths in the family
Elephantidae remained abundant and widespread through most of Europe and
North America until as recently as 12 000–16 000 years ago (Sukumar, 2003).
The modern African elephant emerged about 3 million years ago. Hence, its
relationships with other animal and plant species have been an integral part of
the co-evolutionary history of the ecosystems and biodiversity of Africa.

Herbivores, through their consumption of plant tissues, affect the relative
growth, survival and reproductive output of these plants, with consequences
for vegetation structure, community composition and ecosystem processes
(Huntly, 1991). Even relatively small herbivores can have profound effects in
shaping ecosystem structure, particularly when they occur at high densities.

For example, Côté *et al.* (2004), writing about the increase in deer abundance, had the following to say:

> They affect the growth and survival of many herb, shrub and tree species, modifying patterns of relative abundance and vegetation dynamics. Cascading effects on other species extend to insects, birds, and other mammals. Sustained over-browsing reduces plant cover and diversity, alters nutrient and carbon cycling, and redirects succession ... simplified alternative states appear to be stable and difficult to reverse.

Similarly, smaller herbivores with specific manners of feeding can alter ecosystems, although their abundance and overall use of resources are not great. Feeding by porcupines *Hystrix africaeaustralis* on the bark of red syringas *Burkea africana* exposes the xylem to fire, with consequent increases in tree mortality (Yeaton, 1988; De Villiers & Van Aarde, 1994). Granivory and seedling predation by rodents alters many plant communities (Brown & Heske, 1990).

Nevertheless, the feeding and breakage impacts of elephants on plants are greater in magnitude and scale than those of smaller herbivores, particularly through affecting the structural components of the vegetation like canopy trees (Owen-Smith, 1988). From this perspective elephants have been termed 'megaherbivores', along with other species exceeding 1 000 kg in adult body mass with similarly great impacts on ecosystems, including rhinos and hippos (Owen-Smith, 1988). Herbivore species within this size range were a general feature of ecosystems worldwide until modern humans spread their predatory and land-transforming influences worldwide between 50 000 and 12 000 years ago. It has been surmised that the elimination of these megaherbivores through human hunting contributed to the demise of many other large mammal species, and consequent reduction in species diversity outside of Africa and tropical Asia, as a result of the habitat changes that occurred (Owen-Smith, 1987, 1989). This emphasises that the effects of elephants on biodiversity can be positive as well as negative. However, the biodiversity consequences need to be judged not only at the species level, but also in terms of changes in habitat composition and functional processes (Noss, 1990). This diversity is furthermore expressed across a range of organisational levels from genes to landscapes.

Formerly, ecosystem dynamics were viewed largely from a 'balance of nature' perspective, with changes being regarded as threatening the maintenance of the species richness within these systems. Hence, human interventions were largely directed at counteracting or suppressing changes, aimed at maintaining an 'ideal' state generally defined by some historical perspective, e.g. what was described in writing by early European colonists. The modern perspective views disturbance in various forms as being integral to the generation and maintenance of biodiversity, expressed through hierarchical patch dynamics and consequent spatial heterogeneity within landscapes (Pickett & White, 1985). Hence, in this chapter we are concerned with the changes brought about through the presence of elephants on the species composition, vegetation structure and functioning of the ecosystems of which they are a component. These changes are judged within the context of the overriding context of biodiversity conservation, which is a primary aim set by humans for much of the land within which these elephants reside.

We need to distinguish further an 'elephant' effect from an 'elephant density' effect (Cowling & Kerley, 2002). The former reflects the ability of elephants to influence biodiversity, by virtue of the special characteristics of elephants, while the latter reflects the consequences that depend on the abundance of elephants within the area of concern. Bearing in mind the considerations outlined above, this chapter addresses the following specific questions.

- How are elephants special in the nature of their feeding, and hence, the damage to plants they cause, by virtue of features such as body size, the trunk and tusks?
- How are the impacts of elephants on individual plants translated into changes in vegetation composition and structure?
- How do these changes in vegetation and hence, habitat features for other animal species, affect the coexistence of these species?
- How do the presence and activities of elephants influence nutrient cycling, the effects of fire and the productive potential of the ecosystems they inhabit?
- What are the cascading or knock-on effects of elephants on the components of biodiversity?

In addition, we attempt to identify what we still need to find out in order to better understand the impacts of elephants and the implications for management of these impacts. The approach is to use these questions as a framework to guide the contents of this chapter.

Across Africa, elephants occupy a broad range of terrestrial ecosystems, penetrating deserts such as the Namib along seasonal rivers, as well as being found within the tropical rain forests of the Congo basin (Laws, 1970; Boshoff *et al.*, 2002). However, within South Africa, concern is focused on their effects on savanna and subtropical thicket ecosystems, reflecting current elephant distribution.

SPECIAL FEATURES OF ELEPHANTS

The African elephant is the largest herbivore alive today, with females attaining a maximum body mass of over three tons and males over six tons. Coupled with this large size (and hence megaherbivore status) is a fairly simple digestive system with most digestion taking place in the capacious hindgut, comprising the small intestine and colon. Throughput is relatively rapid, with mean retention time of around 24 hours, independent of the daily food intake (Clauss *et al.*, 2007; Davis, 2007). This fast passage (compared with other large herbivores) means that digestive efficiency is quite low, with less than half of the ingested food being assimilated and the remainder passed out as faeces. On the other hand, large amounts of fibre can be ingested without slowing throughput, in contrast to the situation for ruminants (Janis, 1976). Because of their large size (hence, relatively low external surface area to volume ratio) elephants have a low metabolic rate per unit of body mass, which enables them to obtain adequate nutrition from plant material low in nutrient content. Hence, their relative daily food intake (in dry mass terms) is also low, around 1–1.5 per cent of body mass per day (compared with 2–3 per cent for cattle). Nevertheless, as a consequence of their large size, the absolute amount of vegetation that each elephant consumes per day is huge, estimated to be over 60 kg for a fully grown male, weighed as dry mass, or around 180 kg weighed wet (Owen-Smith, 1988).

FEEDING BEHAVIOUR

Elephants display a variety of feeding behaviours, and have long been known as robust and wasteful feeders (Selous, 1881). As with other vertebrate herbivores, they can ingest forage directly by biting with the mouth, although this occurs infrequently – about 10 per cent of browsing events in subtropical thicket (Lessing, 2007). Alternatively, forage is plucked (broken off the plant or the entire plant uprooted) with the trunk and passed to the mouth where it is ingested through a single bite or multiple bites, or material is stripped off a branch with the trunk and passed to the mouth. They also run branch tips between their teeth to strip off the bark, discarding the interior wood. At certain times of the year they strip off and discard leaves before consuming the bark, while at other times they eat the leaves of these same species (Barnes, 1982; Chafota, 2007).

The trunk, a specialised foraging adaptation with surprising dexterity, plays a crucial role in enabling elephants to achieve a high rate of food intake, in part by allowing them to chew and handle material simultaneously. Food intake has been estimated to approach an instantaneous rate of 2 kg.min⁻¹ when feeding on succulent shrubs (Lessing, 2007). The trunk, together with their high shoulder height, also allows them to forage up to 8 m above ground level (Croze, 1974). Elephants can adopt a bipedal stance in order to reach higher food material (Croze, 1974). Most browsing, however, takes place between 0.5 and 2.5 m (Guy, 1976; Jachmann & Bell, 1985; Chafota, 2007; Lessing, 2007).

The tusks are used for specialised feeding, particularly to strip bark off trees, most commonly during the latter part of the dry season and the early growing season (Barnes, 1982). Thereby elephants probably gain from the carbohydrates flowing through this bark prior to leaf flush (Barnes, 1982). When hard pressed for food, elephants will gouge quite deeply into the trunks of soft-stemmed trees like baobabs *Adansonia digitata* (figure 1). They also use the tusks to dig up the roots of some woody and succulent species (Barnes, 1982; Chafota 2007; Lessing, 2007).

Elephants use their feet to dig out (kicking or scraping) geophytes or grass tussocks, and knock grass tussocks held in the trunk against their legs to dislodge soil (Owen-Smith, 1988).

Elephants have been recorded felling or uprooting trees up to 60 cm in basal diameter (Chafota, 2007). Sometimes they feed on the branch tips or roots of these trees, but on other occasions they abandon the fallen tree without feeding on it. It has been suggested that some tree felling may be a social display unrelated to feeding (Hendrichs, 1971; Midgley *et al.*, 2005), but this has not been confirmed. Trees pushed over in Kasungu National Park, Malawi, were taller (4–5 m) for favoured species than for species generally rejected as food (2–3 m) (Jachmann & Bell, 1985).

Unlike most other herbivores, elephants' feeding actions may lead directly to the death of mature trees (through felling or uprooting), or otherwise expose these trees to other processes leading to tree mortality (through bark removal). Most other herbivores simply remove plant tissues, suppressing plant growth and reproductive potential, except in the case of small seedlings. In this sense, the consequences of elephant feeding for tree dynamics are more akin to those of a predator than is the case for other herbivores.

Figure 1: Damage to baobabs by elephants in the Chobe National Park, Botswana (photo: W S W Trollope)

Forage use as a basis for inferring impact

It is generally presumed that elephant herbivory is an important mechanism that structures plant communities (e.g. Laws, 1970; Tafangenyasha, 1997; Stuart-Hill, 1992; Trollope *et al.*, 1998; Mapaure & Campbell, 2002; Conybeare, 2004). Thus, it is important to have an understanding of elephant diet, and particularly their dietary preferences, in order to predict these impacts. However, some plant species that are not browsed by elephants respond to elephants through indirect mechanisms – for example, trampling and associated path formation (Plumptre, 1993; Landman *et al.*, 2008). In addition, the amount of forage ingested by elephants only represents a fraction of their total forage off-take (Guy, 1976; Paley, 1997); hence, impacts on plant communities are not a simple function of food requirements.

Although numerous studies describe the diet of elephant in a range of habitats – wooded savannas, desert shrublands, fynbos and subtropical thicket (Buss, 1961; Jarman, 1971; Barnes, 1982; Kalemera, 1989; Viljoen, 1989; Kabigumila, 1993; Paley & Kerley, 1998; Steyn & Stalmans, 2001; Milewski, 2002; Greyling, 2004; Minnie, 2006; Chafota, 2007), many are not quantitative in terms of species contribution, and for example describe diet at the broad level of growth forms (Koch *et al.*, 1995; Cerling *et al.*, 1999; Codron *et al.*, 2006). In addition, few studies (Guy, 1976; Jarman, 1971; Viljoen, 1989; De Boer *et al.*, 2000; Greyling, 2004; Minnie, 2006; Landman *et al.*, 2008) assess the relative availability of dietary items, and are thus able to quantify preferences for specific species. Moreover, elephant diet is often indirectly inferred from plant-based studies (Penzhorn *et al.*, 1974; Barratt & Hall-Martin, 1991; Midgley & Joubert, 1991; Stuart-Hill, 1992; Moolman & Cowling, 1994; Lombard *et al.*, 2001), assuming that differences between elephant areas and areas where elephants have been excluded are the result of elephant browsing. In this regard, Landman *et al.* (2008) showed that a significant proportion of such species are not eaten by elephants.

Elephants are mixed feeders, consuming a range of plants and plant parts from grasses to browse, bark, fruit, and bulbs. Their large body size and robust feeding allow them to have a broad diet – for example, 146 plant species in subtropical thicket (Kerley & Landman, 2006). Elephant herbivory can, therefore, influence the fate of a considerable number of plant species. However, the bulk of the daily dry matter intake comes from a few species.

Elephants consume varying proportions of browse and grass depending on region, vegetation cover, water availability, soil nutrient composition, and season (Williamson, 1975; Field & Ross, 1976; Owen-Smith, 1988; Koch *et al.*, 1995; Cerling *et al.*, 1999). Grasses are primarily consumed in the rainy season (40–70 per cent of the diet), and trees or shrubs in the dry season, when grass contributes only 2–40 per cent (Buss, 1961; Bax & Sheldrick, 1963; Wing & Buss, 1970; Jarman, 1971; Field, 1971; Laws *et al.*, 1975; Williamson, 1975; Guy, 1976; Barnes, 1982; Lewis, 1986; Kabigumila, 1993; Spinage, 1994; De Boer *et al.*, 2000; Greyling, 2004). When feeding on grasses, elephants favour leaves and inflorescences during the wet season, turning more to leaf bases and roots during the dry season (Owen-Smith, 1988). Forbs (herbaceous plants besides grasses) are also commonly consumed, and elephants may spend much time feeding in reed beds during the dry season. Under dry conditions, wood, bark and roots constitute 70–80 per cent of the material eaten (Barnes, 1982).

Elephants are selective feeders at the plant species level. For example, 40–70 per cent of the seasonal browse intake of elephants feeding in the Chobe River front region of northern Botswana came from just three shrub species: *Baphia massaiensis*, *Bauhinia petersiana* and *Diplorhynchus condylocarpon*, with a wider range of species eaten during the hot-dry season than at other times of the year (Chafota, 2007). A similar pattern was observed in subtropical thicket, where 25 out of 146 species used comprise 71 per cent of the diet (Kerley & Landman, 2005). Common dietary staples elsewhere include species in the genera *Acacia*,[1] *Azima*, *Colophospermum*, *Combretum*, *Commiphora*, *Cordia*, *Cynodon*, *Dichrostachys*, *Grewia*, *Faidherbia*, *Gardenia*, *Portulacaria*, *Premna*, *Schotia*, *Sclerocarya*, *Tamarix*, *Terminalia* and *Ziziphus*. Genera rejected as food, or eaten rarely, include *Baikiaea*, *Burkea*, *Capparis*, *Croton*, *Erythrophleum*, *Euclea*, *Ochna* and *Scolopia* (see diet references above). Several *Combretum* spp. are commonly eaten, others rejected (e.g. *Combretum mossambicense* is noted by Skarpe *et al.*, 2004).

There is conflicting evidence regarding the nutritional characteristics of plants preferred by elephants. Some studies show preferences for plants with higher levels of protein, sodium, calcium and magnesium (Dougall, 1963; Dougall & Sheldrick, 1964; Van Hoven *et al.*, 1981; Jachmann & Bell, 1985; Hiscocks, 1999), lower levels of crude fibre (Field, 1971; Holdo, 2003), secondary compounds and lignin (Jachmann, 1989). In contrast, Thompson (1975) could not show any differences in mineral or crude protein content between the bark of five species of trees with differing apparent preference. Calcium, magnesium, sodium, potassium, total salts and crude protein apparently do not determine elephant use among 16 species assessed by Anderson & Walker (1974) in Zimbabwe. These relationships are confounded by factors such as soil nutrients, rainfall, plant availability and so on, and need to be further researched.

It has been hypothesised that because of their simple digestive system, involving rapid throughput, elephants are less readily able than ruminants to handle plant secondary chemicals (e.g. resins, tannins and other phenolics), which tend to be concentrated in leaves (Olivier, 1978; Langer, 1984).

Discarded forage

Besides trees felled, elephants also break off and discard plant parts (Ishwaran, 1983). The discarded material could represent as much as a quarter to a half of the mass consumed in the Addo Elephant National Park (Addo) (Paley, 1997; Lessing, 2007). This discarded material could alter the size, distribution, nutrient levels and hence dynamics of litter in subtropical thicket ecosystems (Kerley & Landman, 2006). Elephants are not unique in this behaviour, as for example, kangaroo rats (*Dipodomus* sp.) also discard a large proportion of the forage they harvest (Kerley *et al.,* 1997). This aspect of elephant foraging is poorly described and understood, but may have profound cascading effects on ecosystem function and biodiversity patterns.

Ecological consequences of sexual dimorphism

Male elephants attain a body mass twice that of adult females (Lee & Moss, 1995), leading to differences in feeding behaviour and energetic and nutritional demands besides those associated with reproduction (Stokke & Du Toit, 2000;

Greyling, 2004; Lagendijk *et al.*, 2005; Shannon *et al.*, 2006a). In addition, differences in social structure (group-living cows vs. largely solitary bulls) influence foraging (Dublin, 1996). In savanna, bulls feed more robustly on fewer plant species, but a wider range of plant parts (Stokke & Du Toit, 2000), and consume more low-quality items. Family units more frequently debark and defoliate woody plants, while bulls fell trees and dig up roots more frequently (Greyling, 2004). Males also consume a higher proportion of grass than females. The rate of tree felling by males is much greater than that of females (Guy, 1976), and males also fell substantially larger trees than females. Accordingly, the consequences of the feeding and breakage impacts of the adult male segment of the population are relatively much greater than those of family units. In contrast, in subtropical thicket, males and females show large overlaps in feeding height, pluck size and foraging rates, which do not differ between sexes (Lessing, 2007). Males, however, do access the largest biomass (branch size) per pluck, and tend to harvest more multiple stem portions per pluck (compared to the females who tend to use single stem plucks).

Furthermore, differences in habitat use between sexes have been ascribed to the differential need to access water, with breeding females being found closer to water (Stokke & Du Toit, 2002). There have, therefore, been suggestions that elephant sexes occupy different ecological niches (Stokke & Du Toit, 2000; Shannon *et al.*, 2006a) in savanna. However, Shannon *et al.* (2006b) found no sex-based habitat selection in areas where water was spatially limited.

ECOLOGICAL PROCESSES INFLUENCED BY ELEPHANTS

Elephants affect a broad variety of ecological processes through their feeding, digging and movement. For example in subtropical thicket, Kerley & Landman (2006) showed that the role of elephants (15 broad processes) was comparable to that of the balance of the vertebrate herbivore community (21 species) in terms of the number of ecological processes (table 1). In addition, by virtue of their killing, through aggressive competition, of other herbivore species such as white rhinoceros *Ceratotherium simum* and black rhinoceros *Diceros bicornis* (Slotow *et al.*, 2001; Kerley & Landman, 2006), elephants also play a role analogous to predation. The significance of elephants in all these roles, and how this differs between landscapes, has yet to be quantified. The focus on a few effects such as tree mortality may, therefore, mask both the extent and the mechanisms of elephant impacts (Landman *et al.*, 2008).

Elephant formation of 'browsing lawns', where they reduce the height of mopane veld and increase the quality of forage, is considered to be 'gardening', analogous to the formation of 'grazing lawns' by other herbivores including snails, tortoises, geese and wildebeest (McNaughton, 1984). This shrub coppice state is advantageous for elephants through providing more food and better quality re-growth within the 2–5 m height range favoured by elephants (Jachmann & Bell, 1985). There are also increases (provided the overall cover is not lost) in the availability of forage for other herbivores (Guy, 1981; Smallie & O'Connor, 2000; Styles & Skinner, 2000; Rutina *et al.*, 2005; Makhabu *et al.*, 2006). In addition, they will excavate waterholes in dry riverbeds (Owen-Smith, 1988; Selous, 1881). The paths that they develop in travelling to and from water, and around obstacles such as mountainous ridges, can facilitate movements by other species (e.g. Skead, 2007). Elephants also function as keystone species (Paine, 1969), as shown for example by their dispersal of seeds of a specific range of plant species (Kerley & Landman, 2006). These observations appear to be consistent with the 'keystone herbivore' concept, invoked to explain how the elimination of similar megaherbivores elsewhere (through hunting by early human colonists in the late Pleistocene) contributed to a cascading sequence of extinctions among other large mammal species (Owen-Smith, 1987, 1989; Koch & Barnovsky, 2006).

EFFECTS OF ELEPHANTS ON BIODIVERSITY

If we are to understand the impacts of elephants, it is critical that the connections between elephants and the assumed impacts (defined here as changes brought about by elephants) are clearly understood and demonstrated. Elephant impacts are observed at a range of levels, from soils to coexisting mammals (reviewed below), and in all instances of such impacts, the mechanisms need to be clearly identified.

Individual plants and species

Elephants impact on plants by breaking branches/stems, stripping bark, uprooting plants and toppling trees. The persistence of plant species eaten by elephants is dependent on whether they can cope with herbivory of this nature (i.e. the relative capacity of these species to restrict, resist or compensate for the damage inflicted by resprouting and/or regrowth), or whether mortality is balanced or exceeded by recruitment and regeneration. The ability to resprout is taxon-specific: a range of species coppice readily, whereas *Aloe* spp., *Acacia goetzii, Acacia nigrescens, Acacia nilotica, Acacia polyacantha, Dalbergia melanoxylon* (Luoga *et al.*, 2004; Kruger *et al.*, 2007) and various *Commiphora* spp. (Kruger *et al.*, 2007) have all been reported to be poor resprouters following either cutting or elephant damage.

Broad ecological process	Megaherbivores		Meso-herbivores	Omnivores	Carnivores
	Elephant	Black rhinoceros & hippopotamus			
No. of species in category	1	2	19	3	18
Trophic processes					
Bulk grazing	1	1	3		
Concentrate grazing	1		9		
Browsing	1	1	7	3	
Frugivory	1	1	17	3	6
Predation				2	18
Scavenging				2	9
Transport processes					
Seed dispersal	1	2	19	3	6
Nutrient dispersal	1	2	19	3	18
Habitat architecture processes					
Plant form	1	2	7		
Grazing lawns		1	5		
Path opening	1	2	5	1	
Bipedturbation processes					
Wallowing formation	1	1	1	1	
Soil movement through dust bathing	1		5		
Digging	1		1	2	6
Hoof action	1		19	1	
Geophagy	1				1
River-bed configuration	1	1			
Other processes					
Litter production	1	1		2	
Germination facilitation	1	2	19	2	6
Total no. of processes affected	15	12	14	12	8

Table 1: The relative role of elephants in broad ecological processes (*n* = 19), modified from Kerley & Landman (2006), operating in subtropical thicket in relation to other megaherbivores (2 spp.), mesoherbivores (19 spp.), omnivores (3 spp.) and carnivores (18 spp.)

Responses to bark stripping also vary across taxa, e.g. *Acacia xanthophloea* in Amboseli, Kenya, are relatively tolerant of bark stripping and branch removal by elephants (Young & Lindsay, 1988). *Brachystegia* spp. seem to be highly susceptible to elephant damage, despite their high coppicing ability, resulting in stands of tall trees being converted to shrubby coppice regrowth (Thompson, 1975; Guy, 1989). O'Connor *et al.* (2007) suggest that the sensitivity of woody species to elephant browsing is a function of plant and landscape features.

Through their feeding, elephants can 'negatively' impact plant species and cause extirpation (localised plant species extinction) (Penzhorn *et al.*, 1974; Western, 1989; O'Connor *et al.*, 2007) or conversely, trigger plant growth and regeneration (Stuart-Hill, 1992).

Mechanisms of impact on individual plants

Toppling effects

The ecological effects of pollarding (total breaking of the stem) differ from toppling, where the roots may be removed from the soil, which usually kills the plant. However, if the roots remain in the soil, many species can resprout quite effectively (e.g. *Combretum apiculatum* – Eckhardt *et al.*, 2000). Factors that influence vulnerability to being toppled include strength of the wood, the depth and extensiveness of the root system and substrate stability (O'Connor *et al.*, 2007). Shallow-rooted shrubs (e.g. *Commiphora* spp.) that are uprooted completely by elephants are greatly reduced in their prevalence by elephants, as has happened in sections of Tsavo East National Park, Kenya (Leuthold, 1977), and in Ruaha National Park, Tanzania (Barnes, 1985).

Bark stripping

The impact of stripping on a plant species is dependent on the degree to which the bark is stripped. Ring barking will kill the plant, but if some phloem remains intact, the bark may re-grow (Buechner & Dawkins, 1961; Laws *et al.*, 1975). This may vary between species – mopane can lose up to 95 per cent of the bark without visible signs of stress (Styles, 1993). Features of the tree influence its vulnerability to being stripped, for example, elephants can cause more damage to trees with stringy bark (e.g. *Acacia* spp.) than those with bark that breaks off in chunks (e.g. *Sclerocarya birrea*) (O'Connor *et al.*, 2007). Furthermore, toxins in the bark or stem spinescence reduce preference for bark stripping (Sheil & Salim, 2004; Morgan, 2007). Fluted or multistemmed trunks are better protected against stripping (Sheil & Salim, 2004): in *Balanites maughamii* two-thirds of the bark is protected on account of fluting; while multistemmed trees that avoid total stripping (O'Connor *et al.*, 2007) include various *Combretum* and *Gymnosporia* spp. Further, Sheil & Salim (2004) found that elephants selectively stripped larger trees.

The effects of stripping are exacerbated by borer infestation, rot and fire (Laws *et al.*, 1975; Thompson, 1975). Elephant bark stripping facilitates insect and fungal attacks in *Brachystegia boehmii* woodlands in northern Zimbabwe (Thompson, 1975). However, Smith & Shah-Smith (1999) found no relationship between elephant damage and fungal infection. Van Wilgen *et al.* (2003) suggest that it is highly likely that fire in conjunction with elephant impacts may have resulted in the loss of large trees in Kruger between 1960 and 1989 (see Eckhardt *et al.*, 2000).

Vulnerability of seedlings

Few studies explore elephant impact on seedlings (but see Jachmann & Bell, 1985; Kabigumila, 1993; Barnes, 2001), though there is evidence for species-specific impacts. Examples are baobabs (Edkins *et al.*, 2007), and about 35 per cent mortality in *Acacia erioloba* in Chobe National Park, Botswana (Barnes, 2001). Elephants cause mortality by ripping seedlings from the soil, or prevent recruitment into adult size classes through top kill, maintaining the plants in a size class where they are caught in the 'fire trap' (Barnes, 2001).

Case studies of species-specific impacts

Baobab *Adansonia digitata*

Elephants are the only herbivores that can kill adult baobabs, and are frequently linked to the reduction in baobab densities, e.g. Mana Pools (Swanepoel, 1993), Tanzania (Barnes *et al.*, 1994) and Kruger (Whyte *et al.*, 1996). Barnes *et al.* (1994), in a 10-year study in Tanzania, found that baobab populations declined as elephant numbers increased and that the baobabs recovered when elephant populations declined due to poaching.

As with other species, the impact of elephants on baobabs is confounded by interactions with drought (Whyte *et al.*, 1996), other herbivores (Edkins *et al.*, 2007), and fire. Furthermore, the pattern of elephant effects on baobabs is inconsistent across size-classes, either showing selection against small trees (Weyerhaeuser, 1985; Barnes, 1985), or no size-class selection (Swanepoel, 1993).

Spatial refuges for baobabs occur on steep slopes inaccessible to elephants (figure 2; Edkins *et al.*, 2007). Consequently, it is unlikely that elephants can remove all baobabs from areas that include sufficient topographic relief (Whyte *et al.*, 1996; Edkins *et al.*, 2007).

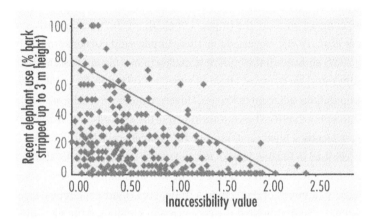

Figure 2: Regression analysis at the 90th quantile of recent elephant use of baobabs in the Kruger National Park and the inaccessibility value calculated for these. Elephant browsing drops below 100 per cent at the 7° slope and below 20 per cent at the 18° slope cut-off (Edkins *et al.*, 2007)

Acacia spp.

Because *Acacia* spp. are commonly selected by elephants (Calenge *et al.*, 2002), and show little or no resprouting once mature, their densities decline under high elephant browsing pressure, e.g. *Acacia tortilis, A. xanthophloea, A. nigrescens, A. senegal* or *A. erioloba* (Van Wyk & Fairall, 1969; Pellew, 1983; Ruess & Halter, 1990; Barnes, 2001). However, *Acacia* spp. have the capacity to regenerate rapidly from seedlings (Western & Maitumo, 2004), and elephants tend to ignore early stage and regenerating trees (Okula & Sise, 1986; Mwalyosi, 1987, 1990; Pellew, 1983; Calenge *et al.*, 2002). Thus, elephant damage may not affect *Acacia* populations overall (Balfour, 2005). In a comparative study of eight co-occurring *Acacia* spp. in Hluhluwe-Umfolozi Park, while levels of impact varied between the different species, no species were selected for or against (Balfour, 2005). In contrast, Western & Maitumo (2004) showed that elephants have brought about the local loss of swamp-edge *A. xanthophloea* woodlands in Amboseli, Kenya, their impacts overriding those of fire or other processes.

Soil chemistry confounds the latter results, however, as rising salinity levels were clearly linked to *A. xanthophloea* mortality in non-swamp areas in both Amboseli, Kenya (Western & Van Praet, 1973), and Ngorongoro, Tanzania (Mills, 2006).

Marula *Sclerocarya birrea*

Despite concern about of the impacts of elephants on marula, early studies (Coetzee *et al.*, 1979), suggested that these impacts did not constitute a threat. Gadd (2002) showed that elephant impacts on marula are sustainable (low mortality rates, recovery of affected trees, no selection for small trees) in three populations adjacent to Kruger. However, other studies have shown that marula trees have suffered severe attrition due to elephants (e.g. Weaver, 1995). In Kruger, Jacobs & Biggs (2002) showed a 7 per cent mortality of marula trees, mostly ascribed to the breakage of main stems by elephants. They also showed that these impacts varied in terms of the extent (number of trees affected) and severity (amount of damage to a tree) across landscape types. Jacobs & Biggs (2002) also highlighted the concern that elephant damage could lead to increased mortality due to other factors such as insect or pathogen attack and fire.

Mopane *Colophospermum mopane*

Elephants browse intensively on mopane trees, and prefer mopane to many other trees (Ben-Shahar, 1993). However, mopane trees are well adapted to regenerate after elephant browsing, and few are killed by this browsing. While unbrowsed mopane has treelike morphologies, mopane woodlands may be converted to stands of shrubby coppice through the feeding impacts of elephants (Lewis, 1991; Smallie & O'Connor, 2000; Styles & Skinner, 2000; Lagendijk *et al.*, 2005). Elephants inhibit height recruitment by repeatedly breaking leader shoots (Anderson & Walker, 1974). However, elephants have more impact in taller mopane, where ring-barking, heavy browsing and toppling cause mortality (Caughley, 1976a; Lewis, 1991).

Several factors affect the degree of elephant damage on mopane. Proximity to water sources appears, as in many other systems, to have the greatest effect (Styles & Skinner, 2000). Soil type also appears important: soils that promote shrub-like mopane yield less stable woodlands than soils that promote tree-like growth (Lewis, 1991). Elephant browsing intensity also tends to fluctuate with time of year, being greatest after spring rains (Styles & Skinner, 2000).

Spekboom *Portulacaria afra*

Spekboom is generally one of the most abundant species in subtropical thicket, and probably the best studied example of the species-specific impacts of elephants in Addo. The roots, shoots and leaves are utilised extensively (contributing about 9 per cent to the diet), usually in proportion to availability (Landman *et al.*, 2008). Elephants reduce the height of individual plants (Stuart-Hill, 1992) and remove more than 50 per cent of the biomass (Penzhorn *et al.*, 1974). Despite these high levels of utilisation (and thus large impacts), *P. afra* persists in the presence of elephants, except in areas with extremely high elephant densities (Barratt & Hall-Martin, 1991). Stuart-Hill (1992) argued that the species is adapted to the 'top-down' browsing by elephants, whereby the lower rooted branches escape elephant browsing impacts, which facilitates vegetative reproduction. The 'top-down' hypothesis is supported by observed elephant browsing heights of above 50 cm in Addo. However, this hypothesis fails when the plants are uprooted and the roots are consumed (Stuart-Hill, 1992; Lessing, 2007).

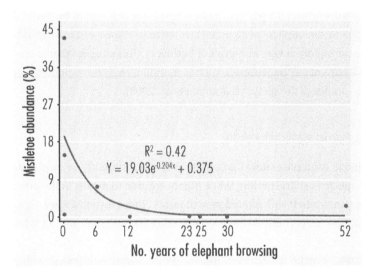

Figure 3: Exponential decline in the abundance of mistletoes (*Viscum rotundifolium, Viscum crassulae, Viscum obscurum*) in the presence of elephants in the Addo Elephant National Park (Magobiyane, 2006)

Mistletoes

Mistletoes (comprising *Viscum rotundifolium, Viscum crassulae, Viscum obscurum, Moquinella rubra*) are highly nutritious (Midgley & Joubert, 1991) and are preferred food items for elephants in Addo (Landman *et al.*, 2008). This guild is treated as an entity here. Mistletoes show an exponential decline in abundance (figure 3) and richness with increasing levels of elephant browsing, with *V. crassulae* disappearing in the presence of elephants (Magobiyane, 2006). *V. rotundifolium*, however, persists at very low densities in elephant habitat. These responses are rapid (a 60 per cent decline in abundance within six years), and after a decade of elephant browsing, mistletoe densities are too low to be used as measures of elephant impact (Magobiyane, 2006).

Aloe **spp.**

Aloes, in particular *A. africana*, have long been known to disappear from the elephant area of Addo, presumably as a result of elephant browsing (Penzhorn *et al.*, 1974; Barratt & Hall-Martin, 1991). Only recently did Landman *et al.* (2008) show that elephants actually consume *A. africana*, albeit in very small proportions (about 0.1 per cent of the diet). Aloes appear to be particularly sensitive to the impacts of elephants (relative to *P. afra* and mistletoes) and disappear rapidly at very low levels of herbivory. This suggests that alternative mechanisms of elephant impact, such as trampling, may be responsible for the disappearance of the species (Landman *et al.*, 2008).

Assessing species-specific vulnerability

The above examples show that plants respond differently to elephant use. Some species decline rapidly, while others are able to persist in the presence of elephants, albeit with altered growth forms. These responses are, however, difficult to interpret due to the presence of a range of confounding variables such as fire, soil nutrients, other herbivores, and elephant densities. O'Connor *et al.* (2007) provide a theoretical framework for assessing the vulnerability of a plant species to extirpation/extinction. They list a range of plant traits, landscape characteristics that might influence the probability of elephants' selection for these species, and management unit characteristics that exacerbate these.

Plant traits

A species would be considered vulnerable to extirpation by elephants if it displayed the following characteristics:

- lacks the ability to sprout as adult and/or cannot regrow its bark so that pollarding or ringbarking causes death
- restricted to selected foraging habitats
- highly selected by elephants
- frequently subjected to pollarding and ringbarking
- regenerates infrequently and/or usually in small numbers
- slow growing
- displays episodic recruitment.

Landscape and management unit characteristics
Vulnerability to extirpation is exacerbated if:

- terrain lacks topographical refuges
- there are no spatial refuges from elephant because distance from water is not a foraging constraint
- reserves are small
- reserve is located in a semi-arid region with variable grass production, hence heightened utilisation of woody material
- reserve is a degraded semi-arid savanna in which suitable grass is no longer available and woody plants form the bulk of the diet.

Fauna

The direct effects of elephants on other animals include direct mortalities and interference competition (as opposed to resource competition). Thus, elephants temporally exclude other species from resources such as waterholes or other resources by actively chasing them away (Owen-Smith, 1996). Alternatively, elephants may also facilitate access to resources through, for example, excavating waterholes (Owen-Smith, 1988) and increasing the availability and quantity of forage (e.g. Skarpe *et al.*, 2004). The understanding of these interactions is again limited due to confounding factors, and the fact that these are normally cascading effects.

Invertebrates

There are few studies on the effects of elephants on invertebrates. Cumming *et al.* (1997) found significantly lower richness of ant species in woodlands that had been impacted by elephants than in intact woodlands. Cicadas were only recorded in the intact woodlands, not in the impacted woodlands.

Mantid communities did not respond to changes in woodland structure (Cumming *et al.*, 1997).

Dung beetles are sensitive to habitat change (Klein, 1989). Disturbance in the form of fire or elephants can have a significant effect on dung beetle species' diversity and biomass (Botes *et al.*, 2006). In Tembe Elephant Park, Maputaland, dung beetle assemblages (Botes *et al.*, 2006) differ between elephant impacted sand forest (a key endemic habitat type) and undisturbed sand forest sites (including the loss of some forest specialist species). Elephants may provide refugia for other species, particularly ground-living invertebrates, under dung and trunks of toppled trees (Govender, 2005).

Musgrave & Compton (1997) demonstrated a significant increase in phytophagous insect feeding damage in the presence of elephants in Addo, and attributed this to an increase in the quality of browsed plants through a decline in secondary chemical compounds (e.g. tannins). This hypothesis has yet to be tested, nor has it been shown which insect species were involved, and what their population or overall insect biodiversity responses were. This apparent increase in nutritional quality of plants needs to be weighed up against the significant decline in overall plant phytomass (Kerley & Landman, 2006).

Reptiles and amphibians

In an attempt to explain high tortoise abundance in Addo, Kerley *et al.* (1999) hypothesise that elephant alteration of subtropical thicket habitat (through their creation of open habitat patches and paths) may favour increased access for tortoises (i.e. leopard tortoises *Stigmochelis pardalis* and angulate tortoises *Chersina angulata*).

Birds

Cummings *et al.* (1997) found a drop in species richness of birds and changes in bird communities (from woodland species to non-woodland species) in response to changes caused by elephants in Miombo woodlands, Zimbabwe. Reduced vertical and horizontal heterogeneity in the elephant-impacted woodlands probably accounts for their observed loss of species richness (c.f. MacArthur, 1964).

In contrast, Herremans (1995), assessing bird community species shifts in riverine forest and Mopane woodland in northern Botswana, found that dramatic woodland change associated with the high abundance of elephants did not result in a reduction in bird diversity. This was possibly due to the fact that woodland conversion was spatially restricted. However, gallinaceous birds were more abundant in areas heavily impacted by elephants than elsewhere in the Chobe River region (Motsumi, 2002).

Elephant removal of large standing trees in savanna (e.g. Eckhardt *et al.*, 2000), may decrease the availability of nesting sites for raptors, especially vultures and other rare, open-savanna species (Monajem & Garcelon, 2005). Little is available in the scientific literature on the nesting requirements of savanna raptors. More research is needed to determine the outcomes of elephant-raptor interactions.

Chabie (1999) showed that in transformed thicket in Addo, there were significant changes in the bird communities. At the guild level, there was a shift from frugivores in intact thicket to a community dominated by insectivores and granivores in opened-up thicket. In addition, there was a shift to larger bodied species in transformed thicket. The hypothesis that elephants drive these changes needs to be further tested.

Bats

The expected loss of large trees and snags due to elephants may decrease both roosting sites of bats and available habitat for species that specialise on feeding within dense vegetation (Fenton *et al.*, 1998). However, Fenton *et al.* (1998) found no decrease in Vespertilionid and Molossid (airborne insectivores) bat species richness, or a loss in specialists, with a reduction in woodland canopy cover. Similar results were observed by Cumming *et al.* (1997) in Miombo woodlands.

Small terrestrial mammals

There are few studies on the impacts of elephants on small mammals. Keesing (2000) showed that the presence of elephants in East African savannas results in an increase in species richness of small mammals, through habitat alteration.

Large terrestrial mammals

Browsers

There is a general negative correlation between elephant biomass and the biomass of browsers and medium-sized mixed feeders across ecosystems (Fritz *et al.*, 2002). A number of mechanisms for this have been proposed, including

(1) the reduction in resources through direct competition, (2) the alteration of habitats for browsers and other ungulates, (3) increase in visibility resulting in higher predation levels, and (4) competition for water (Owen-Smith, 1988; Skarpe *et al.*, 2004; Valeix *et al.*, 2007). While the patterns are significant, and sometimes obvious, the mechanisms are not yet clear: a possible explanation is that elephants reach highest abundances in areas of mopane and other vegetation types which they exploit more effectively than other browsers.

The structural transformation from more wooded to more open habitat conditions benefits some browser species, but leads to a decline in others. The persistent abundance of elephants along the Chobe River and in Hwange National Park has been associated with an increase in kudu *Tragelaphus strepsiceros* and impala *Aepyceros melampus* (Skarpe *et al.*, 2004). The mechanism for this is not clear, however; on the Chobe River, it may reflect the increase in *Capparis tomentosa* vines and *C. mossambicensis* shrubs, which are readily consumed by kudu and impala, but not elephants. In contrast, along the Chobe River, the abundance of bushbuck *Tragelaphus scriptus* has declined substantially following the opening of the riparian woodland by elephants (Addy, 1993).

In Addo, the opening of the succulent thicket vegetation by elephants brought about a decline in bushpig *Potamochoerus larvatus*, Cape grysbok *Raphicerus melanotis* and bushbuck abundance (Novellie *et al.*, 1996; Castley & Knight, 1997). However, it is not known whether populations of these species outside the elephant enclosure have remained unchanged over this period, or whether putative changes in habitat structure are the consequences of elephant impacts (reasonably likely given the trends reviewed here) or some other process such as global climate change (Kerley & Landman, 2006).

The reduction of vegetation cover and density by elephants in Addo results in a change in potential browse availability for black rhinoceros (Kerley & Landman, 2006). The increase in elephant paths, associated with increases in elephant densities, initially facilitates access to browse by black rhinoceros, but the subsequent dominance of the landscape by these paths results in a loss of foraging opportunities.

Sigwela (1999) compared the diet of kudu in the elephant enclosure and botanical reserves of Addo, and showed that elephants had no apparent effect on kudu diet selection. This is surprising given that (1) extensive vegetation changes have occurred in the elephant enclosure, (2) kudu diet (28 species) includes many of the plant species recorded as being impacted by elephants, and (3) elephants consume all the plant species recorded in the diet of kudu here. This suggests that food availability is not limiting to either kudu or elephant at the present densities of vegetation and browsers at these sites (Kerley & Landman, 2006).

Grazers

Given that grass forms a substantial part of the diet of elephants for much of the year (Owen-Smith, 1988), elephants are expected to compete with grazing ungulates if forage is limited. On the other hand, elephants are able to open up the woodland and increase the grass cover (Caughley, 1976b). However, in their broad-scale analysis, Fritz *et al.* (2002) could not detect any effect of elephants on grazers. Western (1989) highlighted the role of elephants in East Africa in facilitating pasture for medium and small ungulates, including domestic livestock.

In several cases, the decline of grazing species has been linked to the encroachment of woody vegetation in the absence of elephants (Owen-Smith, 1988), for example wildebeest *Connochaetus taurinus*, plains zebra *Equus burchelli*, waterbuck *Kobus ellipsiprymnus*, and reedbuck *Redunca arundinum* in Hluhluwe-Umfolozi Park (Owen-Smith, 1989). In Tsavo East National Park, Parker (1982) reported an increase in abundance of several grazing species, including oryx *Oryx gazella*, warthog *Phacochoerus africanus*, and zebra, following the opening of shrubland by the increasing elephant population. Young *et al.* (2004) found that by decreasing cattle grazing in a grassland area, elephants reduced the effects of competition between livestock and zebra.

Not all grazers benefit; for example, the conversion of tall woodlands into shrub coppice is likely to be adverse for sable antelope *Hippotragus niger*, although possibly not for roan antelope *Hippotragus equinus* (Bell, 1981).

Buffalo *Syncerus caffer* show a variety of responses to elephants. In the Chobe region, buffalo herds favoured areas recently grazed by elephants, suggesting facilitation rather than competition (Halley *et al.*, 2003). Skarpe *et al.* (2004) suggested that there is no evidence for competition between buffalo and elephants in Chobe; however there is some evidence for competition between buffalo and elephants in Tanzania (De Boer & Prins, 1990).

Ecosystem patterns and processes

The population and species level impacts brought about by elephants (documented in part above) will be expressed at the community and ecosystem level, including emergent properties of such systems, such as nutrient cycling, vegetation structure and dynamics.

Nutrient cycling

Elephants typically constitute 30–60 per cent of the large herbivore biomass in savanna ecosystems, and are thus responsible for 25–50 per cent (allowing for metabolic scaling) of the plant biomass consumption by herbivores (Owen-Smith, 1988; Fritz *et al.*, 2002). About 50 per cent of the material eaten passes through the gut undigested. Furthermore, elephants process fibrous plant parts such as bark and roots (which are generally not eaten by other herbivores) and thereby accelerate biomass recycling. Their importance for biomass cycling is further enhanced through wasteful feeding (Paley, 1997; Lessing, 2007) and the toppling of trees (Owen-Smith, 1988).

This contribution by elephants to biomass recycling tends to be greater in nutrient-poor than in nutrient-rich ecosystems because of their capacity to exploit vegetation components of low nutritional value. The removal of branch ends as well as leaves, plus felling of mature trees, promotes compensatory regeneration by these plants (Pellew, 1983; Fornara & Du Toit, 2007: Makhabu *et al.*, 2006) and, hence, greater primary production and rates of nutrient recycling than would occur in the absence of elephants. Termites contribute to the release of the nutrients in the fibrous tissues in elephant dung, and fire to releasing the minerals held in the stems of trees toppled by elephants. It has been hypothesised that, in the nutrient-deficient savanna woodlands prevalent on Kalahari sands (with little capacity to retain nutrients), much of the biologically available nitrogen and sodium pool is held within elephant biomass (Botkin *et al.*, 1981).

Elephants play a variety of roles in nutrient cycling, especially in nutrient-deficient ecosystems. They may release the nutrients locked up in tree trunks and roots (Botkin *et al.*, 1981). By removing large trees, they reduce the role that these trees play in extracting mineral nutrients from deep soil layers (Treydte *et al.*, 2007), and also the contribution of these trees to small-scale heterogeneity in soil nutrients through the nitrogen-enrichment promoted by fallen leaves. This generally decreases the availability of high-quality forage resources beneath tree canopies, and could indirectly affect the persistence of grazers (Ludwig, 2001). By reducing the prevalence of nitrogen-fixing legumes such as many *Acacia* spp., elephants suppress the role that these species play in nitrogen enrichment (Treydte *et al.*, 2007), although the absolute and relative extent of this effect has not been quantified.

Soil resources

Because of their large biomass, the trampling effects of elephants on soil compaction can also be substantial, with unclear consequences for vegetation (Plumptre, 1993). The large increase in woody cover associated with the exclusion of elephants in the experimental plots in Uganda dramatically increased soil organic matter and thereby pH, as well as extractable calcium, potassium, and magnesium levels. Organic carbon and nitrogen also increased, but total phosphorus declined slightly (Hatton & Smart, 1984).

Kerley *et al.* (1999) showed that in the Addo elephant enclosure the proportion of the landscape that represented run-on zones (i.e. where resources such as water, litter, soil, and nutrients are trapped during overland flow) declined, while the proportion of run-off zones (i.e. where these resources are lost) increased. The consequence of this was a decline in soil nutrients. Kerley *et al.* (1999) suggested that elephant impacts were less deleterious than goat impacts, but that these studies must be replicated.

Seed dispersal

Elephants play an important role in facilitating the dispersal and germination, and hence regeneration, of a large variety of plant species through endozoochory. Elephants are considered to be the only foragers (and hence dispersers) of the large-fruited *Balanites wilsoniana*, a canopy tree dominant in Kibale Forest, Uganda, as well as other large-fruited forest species (Chapman *et al.*, 1992; Babweteera *et al.*, 2007). Elephants enhance seedling germination (Cochrane, 2003) and increase seedling survival and growth by dispersing propagules far from adult trees (Babweteera *et al.*, 2007). In savanna, seed germination and seedling survival of *Sclerocarya birrea* are also enhanced following fruit ingestion by elephants (Lewis, 1987).

Despite their dietary breadth in subtropical thicket (146 plant species – Kerley & Landman, 2006), elephants are relatively poor seed dispersers in Addo, dispersing only 21 plant species through endozoochory (Mendelson, 1999; Sigwela, 2004), comparable to black rhinoceros and eland (both 20 species – Mendelson, 1999). Why so few species are dispersed is not clear, but may reflect the rarity of most plant species in the diet (25 out of 146 species comprise 71 per cent of the diet – Kerley & Landman, 2005), selective foraging behaviour in terms of plant phenology, complete loss of propagules during digestion, or inadequate sampling. The large volume of forage intake (and faecal output) by elephants (Owen-Smith, 1988), however, allows them to disperse large numbers of seeds (Sigwela, 2004), but their role in plant regeneration through this process needs to be quantified. Levels of zoochory vary between locations: for example, Robertson (1995) recorded 32 dicotyledonous species that were dispersed by elephants in nearby Shamwari Private Game Reserve.

Mortality of seeds during passage through the digestive tract was significantly lower in elephant compared to the goat *Capra hircus*, which served as a model ruminant (Davis, 2007). The effects of passage through the elephant digestive tract on germination differed between plant species (e.g. *Acacia karroo* germination declined, while *Azima tetracantha* germination improved). In addition, patterns of germination after ingestion differed between elephants, goats and pigs (Davis, 2007). This suggests that elephant effects on endozoochory will not be replaced by other herbivores.

Comparison among ecosystems

Perceptions of the extreme vegetation transformation that can be brought about by burgeoning elephant populations have been strongly influenced by particular case studies from outside South Africa. These include the situations in Murchison Falls National Park, Uganda, which led to the first major elephant culling operation implemented in Africa; Tsavo East National Park, Kenya, where a need for drastic culling was proposed but not implemented in the face of opposition; and Chobe National Park, Botswana, where high elephant concentrations have developed in the vicinity of the Chobe River, and culling has been repeatedly advocated but not undertaken because of practical considerations. Most recently, drastic vegetation changes ascribed to elephants have been documented for Amboseli National Park, Kenya. A critical appraisal of the ecological context and what these particular examples show (or do not show) is helpful, before turning to a broader assessment of ecosystem differences.

Illustrative case studies

Murchison Falls in Uganda

Murchison Falls National Park covers a 2 400 km² section of the northern part of the Bunyoro district in western Uganda, divided into southern and northern sections by the Nile River. Elephants were spread more widely over a 3 200 km² range at the time of the study (Laws & Parker, 1968; Laws *et al.*, 1975). The annual rainfall of 1 250 mm supported a *Terminalia glaucescens/Combretum binderanum* savanna woodland, plus open grassland areas with scattered

Acacia sieberiana trees. Also present were patches of closed-canopy forest (including the Budongo Forest), which historically had been more widespread, plus a limited area of bushland. Soils are underlain by basement igneous rocks, with volcanic influences from the adjoining Rift Valley. Annual burns generally occurred early in the dry season. A population approaching 10 000 elephants had become compressed inside the park by surrounding human settlements, creating an effective regional elephant density of around 3 elephant.km^{-2}. The park also supported 6 000 hippos and 14 000 buffaloes, plus numerous kob *Kobus kob*, hartebeest *Alcelaphus buselaphus* and warthog, so that the total large herbivore biomass amounted to 12 000 kg.km^{-2}. Much of the central region had been transformed into treeless *Hyparrhenia* grassland with just tree stumps remaining.

Vegetation changes were documented from aerial photographs (Laws & Parker, 1968; Laws *et al.*, 1975). One section of woodland, covering 5 300 km^{-2} in 1958, in which 24 per cent of trees were dead (Buechner & Dawkins, 1961), had been reduced to 1 060 km^{-2} in 1967, with 98 per cent of trees dead. The radial pattern of damage diminishing outwards from the centre of the park indicated that fire was not the major cause of the tree mortality. In some areas woodland had been replaced by dense *Lonchocarpus taxiflorus* shrubland, apparently resistant to both heavy browsing and fire. Two exclosures established in 1967 had become transformed to closed canopy *A. sieberiana* woodland, 7–10 m high by 1981 (Smart *et al.*, 1985). However, plant species richness had dropped to almost half of that recorded in 1967, especially in the herbaceous layer. Following the build-up of soil organic matter, there was a dramatic increase in extractable cations associated with an elevated soil pH (Hatton & Smart, 1984). Although total soil phosphorus declined, available phosphorus and nitrogen both showed increases. Following a massive reduction of the elephant population during the 1978 civil war, abundant regeneration of dense *Acacia* scrub occurred through much of the formerly open grassland areas of the park and extended into formerly *Terminalia* woodland. However, fire frequency was also reduced during this period.

A point to note in this case history is evidence that elephant damage was the primary factor, and fire secondary in the woodland transformations that occurred. It is also noteworthy that floristic diversity was reduced when elephants were excluded, at least in the herbaceous layer. Furthermore, tree regeneration took place rapidly when elephant impacts were reduced, although not back towards the former woodland composition.

Tsavo in Kenya

Tsavo East and West National Parks cover a combined area exceeding 20 000 km^2 in south-eastern Kenya, divided by the Mombassa road and railway line. Annual rainfall averages around 400 mm in central Tsavo East. Here, the vegetation consists predominantly of *Commiphora* shrubland on acid alluvial soils, with bands of tall trees and other species flanking rivers. Woodland decline had become a source of concern by 1967, at which stage the elephant population had reached at least 24 000 animals (Glover, 1963; Agnew, 1968). Severe drought conditions with rainfall amounting to less than half of the long-term mean prevailed during 1971, resulting in the deaths of at least 7 000 elephants (Corfield, 1973), representing 15–20 per cent of the pre-drought population (Cobb, 1976). *A. tortilis* plants taller than 1 m declined in density by 65 per cent between 1970 and 1974, while baobab trees had been virtually eliminated by 1974 (Leuthold, 1977). Mature *Commiphora* shrubs were reduced in density from 90 plants.ha^{-1} in 1970, to 5 plants.ha^{-1} by 1974 in a 4 400 km^{-2} section of Tsavo East (the rest of the park showed far less change – Myers, 1973). The opening of the woodland, promoted further by fires, led to increases in the abundance of grazers such as Burchell's zebra and oryx, while browsers including lesser kudu *Tragelaphus buxtoni*, gerenuk *Litocranius walleri* and giraffe *Giraffa camelopardalis* declined (Parker, 1982). Black rhino numbers also fell drastically, with poaching responsible for most of the losses.

Poachers also reduced the elephant population within the park to around 6 000 animals by 1994. This lowered density, then allowed abundant woodland regeneration to occur, especially of *A. tortilis* in riparian fringes (Van Wijngaarden, 1985; Leuthold, 1996). *Commiphora* shrubs that had been pushed over resprouted profusely from the base of the stem or roots. Some tree species not eaten by elephants survived virtually unchanged from 1970. Associated with the recovery of woody vegetation, the abundance of lesser kudu and gerenuk increased while the grazers that had shown increases decreased in numbers (Inamdar, 1996).

The Tsavo case illustrates drastic vegetation transformation by elephants during a severe drought followed by the rapid recovery of this vegetation after the abundance of elephants had been reduced to a density of around 0.3 animals.km^{-2}. These changes occurred mostly in the more arid region of the park. Populations of other large herbivores were affected to a relatively minor extent. Hence, no biodiversity losses occurred, apart from the near-extirpation of baobab trees (which occur abundantly outside the park). The major uncertainty is what would have happened had the peak density level of around 2 animals.km^{-2} been maintained for longer.

Chobe River front and adjoining areas in northern Botswana

The 80 000 km² region of northern Botswana within which Chobe National Park lies supported an elephant population which had reached 40 000 animals in 1980 and 140 000 animals by 2006 (Spinage, 1990; Skarpe *et al.*, 2004). Recent dry season densities along the Chobe River front region average around 4 elephants.km⁻², decreasing to 0.5 elephants.km⁻² when these animals disperse during the wet season (rainfall is around 700 mm per year). A narrow strip of riparian forest persisted along the Chobe River front in 1970, although many of the large *Acacia* trees appeared to be dying (Simpson, 1975). By 1980 most trees near the river, mainly *A. nigrescens* and *A. tortilis*, had been reduced to standing dead trunks, while two species unpalatable to elephants (i.e. *Combretum tomentosa*, *C. mossambicense*) had become predominant in the shrub understorey. Further back from the river, a shrubland including *C. eleagnoides*, *Baphia massaiensis* and *Bauhinia petersiana* prevailed on the alluvial terrace, while 3–5 km away from the river the vegetation changed to sandveld woodland with *Burkea africana* predominant on shallower sandy soils and *Baikiaea plurijuga* on deeper sands. Aerial photographs indicated that the area covered by woodland decreased from 60 per cent to 30 per cent between 1962 and 1998, while the area of shrubland expanded from 5 per cent to 33 per cent (Mosugelo *et al.*, 2002). In 1874, before elephants were exterminated from the region by ivory hunters, the vegetation adjoining the Chobe River had appeared quite open (Selous, 1881). Vegetation on the alluvial terrace remained open through the 1930s, with grazing by cattle plus exclusion of fires before the national park was established, contributing to the thicket development (Simpson, 1978).

A study on the ecosystem consequences of these vegetation changes (Skarpe *et al.*, 2004) found little regeneration of the tree species reduced in abundance by elephants, largely due to intense browsing pressure on seedlings by a high density of impala (locally >150 animals km⁻²). The shrub species avoided by elephants were commonly browsed by ruminants (Makhabu *et al.*, 2006), while buffalo appeared to be more abundant in areas of the floodplain where elephants had been feeding than elsewhere. Both small mammals and gallinaceous birds (guinea fowl and spur fowl) appeared more abundant in places that had incurred severe elephant impacts. The Chobe River front retained an exceptionally high density of land birds, especially of migrants (Herremans, 1995). Nevertheless, the opening of the woody vegetation cover by elephants was associated with a substantial reduction in the abundance of bushbuck, to a third or less of their former abundance (Addy, 1993). Fire was not a factor in the river front region, being blocked by the main road paralleling the river.

Further west along the Linyanti River, a similar pattern of woodland conversion is in progress, mostly outside the national park. Extremely high local concentrations of elephants develop here during the late dry season, up to 20 elephants.km^{-2}. By 1991 over 40 per cent of the trees in the riparian fringe were dead (Coulson, 1992; Wackernagel, 1992). *Acacia* spp. were most severely affected, with two-thirds of *A. erioloba* and 45 per cent of *A. nigrescens* trees dead, in many cases due to debarking by elephants. Wind-throw and natural senescence were additional factors contributing to this mortality, and other species such as *Diospyros mespiliformis* and *Combretum imberbe* growing in the riparian woodland showed much less elephant damage. Repeated aerial photographs indicated a net loss rate of canopy trees of only 2 per cent per year between 1992 and 2001, but tree felling was patchy and much of this loss was concentrated in patches where *Acacia* spp. were prevalent (Bell, 1985). In compensation, an expanding shrub layer, largely of *C. mossambicense*, had developed by 2001.

While the vegetation changes brought about by elephants along the Chobe River are extremely severe, the area affected is restricted to a 20–30 km section by human settlements to the east (Kasane town) and west (Kachikau enclave). Animal populations seem to have benefited rather than being adversely affected, apart from bushbuck. Browsing pressure from impala would suppress woodland recovery even if elephants were greatly reduced in abundance. Of greater concern are the trends towards elimination of the *Acacia* component of the woodland plus severe impacts on certain other woody species developing along the Linyanti River. Biodiversity losses are not yet of major concern because of the restricted extent of these vegetation changes within the greater ecosystem context.

Amboseli National Park in Kenya
The Amboseli ecosystem covers 8 500 km^2 in southern Kenya, while Amboseli National Park occupies 388 km^2 within the central basin (Western, 2007). The present-day remnant of a formerly much larger lake generated by drainage from the slopes of Kilimanjaro holds water usually for only a few weeks after heavy rains. Soils derived from volcanic deposits are alkaline and locally saline because of the closed drainage, except around the swamp margins. Further back the vegetation grades into bushland or open woodland with *Acacia tortilis*, *A. mellifera* and *Commiphora* spp. predominating. The mean annual rainfall is 340 mm. The region currently supports a population of 1 400 elephants, with the local density within the park amounting to 2–3 animals.km^{-2}. Elephants formerly migrated seasonally between the basin and surrounding bushland,

and their concentration within the park increased during the late 1970s after Maasai pastoralists and their livestock were excluded from this area.

Die-offs of extensive areas of *A. xanthophloea* (fever tree) woodland that became apparent during the late 1960s were ascribed to a rising water table and consequently increased salinity in the rooting zone (Western & Van Praet, 1973), as documented also in the Ngorongoro caldera (Mills, 2006). However, exclosure plots suggest that elephant damage was the primary contributor to the demise of these woodlands (Western & Maitumo, 2004), although the contributory role of water level and salinity changes cannot be excluded. Within areas fenced off in 1981, dense stands of *A. xanthophloea* had established and reached a height of 7–10 m by 1988, while *Acacia* seedlings outside the exclosures failed to grow and declined in abundance. This indicates the potential of the *Acacia* woodland for rapid recovery in the absence of browsing pressure and other damage by elephants. The total area covered by fever tree woodlands within a 700 km² region declined from 125 km² in 1950 to 2 km² by 2002, coupled with an expansion by alkaline grasslands and scrubland of salt-tolerant *Suaeda monoica* and *Salvadora persica* (Western, 2007). Stands of palms *Phoenix reclinata* have replaced the woodland in some localities. Associated with the woodland decline has been a decrease in the abundance of browsing ungulates within the national park, although these species remain abundant outside the park. Historical records suggest that woodlands were absent from the Amboseli basin in the late 1800s and that the presence of pastoralists with their cattle had contributed to the development of the *A. xanthophloea* stands within the basin. Woodlands outside the park boundary have mostly recovered since the 1970s following the establishment of pastoralist settlements, which are largely avoided by elephants.

This case study illustrates the potential for elephants to largely eliminate a tree species forming a monospecific woodland from a region, as well as the potential of this *Acacia* species for rapid regeneration once protected from elephants. Other factors contributed to both the establishment and demise of the woodlands, and the area affected was a fairly small section of the regional ecosystem.

Subtropical thicket

Research on the impacts of elephants on the plant communities of Addo has followed a tradition of comparing elephant-occupied areas with areas where elephants have been excluded (i.e. botanical reserves). This assumed that any difference in vegetation was due to the influence of elephants. Elephants have been shown to reduce plant species richness, plant biomass, canopy height and volume and density (Penzhorn *et al.*, 1974; Barratt & Hall-Martin, 1991; Stuart-Hill, 1992; Moolman & Cowling, 1994; Lombard *et al.*, 2001). Stuart-Hill (1992) argued that succulent thicket is adapted to the 'top-down' browsing by elephants, which maintains thicket regeneration by protecting canopy cover at ground level. In general, species abundance and richness of 75 special species (endemic-rich geophytes and low succulents – Johnson *et al.*, 1999) and two indicator species (*V. rotundifolium, V. crassulae* – Midgley & Joubert, 1991) declined exponentially with length of exposure to elephant browsing, halving approximately every 7 years (Lombard *et al.*, 2001). An important point is that 168 plant species identified as being entirely reliant on Addo for their conservation (Johnson *et al.*, 1999), are potentially vulnerable to elephant-driven extinction (Kerley & Landman, 2006).

The absence of effective density dependence in subtropical thicket (Gough & Kerley, 2006) is interpreted as a consequence of the aseasonal availability of high-quality forage, and it is predicted that the forage resource (and associated biodiversity) will collapse before density dependence emerges (Kerley & Landman, 2006).

Contrasts across biomes and ecosystems

The above savanna case studies span a rainfall range from arid (Amboseli, Tsavo East) to moist (Murchison Falls) savanna, and in soil fertility from fairly poor (the juxtaposition of Kalahari Sand with the Chobe riparian zone) to excessively eutrophic (the Amboseli basin). In all cases an extreme conversion of savanna structure occurred, associated with local elephant densities ranging between 2 and 4 animals.km^{-2}. The severe effect was limited in its extent to areas between approximately 100 km^2 along the Chobe River front and 4 400 km^2 in Tsavo East, and exacerbated in all cases by other factors compressing the elephant population within this area. The consequences for biodiversity as assessed through changes in habitat composition or species representations have not been quantified. A reduction in plant species diversity must surely have occurred locally, but not necessarily regionally. In some cases the vegetation showed its capacity to recover rapidly once the elephant pressure was reduced substantially; no irreversible threshold was passed, and the recovery time seemed to be merely 2–3 decades. Changes in animal populations appeared to be mostly relatively minor or locally restricted.

The transformation of savanna woodland into open grassland appears most typically as a feature of clayey soils where dense grass cover promotes hot fires (Bell, 1981). Examples include the Rwindi-Rutshuru plains in Kivu National

Park, Congo (Bourliere, 1965), and Maasai Mara Reserve in Kenya (Dublin *et al.*, 1990). On sandy soils allowing deeper water infiltration, many tree and shrub species have the capacity to resprout strongly from underground parts, so that the destruction of canopy trees by elephants leads to the development of a shrub coppice state (Bell, 1984, 1985; McShane, 1989). Examples of this include the Sengwa Research Area (Guy, 1989; Mapaure & Campbell, 2002) and Chizarira National Park (Thompson, 1975) and elsewhere in Zimbabwe (Holdo, 2006), as well as sections of Murchison Falls National Park and the Chobe River region, as described above. A similar conversion to a hedged or shrub coppice state has been documented for mopane woodlands, despite their prevalence on clay soils (Lewis, 1991). In South Africa the contrasts in woodland change between eastern and western regions of Kruger are consistent with this pattern (Eckhardt *et al.*, 2000). Thus, on the eastern basalts a substantial opening of the tree canopy has occurred, while on the western granites the overall woody plant cover did not change although the presence of tall trees decreased.

The studies outlined above have described general features of the consequences of elephant impacts for vegetation structure and composition, for the regions or ecosystems concerned, but some caveats should be noted. All areas show high spatial variability in these impacts as well as temporal variability. It is easy photographically to contrast local devastation with intact woodlands remaining nearby. The causes of this intense localised damage remain unknown, although Chafota's (2007) observations on interactions involving fire, frost, and the persistence of surface water shed some light on possible mechanisms. It is possible that the former pattern was a mosaic cycle of intense utilisation, with elephants moving elsewhere until areas previously heavily impacted had recovered. The extent of the area required for such a spatial pattern of utilisation to be maintained is unknown. Movement studies have merely documented opportunistic concentrations in areas where rainfall has promoted new growth, plus dry season concentrations around remaining sources of water for drinking. Tree populations within semi-arid environments seem also to recruit episodically at long intervals, during rare sequences of years with high rainfall, low fire frequency and low browsing impacts (Young & Lindsay, 1988; Walker 1989).

SPATIAL AND TEMPORAL PERSPECTIVES

The disturbing impacts on vegetation imposed by elephants are not only greater in magnitude than those due to other large herbivore species, but also extend over broader areas. The time taken by canopy tree populations to recover is also correspondingly longer than that for grasses and other herbaceous plants. While it has been proposed that 'intermediate' disturbances are associated with the highest species diversity (Connell, 1978), defining what is intermediate is problematic. It is not only the magnitude of the effect that is important, but also its spatial extent and frequency. Severe plant mortality imposed over the whole extent of a protected area and sustained for longer than the persistence of seed or seedling banks would obviously be disastrous. On the other hand, clearing of the existing vegetation from some areas by elephants potentially opens opportunities for plant species poorly represented elsewhere to colonise, potentially enhancing overall species diversity, but only if these plants are allowed sufficient time to establish. These concepts have not been rigorously applied to the elephant–vegetation interaction.

Temporal perspectives of elephant impact are generally poorly studied. Impacts over time will have two components – that of seasonal/interannual variation in impacts, and that of the actual rate of impacts. Given the seasonal variation in grass availability, and hence diet, it is predicted that elephant impacts on woody vegetation will be higher during winter than summer. This was confirmed for Madikwe (Govender, 2005). Similarly, elephant browsing intensity on mopane is greatest after spring rains (Styles & Skinner, 2000). On an interannual scale, the Tsavo elephant impacts saga is strongly linked to drought conditions (e.g. Leuthold, 1977). In contrast, the lack of seasonal variation in diet composition (and hence presumably impacts) in Addo (Davis, 2007) reflects the evergreen nature of subtropical thicket.

Rates of change are similarly poorly studied, the best documented being for Addo. The elephant enclosure of Addo was enlarged on a number of occasions, providing areas with different periods of elephant occupancy. Using these variations in elephant density and time since exposure to elephants, Barratt & Hall-Martin (1991) showed changes in plant architecture, Lombard *et al.* (2001) showed changes in the regionally rare and endemic small succulent shrubs and geophytes, and Magobiyane (2006) estimated the rate of impact on mistletoes. These studies in subtropical thicket show that some species respond very rapidly to elephant impacts.

In the Sengwe Wildife Research Area, Zimbabwe, annual loss of trees in the height class >5.0 m varied from 3 per cent for *Brachystegia speciformis* to 100 per cent for *Diplorynchus condylocarpon*, in the presence of elephants (Martin *et al.*, 1996), and on average tree loss rates were in the region of 22 per cent across species within Mopane and Miombo woodland (Martin *et al.*, 1996). In the Matusadona Highlands, Zimbabwe, tree loss rates of 21 per cent occurred even at low (<1 elephant.km^{-2}) densities. Modelling of tree loss and recruitment as a function of elephant density shows that very low elephant densities (0.1–0.5 elephant.km^{-2}) are required to achieve equilibrium between tree loss and recruitment (Martin *et al.*, 1996).

Spatial perspectives are better understood. Elephants, like other animals, do not use the landscape in a uniform fashion and hence vary their impacts across landscapes, producing heterogeneity in biodiversity patterns. One of the major factors influencing space use by elephants is topographical relief, and Wall *et al.* (2006) showed that elephants are reluctant to climb slopes. This is expressed in reduced elephant impact in relation to topographic relief (figure 2). The consequences are that tree species that seem most susceptible to elephant impacts, such as marula and baobab, tend to be prevalent in upland regions of the landscape (Weyerhaeuser, 1985; Edkins *et al.*, 2007).

Although Tsavo East National Park is commonly advanced as an example of the devastation potentially brought about by elephants, less than a quarter of its 20 000 km^2 extent was severely affected. Furthermore, this was largely in the lowest rainfall region, where the effect of drought conditions was most severe (Myers, 1973). Likewise, the zone of severe impact on riparian vegetation along the Chobe River spans less than 20 km (Skarpe *et al.*, 2004).

Elephants use vegetation types differently (e.g. Guldemond & Van Aarde, 2007). Despite their reliance on grass in the diet, there is a poor understanding of their use of grasslands, with most studies comparing woodland types, largely in terms of impacts. In Madikwe (Govender, 2005) and Pilanesberg (Moolman, 2007), elephants impacted *Acacia* woodland types significantly more than *Combretum* woodland types. In Phinda, two of the top three impacted habitats were *Acacia* dominated (the other was threatened sand forest), while in Mkhuze one of the top three impacted habitats was *Acacia* dominated (Repton, 2007). Further, some tree species were heavily used at some sites, but the same species was not heavily used at other sites (e.g. Madikwe – Page & Slotow, 2001; Pilanesberg – Moolman, 2007). In the Eastern Cape, elephants avoided karroid shrublands in Kwandwe (Roux, 2006).

The above patterns show that refugia from elephant impacts occur at a variety of spatial and possibly temporal scales, and these patterns need to be better understood. There are two further important aspects of such heterogeneous spatial patterns, firstly where elephants impact the areas around water (see Piosphere effects below), and secondly where their impacts are confined within small areas (see below).

Piosphere effects

Particularly relevant within this context is the abundance and spatial distribution of perennial surface water sources. Being water-dependent, elephants generally drink every 1–2 days (Owen-Smith, 1988), and typically forage up to about 16 km from water, although this extends up to 60 km in extreme cases (Laws, 1970; Western, 1975; Leggett *et al.*, 2003, 2004, 2006a & b). Accordingly, they concentrate near rivers or other sources of drinking water during the dry season, and disperse through a wider area during the wet season when pools are more widely distributed (Western, 1975; Thrash *et al.*, 1995; Owen-Smith, 1996; Leggett *et al.*, 2003, 2004, 2006a & b; Chamaillé-Jammes *et al.*, 2007; Smit *et al.*, 2007). The dry season concentration of elephants near surface water contributes to a gradient of intensifying impacts on vegetation, termed a piosphere, with the sacrifice zone in close proximity to the water source (Andrew, 1988). This region shows increases in soil nutrients, dung deposition, and trampling, decreases in trees and palatable perennial herbs, and increases in annual and unpalatable herbs and the amount of bare ground, soil compaction, and increased erosion (Bax & Sheldrick, 1963; Van Wyk & Fairall, 1969; Weir, 1971; Tolsma *et al.*, 1987; Thrash *et al.*, 1991, 1995; Ben-Shahar, 1993; Belsky, 1995; Owen-Smith, 1996; Thrash, 1998; James *et al.*, 1999). Piospheres may become especially intense around point sources of water such as those provided by boreholes, feeding troughs or artificial pools (Ben-Shahar, 1993; Conybeare, 1991; Owen-Smith, 1996). The availability and distribution of water sources can influence ecosystem structure and function at a range of scales and organisational levels, through its influence on various processes and feedbacks affecting both animals and plants (Gaylard *et al.* 2003; De Beer *et al.*, 2006)

Piospheres are manifested in woody vegetation primarily through changes to local structural heterogeneity by elephant browsing. Documented effects of elephants include a decrease in the density of *C. mopane* shrubs within 100–200 m of borehole sources (Fruhauf, 1997), and declines in plant species composition, density and diversity in areas close to pumped pans (Conybeare, 1991). With close spacing of water points, the regions severely affected tend to coalesce, restricting the opportunity for vegetation to recover when elephants move away, since their presence becomes effectively year-round (Owen-Smith, 1996). Waterpoints established in upland areas of the landscape may be especially detrimental, because tree species, such as marula and baobab, prevalent in these regions appear to have less capacity to recover from elephant damage (Weyerhaeuser, 1985; Edkins *et al.*, 2007). On the other hand, trees growing along river margins have a substantial capacity to recover from floods, let alone elephant damage (Rountree *et al.*, 2000; Rogers & O'Keefe, 2003).

Episodic severe damage and patch dynamics

Much of the extreme damage by elephants to canopy trees will be imposed during restricted periods when elephants experience an acute shortage of food. In northern Botswana, three documented instances related to events associated with fire, frost and extended lack of rainfall (Chafota, 2007). In one instance elephants moved 40 km away from the Chobe River following the first spring rains, to encounter an area that had recently been burnt. In the absence of much accessible forage, over 25 per cent of trees exceeding 10 cm in basal diameter were felled within a brief period, largely by female elephants. In a second case, severe frost eliminated much of the accessible browse in the Kazuma Forest Reserve. Within a few weeks, over 50 per cent of *Brachystegia africana* and *B. boehmii* trees had been felled by a group of bull elephants frequenting this region. In the third instance, early cessation of the summer rains led to greater damage by elephants to mopane and riparian woodland trees near the Linyanti River.

These instances of severe mortality of canopy trees imposed within a limited area over a restricted period could lead to the development of a mosaic interspersion of patches at different stages of recovery. The generation of such patch dynamics through wind-throw has been recognised as contributing to the dynamics of temperate woodlands (Pickett & White, 1985), but explored little for savanna woodlands. The potential consequences of such heterogeneity in vegetation structure and composition will be considered below.

In savanna woodlands, opportunities for successful tree seedling establishment may occur at long intervals when conditions of high rainfall, low fire incidence and low browser pressure are experienced (Young & Lindsay, 1988). Dense stands of regenerating *Faidherbia albida* trees developed on islands and sandbanks in the Zambezi River in 1985 due to some unidentified circumstances, despite an abundance of elephants and other large herbivores (Dunham, 1994). The development of the riparian woodland along the Chobe River has been ascribed to the low abundance of browsing ungulates following the rinderpest epizootic towards the end of the nineteenth century coupled with the elimination of elephants by hunters (Walker, 1989). The rarity of conditions enabling tree seedling recruitment will slow the recovery of woodlands after elephants are removed or reduced in abundance.

Impacts in confined areas/small reserves

Although it may be expected that elephants will utilise confined areas in a uniform fashion, there are limited data to support this. Roux (2006) showed that for smaller reserves (<1 000 km^2) range size was a function of reserve size, but not for larger systems, suggesting that smaller reserves would be used more comprehensively. Nevertheless, even within the Ithala Game Reserve (300 km^2), about 50 per cent of the reserve is not used by elephants because of topography, habitat and behaviour (Wiseman *et al.*, 2004). Within the Songimvelo Reserve (310 km^2), elephants use only a 120 km^2 section at an effective local density of 2.75 elephant.km^{-2} (Steyn & Stalmans, 2001). Elephants are restricted to the eastern half of Pongola Game Reserve by a railway track bisecting the reserve (Shannon *et al.*, 2006a), and hence have an effective density of 1 elephant.km^{-2}. Similarly, although the entire Phinda Reserve (150 km^2) is used by elephants, not all parts are used with the same intensity (Druce *et al.*, 2006). These patterns may in part be due to the relatively short periods that elephants have been confined in some small areas, as well as variations in density within reserves. In contrast, elephants have been confined to Addo for over 50 years, and despite the addition of new areas (growing from 27 to 120 km^2), a clear pattern of homogeneous impacts (i.e. decline in plant richness, taking period of occupation into account) can be seen (e.g. Barratt & Hall-Martin, 1991; Lombard *et al.*, 2001; Magobiyane, 2006).

MEGAHERBIVORE RELEASE

The absence of elephants will bring about changes to ecosystems (e.g. Kerley & Landman, 2005), which is known as megaherbivore release. This complicates the interpretation of elephant impacts, as it has been argued that where elephants have been reintroduced into an area, observed changes are a return to the situation prior to elephant removal (c.f. Conybeare, 2004). Kamineth (2004) showed that in the absence of megaherbivores (including areas with historical megaherbivore records) tree *Euphorbia* populations were dominated by younger plants (<100 years), with few adults (i.e. recruiting populations). In the presence of megaherbivores (historical and current), however, *Euphorbia* populations were characterised by individuals in younger and intermediate (100–150 years) age classes (i.e. irregular age distributions). No recruiting populations were observed in the presence of megaherbivores. Thus, the presence of megaherbivores has resulted in a high incidence of adult tree *Euphorbia* mortality, and may have controlled tree numbers. This suggests that the local abundance of tree *Euphorbias* is an artefact of relaxation from browsing or other effects provided by megaherbivores.

Skarpe *et al.* (2004) also suggest that the large populations of *Acacia* and *Faidherbia* in the Chobe area were established during periods of low herbivore biomass. The mechanisms of megaherbivore release extend beyond direct herbivory, as the absence of elephants will influence a number of ecological processes (Kerley & Landman, 2005).

CONSTRAINTS TO IDENTIFYING ELEPHANT EFFECTS

The interpretation of elephant impacts is rarely possible to do in isolation of possible confounding or synergistic effects such as fire (Trollope *et al.*, 1998; Bond & Keeley, 2005; Chafota, 2007), other herbivores (Cowling & Kerley, 2002; Skarpe *et al.*, 2004), drought (Wiseman *et al.*, 2004), wind toppling (Bell, 1985), soil chemistry and water table (Western & Van Praet, 1973; Mills, 2006), and frost (Holdo, 2007). Specifically, in Ithala, other browsers (black rhino = 13 per cent of individuals; other browsers about 30 per cent) had almost a three-fold higher effect on woody vegetation than did elephants (16 per cent). Of the top 20 plant species by canopy removed, 12 were more heavily impacted by other browsers than by elephants (Wiseman *et al.*, 2004). Note, however, that these relative impacts are not expressed in relation to browser biomass.

There are few studies that show no changes or increases in species richness or numbers of particular species (i.e. so-called 'positive effects') in response to elephants. It is, therefore, not clear as to how much our understanding of elephant impacts is biased by the possible under-reporting of such effects.

Furthermore, the studies on confined populations are complicated by the inability to control for elephant density, as opposed to elephant presence (Cowling & Kerley, 2002). Benchmarking elephant impacts is also complicated by the absence of a 'natural state' yardstick, as well as the consequences of megaherbivore release (see above). The measurement and interpretation of elephant impacts, therefore, needs to be undertaken in a rigorous fashion such that confounding effects are controlled for (Cowling & Kerley, 2002). A useful approach is to quantify impacts on a gradient of elephant density or period of occupation (Barratt & Hall-Martin, 1991). The interpretation of impacts should be based on a sound understanding of the mechanisms of such putative impacts in order to avoid the risk of incorrectly assigning impacts to elephants (Landman *et al.*, 2008).

Given the longevity of elephants, the scales at which they use landscapes, as well as the temporal and spatial scales of responses of ecosystems affected by elephants, there are further constraints on our understanding of elephant-ecosystem interactions. Thus, the typical study (~1 year) on elephant impacts is of too short a duration (Kerley & Shrader, 2007), or too spatially restricted (e.g. Cumming *et al.*, 1997) to provide a real view of the effects. It can be predicted that elephant effects will be further confounded by the effects of climate change (Kerley & Landman, 2006), and this should be borne in mind when designing elephant-effect studies.

CONCEPTUAL/MODELLING FRAMEWORK FOR CONTEXTUALISING ELEPHANT EFFECTS

Caughley's (1976b) model describing the eruptive dynamics of a herbivore population introduced into a new environment, developing through the interaction with vegetation, has been highly influential in guiding thinking about possible long-term trajectories of elephant numbers and vegetation. The fundamental feature underlying these dynamics is the delay in the response of the vegetation to increasing levels of consumption by the herbivores. In suggesting the possible relevance of this model for elephant dynamics, Caughley (1976a) emphasised how the delayed recovery of woodlands following their depression by elephants, coupled with the delayed response of elephants to the woodland reduction (because of their capacity to use grass as an alternative food source), could lead to reciprocal cycling in abundance with a period of around 100 years. Duffey *et al.* (1999) suggested that more realistic parameter values for elephants could lead to stability rather than cycling, but incorporated a stabilising density feedback by basing the functional response on a consumer-resource ratio rather than simply resource abundance.

However, neither of these models accommodates heterogeneity in vegetation structure or composition, or temporal variability in conditions, not even the seasonal cycle of production and decay by plants, nor do they address biodiversity *per se*. Owen-Smith (2002a) demonstrated that effective functional heterogeneity in vegetation quality, coupled with adaptive resource selection by herbivores, could promote stability rather than cycling, and suggested that this finding might have some relevance for the dynamics of elephants and woodlands in terms of achieving a stable state (Owen-Smith, 2002b). However, of most relevance is the potential recovery rate of tree populations.

Baxter & Getz's (2005) model provides a foundation for contextualising the relative effects of elephants, fire and climatic variability on likely trends. This represented a 1 km² cell with woody plant growth dynamics parameters specifically based on mopane, with a relatively simple age structure of the

Box 1: Research needs

There is an urgent need to study the effect of elephants on biodiversity, specifically those aspects which are considered critical for ecosystem integrity (e.g. species level effects), or which are featured in the management objectives for specific protected areas (e.g. landscape level effects such as presence of large trees), as a function of elephant density. The observation that such impacts are often scale- and site-specific or episodic requires that this be undertaken at a range of spatial and temporal scales and at different sites varying in climate and soil features. Sampling should be designed to detect episodic effects.

The rate of change brought about by elephants as a function of elephant density is key to managing biodiversity in elephant areas, and this needs to be specifically quantified. Of value here may be the areas to which elephants have recently been reintroduced.

The mechanisms of elephant impacts need to be more clearly researched, in order to predict the consequences of increased elephant density and to ensure that management responses are appropriate. This is particularly important since interactions with other ecosystem drivers (fire, drought, other herbivores, disease) may be confounding.

It has been shown that different habitats respond differently to elephant impacts and it may be hypothesised that elephant impacts are greater in habitats where they are resource limited. Research is needed to quantify elephant resource requirements and to establish how these may be provided in different habitats in order to guide the introduction of elephants into new locations and predict risks to biodiversity and identify spatial and temporal refuges from elephant impacts.

The response of biodiversity to management interventions to reduce elephant impacts (fencing, habitat expansion, etc) is key to assessing the effects of such interventions. Research is needed to provide evidence for the success or failure of such interventions.

The effects of the absence of elephants (megaherbivore release) need to be further researched, as across South Africa elephants are no longer a functional part of most ecosystems which may be dependent on the process provided by elephants.

elephant population. They suggested that a decline in woody vegetation might occur once effective local elephant densities exceeded 1–2 elephant.km^{-2}. This model needs to be expanded to take into account other woody species with different growth characteristics, as well as seasonal and spatial variation in the local presence of elephants, and a better elephant population model. A model developed by Holdo (2007), specifically for miombo woodland, indicated a likely decline in woody vegetation with elephant densities of around 2 elephant.km^{-2}.

ASSESSMENT

1. That elephants at high densities are having an impact on plant communities, with consequent changes in vegetation structure and species composition, is undeniable. However, such changes vary in extent, rate and severity between ecosystems. There is currently no recommended density for elephants to manage such changes, and the desirability of such changes will depend on the management objectives.

2. Some plant species can cope with elephant browsing, stripping or toppling, although this varies substantially with circumstance (e.g. xeric vs. mesic savannas). Therefore, aside from a number of instances where local extirpation has occurred, the most significant impact that elephants will have is the changing of vegetation structure.

3. There are very few data on rates of change in response to elephants. This will be a function of the density of elephants, the availability of alternative resources and the nature (e.g. life history) of the component of biodiversity of interest, as well as other ecosystem drivers that are involved.

4. It is difficult to untangle the effects of elephants and confounding factors such as fire, natural plant senescence and episodic recruitment events (e.g. Skarpe *et al.*, 2004). These levels of interactions will be exacerbated by climate change.

5. Many plant populations will recover once the pressure of high elephant densities has been released; however, these rates will vary between species and landscapes and the extent of change; animal populations will respond faster, unless they are dependent on the habitat provided by the plants.

6. While extensive data are available from elsewhere in Africa, a paucity of data of elephant effects exists in South Africa. The most comprehensive data are for Addo, with limited information in northern KwaZulu-Natal and Kruger.

CONCLUDING COMMENTS

We conclude that elephants are special in the nature of their feeding, and hence their impacts, by virtue of features such as body size, the trunk and tusks. Overall, our Assessment is that while the impacts of high elephant concentrations may bring about local changes in vegetation and associated animal species, and hence local biodiversity, this need not be the case at the wider ecosystem level. Moreover, unless extreme, the consumption and breakage of woody plants and uprooting of grass tufts by elephants promotes compensatory regeneration and hence probably enhanced ecosystem productivity, as has been demonstrated for grazing systems. The concern is not the local severity of elephant impacts, which could be adverse for both productivity and diversity if extreme, but rather the persistence and extent of such pressure on plants, and the cascading or knock-on effects of elephants on other elements of biodiversity.

Transformation brought about by elephants is restricted in extent by the spatial dispersion of natural perennial surface water, where such dispersion is greater than the average daily foraging distance of elephants (c. 16 km). This is altered by the extent to which water is augmented by dams and boreholes. Elephant feeding on woody plants and grasses can facilitate feeding by other large herbivore species. Adverse consequences for these species arise through habitat transformations rather than direct competition. Prior to the large-scale changes in elephant abundance and distribution, it was recognised that elephants impacted landscapes (Selous, 1881), but unfortunately there are no benchmarks of elephant-landscape interactions in the absence of humans. This is further complicated by the recognition that elephant impacts varied in space and time. Defining the severity of impacts, and hence managing impacts, therefore will depend on management objectives for a particular system.

ENDNOTE

1. For ease of reference, we have retained the genus *Acacia*, but note that the nomenclature is under revision.

REFERENCES

Addy, J.E. 1993. Impact of elephant induced vegetation change on the Chobe bushbuck (*Tragelaphus scriptus ornatus*) along the Chobe River, northern Botswana. MSc thesis, University of the Witwatersrand, South Africa.

Agnew, A.D.Q. 1968. Observations on the changing vegetation of Tsavo National Park (East). *East African Wildlife Journal* 6, 75–80.

Anderson, G.D. & B.H. Walker 1974. Vegetation composition and elephant damage in the Sengwa Wildlife Research Area, Zimbabwe. *Journal of South African Wildlife Management Association* 4, 1–14.

Andrew, M.H. 1988. Grazing impact in relation to livestock watering points. *Trends in Ecology and Evolution* 3, 336–339.

Babweteera, F., P. Savill & N. Brown 2007. *Balanites wilsoniana*: Regeneration with and without elephants. *Biological Conservation* 134, 40–47.

Balfour, D.A. 2005. *Acacia* demography, fire and elephants in a South African savanna. Ph.D. thesis, University of Cape Town, South Africa.

Barnes, M.E. 2001. Effects of large herbivores and fire on the regeneration of *Acacia erioloba* woodlands in Chobe National Park, Botswana. *African Journal of Ecology* 39, 340–350.

Barnes, R.F.W. 1982. Elephant feeding behaviour in Ruaha National Park, Tanzania. *African Journal of Ecology* 20, 123–136.

Barnes, R.F.W. 1985. Woodland changes in Ruaha National Park (Tanzania) between 1976 and 1982. *African Journal of Ecology* 23, 215–222.

Barnes, R.F.W., K.L. Barnes & E.B. Kapela 1994. The long term impact of elephant browsing on baobab trees at Msembe, Ruaha National Park, Tanzania. *African Journal of Ecology* 32, 177–184.

Barratt, D.G. & A.J. Hall-Martin 1991. The effects of indigenous herbivores on Valley Bushveld in the Addo Elephant National Park. In: P.J.K. Zacharias & G.C. Stuart-Hill (eds) *Proceedings of the First Valley Bushveld/Subtropical Thicket Symposium*. Grassland Society of Southern Africa, Howick, 14–16.

Bax, P.N. & D.L.W. Sheldrick 1963. Some preliminary observations on the food of elephant in the Tsavo Royal National Park (East) of Kenya. *East African Wildlife Journal* 1, 40–53.

Baxter, P.W.J. & W.M. Getz 2005. A model-framed evaluation of elephant effects on tree and fire dynamics in African savannas. *Ecological Applications* 15, 1331–1241.

Bell, R.H.V. 1981. Outline of a management plan for Kasungu National Park, Malawi. In: P.A. Jewell, S. Holt & D. Hart (eds) *Problems in the management of locally abundant wild mammals*. Academic Press, New York, 69–89.

Bell, R.H.V. 1984. Notes on elephant-woodland interactions. In: D.H.M. Cumming & P. Jackson (eds) *The status and conservation of Africa's elephants and rhinos,* Proceedings of the joint meeting of IUCN/SSC African elephant and African rhino specialist groups, 1981, Zimbabwe. IUCN, Gland, 98-103.

Bell, R.H.V. 1985. Elephants and woodland – a reply. *Pachyderm* 5, 17-18.

Belsky, J. 1995. Spatial and temporal patterns in arid and semi-arid savannas. In: L. Hansson, L. Fahrig & G. Merriam (eds) *Mosaic landscapes and ecological processes.* Chapman and Hall, London, 31-56.

Ben-Shahar, R. 1993. Patterns of elephant damage to vegetation in northern Botswana. *Biological Conservation* 65, 249-256.

Bond, W.J. & J.E. Keeley 2005. Fire as a global 'herbivore': the ecology and evolution of flammable ecosystems. *Trends in Ecology and Evolution* 20, 387-394.

Boshoff, A.F., J. Skead & G.I.H. Kerley 2002. Elephants in the broader Eastern Cape – a historical overview. In G.I.H. Kerley, S. Wilson & A. Massey (eds) *Proceedings of a workshop on Elephant Conservation and Management in the Eastern Cape,* Terrestrial Ecology Research Unit Report 35, University of Port Elizabeth, South Africa, 3-15.

Botes, A., M.A. McGeoch & B.J. van Rensburg 2006. Elephant- and human-induced changes to dung beetle (Coleoptera: Scarabaeidae) assemblages in the Maputaland Centre of Endemism. *Biological Conservation* 130, 573-583.

Botkin, D.B., J.M. Mellilo & L.S-Y. Wu 1981. How ecosystem processes are linked to large mammal population dynamics. In C.W. Fowler & T.D. Smith (eds) *Dynamics of large mammal populations.* John Wiley, New York, 373-388.

Bourliere, F. 1965. Densities and biomasses of some ungulate populations in eastern Congo and Ruanda, with notes on population structure and lion/ungulate ratios. *Zoological Africana* 1, 199-207.

Brown, J.H. & E.J. Heske 1990. Control of a desert-grassland transition by a keystone rodent guild. *Science* 250, 1705-1708.

Buechner, H.K. & H.C. Dawkins 1961. Vegetation change induced by elephants and fire in Murchison Falls National Park, Uganda. *Ecology* 42, 752-766.

Buss, I.O. 1961. Some observations on food habits and behavior of the African elephant. *Journal of Wildlife Management* 25, 131-148.

Calenge, C., D. Maillard, J.-M. Gaillard, L. Merlot & R. Peltier 2002. Elephant damage to trees of wooded savanna in Zakouma National Park, Chad. *Journal of Tropical Ecology* 18, 599-614.

Castley, J.G. & M.H. Knight 1997. *Helicopter-based surveys of Addo Elephant National Park: March 1996 and February 1997*. South African National Parks, Scientific Services Report, Kimberley, South Africa.

Caughley, G. 1976a. The elephant problem – an alternative hypothesis. *East African Wildlife Journal* 14, 265–283.

Caughley, G. 1976b. Plant-herbivore systems. In R.M. May (ed.). *Theoretical Ecology*. Blackwell Scientific Publications, Oxford, 94–113.

Cerling, T.E., J.M. Harris & M.G. Leakey 1999. Browsing and grazing in elephants: the isotope record of modern and fossil Proboscideans. *Oecologia* 120, 364–374.

Chabie, B.P.M. 1999. Avifauna community turnover between transformed and untransformed Thicket vegetation. Honours thesis, University of Port Elizabeth, South Africa.

Chafota, J. 2007. Factors governing selective impacts of elephant on woodland. Ph.D. thesis, University of the Witwatersrand, South Africa.

Chamaillé-Jammes, S., M. Valeix & H. Fritz 2007. Managing heterogeneity in elephant distribution: interactions between elephant population density and surface-water availability. *Journal of Applied Ecology* 44, 625–633.

Chapman, L.J., C.A. Chapman & R.W. Wrangham. 1992. *Balanites wilsoniana* – elephant dependent dispersal. *Journal of Tropical Ecology* 8, 275–283.

Clauss, M., W.J. Streich, A. Schwarm, S. Ortmann & J. Hummel 2007. The relationship of food intake and ingesta passage predicts feeding ecology in two different megaherbivore groups. *Oikos* 116, 209–216.

Cobb, S. 1976. The distribution and abundance of the large herbivore community of Tsavo National Park, Kenya. D.Phil thesis, University of Oxford, Oxford.

Cochrane, E.P. 2003. The need to be eaten: *Balanites wilsoniana* with and without elephant seed-dispersal. *Journal of Tropical Ecology* 19, 579–589.

Codron, J., J.A. Lee-Thorp, M. Sponheimer, D. Codron, C.C. Grant & D.J. de Ruiter 2006. Elephant (*Loxodonta africana*) diets in Kruger National Park, South Africa: spatial and landscape differences. *Journal of Mammalogy* 87, 27–34.

Coetzee, B.J., A.H. Engelbrecht, S.C.J. Joubert & P.F. Retief 1979. Elephant impact on *Sclerocarya caffra* trees in *Acacia nigrescens* tropical plains thornveld of the Kruger National Park. *Koedoe* 22, 39–60.

Connell, J.H. 1978. Diversity in tropical rain forests and coral reefs. *Science* 199, 302–1310.

Conybeare, A.M. 1991. Elephant occupancy and vegetation change in relation to artificial water points in a Kalahari sand area of Hwange National Park. DPhil thesis, University of Zimbabwe, Harare.

Conybeare, A.M. 2004. Elephant impacts on vegetation and other biodiversity in the broadleaved woodlands of south-central Africa. In J.R. Timberlake & S.L. Childes (eds) *Biodiversity of the Four Corners Area: Technical Reviews*, Volume Two (Chapters 5-15), Occasional Publications in Biodiversity No 15, Biodiversity Foundation for Africa, Bulawayo/Zambezi Society, Harare, Zimbabwe, 1-35.

Corfield, T.F. 1973. Elephant mortality in Tsavo National Park, Kenya. *East African Wildlife Journal* 11, 339-368.

Côté, S.D., T.P. Rooney, J-P. Tremplay, C. Dussault & D.M. Waller 2004. Ecological impacts of deer overabundance. *Annual Review of Ecology and Systematics* 35, 113-147.

Coulson, I.M. 1992. *Linyanti/Kwando vegetation survey: March 1992*. Report to Kalahari Conservation Society, Botswana.

Cowling, R. & G.I.H. Kerley 2002. Impacts of elephants on the flora and vegetation of subtropical thicket in the Eastern Cape. In G.I.H. Kerley, S. Wilson & A. Massey (eds) *Proceedings of a workshop on Elephant Conservation and Management in the Eastern Cape*, Terrestrial Ecology Research Unit Report 35, University of Port Elizabeth, South Africa, 55-72.

Croze, H. 1974. The Seronera bull problem. I. The elephants. *East African Wildlife Journal* 12, 1-27.

Cumming, D.H.M., M.B. Fenton, I.L. Rautenbach, R.D. Taylor, G.S. Cumming, M.S. Cumming, J.M. Dunlop, G.A. Ford, M.D. Hovorka, D.S. Johnston, M. Kalcounis, Z. Mahlangu & C.V.R. Portfors 1997. Elephants, woodlands and biodiversity in southern Africa. *South African Journal of Science* 93, 231-236.

Davis, S. 2007. Endozoochory in Subtropical Thicket: comparing effects of species with different digestive systems on seed fate. MSc thesis, Nelson Mandela Metropolitan University, South Africa.

De Beer, Y., W. Kilian, W. Versfeld & R.J. van Aarde 2006. Elephants and low rainfall alter woody vegetation in Etosha National Park, Namibia. *Journal of Arid Environments* 64, 412-421.

De Boer, W.F. & H.H.T. Prins 1990. Large herbivores that strive mightily but eat and drink as friends. *Oecologia* 82, 264-274.

De Boer, W.F., C. Ntumi, C. Correia & J. Mafuca 2000. Diet and distribution of elephant in the Maputo Elephant Reserve, Mozambique. *African Journal of Ecology* 38, 188-201.

De Villiers, M.S. & R.J. van Aarde 1994. Aspects of habitat disturbance by Cape porcupine in a savanna ecosystem. *South African Journal of Zoology* 29, 217-220.

Dougall, H.W. 1963. On the chemical composition of elephant faeces. *East African Wildlife Journal* 1, 123.

Dougall, H.W. & D.L.W. Sheldrick 1964. The chemical composition of a days' diet of an elephant. *East African Wildlife Journal* 2, 51–59.

Druce, H., K. Pretorius, D. Druce & R. Slotow 2006. The effect of mature elephant bull introductions on the spatial ecology of resident bull's: Phinda Private Game Reserve, South Africa. *Koedoe* 49, 77–84.

Dublin, H.T. 1996. Elephants of the Masai Mara, Kenya: seasonal habitat selection and group size patterns. *Pachyderm* 22, 25–35.

Dublin, H.T., A.R.E. Sinclair & J. McGlade 1990. Elephants and fire as causes of multiple stable states in the Serengeti-Mara woodlands. *Journal of Animal Ecology* 59, 1147–1164.

Duffy, K.J., B.R. Page, J.H. Swart & V.B. Bajic 1999. Realistic parameter assessment for a well known elephant-tree ecosystem model reveals that limit cycles are unlikely. *Ecological Modeling* 121, 115–125.

Dunham, K.M. 1994. The effect of drought on the large mammal populations of Zambezi riverine woodlands. *Journal of Zoology, London* 234, 489–526.

Eckhardt, H.C., B.W. van Wilgen & H.C. Biggs 2000. Trends in woody vegetation cover in the Kruger National Park, South Africa, between 1940 and 1998. *African Journal of Ecology* 38, 108–115.

Edkins, M.T., L.M. Kruger, K. Harris & J.J. Midgley 2007. Baobabs and elephants in Kruger National Park: nowhere to hide. *African Journal of Ecology* 38, 108–115.

Fenton, M.B., D.H.M. Cumming, D.R. Taylor, L.I. Rautenbach, S.G. Cumming, S.M. Cumming, G. Ford, J. Dunlop, D.M. Hovorka, S.D. Johnston, V.C. Portfors, C.M. Kalcounis & Z. Mahlanga 1998. Bats and the loss of tree canopy in African woodlands. *Conservation Biology* 12, 399–407.

Field, C.R. 1971. Elephant ecology in the Queen Elizabeth National Park, Uganda. *East African Wildlife Journal* 9, 99–123.

Field, C.R. & I.C. Ross 1976. The Savanna ecology of Kidepo Valley National Park: II Feeding ecology of elephant and giraffe. *East African Wildlife Journal* 14, 1–15.

Fornara, D.A. & J.T. du Toit 2007. Browsing lawns? Responses of *Acacia nigrescens* to ungulate browsing in an African savanna. *Ecology* 88, 200–209.

Fritz, H., P. Duncan, I.J. Gordon & A.W. Illius 2002. The influence of megaherbivores on the abundance of the trophic guilds in African ungulate communities. *Oecologia* 131, 620–625.

Fruhauf, N. 1997. Pattern of elephant impacts around artificial waterpoints in the northern region of the Kruger National Park. BSc Honours thesis, University of the Witwatersrand, South Africa.

Gadd, M.E. 2002. The impact of elephants on the marula tree *Sclerocarya birrea*. *African Journal of Ecology* 40, 328–336.

Gaylard, A., N. Owen-Smith & J.V. Redfern 2003. Surface water availability: implications for heterogeneity and ecosystem processes. In J.T. du Toit, K.H. Rogers & H.C. Biggs (eds) *The Kruger Experience: Ecology and Management of Savanna Heterogeneity*. Island Press, Washington, DC, 171–188.

Glover, J. 1963. The elephant problem at Tsavo. *East African Wildlife Journal* 1, 30–39.

Gough, K.F. & G.I.H. Kerley 2006. Demography and population dynamics in the elephants *Loxodonta africana* of Addo Elephant National Park, South Africa: is there evidence of density-dependent regulation? *Oryx* 40, 434–441.

Govender, N. 2005. The effect of habitat alteration by elephants on invertebrate diversity in two small reserves in South Africa. MSc thesis, University of KwaZulu-Natal, Pietermaritzburg, South Africa.

Greyling, M.D. 2004. Sex and age related distinctions in the feeding ecology of the African elephant, *Loxodonta africana*. Ph.D. thesis, University of the Witwatersrand, South Africa.

Guldemond, R. & R.J. van Aarde 2007. The impacts of elephants on plants and their community variable in South Africa's Maputaland. *African Journal of Ecology* 45, 327–335.

Guy, P.R. 1976. The feeding behaviour of elephant (*Loxodonta africana*) in the Sengwa area, Rhodesia. *South African Journal of Wildlife Research* 6, 55–63.

Guy, P.R. 1981. Changes in the biomass and productivity of woodlands in the Sengwa Wildlife Research Area, Zimbabwe. *Journal of Applied Ecology* 18, 507–519.

Guy, P.G. 1989. The influence of elephants and fire on a *Brachystegia-Julbernardia* woodland in Zimbabwe. *Journal of Tropical Ecology* 5, 215–226.

Halley, D.J., C.L. Taolo, M.E.J. Vandewalle, M. Mari 2003. Buffalo research in Chobe National Park. In M. Vandewalle (ed.) *Effects of fire, elephants and other herbivores on the Chobe riverfront ecosystem*. Proceedings of a Conference Organised by The Botswana-Norway Institutional Co-Operation and Capacity Building Project (BONIC), Botswana Government Printer, Gaborone, 73–79.

Hatton, J.C. & N.O.E. Smart 1984. The effect of long-term exclusion of large herbivores on soil nutrient status in Murchison Falls National Park, Uganda. *African Journal of Ecology* 22, 23–30.

Hendrichs, H. 1971. Freilandbeobachtungen zum Sozialsystem des afrikanishen Elefanten, *Loxodonta africana* (Blumenbach, 1797). In H. & U. Hendrichs (eds) *Dikdik und Elefanten*. R. Piper, Munich, 77–173.

Herremans, M. 1995. Effects of woodland modification by African elephant *Loxodonta africana* on bird diversity in northern Botswana. *Ecography* 18, 440–454.

Hiscocks, K. 1999. The impact of increasing elephant population on the woody vegetation in southern Sabi Sand Wildtuin, South Africa. *Koedoe* 42, 47–55.

Holdo, R.M. 2003. Woody plant damage by African elephants in relation to leaf nutrients in western Zimbabwe. *Journal of Tropical Ecology* 19, 189–196.

Holdo, R.M. 2006. Elephant herbivory, frost damage, and topkill in Kalahari sand woodland savanna trees. *Journal of Vegetation Science* 17, 509–518.

Holdo, R.M. 2007. Elephants, fire, and frost can determine community structure and composition in Kalahari woodlands. *Ecological Applications* 17, 558–568.

Huntly, N. 1991. Herbivores and the dynamics of communities and ecosystems. *Annual Review of Ecology and Systematics* 22, 477–503.

Inamdar, A. 1996. The ecological consequences of elephant depletion. Unpublished Ph.D. thesis, University of Cambridge, Cambridge.

Ishwaran, N. 1983. Elephant and woody-plant relationships in Gal Oya, Sri Lanka. *Biological Conservation* 3, 225–270.

Jachmann, H. 1989. Food selection by elephants in the Miombo Biome, in relation to leaf chemistry. *Biochemical Systematics and Ecology* 17, 15–24.

Jachmann, H. & R.H.V. Bell 1985. Utilization by elephants of the *Brachystegia* woodlands of the Kasungu National park, Malawi. *African Journal of Ecology* 23, 245–258.

Jacobs, O.S. & R. Biggs 2002. The impact of the African elephant on marula trees in the Kruger National Park. *South African Journal of Wildlife Research* 32, 13–22.

James, C.D., J. Landsberg & S.R. Morton 1999. Provision of watering points in the Australian arid zone: a review of effects on biota. *Journal of Arid Environments* 41, 87–121.

Janis, C. 1976. The evolutionary strategy of the Equidae and the origins of rumen and cecal digestion. *Evolution* 30, 757–774.

Jarman, P.J. 1971. Diets of large mammals in the woodlands around Lake Kariba, Rhodesia. *Oecologia* 8, 157–178.

Johnson, C.F., R.M. Cowling & P.B. Phillipson 1999. The flora of the Addo Elephant National Park, South Africa: are threatened species vulnerable to elephant damage? *Biodiversity Conservation* 8, 1447–1456.

Kabigumila, J. 1993. Feeding habits of elephants in Ngorongoro Crater, Tanzania. *African Journal of Ecology* 31, 156–164.

Kalemera, M.C. 1989. Observations on feeding preference of elephants in the *Acacia tortilis* woodland of Lake Manyara National Park, Tanzania. *African Journal of Ecology* 27, 325–333.

Kamineth, A.I. 2004. The population dynamics and distribution of Tree Euphorbias in Thicket. MSc thesis, University of Port Elizabeth, South Africa.

Keesing, F. 2000. Cryptic consumers and the ecology of an African savanna. *Bioscience* 50, 205–215.

Kerley, G.I.H. & M. Landman 2005. Gardeners of the gods: the role of elephants in the Eastern Cape Subtropical Thickets. In C.C. Grant (ed.) *A compilation of contributions by the Scientific Community for SANParks, 2005. Elephant effects on biodiversity: an assessment of current knowledge and understanding as a basis for elephant management in SANParks.* Scientific Report 3/2005, South African National Parks, Scientific Services, Skukuza, 173–190.

Kerley, G.I.H. & M. Landman 2006. The impacts of elephants on biodiversity in the Eastern Cape Subtropical Thickets. *South African Journal of Science* 102, 395–402.

Kerley, G.I.H. & A. Shrader 2007. Elephant contraception: silver bullet or a potentially bitter pill. *South African Journal of Science* 103, 181–182.

Kerley, G.I.H., D. Tongway & J. Ludwig 1999. Effects of goat and elephant browsing on soil resources in Succulent Thicket, Eastern Cape, South Africa. *Proceedings of the VI International Rangeland Congress* 1, 116–117.

Kerley, G.I.H., W.G. Whitford & F.R. Kay 1997. Mechanisms for the keystone status of kangaroo rats: graminivory rather than granivory? *Oecologia* 111, 422–428.

Klein, B.C. 1989. Effects of forest fragmentation on dung and carrion beetle communities in central Amazonia. *Ecology* 70, 1715–1725.

Koch, P.L., J. Heinsinger, C. Moss, R.W. Carlson, M.L. Fogel & A.K. Behrensmeyer 1995. Isotopic tracking of change in diet and habitat use in African elephants. *Science* 267, 1340–1343.

Koch, P.L. & A.D. Barnovsky 2006. Late Quaternary extinctions: state of the debate. *Annual Review in Ecology, Evolution and Systematics* 37, 215–250.

Kruger, L.M., J.A. Coetzee & K. Vickers 2007. *The impacts of elephants on woodlands and associated biodiversity.* Summary report to South African National Parks, Organization of Tropical Studies.

Lagendijk, D.D.G., W.F. de Boer & S.E van Wieren 2005. Can African elephants (*Loxodonta africana*) survive and thrive in monostands of *Colophospermum mopane* woodlands? *Pachyderm* 39, 43–49.

Landman, M., G.I.H. Kerley & D.S. Schoeman 2008. Relevance of elephant herbivory as a threat to important plants in the Addo Elephant National Park, South Africa. *Journal of Zoology, London* 274, 51–58.

Langer, P. 1984. Anatomical & nutritional adaptations in wild herbivores. In: F.M.C. Gilchrist & R.I. Mackie (eds) *Herbivore nutrition in subtropics and tropics.* Science Press, Johannesburg.

Laws, R.M. 1970. Elephants as agents of habitat and landscape change in East Africa. *Oikos* 21, 1–15.

Laws, R.M. & I.S.C. Parker 1968. Recent studies on elephant populations in East Africa. *Symposium of the Zoological Society of London* 21, 319–359.

Laws, R.M., I.S.C. Parker & R.C.B. Johnstone 1975. *Elephants and their habitats: The ecology of elephants in North Bunyoro, Uganda.* Clarendon Press, Oxford.

Lee, P.C. & C. Moss 1995. Satural growth in known-age African elephants (*Loxodonta africana*). *Journal of Zoology, London* 236, 29–41.

Leggett, K.E.A. 2006a. Effect of artificial water points on the movement and behaviour of desert-dwelling elephants of north-western Namibia. *Pachyderm* 40, 24–34.

Leggett, K.E.A. 2006b. Home range and seasonal movement of elephants in the Kunene Region, northwestern Namibia. *African Zoology* 41, 17–36.

Leggett, K.E.A., J. Fennessy & S. Schneider 2003. Seasonal distributions and social dynamics of elephants in the Hoanib River catchment, northwestern Nambia. *African Zoology* 38, 305–316.

Leggett, K.E.A., J. Fennessy & S. Schneider 2004. A study of animal movement in the Hoanib River catchment, northwestern Nambia. *African Zoology* 39, 1–11.

Lessing, J. 2007. Elephant feeding behaviour and forage offtake implications in the Addo Elephant National Park. MSc thesis, Nelson Mandela Metropolitan University, South Africa.

Leuthold, W. 1977. Changes in tree populations of Tsavo East National Park, Kenya. *East African Wildlife Journal* 15, 61–69.

Leuthold, W. 1996. Recovery of woody vegetation in Tsavo National Park, Kenya, 1970-1994. *African Journal of Ecology* 34, 101–112.

Lewis, D.M. 1986. Disturbance effects on elephant feeding: evidence for compression in Luangwa Valley, Zambia. *African Journal of Ecology* 24, 227–241.

Lewis, D.M. 1987. Fruiting patterns, seed germination, and distribution of *Sclerocarya caffra* in an elephant-inhabited woodland. *Biotropica* 19, 50–56.

Lewis, D.M. 1991. Observations on tree growth, woodland structure and elephant damage on *Colophospermum mopane* in Luangwa Valley, Zambia. *African Journal of Ecology* 24, 227–241.

Lombard, A.T., C.F. Johnson, R.M. Cowling & R.L. Pressey 2001. Protecting plants from elephants: botanical reserve scenarios within the Addo Elephant National Park, South Africa. *Biological Conservation* 102, 191–203.

Ludwig, F. 2001. Tree-grass interaction on an East African savanna: the effects of competition, facilitation and hydraulic lift. Ph.D. thesis, Wageningen University, The Netherlands.

Luoga, E.J., E.T.F. Witkowski & K. Balkwill 2004. Regeneration by coppicing (resprouting) of Miombo (African savanna) trees in relation to land use. *Forest Ecology and Management* 189, 23–35.

MacArthur, R.H. 1964. Environmental factors affecting bird species diversity. *American Naturalist* 98, 387–397.

Magobiyane, B. 2006. The effects of elephants on mistletoes in the Eastern Cape. Honours thesis, Nelson Mandela Metropolitan University, South Africa.

Makhabu, S.W., C. Skarpe & H. Hytteborn 2006. Elephant impact on shoot distribution on trees and on rebrowsing by smaller herbivores. *Acta Oecologica* 30, 136–146.

Mapaure, I.N. & B.M. Campbell 2002. Changes in miombo woodland cover in and around Sengwa Wildlife Research Area, Zimbabwe, in relation to elephants and fire. *African Journal of Ecology* 40, 212–219.

Martin, R.B., G.C. Craig & V.R. Booth (eds) 1996. *Elephant management in Zimbabwe*, Third edition. Department of National Parks and Wildlife Management, Harare, Zimbabwe.

McNaughton, S. J. 1984. Grazing lawns: animals in herds, plant form, and co-evolution. *The American Naturalist* 124, 863–886.

McShane, T.O. 1989. Some preliminary results of the relationship between soils and tree response to elephant damage. *Pachyderm* 11, 29–31.

Mendelson, S.S. 1999. The role of thicket herbivores in seed dispersal within restoration sites in Valley Bushveld. MSc thesis, University College, London.

Midgley, J.J. & D. Joubert 1991. Mistletoes, their host plants and the effects of browsing by large mammals in the Addo Elephant National Park. *Koedoe* 34, 149–152.

Midgley, J.J., D. Balfour & G.I.H. Kerley 2005. Why do elephants damage savanna trees? *South African Journal of Science* 101, 213–215.

Milewski, A.V. 2002. Elephant diet at the edge of the Fynbos Biome, South Africa. *Pachyderm* 32, 29–38.

Mills, A.J. 2006. The role of salinity and sodicity in the dieback of *Acacia xanthophloea* in Ngorongoro Caldera, Tanzania. *African Journal of Ecology* 44, 61–71.

Minnie, L. 2006. The dietary shift of African elephant (*Loxodonta africana*) in the Karoo vegetation of Asante Sana Private Game Reserve. Honours thesis, Nelson Mandela Metropolitan University, South Africa.

Monajem, A. & D.K. Garcelon 2005. Nesting distribution of vultures in relation to land use in Swaziland. *Biodiversity and Conservation* 14, 2079–2093.

Moolman, L. 2007. Elephant browsing patterns in Pilanesberg National Park. MSc thesis, University of KwaZulu-Natal, Durban, South Africa.

Moolman, H.J. & R.M. Cowling 1994. The impact of elephant and goat grazing on the endemic flora of South African Succulent Thicket. *Biological Conservation* 68, 53–61.

Morgan, B.J. 2007. Unusually low incidence of debarking by forest elephants in the Réserve de Faune du Petit Loango, Gabon. *African Journal of Ecology*.

Mosugelo, D.K., S.R. Moe, S. Ringrose & C. Nellemann 2002. Vegetation changes during a 36-year period in northern Chobe National Park, Botswana. *African Journal of Ecology* 40, 232–240.

Motsumi, S.S. 2002. Seasonal population densities and distribution of gallinaceous birds in relation to habitat types and large herbivore impact in north east Chobe National Park, Botswana. MSc thesis, Agricultural University of Norway, Ås.

Musgrave, M.K. & S.G. Compton 1997. Effects of elephant damage to vegetation on the abundance of phytophagous insects. *African Journal of Ecology* 35, 370–373.

Mwalyosi, R.B.B. 1987. Decline of *Acacia tortilis* in Lake Manyara National Park, Tanzania. *African Journal of Ecology* 25, 51–53.

Mwalyosi, R.B.B. 1990. The dynamics ecology of *Acacia tortilis* woodland in Lake Manyara National Park, Tanzania. *African Journal of Ecology* 28, 189–199.

Myers, N. 1973. Tsavo National Park, Kenya, and its elephants: an interim appraisal. *Biological Conservation* 5, 123–132.

Noss, R.F. 1990. Indicators for monitoring biodiversity: a hierarchical approach. *Conservation Biology* 4, 355–364.

Novellie, P.A., M.H. Knight & A.J. Hall-Martin 1996. Sustainable utilization of Valley Bushveld: an environmental perspective. In G.I.H. Kerley, S.L. Haschick, C. Fabricius & G. La Cock (eds) *Proceedings of the Second Valley Bushveld Symposium*, Grassland Society of Southern Africa Special Publication, 8–10.

O'Connor, T.G., P.S. Goodman & B. Clegg 2007. A functional hypothesis of the threat of local extirpation of woody plant species by elephant in Africa. *Biological Conservation* 136, 329–345.

Okula, J.P. & W.R. Sise 1986. Effects of elephant browsing on *Acacia seyal* in Waza National Park, Cameroon. *African Journal of Ecology* 24, 1–6.

Olivier, R.C.D. 1978. On the ecology of the Asian elephant. Ph.D. thesis, University of Cambridge, Cambridge.

Owen-Smith, N. 1987. Pleistocene extinctions: the pivotal role of megaherbivores. *Paleobiology* 13, 351–362.

Owen-Smith, N. 1988. *Megaherbivores. The influence of very large body size on ecology*. Cambridge University Press, Cambridge.

Owen-Smith, N. 1989. Megafaunal extinctions: the conservation message from 11 000 years B.P. *Conservation Biology* 3, 405–412.

Owen-Smith, N. 1996. Ecological guidelines for waterpoints in extensive protected areas. *South African Journal of Wildlife Research* 26, 107–112.

Owen-Smith, N. 2002a. A metaphysiological modelling approach to stability in herbivore-vegetation systems. *Ecological Modelling* 149, 153–178.

Owen-Smith, N. 2002b. Credible models for herbivore-vegetation systems: towards an ecology of equations. *South African Journal of Science* 98, 445–449.

Page, B.R. & R. Slotow 2001. The influence of browsers, particularly elephants, on vegetation (diversity) at Madikwe Game Reserve. Preliminary unpublished project report, University of KwaZulu-Natal, South Africa.

Paine, R.T. 1969. A note on trophic complexity and community stability. *American Naturalist* 103, 91–93.

Paley, R.G.T. 1997. The feeding ecology of elephants in the Eastern Cape Subtropical Thicket. MSc thesis, Imperial College of Science, Technology and Medicine, University of London, London.

Paley, R.G.T. & G.I.H. Kerley 1998. The winter diet of elephant in Eastern Cape Subtropical Thicket, Addo Elephant National Park. *Koedoe* 41, 37–45.

Parker, I.S.C. 1982. The Tsavo story: an ecological case history. In R.N. Owen-Smith (ed.) *Management of Large Mammals in African Conservation Areas.* Haum, Pretoria, 37–50.

Pellew, R.A.P. 1983. The impacts of elephant, giraffe and fire upon the *Acacia tortilis* woodlands in the Serengeti. *African Journal of Ecology* 21, 41–74.

Penzhorn, B.L., P.J. Robbertse & M.C. Olivier 1974. The influence of the African elephant on the vegetation of the Addo Elephant National Park. *Koedoe* 17, 137–158.

Pickett, S.T. & P.S. White 1985. *The ecology of natural disturbance and patch dynamics.* Academic Press, New York.

Plumptre, A.J. 1993. The effects of trampling damage by herbivores on the vegetation of the Parc National des Volcans, Rwanda. *African Journal of Ecology* 32, 115–129.

Repton, M. 2007. Vegetation utilization by elephants in the Mkhuzi and Phinda Game Reserves. MSc thesis, University of KwaZulu-Natal, South Africa.

Robertson, M. 1995. Seed dispersal by Eastern Cape elephant (*Loxodonta africana*) in Valley Bushveld. Unpublished project. Rhodes University, South Africa.

Robinson, T.J. & E.R. Seiffert. 2003. Afrotherian origins and interrelationships: new views and future prospects. *Current Topics in Development Biology* 63, 37–60.

Rogers, K.H. & J. O'Keefe 2003. River heterogeneity: ecosystem structure, function and management. In J.T. du Toit, K.H. Rogers & H.C. Biggs (eds) *The Kruger Experience: Ecology and Management of Savanna Heterogeneity.* Island Press, Washington, DC, 189–218.

Rountree, M.W., K.H. Rogers & G.L. Heritage 2000. Landscape state change in the semi-arid Sabie River, Kruger National Park, in response to flood and drought. *South African Geographical Journal* 82, 173–181.

Roux, C. 2006. Feeding ecology, space use and habitat selection of elephants in two enclosed game reserves in the Eastern Cape Province, South Africa. MSc thesis, Rhodes University, South Africa.

Ruess, R.W. & F.L. Halter 1990. The impact of large herbivores on the Seronera woodlands, Serengeti National Park, Tanzania. *African Journal of Ecology* 28, 259–275.

Rutina, L.P., S.R. Moe & J.E. Swenson 2005. Elephant *Loxodonta africana* driven woodland conversion to shrubland improves dry-season browse availability for impala *Aepyceros melampus*. *Wildlife Biology* 11, 207–213.

Selous, F.C. 1881. *A hunter's wanderings in Africa, being a narrative of nine years spent amongst the game of the far interior of South Africa.* Richard Bentley, London.

Shannon, G., B.R. Page, K.J. Duffy & R. Slotow 2006a. The role of foraging behaviour in the sexual segregation of the African elephant. *Oecologia* 150, 344–354.

Shannon, G., B. P. Page, K. J. Duffy, & R. Slotow 2006b. The consequences of body size dimorphism: are African elephants sexually segregated at the habitat scale? *Behaviour* 143, 1145–1168.

Sheil, D. & A. Salim 2004. Forest tree persistence, elephants and stem scars. *Biotropica* 36, 505–521.

Sigwela, A.M. 1999. Goats and kudu in Subtropical Thicket: dietary competition and seed dispersal efficiency. MSc thesis, University of Port Elizabeth, South Africa.

Sigwela, A.M. 2004. Animal seed interactions in the thicket biome: consequences of faunal replacements and land use for seed dynamics. Ph.D. thesis, University of Port Elizabeth, South Africa.

Simpson, C.D. 1975. A detailed vegetation study on the Chobe River in north-east Botswana. *Kirkia* 10, 185–227.

Simpson, C.D. 1978. Effects of elephant and other wildlife on vegetation along the Chobe River, Botswana. *Occasional Papers of the Museum of Texas Tech University* No. 48.

Skarpe, C., P.A. Aarrestad, H.P. Andreassen, S.S. Dhillion, T. Dimakatso, J.T. du Toit, D.J. Halley, H. Hytteborn, S. Makhabu, M. Mari, W. Marokane, G. Masunga, D. Modise, S.R. Moe, R. Mojaphoko, D. Mosugelo, S. Mptsumi, G. Neo-Mahupeleng, M. Ramotadima, L. Rutina, L. Sechele, T.B. Sejoe, S. Stokke, J.E. Swenson, C. Taolo, M. Vandewalle & P. Wegge 2004. The return of the giants: ecological effects of an increasing elephant population. *Ambio* 33, 276–282.

Skead, C.J. 2007. *Historical incidence of the larger land mammals of the broader Eastern Cape.* Second edition, A.F. Boshoff, G.I.H. Kerley & P. Lloyd (eds) Centre for African Conservation Ecology, Nelson Mandela Metropolitan University, South Africa.

Slotow, R., D. Balfour & O. Howison 2001. Killing of black and white rhinoceros by African elephant in Hluhluwe-Umfolozi Park, South Africa. *Pachyderm* 31, 14–20.

Smallie, J.J. & T.G. O'Connor 2000. Elephant utilization of *Colophospermum mopane*: possible benefits of hedging. *African Journal of Ecology* 38, 352–359.

Smart, N.O.E., J.C. Hatton & D.H.N. Spence 1985. The effect of long-term exclusion of large herbivores on vegetation in Murchison Falls National Park, Uganda. *Biological Conservation* 33, 229–245.

Smit, I.P.J., C.C. Grant & B.J. Devereux 2007. Do artificial waterholes influence the way herbivores use the landscape? Herbivore distribution patterns around rivers and artificial surface water sources in a large African savanna park. *Biological Conservation* 136, 85–99.

Smith, P.P. & D.A. Shah-Smith 1999. An investigation into the relationship between physical damage and fungal infection in *Colophospermum mopane*. *African Journal of Ecology* 37, 27–37.

Spinage, C.A. 1990. Botswana's problem elephants. *Pachyderm* 13, 14–19.

Spinage, C.A. 1994. *Elephants*. T. Poyser & A.D. Poyser Ltd., London.

Steyn, A. & M. Stalmans 2001. Selective habitat utilisation and impact on vegetation by African elephant within a heterogenous landscape. *Koedoe* 44, 95–103.

Stokke, S. & J.T. du Toit 2000. Sex and size related differences in the dry season feeding patterns of elephants in Chobe National Park, Botswana. *Ecography* 23, 70–80.

Stokke, S. & J.T. du Toit 2002. Sexual segregation in habitat use by elephants in Chobe National Park, Botswana. *African Journal of Ecology* 40, 360–371.

Stuart-Hill, G.C. 1992. Effects of elephants and goats on the Kaffrarian succulent thicket of the Eastern Cape, South Africa. *Journal of Applied Ecology* 29, 699–710.

Styles, C.V. 1993. Relationships between herbivores and *Colophospermum mopane* of the northern Tuli Game Reserve, Botswana. MSc thesis, University of Pretoria, South Africa.

Styles, C.V. & J.D. Skinner 2000. The influence of large mammalian herbivores on growth form and utilization of mopane trees, *Colophospermum mopane*, in Botswana's Northern Tuli Game Reserve. *African Journal of Ecology* 38, 95–101.

Sukumar, R. 2003. *The living elephants: evolutionary ecology, behaviour and conservation*. Oxford University Press, New York.

Swanepoel, C.M. 1993. Baobab damage in Mana Pools National Park, Zimbabwe. *African Journal of Ecology* 31, 220–225.

Tafangenyasha, C. 1997. Tree loss in the Gonarezhou National Park (Zimbabwe) between 1970 and 1983. *Journal of Environmental Management* 49, 355–366.

Thompson, P.J. 1975. The role of elephants, fire and other agents in the decline of *Brachystegia* woodland. *Journal of the South African Wildlife Management Association* 5, 11–18.

Thrash, I. 1998. Impact of large herbivores at artificial watering points compared to that at natural watering points in Kruger National Park, South Africa. *Journal of Arid Environments* 38, 315–324.

Thrash, I., P.J. Nel, G.K. Theron & J. du P. Bothma 1991. The impact of the provision of water for game on the woody vegetation around a dam in the Kruger National Park. *Koedoe* 34, 131–148.

Thrash, I., G.K. Theron & J. du P. Bothma 1995. Dry season herbivore densities around drinking troughs in the Kruger National Park. *Journal of Arid Environments* 29, 213–219.

Tolsma, D.J., W.H.O. Ernst & R.A. Verwey 1987. Nutrients in soil and vegetation around two artificial waterpoints in eastern Botswana. *Journal of Applied Ecology* 24, 991–1000.

Treydte, A.C., I.M.A. Heitkonig, H.H.T. Prins & F. Ludwig 2007. Trees improve grass quality for herbivores in African savannas. *Perspectives in plant ecology, evolution and systematics* 8, 197–205.

Trollope, W.S.W., L.A. Trollope, H.C. Biggs, D. Pienaar & A.L.F. Potgieter 1998. Long-term changes in the woody vegetation of the Kruger National Park, with special reference to the effects of elephants and fire. *Koedoe* 41, 103–112.

Valeix, M., H. Fritz, S. Dubois, K. Kanengoni, S. Alleaume & S. Saïd 2007. Vegetation structure and ungulate abundance over a period of increasing elephant abundance in Hwange National Park, Zimbabwe. *Journal of Tropical Ecology* 23, 87–93.

Van Hoven, W., R.A. Prins & A. Lankhorst 1981. Fermentative digestion in the African elephant. *South African Journal of Wildlife Research* 11, 78–86.

Van Wijngaarden, W. 1985. *Elephants – trees – grass – grazers*. International Institute for Aerospace Survey and Earth Sciences, Netherlands. ITC publication No. 4, 1–159.

Van Wilgen, B.W., W.S.W. Trollope, H.C. Biggs, A.L.F. Potgieter & B.H. Brockett 2003. Fire as a driver of ecosystem variability. In J.T. du Toit, K.H. Rogers & H.C. Biggs (eds). *The Kruger experience: Ecology and management of savanna heterogeneity*. Island Press, Washington, DC, 149–170.

Van Wyk, P. & N. Fairall 1969. The influence of the African elephant on the vegetation of the Kruger National Park. *Koedoe* 12, 57–89.

Viljoen, P.J. 1989. Habitat selection and preferred food plants of a desert-dwelling elephant population in the northern Namib Desert, South West Africa/Namibia. *African Journal of Ecology* 27, 227–240.

Wackernagel, A. 1993. Elephants and vegetation: severity, scale and patchiness of impacts along the Linyanti River, Chobe District, Botswana. MSc thesis, University of the Witwatersrand, South Africa.

Walker, B. 1989. Diversity and stability in ecosystem conservation. In: C. Western & M. Pearl (eds) *Conservation for the twenty-first century*. Oxford University Press, Oxford, 121–130.

Wall, J., I. Douglas-Hamilton & F. Vollrath 2006. Elephants avoid costly mountaineering. *Current Biology* 16, R527–R529.

Weaver, S.M. 1995. Habitat utilization by selected herbivores in the Klaserie Private Nature Reserve, South Africa. MSc thesis, University of Pretoria, South Africa.

Weir, J.S. 1971. The effect of creating additional water supplies in a central African National Park. In E. Duffey & A.S. Watt (eds) *The scientific management of animal and plant communities for conservation*. Blackwell Scientific, London, 367–376.

Western, D. 1975. Water availability and its influence on the structure and dynamics of a savanna large mammal community. *East African Wildlife Journal* 13, 265–286.

Western, D. 1989. The ecological role of elephants in Africa. *Pachyderm* 12, 43–46.

Western, D. 2007. A half century of habitat change in Amboseli National Park, Kenya. *African Journal of Ecology* 45, 302–310.

Western, D. & D. Maitumo 2004. Woodland loss and restoration in a savanna park: a 20-year experiment. *African Journal of Ecology* 42, 111–121.

Western, D. & C. Van Praet 1973. Cyclical changes in the habitat and climate of an East African ecosystem. *Nature* 241, 104–106.

Weyerhaeuser, F.J. 1985. Survey of elephant damage to baobabs in Tanzania's Lake Manyara National Park. *African Journal of Ecology* 23, 235–243.

Whyte, I.J., P.J. Nel, T.M. Steyn & N.G. Whyte 1996. *Baobabs and elephants in the Kruger National Park*. Preliminary Report, Kruger National Park.

Williamson, B.R. 1975. The condition and nutrition of elephant in Wankie National Park, Rhodesia. *Arnoldia* 7, 1–20.

Wing, L.D. & I.O. Buss. 1970. Elephants and forest. *Wildlife Monographs* 19, 1–92.

Wiseman, R., B.R. Page & T.G. O'Connor 2004. Woody vegetation change in response to browsing in Ithala Game Reserve, South Africa. *South African Journal of Wildlife Research* 34, 25–37.

Yeaton, R.I. 1988. Porcupines, fires and the dynamics of the tree layer of the *Burkea africana* savanna. *The Journal of Ecology* 76, 1017–1029.

Young, T.P. & W.K. Lindsay 1988. Role of even aged populations in the disappearance of *Acacia xanthophloea* woodlands. *African Journal of Ecology* 26, 69–71.

Young, T.P., T.M. Palmer & M.E. Gadd 2004. Competition and compensation among cattle, zebras, and elephants in a semi-arid savanna in Laikipia, Kenya. *Biological Conservation* 122, 351–359.

4

INTERACTIONS BETWEEN ELEPHANTS AND PEOPLE

Lead author: Wayne Twine
Author: Hector Magome

INTRODUCTION

MOST PRESENT-DAY interactions between elephants and people in South Africa occur within conservation areas, and are predominantly positive. However, human–elephant interactions in Africa have received increasing attention in the scientific literature in the last decade because of a perceived rise in levels of conflict between the two (Hough, 1988; Thouless & Sakwa, 1995; Tchamba, 1996; Naughton *et al.*, 1999; O'Connell-Rodwell *et al.*, 2000; Dublin & Hoare, 2004). Important advances have been made in researching, responding to, and reducing conflict between elephants and people across the continent (Hoare, 2000; Dublin & Hoare, 2004; Sitati, 2007), and are discussed later in this chapter. However, the 'conflict paradigm' (Lee & Graham, 2006) has presented an unbalanced perspective on the way elephants and humans interact by overlooking positive interactions. This chapter seeks to redress this oversight by additionally assessing a range of positive types of interactions, such as between tourists and elephants.

Because human–elephant conflict (HEC) is a highly emotive and politicised issue, it is reviewed in this chapter in detail. The attention it receives is not meant to reflect the current scale of the problem in South Africa. Rather, it provides a comprehensive assessment to aid formulation of policy in the current context, while recognising that HEC could become an increasingly important political issue in South Africa with the expansion of conservation areas, growing elephant populations, a burgeoning tourism industry, and greater participation by rural communities in resource management.

Elephants and people have interacted in Africa for thousands of years. Humans have preyed on elephants since the Stone Age, as evidenced by rock art depicting elephant hunts (Carrington, 1958). However, the advent of cultivation probably changed the relationship between the two species from one of 'a mild predator/prey interaction' to one that was 'fundamentally competitive' (Parker & Graham, 1989). Humans and elephants have the same habitat preferences, and this would have given rise to localised competition between the two for space, probably resulting in elephants raiding people's crops from time to time (Parker & Graham, 1989). It has been speculated that wide stone walls constructed around ancient villages may have been to deter crop-raiding elephants (Clutton-Brock, 2000). However, some indirect interactions between elephants and people in pre-colonial times were also positive. Localised bush encroachment caused by overgrazing by livestock may have favoured elephants (Parker & Graham, 1989). Conversely, elephants could have caused the local disappearance of tsetse fly, which is a vector for sleeping sickness, by opening up thickets and woodlands (Ford, 1966), and thus creating new areas suitable for human habitation (Parker & Graham, 1989). Various African societies have totems and folklore about elephants which are indicative of respect (Mutwa, 1997).

Human–elephant interactions during the colonial period were characterised by the decimation of elephant populations by sport and ivory hunters (see Chapter 1). However, since the late colonial period, and into the post-colonial era, interactions between people and elephants have intensified and diversified with the growth of human populations, expansion of conservation areas, and localised increases in elephant populations (Tchamba, 1996; Hoare & Du Toit, 1999; Smith & Kasiki, 2000; O'Connell-Rodwell *et al.*, 2000; Sitati, 2007). This is particularly relevant to South Africa, where elephant conservation and the unprecedented expansion of both state and private protected areas has resulted in a dramatic recovery of elephant populations since the 1960s (Hall-Martin, 1992), drawing in a greater range of role players and creating new types of human–elephant interactions.

SOCIETAL VALUES AND ATTITUDES TO ELEPHANTS

Many controversies over wildlife management become acrimonious because they are either conducted at cross-purposes or reflect fundamental differences in values and attitudes that cannot be changed through argument (Bell, 1983; see also *Conservation and Society* 4(3) 2006 on evictions from national parks). Also, there is a tendency to find short-term solutions to issues that derive from distal or long-term causes. Conflict arises when management decisions do not suit all stakeholders and when unpopular decisions are taken (Caughley & Sinclair, 1994). For example, Leakey (2001) cites the conflict that can arise over nature conservation when wealthy tourist demands are contrasted with those of the poor and hungry. In many instances there are no clear 'right' or 'wrong' answers to the difficult questions arising from the complex socio-economic dimensions of wildlife management. Rather than discussing these further, this chapter describes how elephants are perceived and lists the values that tend to shape attitudes and perceptions.

People view elephants variously as beautiful and charismatic icons of conservation, dangerous and destructive pests, a valuable and exploitable resource, and as keystone species in ecosystems (Hoare, 2000; Dublin & Hoare, 2004). These different attitudes reflect different societal values, which are defined as 'conception[s] of what is good' (Rokeach, 1973). People's values are socially constructed and are shaped by factors such as personal experience, ethnicity, culture, gender, age, socio-economic context, and political orientation (Steel *et al.*, 1994; Manfredo & Zinn, 1996; Vaske & Donnelly, 1999; Dougherty *et al.*, 2003; Lockwood 2006). A range of different types of societal values of wildlife have been recognised (Giles, 1978; Rolston, 1988; Gilbert & Dodds, 2001; Conover & Conover, 2003), and key categories relevant to human–elephant interactions are summarised in table 1. These form the basis of the components of total economic value of elephants discussed in Chapter 10.

According to the cognitive hierarchy model of human behaviour, values held by individuals underpin their attitudes (Rokeach, 1973; Fulton *et al.*, 1996; Tarrant *et al.*, 1997; Vaske & Donnelly, 1999; Tarrant & Cordell, 2002). Specific patterns of values held by a person create 'value orientations' or basic belief patterns, which shape the way the individual interprets and understands the world. This influences the attitudes and opinions held by the person on particular objects or issues. Attitudes, in turn, influence people's behavioural intentions, and ultimately, their behaviour (Fulton *et al.*, 1996; Vaske & Donnelly, 1999).

Sociologists recognise a continuum of environmental value orientations in society (Vaske & Donnelly, 1999). Glaser (2006) suggests that this variation can be represented by human-nature mind-maps. At the one end is the *anthropocentric* value orientation, based on a definition of nature through a 'social lens' (Glaser, 2006), focusing on human uses and benefits from nature (Vaske & Donnelly, 1999). Society and nature are conceived of as two separate systems. At the other end is the *biocentric* or *ecocentric* value orientation which considers society as part of nature (Glaser, 2006), and places greater emphasis on the non-use values of biodiversity (Vaske & Donnelly, 1999). Mental models in this value orientation include the traditional African world-view (Mutwa, 1997), as well as western notions of 'pristine nature' impacted by society, and absolute biocentrism typified by the 'deep ecology' model, which regards humans and their needs as no more important than those of any other species (Glaser, 2006).

Value	Definition	Relevance to elephants
Aesthetic	Appreciation through the senses	People enjoy observing elephants because of their size and power
Commercial	Importance for generating income	Non-consumptive use, such as tourism, and consumptive use, such as trophy hunting, use of meat, hides and ivory
Cultural	Importance as cultural symbols	Associated with power and royalty, and used as clan totems and names
Ecological	Role in contributing to ecosystem composition, structure and function	Valued for their role as ecosystem engineers or keystone species
Empathetic	Satisfaction from being able to emotionally relate to another species	General public empathise with elephants as intelligent, social and long-lived creatures
Existence	Sense of wellbeing from knowledge of their existence	Most South Africans have not seen an elephant in the wild but many still care what happens to them
Historical	Symbols of a past era	Nostalgic appreciation of elephants as symbols of 'wild Africa'. 'Big Five' status harks back to the days of the great game hunters and explorers
Recreational	Enjoyment of experience from recreational activities	Tourists enjoy the thrill of finding and observing elephants in the wild and experiences such as elephant-back safaris
Scientific	Importance for the advancement of knowledge and understanding	Great scientific interest in the complex challenge of solving the 'elephant problem'
Subsistence	Used for purposes of non-commercial consumption	Consumption of elephant meat or use of dung for medicinal purposes

Table 1: Key ways in which society values wildlife, with examples specific to elephants

The anthropocentric-biocentric continuum is similar to the utilisation-protection or benefits-existence continua in the wildlife management literature (Fulton *et al.*, 1996; Vaske & Donnelly, 1999). In reality, these value orientations are not mutually exclusive, and individuals or societies may exhibit a combination of values. In fact, Glaser (2006) proposes that 'interdisciplinary' and 'complex systems' mind-maps of human-nature interactions have emerged in the scientific community as a result of dissatisfaction with the reductionism inherent in both anthropocentric and biocentric mind-maps.

The environmental value orientation of an individual will influence his or her attitude to elephants and opinions on issues relating to their management. This helps to explain why two people who are equally passionate about elephants can have diametrically opposing beliefs, attitudes and opinions on controversial topics such as culling. The biocentric orientation is broad, and underpins both the perspective of animal rights groups who place great importance on the existence value and rights of individual elephants, as well as the perspective of people who believe that the control of elephant populations may be necessary for the greater good of ecosystems and other species. The anthropocentric orientation underlies the perspectives of those who support sustainable utilisation of elephants as a valuable consumptive resource. The deep moral issues associated with these differences in orientations are dealt with in Chapter 9. The perspective of conservation authorities in South Africa today probably reflects a combination of value orientations, simultaneously valuing elephants for their existence value, their ecological role, and their economic importance for tourism. Similarly, tourists may also value elephants for their intrinsic beauty as well as the enjoyment they get from recreation associated with them.

People's values, value orientations and attitudes strongly influence the way they interact with elephants and issues relating to them. For example, people may choose to visit a game reserve to see elephants or pay to hunt them for recreation, become passionately involved in the 'elephant debate' or totally ignore it. Apathy towards issues such as culling may not necessarily be indicative of an uncaring society, but may rather reflect the fact that many people do not consider them of particular concern to society, especially if they are not directly affected materially by what happens to elephants. It is important to note that in addition to being shaped by values, interactions between people and elephants may also themselves shape values. For example, a person who has suffered loss due to elephants, such as through crop damage, may value elephants less for their existence or aesthetic value than somebody who has only had meaningful positive interactions with them.

HUMAN–ELEPHANT INTERACTIONS IN PROTECTED AREAS

Effectively all wild elephant populations in South Africa are confined to fenced national parks, game reserves and privately owned ranches. Particular types of interactions thus occur between elephants and a range of human role-players within clearly defined land use types, or spatial domains. We use this concept as an organising framework to assess human–elephant interactions in South Africa (figure 1).

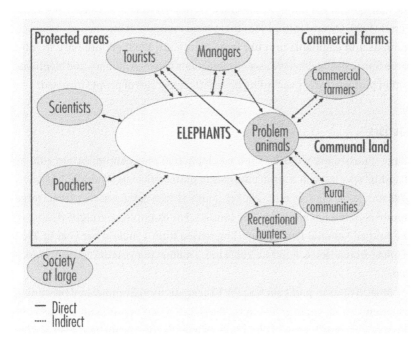

Figure 1: Key human–elephant interactions in different land-use settings

Elephants occur in both privately-owned and state-owned protected areas in South Africa. The primary objective of most private reserves is to generate profits through ecotourism or recreational hunting. Elephants, as one of the 'Big Five', are a huge marketing tool in this regard. Although state-owned reserves also generate revenue, this is to subsidise their primary activity of conserving biodiversity.

Protected areas containing elephants fall into two management categories: 'small' reserves covering less than 1 000 km² (i.e. smaller than the typical elephant home range size), and 'large' reserves greater than 1 000 km² in extent (Owen-Smith *et al.*, 2006). Wider elephant movement is limited within small reserves, and they are therefore usually characterised by higher elephant densities and levels of impact across the landscape and throughout the year (Owen-Smith *et al.*, 2006). This implies that the intensity of encounters between people and elephants, and levels of management interventions needed, are usually higher in smaller protected areas. Exceptions include large reserves with high elephant densities due to an abundance of artificial water points.

There is little readily available literature on human–elephant interactions in protected areas. Even in books or documents ostensibly dedicated to addressing the management of mammals in conservation areas (see Jewell & Holt, 1981; Ferrar, 1983; Owen-Smith, 1983) there is no mention of interactions between humans and elephants or other large mammals. We therefore have to infer these from existing data. We assess interactions between humans and elephants within protected areas according to various categories of people involved.

Tourists

Large charismatic wildlife, such as elephants, plays an important role as 'flagship' species which attract tourists to protected areas (Walpole & Leader-Williams, 2002; Lindsey *et al.*, 2007a). South African game reserves attract large numbers of foreign and domestic tourists. For example, overnight visitors to the Kruger National Park (Kruger) increased from 1 million per year in 2002 (Freitag-Ronaldson & Foxcroft, 2003) to 1.3 million per year in 2006 (SANParks, 2007).

Most visitors to parks such as the Kruger are strongly motivated by wildlife experiences (Saayman & Slabbert, 2004), but it is often difficult to tease out the specific contribution of elephants as a motivating factor. However, Kerley *et al.* (2003) found that 77 per cent of tourists visiting Addo Elephant National Park came mainly to see elephants, and most were satisfied if they were successful in that goal, even if they saw little else. Addo would probably enjoy little mainstream tourist attention were it not for the presence of elephants, due to the dense thicket vegetation which makes game viewing difficult. In their assessment of viewing preferences of tourists in four South African savanna game reserves (Kruger National Park, Pilanesberg National Park, Djuma Game Reserve and Ngala Game Reserve), Lindsey *et al.* (2007a) found that elephants were among the most popular species, especially among first-time and overseas visitors. Regardless of the primary motivation for visiting a protected area, positive interactions between tourists and elephants generate public support and goodwill for elephants and conservation in general. Such experiences may galvanise popular opinion on issues relating to elephant management.

Although the overwhelming majority of interactions between tourists and elephants in protected areas are positive, negative interactions also do occur. Elephants may become stressed under conditions of high tourist activity (Pretorius, 2004) or if tourists get too close (Pretorius, 2004; Burke, 2005). Burke (2005) proposed the '50 m rule' based on her observation in Pilanesberg National Park that stress in elephants due to the presence of tourists was substantially reduced at distances greater than 50 m. This needs further investigation in other settings. In rare instances, elephants injure or even kill tourists. This is discussed in more detail in the next section which deals with human–elephant conflict.

Managers

Interactions between managers and elephants are both direct and indirect, and are covered in detail in Chapters 5–8 and 12. The important point here is that tensions exist between (1) conservation management objectives, which include but also extend beyond elephants (e.g. conserving biodiversity at various scales), (2) the expectations of tourists (either to see more elephants or to see less destruction of vegetation by elephants), and (3) the sentiment of the general public.

Scientists

Elephants generate much interest in the scientific community because of their advanced social behaviour and the substantial impacts they have on the vegetation and other species in the ecosystem. South African scientists have authored or co-authored roughly 150 scientific journal articles dealing with elephant physiology, anatomy, social behaviour, ecology, impacts, and management. Interactions between scientists and elephants are both direct, such as observation of social behaviour, and indirect, such as measurement of elephant damage.

Recreational hunters

Elephants are prized as targets by recreational hunters. Their status as one of the 'Big Five' game species in the contemporary tourism industry originates from their reputation as one of the five most dangerous but highly desirable species during the era of sport hunting in the nineteenth century. Big game trophies, such as elephants, are highly sought after by foreign hunters, particularly from North America and Europe (Taylor, 1993). In South Africa, approximately 30 elephants are hunted annually, mainly on private land (Lindsey *et al.*, 2007b). Taylor (1993) suggests that professional and sports hunters have a potential role to play in the control of problem elephants and in providing an opportunity for rural communities to generate money from elephants on their land. This is discussed further under HEC mitigation strategies. Negative impacts of elephant hunting, other than on the individual killed, include stress in the rest of the proximate population (Burke, 2005, Bradshaw & Schore, 2007), and genetic shifts in extreme cases of selective hunting (Nyakaana *et al.*, 2001). Attitudes to trophy hunting in South Africa are polarised between those who find it morally reprehensible, and those who support it for the contribution it makes to conservation by generating revenue (Lindsey *et al.*, 2007b).

Poachers

Although it remains a factor in elephant management, poaching has significantly diminished as a concern in South Africa. For example, poaching of elephants in the Kruger National Park has dropped substantially since the early 1980s, when over 100 elephants were poached in 1981, and has been consistently five or less per year over the last decade (Freitag-Ronaldson & Foxcroft, 2003). The general decline in elephant poaching in South Africa is largely attributed to effective anti-poaching enforcement (Lee & Graham, 2006). The contribution of the CITES ban on ivory trade appears to be minor in comparison to effective management in southern Africa (Stiles, 2004). Most elephant poaching in South Africa, whether by locals or cross-border raids, is commercial rather than subsistence, targeting ivory rather than meat.

Rural communities

Most positive interactions between rural communities and elephants in protected areas are indirect, and include cultural values and economic benefits. Elephants are prominent in African folklore and have particular significance for clans for whom they are a totem, such as the Batloung and Ndlovu clans (Mutwa, 1997). Rural communities thus value elephants in protected areas as part of their cultural heritage. However, although communities living adjacent to South African protected areas allude to the cultural value of elephants and the historical relationships between them and people (see SANParks, 2005), little has been published or is known about the details of these. This needs further research.

Conservation, tourism and recreational hunting create secondary benefits from elephants for local rural communities, such as employment and training. This, in turn, has a positive influence on local attitudes towards protected areas (Anthony, 2007). However, since parks employ a relatively small proportion of the neighbouring population, rural communities have expectations of greater access to other economic benefits from parks. Parks like Pilanesberg and Madikwe in North-West Province outsource management activities such as fence maintenance to contractors in adjacent communities. In the context of elephant management, community representatives at the Great Elephant Indaba (Berg-en-Dal, 19–21 October 2004) expressed their desire to benefit economically from elephant culling operations in Kruger (SANParks, 2005). This included outsourcing of functions such as processing, marketing and selling elephant by-products like meat and hides (Mabunda, 2005). Community expectations of benefits from culling add another layer of complexity to the culling debate, and therefore need to be investigated further.

Land restitution and the emergence of community-based natural resource management (CBNRM) initiatives provide opportunities for communities to benefit from revenues and employment from tourism and recreational hunting on their own conserved communal land which has been restored to them (see Chapter 10). However, as exemplified by the case of the Makuluke land claim, commercial hunting of elephants in restored communal land within national parks can be politically complex and controversial (Steenkamp & Grossman, 2001). The possibility exists for communal land adjacent to protected areas to serve as 'sink' areas for elephants to disperse from high density areas, within a metapopulation management approach (Van Aarde & Jackson, 2007). This would provide tangible benefits for communities, such as meat and revenue from controlled sports hunting on their land.

Positive direct interactions between rural communities and elephants are limited. Communities get access to meat when problem animals are destroyed, and some local residents visit neighbouring parks as tourists. Anthony & Bellinger (2007) found that rural residents adjacent to the Kruger value elephants for meat and recreation, as well as for ornaments (ivory) and religious purposes. Negative interactions with elephants are discussed under human–elephant conflict.

Society at large

Most people will never see elephants in the wild, but they interact indirectly with images of elephants through the media. The media play a powerful role in shaping public opinion, both domestically and overseas, on controversial issues such as hunting of elephants in contractual parks (Steenkamp & Grossman, 2001). In a modern democracy like South Africa, society at large will ultimately decide on the objectives and desired course of action in the management of the nation's elephants (Owen-Smith *et al.*, 2006).

HUMAN–ELEPHANT CONFLICT IN COMMUNAL LANDS, COMMERCIAL FARMS AND PROTECTED AREAS

Defining HEC and 'problem elephants'

The negative impacts of elephants on humans have replaced the concern over poaching (since the 1970s and 1980s) as the main source of conflict between elephants and humans (Kangwana, 1995; Sitati, 2007). Although some would consider the issue of culling to be a prominent HEC, it is not covered in this chapter, as it is dealt with in Chapters 8 and 9. Current consideration of HEC in the literature typically refers to those interactions between people and elephants which threaten human lives and livelihoods (Hillman Smith *et al.*, 1995; Smith & Kasiki, 2000; Dublin & Hoare, 2004; Thirgood *et al.*, 2005). Such conflicts emerge where human and elephant ranges coincide, either in unprotected landscapes or in land-use mosaics of protected areas and human settlement (Hoare, 1999; 2000). Elephants come into conflict with humans, particularly subsistence farmers, because they are large, strong, social, intelligent, long-lived, require large amounts of food and water, are destructive feeders, can move silently, and move over large home ranges (Smith, 1989; Lee & Graham, 2006).

HEC takes the form of direct and indirect impacts or costs to those affected. Direct costs to humans include destroyed crops, raided food stores, damaged infrastructure and water sources, disturbed or killed livestock, injury, and loss of human life (Thouless, 1994; Tchamba, 1996; Hoare, 1999; Naughton *et al.*, 1999; Hoare, 2000; Dublin & Hoare, 2004; Gadd, 2005; Lee & Graham, 2006). Indirect or social costs include disturbance of normal human activities, such as interference with school attendance (Kangwana, 1995; Kiiru, 1995; Malima *et al.*, 2005), disruption of household chores like collecting water and firewood (Lee & Graham, 2006), loss of time due to guarding fields (Lee & Graham, 2006), and loss of productivity due to sleepless nights guarding fields (Kangwana, 1995; Kiiru, 1995). Injury or death due to retribution by humans are direct costs to elephants, while indirect costs include disturbance and denial of habitat.

In the prevailing context of HEC in Africa, the term 'problem elephants' is typically applied to those individuals or groups which temporarily extend their range into human settlements and engage in activities which negatively impact on humans (Hoare, 1999; 2000). However, in a more general sense, the term also includes elephants which exhibit deviant behaviour that frustrates management activities or objectives *within* protected areas. Examples of these include elephants which habitually damage infrastructure, threaten the lives of staff (Whitehouse & Kerley, 2002) and tourists (Nel, 2004), or kill other wildlife (Slotow *et al.*, 2000; 2001). These are largely atypical behaviours that are considered unprecedented. It has been suggested that such deviant behaviour is a result of stressors such as culling, hunting, poaching, translocation, habitat fragmentation, and high tourist pressure, which disrupt social processes (Slotow *et al.*, 2000; Nel, 2004; Bradshaw *et al.*, 2005; Bradshaw & Schore, 2007). It could thus be argued that 'problem elephant' behaviour reflects changes in human behaviour.

Hoare (2001) observed that removing individual problem animals frequently does not solve the problem, and he proposed the idea of a 'problem component' within elephant populations. The implication of this untested hypothesis is that as problem animals are removed, others take their place. However, the notion of a 'problem component' may be inappropriate in contexts where human behaviour causing the problem persists, such as cultivating highly desirable foods like maize in unprotected fields close to reserve boundaries. In such cases, the unprotected temptation could be regarded as the problem, not the elephants.

HEC in Africa has received increasing attention in the scientific literature in the last decade (e.g. Kangwana, 1995; Thouless & Sakwa, 1995; Tchamba, 1996; Naughton-Treves, 1998; Hoare, 1999; Hoare & Du Toit, 1999; Naughton et al., 1999; Hoare, 2000; Smith & Kasiki, 2000; De Boer & Ntumi, 2001; Sitati et al., 2003; Dublin & Hoare, 2004; Barnes et al., 2005; Gadd, 2005; Lee & Graham, 2006; Sitati, 2007). Naughton et al. (1999) attribute the perceived intensification of conflicts to a combination of changes in (1) land use, (2) elephant behaviour and socio-ecology due to human intervention, and (3) socio-economic changes in rural communities which bring elephants and humans into closer contact and reduce human tolerance of elephants. However, Lee and Graham (2006) challenge the assertion of intensification of HEC on the basis that it has not been adequately substantiated. Other authors note that reports of HEC may be sensationalised or inflated by the media (Kangwana, 1995; Lee & Graham, 2006).

How much of a problem is HEC?

Direct conflicts between elephants and rural populations in southern Africa are comparatively few (Sitati, 2007). This is perhaps because of the hard-boundary effect created by fencing of protected areas in the region. Although no fence is totally elephant proof (Thouless & Sakwa, 1995), there is strong evidence that electric fences dramatically curb the incidence of elephants leaving protected areas, and they thus substantially reduce the levels of conflict between elephants and adjacent human populations (Taylor, 1993; O'Connell-Rodwell et al., 2000; Omondi et al., 2004; Kioko et al., 2006). Nevertheless, localised problems with elephants escaping protected areas and causing damage do occur in South Africa. In one of the very few studies quantifying this in South Africa, Anthony (2007) found that 12.1 per cent of households (n = 240) in 38 rural communities along the western boundary of the Kruger claimed to have experienced damage from wildlife in the last two years (mid-2002 to mid-2004). Of the 386 reported incidents concerning damage-causing animals between October 1998 and October 2004, 14.5 per cent involved elephants, all of which came from the park.

Damage to crops

Crop-raiding is by far the most common source of HEC in Africa (Newmark *et al.*, 1994; Osborn & Parker, 2003; Sitati *et al.*, 2003; Malima *et al.*, 2005). While impacts of crop-raiding may be catastrophic for individual households, these forays by problem elephants are generally uncommon, localised, and seasonal (Thouless, 1994; Lahm, 1996; Naughton-Treves, 1998; Naughton *et al.*, 1999; Hoare, 2000; De Boer & Ntumi, 2001; Dublin & Hoare, 2004; Sitati *et al.*, 2003; Adjewodah *et al.*, 2005). Although some farmers may experience near-total destruction of their crops, this is exceptional, and damage is usually medium- to low-level (Naughton *et al.*, 1999; Adjewodah *et al.*, 2005; Malima *et al.*, 2005). Caution should be exercised when interpreting data on levels of crop damage in the literature because some studies disproportionately sample areas hard-hit by elephants, making the data difficult to extrapolate (Naughton *et al.*, 1999).

Crop damage by elephants is often less than that caused by livestock (Naughton-Treves, 1998) or other wildlife pests such as insects, birds, rodents, primates, antelope, and bushpigs (Newmark *et al.*, 1994; Lahm, 1996; De Boer & Baquete, 1998; Naughton-Treves, 1998; Naughton *et al.*, 1999; Omondi *et al.*, 2004). In a review of 25 studies of wildlife pests in Africa (South Africa excluded), Naughton *et al.* (1999) found that across all studies, elephants accounted for less than 10 per cent of total crop damage. They concluded that elephants may be a significant pest locally, but not nationally.

Despite their relatively modest impact, elephants are less tolerated than most other wildlife pest species because of their size, which makes them more obvious, and the danger they pose (Naughton *et al.*, 1999; Sitati *et al.*, 2003). It is also widely reported that complaints of crop damage by elephants are usually disproportionate to the actual damage (De Boer & Baquete, 1998; Hoare, 2000; Dublin & Hoare, 2004; Lee & Graham, 2006). Farmers may inflate estimates of crop damage by up to 30–40 per cent (Tchamba, 1996) in anticipation of compensation (Tchamba, 1995, 1996) or meat from shot problem elephants (Taylor, 1993). Regardless of the level of impact, crop-raiding has a significant negative impact on local people's attitude towards conservation (De Boer & Baquete, 1998; Naughton *et al.*, 1999; O'Connell-Rodwell *et al.*, 2000; Dublin & Hoare, 2004; Gadd, 2005; Sitati *et al.*, 2005) and provides a convenient avenue for them to vent other grievances about neighbouring conservation areas (Naughton *et al.*, 1999; Lee & Graham, 2006).

Our assessment of the South African situation is that crop-raiding is relatively rare, primarily because protected areas are fenced. Nevertheless, elephants do break out and damage crops from time to time (Anthony, 2007). These events are likely to be in localised 'hot-spots' associated with very particular situations, such as where fences are not maintained (Anthony, 2007) or where erecting fences may be difficult, such as along or across rivers (see Chapter 7).

Most of the elephant populations in reserves in Limpopo, Mpumalanga, North-West and KwaZulu-Natal provinces are adjacent to densely populated communal lands. By contrast, the elephant populations in the Eastern Cape (and Welgevonden Private Game Reserve in Limpopo Province) mainly abut privately-owned commercial farms. The human density in communal lands adjacent to protected areas such as the Kruger can be as high as 300 people. km^{-2} (Pollard *et al.*, 2003). This is a legacy of the forced removals during the apartheid era. Black people were displaced from land earmarked for white-owned agriculture or conservation areas, and were crammed into 'homelands'. This resulted in 74 per cent of the population being allocated a mere 13 per cent of the land surface of the country (Anderson *et al.*, 2002). Such human densities greatly exceed the threshold density of around 16 people.km^{-2} beyond which elephants rapidly disappear from savanna landscapes due to insufficient habitat (Hoare & Du Toit, 1999). Human density in the communal lands of the former homelands thus acts as an effective barrier to elephants, and elephant incursions into rural communities are therefore likely to be of short duration and distance.

Even rare, brief crop-raiding incursions by elephants can wreak havoc locally and have a substantial impact on community perception of conservation areas. This is illustrated by a South African study in the lowveld in which 49–80 per cent of respondents in villages bordering protected areas felt that benefits from tourism were not enough to make up for problems with wildlife, including crop-raiding by baboons and elephants, and stock losses to predators (Spenceley, 2005). The bitter association between conservation and displacement, loss of land and exclusion has bred hostility among rural populations towards protected areas (Fabricius *et al.*, 2001), which is also likely to taint local attitudes to elephants and isolated fence-breaking incidents. Slow response by authorities when elephants break out of protected areas and lack of compensation for damage caused by problem elephants are contentious issues which contribute further to these negative attitudes (Anthony, 2007).

Direct livelihood impacts of crop damage by elephants are likely to be very modest in the communal lands of South Africa, although very poor and vulnerable households will be disproportionately affected. In the rest of Africa, the livelihoods of *subsistence* farmers are the hardest hit by crop-raiding (Smith & Kasiki, 2000; Osborn & Parker, 2002; Dublin & Hoare, 2004). Direct livelihood impacts of crop damage include loss of food sources and income (Osborn & Parker, 2003). However, in South Africa, the subsistence farming peasantry had been virtually eliminated by the end of the 1950s (Seekings, 2000). Small-scale agriculture thus contributes less than a third of total household income in the former homelands (Seekings, 2000; Leroy *et al.*, 2001; Crookes, 2003), and nearly two thirds of rural African households earn nothing at all from agriculture (Seekings, 2000). Rural livelihoods in South Africa are primarily cash-based, with a high reliance on income from migrant labour (May 1990) and government social grants (Carter & May, 1999). Agriculture in communal areas of South Africa therefore plays a safety-net function, rather than being the mainstay of rural livelihoods (Shackleton *et al.*, 2001). However, in relative terms, agriculture makes a greater contribution to the livelihoods of the poorest and most marginalised households in these rural communities (Carter & May, 1999).

Crop-raiding on commercial farms is rare, the damage is probably very localised, and the levels of damage are medium to low. Direct impacts on the livelihoods of commercial farmers are thus likely to be negligible, although they may negatively impact on the attitudes of farmers to elephants and conservation areas. Crop-raiding on commercial farms was largely eliminated around the Addo Elephant National Park with the construction of the elephant-proof Armstrong fence in the 1950s (Woodd, 1999). However, raiding of neighbouring citrus farms may become a problem in the future as the elephant population expands into new sections of the park fenced with more conventional game fencing. In the lowveld, elephants used to make frequent forays into commercial sugar cane farms south of the Kruger in the dry season. This was a major reason for electrifying the southern boundary of the park (Bigalke, 2000).

Conflicts with livestock

Few studies have quantified the impacts of elephants on livestock in Africa. Elephants may chase or even occasionally kill livestock (Gadd, 2005; Thouless, 1994), and in situations where elephants live outside of parks, they may also compete with livestock for food (Gadd, 2005; Young *et al.*, 2005) and water (Kuriyan, 2002; Gadd, 2005). However, the scant evidence which exists suggests that conflict with livestock is a minor issue, especially when compared to crop-raiding and the social impacts of HEC (Gadd, 2005; Malima *et al.*, 2005).

We found no documented evidence of disturbance of livestock by elephants in South Africa in recent times. Given that the minority of rural households in the former homelands own cattle (Shackleton *et al.*, 2001), the implications for rural livelihoods of any isolated incidents which might occur are negligible. Similarly, direct impacts on commercial cattle farmers are inconsequential. Implications of fence-breaking for livestock are dealt with under *indirect impacts*.

Damage to property and infrastructure

Property and infrastructure damaged by elephants around human settlements typically includes fences, food stores, and water sources (Kangwana, 1995; Kiiru, 1995; Hoare, 1999; Gadd, 2005; Malima *et al.*, 2005). As in the case of livestock, few studies have quantified this. Destruction of fences is usually collateral damage associated with crop-raiding, while damage to other property or infrastructure appears to be occasional and localised. It is therefore an unimportant issue outside of protected areas at the national level in South Africa. Although the literature focuses on elephants breaking out of protected areas, it should be noted that illegal immigrants and poachers also sometimes cut fences, allowing elephants to leave these areas freely. There is anecdotal evidence that within protected areas, elephants sometimes cause substantial damage to infrastructure, particularly water pipes.

Human injury and loss of life

An objective assessment of the relative impact of injury or loss of human life due to elephants is hard to achieve. Human life should not be lumped together with the value of crops damaged and property destroyed, nor weighed up against the life of an elephant. It could justifiably be argued that one human death is one too many, and the same could be said for an elephant killed. However, human injury and death caused by elephants are very rare events (Tchamba, 1995;

Sitati et al., 2003; Malima et al., 2005), possibly accounting for less than 0.5 per cent of all HEC incidents (Tchamba, 1996; Malima et al., 2005). Incidents resulting in human injury or death are usually 'unfortunate spatial coincidences' when the paths of elephants and people cross (Sitati et al., 2003).

The risk of being killed by an elephant is very low, especially compared to other causes of mortality, such as malaria or motor vehicle accidents (Kuriyan, 2002). Thus, although every injury or death due to elephants is a regrettable tragedy, devastating at the household level, it is not a significant problem at the national level. However, even isolated incidents fuel the pervasive fear of elephants in rural communities, even at some distance from protected areas (Kaltenborn *et al.*, 2006), and sour local perceptions of wildlife and conservation (Thirgood *et al.*, 2005).

Year	Protected areas	Communal land	Enterprises using tame elephants	Total
2002	2	1	0	3
2003	3	2	0	5
2004	2	0	0	2
2005	4	0	1	5
2006	2	0	1	3
2007	0	1	0	1
Total number	13	4	2	19
Per cent	60%	21%	11%	

Table 2: Annual numbers of human deaths caused by elephants in protected areas, communal lands, and enterprises using tame elephants (e.g. elephant theme parks and elephant-back safari operations) in South Africa (from media reports)

In South Africa, based on information gleaned from the media, no more than five people were killed by elephants in any given year over the last five years (table 2). This includes animal handlers killed in enterprises involving tame elephants, such as elephant theme parks and elephant-back safari operations. An important observation is that 72 per cent of all recorded fatalities since 2002 occurred in protected areas, compared with 17 per cent in communal lands and 11 per cent in elephant-based enterprises. It is possible that there may be some minor under-reporting of incidents in remote rural locations outside of protected areas, but it is unlikely that this dramatically alters the picture. These data suggest that staff, tourists, scientists, hunters, and poachers in protected areas are more at risk than neighbouring communities from attack by elephants. The threat particularly to staff and tourists may be increasing due to growing densities of elephant (see Chapter 8) or tourists (Nel, 2004) in many parks.

Indirect costs

The indirect costs of HEC, such as disturbance of normal human activities, are significant and may even outweigh the direct costs in people's experience of conflict with elephants (Hoare, 2000; Dublin & Hoare, 2004; Sitati *et al.*, 2005). Because indirect costs are difficult to quantify, it is not possible to assess their relative impact compared to direct costs. HEC colours rural communities' sentiments towards elephants and conservation in general (Naughton *et al.*, 1999; O'Connell-Rodwell *et al.*, 2000; Dublin & Hoare, 2004; Gadd, 2005), which is a serious indirect cost borne by governments and conservation authorities (De Boer & Baquete, 1998; Naughton *et al.*, 1999; O'Conell-Rodwell *et al.*, 2000; Dublin & Hoare, 2004; Thirgood *et al.*, 2005).

Indirect costs of HEC have not been investigated in South Africa, but based on the relatively low incidence of elephants leaving protected areas, these are likely to be very low for affected communities. An issue not mentioned in studies elsewhere in Africa is the costs associated with other wildlife leaving protected areas when fences are damaged by elephants. This provides opportunities for predators to kill livestock and for the transmission of disease such as foot and mouth disease and corridor disease between wildlife and cattle (see Chapter 7). This has not been quantified, but may be more significant than other social costs to those impacted by sporadic incidents of elephants escaping from reserves. However, the animosity HEC creates towards conservation among rural communities is the most serious indirect cost in South Africa.

Factors determining risk and intensity of conflict

Studies from across Africa reveal a range of spatial, temporal and other factors which influence risk and intensity of HEC. Risk is not evenly distributed and appears to be less predictable in space than in time (Sitati *et al.*, 2003).

Spatial factors

One of the clearest spatial risk factors is distance from the boundary of the protected area. A growing number of studies show that incidence of HEC increases sharply with proximity to protected areas (Barnes *et al.*, 1995; Naughton-Treves, 1998; Naughton *et al.*, 1999; O'Connell-Rodwell *et al.*, 2000; De Boer & Ntumi, 2001; Parker & Osborn, 2001; Barnes *et al.*, 2005; Sam *et al.*, 2005). Exceptions to this pattern are rare (e.g. Hoare, 1999; Smith & Kasiki, 2000; Sitati *et al.*, 2003). Households most affected by damage-causing animals, including elephants, in the communal lands next to the Kruger were within 3 km of the park boundary (Anthony, 2007).

Proximity to rivers in the dry season is another possible distance predicator of risk. Parker & Osborn (2001) showed that lower frequency but higher intensity crop-raiding occurred close (<5 km) to rivers in the dry season. Communities close to a park boundary defined by a river may thus be at significantly greater risk of crop-raiding, particularly in the dry season. This is intuitive, but needs to be validated with more data. Risk of elephant-induced injury or death may be positively correlated with proximity to roads because of the higher probability of human–elephant encounters along transport routes (Sitati *et al.*, 2003), despite elephant densities possibly increasing with distance from roads (Blom *et al.*, 2004).

Some studies have shown that risk of crop-raiding generally increases with field size (Barnes *et al.*, 2005; Sitati *et al.*, 2005). Others have found that total area of land cultivated in a region, rather than area of an individual field, increases the risk of a field being raided (Sam *et al.*, 2005; Sitati *et al.*, 2005). Yet other studies have shown that total area of land cultivated around a settlement is a weak predictor of risk, and that smaller, more isolated farms are more vulnerable (Malima *et al.*, 2005; Lee & Graham, 2006). We conclude that no consistently predictable relationship exists between risk of crop-raiding and area of cultivation. This may be because of confounding factors such as type and number of crops grown. For example, maize, which is the staple food crop in much of the continent including South Africa, is favoured by elephants (Taylor, 1993; Kiiru, 1995; Smith & Kasiki, 2000; De Boer & Ntumi, 2001; Barnes *et al.*, 2005). Crop-raiding may also increase with increasing number of crops grown (Barnes *et al.*, 2005; Sam *et al.*, 2005). These results imply that rural communities growing maize, along with a mix of other crops, in communal lands may be at greater risk than commercial mono-crop farms. However, a lack of comparative studies precludes affirmation of this possibility.

Neither human nor elephant densities appear to be good predictors of the *amount* of direct conflict between people and elephants (Naughton-Treves, 1998; Hoare, 1999; Hoare, 2000; Dublin & Hoare, 2004). However, Nel (2004) observed a correlation between the increasing number of 'serious elephant incidents' and the rising number of tourist beds (an indication of the number of game drives) in Madikwe Game Reserve from 1992 to 2004. As he pointed out, this apparent relationship needs to be explored further.

Temporal factors

The large majority of crop-raiding incidents and elephant attacks on people in communal lands occur at night (Hillman Smith *et al.*, 1995; Hoare, 1999; Smith & Kasiki, 2000; Osborn & Parker 2003; Sitati *et al.*, 2003; Sitati *et al.*, 2005; Kioko *et al.*, 2006). Elephants are most likely to raid crops after dark in order to minimise the risk of being detected. This would account for crop-raiding being lowest during full moon (Barnes *et al.*, 2007). By contrast, almost all elephant attacks on people in protected areas in South Africa occur during the day, when humans are most active in elephant habitat.

Elephant crop-raiding is strongly seasonal, with highest frequency of raids occurring when crops are mature and ready for harvesting (Hillman Smith *et al.*, 1995; Kiiru, 1995; Hoare, 1999; Parker & Osborn, 2001; Adjewodah *et al.*, 2005; Malima *et al.*, 2005; Sam *et al.*, 2005). This has also been noted in communal lands of South Africa (Spenceley, 2005), and is particularly frustrating for farmers.

Behavioural factors

Risk of HEC incidents is also influenced by the behaviour of both elephants and humans. Crop-raiding usually involves female-led mixed groups (Smith & Kasiki, 2000; Sitati *et al.*, 2003; Malima *et al.*, 2005), although lone bulls or small male groups may be the dominant crop-raiders in particular areas (Hoare, 1999; Chiyo & Cochrane, 2005; Kioko *et al.*, 2006). Habitual fence-breakers or crop-raiders are often bulls (Thouless & Sakwa, 1995; Hoare, 1999). Most crop-raiding groups are relatively small, consisting of 10 or fewer individuals (Smith & Kasiki, 2000; Sitati *et al.*, 2003; Malima *et al.*, 2005; Kioko *et al.*, 2006).

Bulls are more risk-tolerant than females and are therefore more likely to be problem animals in risky situations, such as close to towns and roads (Hoare, 1999; Sitati *et al.*, 2003). The physiological state of individual bulls influences risk of life-threatening encounters with them. Males can be extraordinarily aggressive when in musth, a period of heightened testosterone levels indicated by copious secretion from the temporal glands (Poole & Moss, 1981). However, bulls in this state are less likely to engage in crop-raiding as their priorities change from feeding to fighting and breeding (Hall-Martin, 1987; Poole, 1989). Human behaviour is also a key factor, and attacks on people by elephants are usually associated with situations where people get too close to elephants which are traumatised, sick, injured, harassed, bulls in musth, or females with young calves (Leggat *et al.*, 2001). Bradshaw *et al.* (2005) and Bradshaw

& Schore (2007) argue that stress caused by social disruptions associated with culling, translocation and habitat loss underlie such aggressive behaviour.

Mitigation of HEC

Over a decade of research on HEC has yielded insights on mitigation strategies which show potential for reducing conflict between people and elephants, and increasing tolerance of affected communities towards elephants. These lessons are useful for informing policy and national mitigation strategies to address both current and future HEC scenarios. The IUCN African Elephant Specialist Group (AfESG) uses the term 'mitigation' rather than 'prevention', based on the belief that HEC can never be totally eliminated, but should be reduced to local tolerance levels (Dublin & Hoare, 2004). Nelson *et al.* (2003) provide a detailed review of strategies for managing HEC. Options for mitigating HEC are briefly discussed below, and methods for changing elephant behaviour are discussed in more detail in Chapter 7.

Strategies used by rural communities

Traditional methods still used by rural communities include guarding fields, making loud noises, making fires, clearing field boundaries, erecting simple barriers, planting decoy foods or unpalatable crops, and using traps, spikes and home-made weapons (Nelson *et al.*, 2003). These are largely ineffective, especially in the long term (Osborn & Parker, 2002; 2003; Nelson *et al.*, 2003). Elephants often become habituated to some of these methods (Thouless, 1994). Little is known about indigenous knowledge relevant to mitigating elephant impacts in pre-colonial times.

Strategies currently used by conservation authorities

Electric fences are the most effective barrier to elephants (O'Connell-Rodwell et al., 2000; Osborn & Parker, 2003) and are the most important and effective proactive HEC mitigation strategy employed in South Africa. However, although they can substantially reduce incidents of HEC, they are not impregnable to elephants (see Chapter 7) and their effectiveness is highly dependent on regular maintenance (Nelson *et al.*, 2003). Rural communities who experienced damage from elephants escaping from the Kruger blamed the park authorities for not maintaining the boundary fence (Anthony, 2007). Fences may be most effective when combined with punishing offenders (O'Conell-Rodwell *et al.*, 2000). Refer to Chapter 7 for a discussion of the effects of fences on elephant movement.

Reactive strategies currently used by South African conservation authorities are *disturbance methods* (firing weapons to scare off elephants and driving stray elephants back into parks with helicopters, vehicles or people) and *problem animal control* (killing problem elephants). Although firing weapons to scare elephants usually provides initial relief, it is seldom effective in the long term (Nelson *et al.*, 2003), and elephants can become habituated to such techniques (Kangwana, 1995). Stress associated with some of these techniques may exacerbate the problem (Bradshaw & Schore, 2007).

Killing problem animals is a quick-fix solution with high public relations value because authorities are seen to be doing something and communities usually get the meat. Although often regarded as one of the most effective means of controlling problem elephants, limitations of this approach include: (1) it is dangerous and needs to be conducted by well-trained personnel, (2) it is often difficult to identify culprits since elephant forays out of protected areas usually occur at night, and the culprit may rejoin herds once back in the park, (3) it is a poor deterrent to other elephants, and other individuals may move in to replace the culprit as problem elephants, (4) it may cause stress in other elephants, (5) response by centralised authorities is often slow, and (6) it raises difficult ethical questions (Kangwana, 1995; Hoare 2001; Osborn & Parker, 2003; Nelson *et al.*, 2003; Burke, 2005; Bradshaw & Schore, 2007). Refer to Chapter 8 for further discussion of lethal methods of controlling problem elephants.

In the context of South African legislation (*res nullius* principle – see Chapter 11), the authority responsible for destroying a problem animal is usually not the authority managing the protected area from which the elephant escaped. This may result in further delays and cause confusion among affected communities as to who is responsible (SANParks, 2005). Currently, South African communities receive meat from shot animals, but do not get any direct economic benefit from them.

Other strategies for consideration

One of the first responses of rural communities to damage caused by wildlife is to demand *compensation*, especially if the animals are viewed as property of the state (Nelson *et al.*, 2003; Nyhus *et al.*, 2005). A number of African states experiencing HEC, including Kenya, Botswana, Malawi and Zimbabwe, have implemented or experimented with compensation schemes, and most have abandoned them. Major problems with compensation schemes include

(1) high administration costs, (2) lack of funds, (3) challenge of accurately and promptly verifying damage, (4) lodging of fraudulent claims, (5) disincentives for guarding fields, (6) subsidising uneconomical agriculture, and (7) no discernable improvement in relations between communities and conservation authorities (Bell, 1984; Thouless, 1994; Taylor, 1993; Taylor, 1999; Bulte & Rondeau, 2005; Nyhus *et al.*, 2005). Like many of the other strategies, compensation addresses the effects of HEC, not the cause (Hoare, 1995), and should therefore complement proactive measures to reduce HEC incidents (Nyhus *et al.*, 2005). This highly emotive issue needs to be considered as a policy option with caution.

Experimentation with *repellents* has shown that chilli (*Capsicum*) products, such as aerosol sprays, grease or smoke are effective in repelling elephants (Osborn & Rasmussen, 1995; Osborn, 2002; Nelson *et al.*, 2003). However, these are expensive (Osborn, 2002), and therefore not viable on a large scale or in poor rural communities if not subsidised or locally produced using simple technology. Most other repellent techniques, such as auditory repellents (e.g. elephant distress calls) (O'Connell-Rodwell *et al.*, 2000) and bees (Karidozo & Osborn, 2005), are ineffective.

Since rural communities incur costs from elephant damage and mitigation, they should also receive greater benefits, which would increase community tolerance of elephants (Leader-Williams & Hutton, 2005; Walpole *et al.*, 2006). Options for *community beneficiation* range from 'outreach programmes' in which revenues from protected areas are shared with neighbouring communities, to Community Based Natural Resource Management (CBNRM) projects where communities are empowered to manage their natural resources and earn income from elephants on their own land, through tourism or commercial hunting concessions (Nelson *et al.*, 2003). Some of this income could be used to insure households against damages caused by elephants. Ironically, the *res nullius* law (see Chapter 11) which creates ambiguity around managing HEC in South Africa, also provides an opportunity for communities to benefit economically from elephants that wander onto communal land. CBNRM holds much promise for enabling communities to better respond to HEC and realise benefits from elephants (Omondi *et al.*, 2004). A positive spin-off is increased community tolerance of elephants (Taylor, 1993). However, CBNRM is difficult to apply for many reasons, many of which have to do with the complex nature of communities (Nelson *et al.*, 2003; Koch, 2004). CBNRM is in its infancy in South Africa, and the role of elephants in revenue generation in the flagship Makuluke project remains to be seen. For further discussion of the economic benefits of CBNRM for rural communities, see Chapter 10.

One option for generating revenue for communities and conservation is to integrate *commercial hunting* safaris into problem animal control strategies (Leader-Williams & Hutton, 2005). This has been employed in CAMPFIRE projects in Zimbabwe where income from commercial hunts has been shared with local communities as an incentive to conserve wildlife on communal land (Taylor, 1993). However, challenges include: (1) quotas can be manipulated, (2) problem elephants are not always desirable trophy animals, (3) problem elephants are often difficult to identify, and 4) professional hunters seldom have clients ready and waiting to quickly respond to a HEC incident (Taylor, 1993; Nelson *et al.*, 2003). Members of rural communities adjacent to Kruger have expressed interest in being trained as professional hunters (SANParks, 2005), which may improve the efficiency of problem animal control while creating local employment and revenue. Thus, although hunting problem elephants is unlikely to be an effective method for reducing the incidence of elephant break-outs from reserves, it has potential of contributing to poverty mitigation and increasing local tolerance of HEC.

Land-use planning has been identified as being fundamental to managing HEC (Hoare, 2000; Omondi *et al.*, 2004). This can occur at a national level, such as in Namibia, where the entire country was classified into different elephant use zones (Kangwana, 1995), and at local level, such as the Nyaminyami CAMPFIRE project in Zimbabwe, where communal land was zoned into settlements and fields, elephant sanctuary, and safari hunting areas (Taylor, 1993). Buffer zones – areas with low human and elephant density and minimal agriculture – between protected areas and settlements could lessen the incidence of conflict between humans and elephants (Taylor, 1982). However, large areas of high human density and transformed commercial agricultural land abut protected areas in South Africa. These contexts produce hard edges and an inflexible land-use template, which pose a challenge to land-use planning for mitigating HEC. Nevertheless, this needs to be explored further.

HEC policy

Lack of adequate HEC policy leads to crisis management which focuses on the effects instead of the causes of the problem (Kangwana, 1995). The negative political impacts of HEC are usually also disproportionate to the actual impacts on people and their livelihoods. Clear policy on dealing with problem elephants is thus vital for government credibility (Dublin & Hoare, 2004), especially given the emotive nature of the issue. Policy should clearly state who holds responsibility for problem elephants and define appropriate responses for particular situations (Kangwana, 1995). Greater attention needs to be given to involving and empowering local communities in HEC mitigation, including mechanisms for communities to gain more economic benefits from elephants. It is clear that no single strategy on its own will be sufficient and policy will thus need to integrate different approaches to addressing HEC proactively (Omondi *et al.*, 2004; Walpole *et al.*, 2006).

For South Africa, HEC policy is also important within the regional context, given the emergence of transfrontier conservation areas (TFCAs) with neighbouring countries such as Mozambique, Swaziland, Zimbabwe and Botswana. Three existing or potential TFCAs, namely the Great Limpopo Transfrontier Park (GLTP), Lubombo Transfrontier Conservation Area (LTCA), and Limpopo-Shashe Transfrontier Conservation Area (LSTCA), contain populations of elephants. Indeed, providing more habitat for elephants, reconnecting isolated populations, and relieving population pressure in areas with high elephant densities has been an important conservation motivation for creating TFCAs (Hanks, 2000; 2003). A potential policy consideration is South Africa's position in situations where neighbouring states experience high levels of HEC in TFCAs sharing elephant populations with South Africa.

CONCLUSION

Elephants and people interact with each other both directly and indirectly, and positively and negatively. Our Assessment has shown that most of these interactions occur within conservation areas, and are predominantly positive. Further, levels of direct conflict between humans and elephants outside of protected areas are generally low. However, the impact of sporadic conflict incidents outside of protected areas has a negative effect on local attitudes towards elephants and conservation, often disproportionate to the actual damage caused. An important observation is that human–elephant interactions ending in human injury and death are very rare and occur mainly in protected areas. Phenomena such as the expansion of conserved land, growing elephant populations, intensification of ecotourism, and increasing inclusion of rural communities in resource management are expected to intensify interactions between humans and elephants in South Africa. This will necessitate adequate policy and management strategies for mitigating and responding to conflict between humans and elephants. At the same time, efforts need to be made to address some of the misconceptions about human–elephant conflicts.

Key information gaps identified by the authors, particularly for the South African context, include:

- The historical relationship between indigenous human populations and elephants, including indigenous knowledge and cultural beliefs, practices and values.
- Contemporary societal value systems underpinning opinions that give rise to conflicts over elephant management, and the factors shaping these value systems.
- Size, characteristics, and opinions of different elephant stakeholder groups.
- Comprehensive and consistent records of elephant break-outs from protected areas, analysis of factors determining the probability of break-outs, and quantification of direct and indirect impacts on humans.
- The incidence and determinants of human–elephant conflict within protected areas, including attacks on humans, destruction of infrastructure, and tourist pressure on elephants.
- Locally appropriate models for beneficiation of rural communities from elephant conservation and management.

ACKNOWLEDGEMENTS

We would like to thank our reviewers, the editors, and Jane Carruthers for valuable comments and suggestions for this chapter.

REFERENCES

Adjewodah, P., P. Beier, M.K. Sam & J.J. Mason 2005. Elephant crop damage in the Red Volta Valley, north-eastern Ghana. *Pachyderm* 38, 39–48.

Anderson, B.A., J.H. Romani, H.E. Phillips & J.A. van Zyl 2002. Environment, access to health care and other factors affecting infant and child survival among the African and coloured populations of South Africa, 1989-94. *Population and Environment* 23, 349–364.

Anthony, B. 2007. The dual nature of parks: attitudes of neighbouring communities towards Kruger National Park, South Africa. *Environmental Conservation* 34, 236–245.

Anthony, B.P. & E.G. Bellinger 2007. Importance value of landscapes, flora and fauna to Tsonga communities in the rural areas of Limpopo Province, South Africa. *South African Journal of Science* 103, 148–154.

Barnes, R.F.W., S. Azika & B. Asamoah-Boateng 1995. Timber, cocoa and crop-raiding elephants: a preliminary study from southern Ghana. *Pachyderm* 19, 33–38.

Barnes, R.F.W., E.M. Hema, A. Nandjui, M. Manford, U.F. Dubiure, E.K.A. Danquah & Y. Boafo 2005. Risk of crop-raiding by elephants around the Kakum Conservation Area, Ghana. *Pachyderm* 39, 19–25.

Barnes, R.F.W., U.F. Dubiure, E. Danquah, Y. Boafo, A. Nandjui, E. M. Hema, M. Manford 2007. Crop-raiding elephants and the moon. *African Journal of Ecology* 45, 112–115.

Bell, R. 1983. Deciding what to do and how to do it. In: A.A. Ferrar (ed.) *Guidelines for the management of large mammals in African conservation areas.* Report No. 69, South African National Scientific Programmes, Pretoria, 51–75.

Bell, R.H.V. 1984. The man-animal interface: an assessment of crop damage and wildlife control. In: R.H.V. Bell & E. McShane-Caluzi (eds) *Conservation and Wildlife Management in Africa.* Proceedings of a workshop organised by the U.S. Peace Corps at Kasungu National Park, Malawi, October 1984. US Peace Corps, Malawi, 387–416.

Bigalke, R.C. 2000. Functional relationships between protected and agricultural areas in South Africa and Namibia. In: H. H. T. Prins, J. G. Grootenhuis & T. T. Dolan (eds) *Wildlife conservation by sustainable use.* Kluwer Academic Press, Dordrecht, 169–201.

Blom, A., R. van Zalinge, E. Mbea, I.M.A. Heitkönig & H.H.T. Prins 2004. Human impact on wildlife populations within a protected central African forest. *African Journal of Ecology* 42, 23–31.

Bradshaw, G.A., A.N. Schore, J.L. Brown, J. H. Poole & C.J. Moss 2005. Elephant breakdown. *Nature* 433, 807.

Bradshaw, G.A. & A.N. Schore 2007. How elephants are opening doors: developmental neuroethology, attachment, and social context. *Ethology* 113, 426–436.

Bulte, E.H. & D. Rondeau 2005. Why compensation schemes may be bad for conservation. *Journal of Wildlife Management* 69, 14–19.

Burke, T. 2005. The effect of human disturbance on elephant behaviour, movement dynamics and stress in a small reserve: Pilanesberg National Park. MSc thesis, University of KwaZulu-Natal, Durban.

Carrington, R. 1958. *Elephants: A short account of their natural history, evolution and influence on mankind.* Chatto & Windus, London.

Carter, M.R. & J. May 1999. Poverty, livelihood and class in rural South Africa. *World Development* 27, 1–20.

Caughley, G & A.R.E. Sinclair 1994. *Wildlife ecology and management.* Blackwell Science, Oxford.

Chiyo, P.I. & E.P. Cochrane 2005. Population structure and behaviour of crop-raiding elephants in Kibale National Park, Uganda. *African Journal of Ecology* 43, 233–241.

Clutton-Brock, J. 2000. Cattle, sheep and goats south of the Sahara: an archaeozoological perspective. In: R. Blench & K. MacDonald (eds) *The origins and development of African livestock: archaeology, genetics, linguistics and ethnography*. University College London Press, London, 30–37.

Conover, M.R. & D.O. Conover 2003. Unrecognized values of wildlife and the consequences of ignoring them. *Wildlife Society Bulletin* 31, 843–848.

Crookes, D. 2003. The contribution of livelihood activities in the Limpopo Province: case study evidence from Makua and Manganeng. *Development Southern Africa* 20, 144–159.

De Boer, F. & D.S. Baquete 1998. Natural resource use crop damage and attitudes of rural people in the vicinity of the Maputo Elephant Reserve, Mozambique. *Environmental Conservation* 25, 208–218.

De Boer, F. & C. Ntumi 2001. Elephant crop damage and electric fence construction in the Maputo Elephant Reserve. *Pachyderm*, 57–64.

Dougherty, E.M., D.C. Fulton & D.H. Anderson 2003. The influence of gender on the relationship between wildlife value orientations, beliefs, and the acceptability of lethal deer control in Cuyahoga Valley National Park. *Society and Natural Resources* 16, 603–623.

Dublin, H.T. 2005. African elephant specialist report. *Pachyderm* 38, 1–10.

Dublin, H.T. & R.E. Hoare 2004. Searching for solutions: the evolution of an integrated approach to understanding and mitigating human–elephant conflict in Africa. *Human Dimensions of Wildlife* 9, 271–278.

Fabricius, C., E. Koch & H. Magome 2001. Towards strengthening collaborative ecosystem management: lessons from environmental conflict and political change in southern Africa. *Journal of the Royal Society of New Zealand* 31, 831–844.

Ferrar, A.A. (ed.) 1983. *Guidelines for the management of large mammals in African conservation areas*. CSIR, Pretoria.

Ford, J. 1966. The role of elephants in controlling the distribution of tsetse flies. *IUCN Bulletin* 19, 6.

Freitag-Ronaldson, S. & L.C. Foxcroft 2003 Anthropogenic influences at the ecosystem level. In: J.T. du Toit, K. H. Rogers & H. C. Biggs (eds) *The Kruger experience: ecology and management of savanna heterogeneity*. Island Press, Washington, 391–421.

Fulton, D.C., M.J. Manfredo & J. Lipscomb 1996. Wildlife value orientations: A conceptual and measurement approach. *Human Dimensions of Wildlife* 1, 24–47.

Gadd, M.E. 2005. Conservation outside of parks: Attitudes of local people in Laikipia, Kenya. *Environmental Conservation* 32, 50–63.

Gilbert, F.F. & D.G. Dodds 2001. *The philosophy and practice of wildlife management.* Krieger, Florida.

Giles, R.H. 1978. *Wildlife management.* W.H. Freeman, San Francisco.

Glaser, M. 2006. The social dimension in ecosystem management: strengths and weaknesses of human-nature mind maps. *Human Ecology Review* 13, 122–142.

Hall-Martin, A.J. 1987. Role of musth in the reproductive strategy of the African elephant (*Loxodonta africana*). *South African Journal of Science* 83, 616–620.

Hall-Martin, A.J. 1992. Distribution and status of the African elephant *Loxodonta africana* in South Africa, 1652–1992. *Koedoe* 35, 65–88.

Hanks, J. 2000. The role of Transfrontier Conservation Areas in southern Africa in the conservation of mammalian biodiversity. In: A. Entwistle & N. Dunstone (eds) *Priorities for the conservation of mammalian diversity. Has the panda had its day?* Conservation Biology 3, Cambridge University Press, Cambridge, 239–256.

Hanks, J. 2003. Transfrontier Conservation Areas (TFCAs) in southern Africa: their role in conserving biodiversity, socioeconomic development and promoting a culture of peace. *Journal of Sustainable Forestry* 17, 127–148.

Hillman Smith, A.K.K., E. de Merode, A. Nicholas, B. Buls & A. Ndey 1995. Factors affecting elephant distribution at Garamba National Park and surrounding reserves, Zaire, with focus on human–elephant conflict. *Pachyderm* 19, 39–48.

Hoare, R.E. 1995. Options for the control of elephants in conflict with people. *Pachyderm* 19, 54–63.

Hoare, R.E. 1999. Determinants of human–elephant conflict in a land-use mosaic. *Journal of Applied Ecology* 36, 689–200.

Hoare, R.E. 2000. African elephants and humans in conflict: the outlook for co-existence. *Oryx* 34, 34–38.

Hoare, R. 2001. Management implications of new research on problem elephants. *Pachyderm* 30, 44–48.

Hoare, R.E. & J.T. du Toit 1999. Coexistence between people and elephants in African savannas. *Conservation Biology* 13, 633–639.

Hough, J.L. 1988. Obstacles to effective management of conflicts between national parks and surrounding human communities in developing countries. *Environmental Conservation* 15, 129–135.

Jewell, P.A. & S. Holt (eds) 1981. *Problems in management of locally abundant wild mammals.* Academic Press, New York.

Kaltenborn, B.P., T. Bjerke & J. Nyahongo 2006. Living with problem animals – self reporting fear of potentially dangerous species in the Serengeti region of Tanzania. *Human Dimensions of Wildlife* 11, 397–409.

Kangwana, K. 1995. Human–elephant conflict: the challenge ahead. *Pachyderm* 19, 11–14.

Karidozo, M. & F.V. Osborn 2005. Can bees deter elephants from raiding crops? An experiment in the communal lands of Zimbabwe. *Pachyderm* 39, 26–32.

Kerley, G.I.H., B.G.S. Geach & C. Vial 2003. Jumbos or bust: do tourists' perceptions lead to an under appreciation of biodiversity? *South African Journal of Wildlife Research* 33, 13–21.

Kiiru, W. 1995. The current status of elephant-human conflict in Kenya, *Pachyderm* 19, 15–17.

Kioko, J., J. Kiringe & P. Omondi 2006. Human–elephant conflict outlook in the Tsavo-Amboseli ecosystem, Kenya. *Pachyderm* 41, 53–60.

Koch, E. 2004. Putting out fires: does the 'C' in CBNRM stand for community or centrifuge? In: C. Fabricius & E. Koch (eds) *Rights, resources and rural development: Community-based natural resource management in Southern Africa.* Earthscan, London, 78–92.

Kuriyan, R. 2002. Linking local perceptions of elephants and conservation: Samburu pastoralists in northern Kenya. *Society and Natural Resources* 15, 949–957.

Lahm, S.A. 1996. A nationwide survey of crop-raiding by elephants and other species in Gabon. *Pachyderm* 21, 69–77.

Leakey, R.E. 2001. *Wildlife wars: My battle to save Kenya's elephants.* Macmillan, London.

Leader-Williams, N. & J.M. Hutton 2005. Does extractive use provide opportunities to offset conflicts between people and wildlife? In: R. Woodroffe, S. Thirgood & A. Rabinowitz Book (eds) *People and wildlife: conflict or coexistence?* Cambridge University Press, Cambridge, 140–161.

Lee, P.C. & M.D. Graham 2006. African elephants *Loxodonta africana* and human–elephant interactions: Implications for conservation. *International Zoo Yearbook* 40, 9–19.

Leggat, P.A., D.N. Durrheim & L. Braack 2001. Travelling in wildlife reserves in South Africa. *Journal of Travel Medicine* 8, 41–45.

Leroy, J.L.J.P., J. van Rooyen , L. D'Haese & A-M. de Winter 2001. A quantitative determination of the food security status of rural farming households in the Northern Province of South Africa. *Development Southern Africa* 18, 5–17.

Lindsey, P.A., R. Alexander, M.G.L. Mills, S. Romanach & R. Woodroffe 2007a. Wildlife viewing preferences of visitors to protected areas in South Africa: Implications for the role of ecotourism in conservation. *Journal of Ecotourism* 6, 19–33.

Lindsey, P.A., P.A. Roulet, S.S. Romanach 2007b. Economic and conservation significance of the trophy hunting industry in sub-Saharan Africa. *Biological Conservation* 134, 455–469.

Lockwood, M. 2006. Values and benefits. In: M. Lockwood, G.L. Worboys & A. Kothari (eds) *Managing protected areas: a global guide.* Earthscan, 101–115.

Mabunda, D. 2005. *Report on the elephant management strategy.* Report to the Minister: Environmental Affairs and Tourism. South African National Parks, Pretoria.

Malima, C., R. Hoare & J. Blanc 2005. Systematic recording of human–elephant conflict: a case study in south-eastern Tanzania. *Pachyderm* 38, 29–38.

Manfredo, M.J. & H.C. Zinn 1996. Population change and its implications for wildlife management in the new west: a case study of Colorado. *Human Dimensions of Wildlife* 1, 62–74.

May, J. 1990. The migrant labour system: changing dynamics in rural survival. In: N. Nattrass & E. Ardington Book (eds) *The migrant labour system: changing dynamics in rural survival.* Oxford University Press, Cape Town, 175–186.

Mutwa, C. 1997. *Isilwane – the animal. Tales and fables of Africa.* Struik Publishers, Cape Town.

Naughton, L., R. Rose & A. Treves 1999. *The social dimensions of human-elephant conflict in Africa: a literature review and case studies from Uganda and Camaroon.* Report to the African Elephant Specialist Group, Human-elephant Conflict Task Force. IUCN, Gland, Switzerland.

Naughton-Treves, L. 1998. Predicting patterns of crop damage by wildlife around Kibale National Park, Uganda. *Conservation Biology* 12, 156–168.

Nel, P. 2004. *Elephant management in Madikwe Game Reserve – conceptualization of a future approach.* Paper presented at the EMOA Elephant Symposium, 13-16 October 2004, Bakgatla, Pilanesberg National Park.

Nelson, A., P. Bidwell & C. Sillero-Zubiri 2003. *A review of human–elephant conflict management strategies.* People and Wildlife Initiative, Wildlife Conservation Unit, Oxford University.

Newmark, W.D., D.N. Manyanza, D-G.M. Gamassa & H.I. Sariko 1994. The conflict between wildlife and local people living adjacent to protected areas in Tanzania: human density as a predictor. *Conservation Biology* 8, 249–255.

Nyakaana, S., E. L. Abe, P. Arctander & H. R. Siegismund. 2001. DNA evidence for elephant social behaviour breakdown in Queen Elizabeth National Park, Uganda. *Animal Conservation* 4, 231–237.

Nyhus, P.J., S.A. Osofsky, P. Ferraro, F. Madden & H. Fischer. 2005. Bearing the cost off human-wildlife conflict: the challenges of compensation schemes. In: R. Woodroffe, S. Thirgood & A. Rabinowitz Book (eds) *People and wildlife*: *conflict or coexistence?* Cambridge University Press, Cambridge, 107–121.

O'Connell-Rodwell, C., T. Rodwell, M. Rice & L.A. Hart 2000. Living with the modern conservation paradigm: can agricultural communities co-exist with elephants? A five year case study in East Caprivi, Namibia. *Biological Conservation* 93, 381–391.

Omondi, P., E. Bitok & J. Kagiri 2004. Managing human–elephant conflicts: the Kenyan experience. *Pachyderm* 36, 80–86.

Osborn, F.V. 2002. *Capsicum* oleoresin as an elephant repellent: field trials in the communal lands of Zimbabwe. *Journal of Wildlife Management* 66, 674–677.

Osborn, F.V. & G.E. Parker 2002. Community-based methods to reduce crop loss to elephants: experiments in the communal lands of Zimbabwe. *Pachyderm* 33, 32–38.

Osborn, F.V. & G.E. Parker 2003. Towards an integrated approach for reducing the conflict between elephants and people: A review of current research. *Oryx* 37, 80–84.

Osborn, F.V. & L.E.L. Rasmussen 1995. Evidence for the effectiveness of an oleo-resin capsicum against wild elephants. *Pachyderm* 10, 15–22.

Owen-Smith, R.N. (ed.) 1983. *Management of large mammals in African conservation areas.* Haum, Pretoria.

Owen-Smith, N., G.I.H. Kerley, B. Page, R. Slotow & R.J. van Aarde 2006. A scientific perspective on the management of elephants in the Kruger National Park and elsewhere. *South African Journal of Science* 102, 389–394.

Parker, I.S.C. & A.D. Graham 1989. Elephant decline (Part I): downward trends in African elephant distribution and numbers. *International Journal of Environmental Studies* 34, 287–305.

Parker, G.E. & F.V. Osborn 2001. Dual-season crop damage by elephants in the Eastern Zambezi Valley, Zimbabwe. *Pachyderm* 30, 49–56.

Pollard, S., C. Shackleton & J. Carruthers 2003. Beyond the fence: people and the lowveld landscape. In: J.T. du Toit, K. H. Rogers & H. C. Biggs Book (eds) *The Kruger experience: Ecology and management of savanna* heterogeneity. Washington, Island Press, 422–446.

Poole, J.H. 1989. Mate guarding, reproductive success and female choice in African elephants. *Animal Behaviour* 37, 842–849.

Poole, J.H. & C.J. Moss 1981. Musth in the African elephant, *Loxodonta africana*. *Nature* 292, 380–381.

Pretorius, Y. 2004. Stress in the African elephant on Mabula Game Reserve, South Africa. MSc thesis. University of KwaZulu-Natal, Durban.

Rokeach, M. 1973 *The nature of human values*. Free Press, New York.

Rolston, H. 1988. Human values and natural systems. *Society and Natural Resources* 1, 271–283.

Saayman, M. & E. Slabbert 2004. A profile of tourists visiting the Kruger National Park. *Koedoe* 47, 1–8.

Sam, M.K., E. Danquah, S.K. Oppong & E.A. Ashie 2005. Nature and extent of human–elephant conflict in Bia Conservation Area, Ghana. *Pachyderm* 38, 49–58.

SANParks. 2005. *The great elephant indaba: Finding an African solution to an African problem*. South African National Parks, Pretoria.

SANParks, 2007. *The Annual Report 2006/7*. South African National Parks, Pretoria.

Seekings, J. 2000. Visions of society: Peasants, workers and the unemployed in a changing South Africa. *Journal for Studies in Economics and Econometrics* 24, 53–71.

Shackleton, C.M., S.E. Shackleton & B. Cousins. 2001. The role of land-based strategies in rural livelihoods: The contribution of arable production, animal husbandry and natural resource harvesting in the communal areas. *Development Southern Africa* 18, 581–604.

Sitati, N.W., M.J. Walpole, R.J. Smith & N. Leader-Williams 2003. Predicting spacial aspects of human–elephant conflict. *Journal of Applied Ecology* 40, 667–677.

Sitati, N.W., M.J. Walpole & N. Leader-Williams 2005. Factors affecting susceptibility of farms to crop-raiding by African elephants: using a predictive model to mitigate conflict. *Journal of Applied Ecology* 42, 1175–1182.

Sitati, N. 2007. Challenges and partnerships in elephant conservation and conflict mitigation. *Oryx* 41, 135–139.

Slotow, R., G. van Dyk, J.H. Poole, B. Page & A. Klocke 2000. Older bull elephants control young males. *Nature* 408, 425–426.

Slotow, R., D. Balfour, & O. Howison 2001. Killing of black and white rhinoceroses by African elephants in Hluhluwe-Umfolozi, South Africa. *Pachyderm* 31, 14–20.

Smith, D.A. 1989. Elephants and man, a big problem. *Canadian Veterinary Journal* 30, 785–878.

Smith, R.J. & S.M. Kasiki 2000. *A spatial analysis of human-elephant conflict in the Tsavo ecosystem, Kenya.* Report to the African Elephant Specialist Group, Human–elephant Conflict Task Force. IUCN, Gland, Switzerland.

Spenceley, A. 2005. Nature-based tourism and environmental sustainability in South Africa. *Journal of Sustainable Tourism* 13, 136–170.

Steel, B.S., P. List & B. Schindler 1994. Conflicting values about federal forests: a comparison of national and Oregon publics. *Society and Natural Resources* 7, 137–153.

Steenkamp, C. & D. Grossman 2001. *People and parks: Cracks in the paradigm.* Policy Think Tank Series No 10, IUCN, Pretoria.

Stiles, D. 2004. The ivory trade and elephant conservation. *Environmental Conservation* 31, 309–321.

Tarrant, M.A., A.D. Bright & H.K. Cordell 1997. Attitudes toward wildlife species protection: assessing moderating and mediating effects in the value-attitude relationship. *Human Dimensions of Wildlife* 2, 1–20.

Tarrant, M.A. & H.K. Cordell 2002. Amenity values of public and private forests: examining the values-attitude relationship. *Environmental Management* 30, 692–703.

Taylor, R.D. 1982. Buffer zones: Resolving conflict between human and wildlife interests in the Sebungwe region. *Zimbabwe Agricultural Journal* 79, 179–184.

Taylor, R.D. 1993. Elephant management in Nyaminyami district, Zimbabwe: turning a liability into an asset. *Pachyderm* 17, 19–29.

Taylor, R.D. 1999. *A review of problem elephant policies and management policies in southern Africa.* IUCN African Elephants Specialist Group Report.

Tchamba, M.N. 1995. The problem elephants of Kaele: a challenge for elephant conservation in northern Cameroon. *Pachyderm* 19, 26–32.

Tchamba, M.N. 1996. History and present status of the human/elephant conflict in the Waza-Logone region, Cameroon, West Africa. *Biological Conservation* 75, 35–41.

Thirgood, S., R. Woodroffe & A. Rabinowitz 2005. The impact of human-wildlife conflict on human lives and livelihoods. In: R. Woodroffe, S. Thirgood & A. Rabinowitz Book (eds) *People and wildlife: conflict or coexistence?* Cambridge University Press, Cambridge, 13–26.

Thouless, C.R. 1994. Conflict between humans and elephants on private land in northern Kenya. *Oryx* 28, 119–127.

Thouless, C.R. & J. Sakwa 1995. Shocking elephants: fences and crop raiders in Laikipia district, Kenya. *Biological Conservation* 72, 99–107.

Van Aarde, R.J. & T.P. Jackson 2007. Megaparks for metapopulations: addressing the causes of locally high elephant numbers in southern Africa. *Biological Conservation* 134, 298–297.

Vaske, J.J. & M.P. Donnelly 1999. A value-attitude-behavior model predicting wildland preservation voting intentions. *Society and Natural Resources* 12, 523–537.

Walpole, M. & N. Leader-Williams 2002. Ecotourism and flagship species in conservation. *Biodiversity and Conservation* 11, 543–547.

Walpole, M., N. Sitati , B. Stewart-Cox, L. Niskanen & P.J. Stephenson 2006. Mitigating human–elephant conflict in Africa: a lesson-learning and network development meeting. *Pachyderm* 41, 95–99.

Whitehouse, A.M. & G.I.H. Kerley 2002. Retrospective assessment of long-term conservation management of elephants in Addo Elephant National Park, South Africa. *Oryx* 36, 243–248.

Woodd, A.M. 1999. A demographic model to predict future growth of the Addo elephant population. *Koedoe* 42, 97–100.

Young, T.P., T.M. Palmer & M.E. Gadd 2005. Competition and compensation among cattle, zebras and elephants in a semi-arid savanna in Laikipia, Kenya. *Biological Conservation* 122, 251–259.

ELEPHANT TRANSLOCATION

Lead author: Douw G Grobler

Authors: JJ van Altena, Johan H Malan, and Robin L Mackey

INTRODUCTION

The development of elephant translocations in South Africa

THE NUMBER of game reserves and game ranches increased tremendously in South Africa over the past two decades, setting demands on the wildlife translocation industry that spurred the evolution and unique development of elephant translocation to the current level of proficiency. Initially, small groups of juvenile elephants, originating from culling operations in Kruger, were translocated to several game ranches and reserves all over South Africa (Du Toit, 1991). Larger groups were moved to places such as Pilanesberg and Madikwe National Parks in the North West Province, Hluhluwe-Umfolozi in KwaZulu-Natal and Songimvelo in Mpumulanga.

The first adult elephant groups were moved in 1993 from Gonarhezou in Zimbabwe to Madikwe National Park (200 elephants) and Phinda Game Reserve (10 elephants). In the following year, which also marked the end of elephant culling in Kruger, 146 elephants were moved from Kruger into various reserves, with 50 of them going to Welgevonden in Limpopo Province. An important landmark was achieved in 1997 with the first translocations of adult elephant bulls to Pilanesberg from Kruger, which now meant that any size of elephant could be moved, making South Africa a world leader on this front (Slotow & Van Dyk, 2002).

Historical problems and solutions

The translocated juvenile elephants formed large groups, were very secretive and avoided human contact, staying mostly in dense bush and thickets. There were reports of break-outs and abnormal aggression towards humans, and in some instances even fatal attacks (Slotow & Van Dyk, 2002). The introduction of family groups in Madikwe in 1993 had a positive effect on their behaviour and the majority of juveniles integrated with these herds and became less secretive afterwards. An additional dramatic reaction was the killing of black and white rhino by young, rogue elephant bulls coming into musth at an early age, especially in Pilanesberg National Park and Hluhluwe-Umfolozi Game Reserves. While the majority of problem cases were handled by destroying the specific culprits, the translocation of adult elephant bulls into Pilanesberg National Park and Hluhluwe–Umfolozi provided a long-term solution for the rhino killers (Slotow *et al.,* 2002).

Reasons for translocations

The most important reasons are ecological considerations, dealing with overpopulation by elephants at the source site. Additionally, translocation improves eco-tourism at the release sites. Elephants are one of the charismatic animals of the 'Big Five'; they enchant tourists and are perceived to display the spirit of Africa. Every self-respecting private game reserve wants to have elephants in order to attract tourists and provide a special experience.

Small founder population sizes make it probable that the genetic diversity of the elephant population will be reduced (Knight *et al.*, 1995). The introduction of new elephants into existing populations will add new genes to that population, for instance Kruger bulls into the Addo population.

Trophy hunting is practised in KwaZulu-Natal, North West Province and Limpopo Province and has created a market for adult elephant bulls. Animals are often translocated in view of future hunting opportunities. Stipulations from authorities in the field of nature conservation state that an elephant has to have the chance to procreate on a new reserve, or be there for at least three years, before it can be hunted. As with all other species of wildlife in South Africa, the provision of elephants for hunting needs to be sustainable in order to ensure its survival.

SELECTION OF ELEPHANTS FOR TRANSLOCATIONS

At the inception of translocation of adult elephants in 1993 from Kruger, specific herds or individuals were not identified for selection prior to translocation. On any given day of translocation, the first suitable elephant group found in the capture area on the day of translocation was herded to the closest road, captured and loaded. Table 1 indicates all the elephants translocated from Kruger during the period 1994–2006.

The last translocation of elephants from Kruger to private landowners within South Africa took place in 2002. Most of the translocations during that period were from other locations, such as Sabi Sand Game Reserve, Madikwe National Park, Kapama, Shamwari and Phinda Game Reserves (table 2). During this period, translocations from Kruger were mainly done to other parks or contractual parks, such as Marakele, Addo and Limpopo (Mozambique) National Parks, and to the Eastern Shores in KwaZulu-Natal.

It has become apparent over time that habituated and 'well-behaved' elephants at the capture site will display the same behaviour at the point of release. No problems have been encountered with the translocations of seven cohesive groups of elephants originating from the Sabi Sand Game Reserve where they were used to being viewed in close proximity by open game vehicles. Four cohesive groups translocated from Phinda Game Reserve in 2003 and three groups removed from Kapama in 2005 rendered similar results to the Sabi Sand elephants. In all of these cases, active viewing and close proximity of tourists to elephants took place within days after the groups were released into their new locations. The elephants in these translocation cases reacted calmly, without any aggression towards human activities or abnormal behaviour, indicating no evidence of stress or post-translocation trauma. Similar results were obtained by other operators with elephants originating from Shamwari, where 28 elephants have been moved to several destinations over the last four years (pers. comm. Johan Joubert, Shamwari).

Year	Total	Family group	Bulls	Mozambique
1994	146	146	0	0
1995	83	83	0	0
1996	52	52	0	0
1997	46	34	12	0
1998	31	13	18	0
1999	12	0	12	0
2000	49	22	27	0
2001	93	69	24	25
2002	84	67	17	48
2003	51	46	5	38
2004	0	0	0	0
2005	0	0	0	0
2006	2	0	2	0
Total	**741**	**502**	**128**	**111**

Table 1: Number of elephants translocated on a yearly basis from the Kruger National Park (1994–2006)

In three instances where the translocation of problem elephants was attempted, the problematic behaviour continued at the translocation site. (1) A known fence-breaking bull from Sabi Sand Game Reserve continued to break fences in KwaZulu Private Game Reserve and ultimately had to be destroyed in 2004. (2) A wild herd at Shambala was translocated to Rietfontein in 2005, but stayed wild and led to human–elephant conflicts. (3) An aggressive cow in a family group from Madikwe was moved to Bayethe Private Game Reserve in 2001 but had to be euthanised shortly afterwards. It is therefore important not to translocate problem elephants as a means of trying to solve a problem.

Year	Origin	Number	Destination
2001	Madikwe National Park	12	Sandhurst, Tosca
2001	Madikwe National Park	8	Bayete, Eastern Cape
2001	Madikwe National Park	9	Kwandwe, Eastern Cape
2002	Sabie Sand Game Reserve	10	Thanda, KwaZulu-Natal
2003	Phinda Private Game Reserve	7	Kwantu, Eastern Cape
2003	Phinda Private Game Reserve	10	Amakhala, Eastern Cape
2003	Phinda Private Game Reserve	9	Nanbithi, KwaZulu-Natal
2003	Phinda Private Game Reserve	12	Onverwacht, KwaZulu-Natal
2003	Shamwari Private Nature Reserve	9	Bushman Sands, Eastern Cape
2004	Sabie Sand Game Reserve	11	Kariega, Eastern Cape
2004	Sabie Sand Game Reserve	9	Asante Sana, Eastern Cape
2004	Sabie Sand Game Reserve	9	Shambala, Limpopo
2004	Kapama Game Reserve	8	Pumba, Eastern Cape
2004	Shambala	8	Rietfontein, North West Province
2004	Shamwari Private Nature Reserve	6	Hopewell, Eastern Cape
2004	Shamwari Private Nature Reserve	5	Sawubona, Eastern Cape
2005	Sabie Sand Game Reserve	7	Ka'Ingo, Limpopo
2005	Sabie Sand Game Reserve	8	Blaauwbosch, Eastern Cape
2005	Kapama Game Reserve	8	Mziki, North West Province
2006	Welgevonden	6	Lalibela, Eastern Cape
2006	Kapama Game Reserve	7	Sibuya, Eastern Cape
2006	Thukela Biosphere	9	Sanwild, Limpopo
2006	Shamwari	5	Mpongo Game Reserve, Eastern Cape
2006	Shamwari	3	Kqmala, Eastern Cape

Table 2: Elephants (number of individuals in family groups) translocated from other game reserves (2001–2006)

The known mortality rate of all elephants translocated since 1994 is only 2.7 per cent (27 known mortalities out of 1 014 elephants). The reasons for these deaths vary from the transport truck overturning, elephants with predisposed disease problems, elephants falling accidentally in water or down a cliff after being darted, and failure of equipment.

CURRENT TECHNIQUES AND EQUIPMENT

The techniques for elephant translocation have, in general, stayed the same over the past decade. However, various improvements (equipment enlargements and modifications) have been made to facilitate better and safer methods for manoeuvring elephants.

The translocation procedure is divided into six stages: capture (darting), recovery, wake-up, loading, transport, and release.

Figure 1: A wild elephant bull reacting to the effects of M99, fast asleep on its feet!

Capture

The immobilisation of elephants is achieved through the use of Schedule 7 drugs, thus making it imperative that a veterinarian direct the procedures. Immobilisation drugs such as M99 (etorphine hydrochloride) in combination with a tranquilliser such as Stresnil (Azaperone) have made it possible to safely anaesthetise elephants of different sizes and to allow manipulation for translocations (Du Toit, 2001). A strong, turbo-operated four-seat helicopter with an experienced helicopter pilot is an absolute necessity to safely conduct the capture process. The helicopter is used to dart the elephants as well as to steer them to a suitable area where they can be loaded.

Recovery

There are two ways of recovering elephants. The first involves using a powerful winch on a flatbed trailer pulled by a tractor. The elephant is rolled over onto a long stretcher made of a conveyor belt. Lying on its side, the elephant is then winched up onto the trailer (figure 2).

Figure 2: An anaesthetised elephant being winched up on a transport trailer

The second method is a recent development, and requires a strong hydraulic crane and special slings to lift the elephant by its legs (while supporting the head) onto a similar conveyor stretcher placed on the back of a flatbed truck (figure 3). Although this method does look awkward, it is a safe and fast loading method. After loading, the elephant is transported sedated, lying on its side on the conveyor stretcher, to the wake-up area.

Figure 3: Loading an anaesthetised elephant bull by means of a hydraulic crane

Wake-up crate

This is a very important piece of equipment in the loading process; its dimensions are such that it is large enough to accommodate a full-sized, mature elephant bull (up to 7 500 kg and 3.5 m shoulder-height). The transport truck and crates are set up in line with and adjacent to the wake-up container and hydraulic crane truck (figure 4). The elephant is winched or pulled into the wake-up crate with its backside going in first. At this stage the immobilising process is reversed by using antidotes such as M5050 (deprenorphine hydrochloride) and Naltrexone. Generally, the elephant wakes up within two minutes after the intravenous injection of the antidotes; it will stand up, and immediately walk backwards into the transport crate (figure 5).

Figure 4: Elephant on the crane truck just before being winched into the wake-up crate with the transport truck in the background

Transport

The crates and trailers that are used to move elephants must meet certain height and load requirements. A family group of elephants can easily weigh in excess of 30 tonnes, and a moving load of this nature has to be well balanced. A 12-m low-bed trailer is utilised to accommodate the abnormal load as well as to allow for the height restriction of 4.5 m generally in place on South African roads. The trailer is fitted with two crates, each 6 m long x 2.4 m wide x 3.4 m high, that house the elephants for the duration of the trip. The elephants are tranquillised before and during transport.

Family groups

Translocated elephant groups range in size from 7 to 12 elephants, and generally comprise adult cows and their offspring. The group is loaded into one transport unit comprising two crates. These crates have strong sliding doors with duplication locks and safety locks in place, making it impossible for even the strongest elephant to open them. The entire family/cohesive group is then moved together as a unit. During the loading process elephants are marked for identification because it is very important to have a cow and her own calves together in the same crate, otherwise it will lead to injuries or even death of the calves, as cows will only nurse and tend to their own direct offspring and often react aggressively towards other calves. The transport crates have an inside height measurement of approximately 3.4 m in order to accommodate large bulls if necessary.

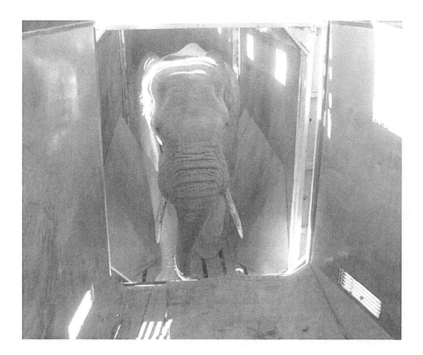

Figure 5: An elephant bull after awakening, walking backwards out of the wake-up crate into the transport truck

Adult bulls

When mature bulls are moved, normally only two individuals are moved at a time, each in its own crate. While mature bulls are transported individually due to their size and weight, younger bulls may be transported together. The crates used to move bulls are the same dimensions as those used to translocate family groups. The crates originally used to translocate Kruger elephant bulls have been made narrower (reduced from 2.4 m to 1.7 m wide) to ensure that the mature bulls stand in the middle line of the trailer to avoid the uneven distribution of weight (figure 6).

Figure 6: Specialised bull transport crates and truck

Importance of release boma and correct fencing

Elephants are intelligent animals and therefore different individuals react differently to the translocation experience, depending on their unique temperament and life experiences. For this reason, the release boma, constructed using electric fences, was introduced to provide translocated elephants with an adaptation or training period in order for them to be introduced to, or become accustomed to electric fences. This adaptation period is necessary for elephants that may never have encountered a fence before, as during this time they are able to develop a lasting respect for electrified fencing. The effectiveness of this boma and the management of this critical period of time are essential to the success of the entire operation.

The management of break-outs has varied in the past. In some cases, authorities have stepped in and destroyed escaped elephants without considering other options (Kruger, 2002). There have been instances where individuals that have repeatedly broken out of the boma and/or reserve, have been moved back to their place of origin. For example, an adult bull moved to Shambala in Limpopo was taken back to Kruger after two break-outs. Additionally a group of eight elephants in Mziki (North West Province) was returned to Marakele National Park after breaking out of the boma and reserve. There have also been instances where the matriarch or other members of the group had to be destroyed, for example in Phinda Game Reserve (1993), Shamwari, and Bayethe. In other cases, elephants that broke out of their new locations were brought back to that reserve without further problems (e.g. Mkuze Falls Game Reserve; Bayethe). It has also been reported that adult bulls are more prone to break out of their new destinations, and that there is a high correlation (80 per cent) between break-outs from the release boma and break-outs from the same reserve (Carr & Garaï, 2002).

The elephants are generally kept in the release boma for 24–48 hours, after which the release gate is opened and they are free to walk out at will.

LIMITATIONS AND COSTS

The size of the elephant translocation equipment, because it can restrict access to locations, is the only important limiting factor associated with elephant translocation. Heavy equipment may have limited access on bad roads. However, access limitations can often be circumvented by experienced helicopter pilots who are able to herd elephants over long distances (up to 20 km) to more accessible locations.

There is no elephant too big, or group too large, to be translocated. Recent techniques, modern equipment and experienced personnel have made it possible to perform any translocation required or desired. The task of moving a large number of elephants, even as many as 1 000 individuals, is purely dependent upon human effort and financial resources.

The cost of a translocation operation depends on several factors. The factors that contribute the greatest cost and greatest variability in cost to translocation deal with the actual transport: distances that equipment (including the helicopter) have to be moved to get to the capture site, and the distance the elephants are moved. Additionally, the travel and accommodation costs of personnel may be high, depending on the location and duration of the operation. In terms of the treatment of elephants, full-grown adults are more expensive to dart than smaller individuals because adults require a larger dosage of drugs for sedation (Du Toit, 2001). The geographic location of target elephants has no effect on the cost of darting (e.g. darting in Kruger costs the same as darting in the Eastern Cape).

To illustrate the reduction in cost per elephant for translocating an increasing number of elephants, an example is presented in table 4. Costs in the presented example are defined by a capture destination 500 km from the translocation's team base, and, for every 10 elephants translocated, two hours of helicopter time and two days' accommodation and expenses for all personnel. Due to high fixed daily expenses, the cost per elephant for moving fewer than 10 individuals is much higher than for moving a greater number of individuals.

Financial parameters	Cost (ZAR)
Bell Jet Ranger helicopter	4 700/hour
Wildlife veterinarian	4 000/day
Crane truck and trailer	12/km
Wake-up crate	12/km
Transport truck and crates	12/km
Minimum personnel (10 people)	3 000/day
Stay and travel costs	150/per person
Personnel transport	4/km
Capture costs	1 500/adult elephant
Loading to wake-up crate	1 000/elephant

Table 3: The cost of a translocation operation depends on several factors

Elephant numbers	Drugs and loading (ZAR)	Fixed costs (ZAR)	Cost per elephant (ZAR)
1	2 500	58 200	60 700
5	12 500	58 200	14 140
10	25 000	58 200	8 320
20	50 000	94 200	7 210
100	250 000	418 200	6 682
500	1 250 000	1 858 200	6 216

Table 4: The costs incurred per elephant decrease with increasing numbers of elephants translocated

LEGAL REQUIREMENTS

An elephant may only be translocated on the following conditions (DEAT, 2008):

a. the translocation must comply with all relevant permitting requirements
b. the translocation must be effected in accordance with the provisions of the Biodiversity Act
c. the translocation must comply with the relevant provisions of the Animal Protection Act, 71 of 1962, and the Translocation of Certain Wild Herbivores (SABS Protocol SABS 0331) as amended
d. if elephants are captured within a protected area for the purpose of translocation, the capture must be in accordance with an approved management plan for the protected area within which the elephants occur

e. if the elephants are to be introduced into a protected area, the introduction must be in accordance with an approved management plan for the protected area to which the elephants are to be introduced

f. at the point of destination, the elephants must initially be released into a specified release camp

g. immediately prior to offloading into a release camp, the matriarch, other adults and juveniles must, if necessary, be tranquillised with short or long-acting tranquillisers.

The current permitting system is an extremely lengthy and laborious process that requires significant input from a variety of sectors. Firstly, an independent environmental consultant must conduct an environmental impact assessment (EIA) of the potential recipient reserve; this must satisfy the local Conservation Authority. The Conservation Authority will then assess the EIA and list any required changes. Thereafter, it will conduct the necessary fence inspections before any attention will be given to the needed legal issues, such as movement permits (export and/or import). All neighbours of the recipient game reserve must give their written permission accepting the elephant translocation and the new owner is advised to have extensive third party insurance in the event of a break-out or any other worst-case scenario.

The various Provincial Nature Conservation bodies have so many different policies and decision processes in place, especially with regard to elephants, that it is very difficult to find out correct procedures and requirements. It is proposed that a national policy for translocating elephants should be developed and put in place to facilitate future decisions, movements, and actions.

CONCLUSIONS

Effects on social organisation and behaviour

There is a need to develop procedures and/or guidelines for correctly identifying an active family group in the field. By definition, a family group is a cohesive group of females and their calves, led by a matriarch or another older female, and generally comprising no fewer than 6–8 individuals, which associate regularly and closely with one another (Dublin & Niskanen, 2003).

To determine the effects of translocation on behaviour we can review several case studies conducted over the past years, ranging from ones that were highly successful to others not so successful. Problems encountered are normally limited to either break-outs, interactions with people or destruction of facilities. Some of these problems have been associated with the translocation of an incomplete family group. It is of vital importance that adult cows are translocated together with all their offspring. Cases associated with abnormal aggressive behaviour were all linked to abnormal aggression from the matriarch's side, and once the matriarch was removed, the entire herd settled in normally (Phinda Game Reserve; Shamwari; Bayethe).

Post-release monitoring

In general, the success of a translocation is measured by the degree of adaptation of the elephants to their new location. All newly translocated elephant groups should be fitted with a VHF radio-collar to facilitate long-term monitoring. The condition and behaviour of individuals, as well as the cohesiveness and geographical position of herds, should be monitored. While many individuals behave the same in their new environment as they did before translocation, this is not always the case. Thus, if elephants have been moved into a reserve to improve eco-tourism, it is important to monitor them after their release to determine whether they have habituated to tourism influences in that reserve. It is also important to monitor individuals to determine whether they exhibit aberrant behaviour that could put themselves, other elephants or people in danger.

Minimising and monitoring stress imposed by capture operations

There is no doubt that any capture process or translocation operation imposes a high level of stress on the captured elephants, and that during the journey to their new location, they will suffer additional stress. Over time, the translocation process has been modified in an attempt to reduce stress on the elephants before, during and after the operation. For example, through our interpretation of stress indicators, combinations of tranquillisers are now administered to control and minimise stress (Du Toit, 2001). Also, selecting a group or individuals that seem less susceptible to stress, based on their behaviour in their original location, may lead to a reduction in overall stress levels.

Physiological responses of elephants to stressful situations can be monitored by measuring glucocorticoid (stress hormone) metabolite levels in dung. These metabolites serve as an indicator of the degree of stress experienced by an elephant in the recent past. Measurement of stress hormones from dung collected throughout and after the translocation process will provide insight into the degree of stress experienced by translocated individuals (Pretorius

& Slotow, 2002). In another more recent study it was shown that transport of captive elephants did increase their secretion of glucocorticoids. However, the proximity of other known herd members and allowing them to interact, did decrease their stress levels (Millspaugh *et al.*, 2007). Such information may assist translocation teams' efforts to reduce stress in future translocations.

Translocation as a tool to reduce elephant numbers

It is possible to use translocation to reduce elephant numbers (Du Toit, 2001; Dublin & Niskanen, 2003). In recent years, game reserves such as Phinda, Kapama and Shamwari have removed elephants through translocation to keep elephant numbers down. Currently, the lack of new areas or reserves is the greatest limitation to using translocation as a means of controlling elephant population size within a reserve.

Genetic intervention in the long run

The present scenario in South Africa for elephant conservation is more positive than ever before, with a great increase in elephant numbers over the past decades, through protection and expansion of their ranges – predominantly through translocation into state and private game reserves. This in itself has generated its own conservation problems, as the once continuous elephant population has now become highly fragmented, scattered in a multitude of varied sized reserves (Knight *et al.*, 1995).

In the greater picture of elephant conservation, these small populations play a relatively insignificant role. As most translocations are to small, fenced reserves, the dissemination of genetic material is restricted by the small numbers of elephants available to form viable breeding nuclei. As a result, these translocated populations are not considered genetically viable. It is imperative that management interventions be focused on genetic diversification; if not, a population bottleneck situation, in terms of reduced genetic diversity, will occur on smaller game reserves. In order to maintain the genetic diversity of the population at acceptable levels, it is recommended that a few animals should be translocated per generation, which would be every 15–20 years (Knight *et al.*, 1995).

REFERENCES

Carr, R.D. & M.E. Garaï 2002. *An investigation into some unsuccessful introductions of adult elephant bulls to confined areas in South Africa.* Proceedings of a workshop on elephant research, Elephant Management and Owners Association, 4–15.

DEAT 2008. Draft national norms and standards for the management of elephants in South Africa. Department of Environment and Tourism. *Government Gazette*, 8 May 2008.

Dublin H.T & L.S. Niskanen (eds) 2003. *The African elephant specialist of the IUCN/SSC. Guidelines for the in situ Translocation of the African Elephant for Conservation Purposes.* IUCN, Gland, Switzerland and Cambridge, UK, 1–54.

Du Toit, J.G. 1991. *The African Elephant (Loxodonta africana) as a game ranch animal.* Proceedings of a Symposium: The African Elephant (*Loxodonta africana*) as a game ranch animal, 37–42.

Du Toit, J.G. 2001. *Veterinary Care of African Elephants.* The SA Veterinary Foundation/Novartis Wildlife Fund, Pretoria.

Knight, M.H., A.S. van Jaarsveld & A.O. Nicholls 1995. *Evaluating population persistence in fragmented South African elephant populations: conservation and management implications.* Proceedings of a workshop on elephant research, Elephant Management and Owners Association, 76–80.

Kruger, J. 2002. *The control of problem elephants in the Limpopo Province.* Proceedings of a workshop on elephant research, Elephant Management and Owners Association, 97–101.

Millspaugh, J.J., T. Burke, D. van Dyk, R. Slotow, G.E. Washburn & R.J. Woods 2007. Stress response of working African elephants to transportation and safari adventures. *Journal of Wildlife Management* 71(4), 2157–2160.

Pretorius, Y. & R. Slotow 2002. *Tourism as a possible cause of stress in the African elephant of Mabula game reserve.* Proceedings of a workshop on elephant research, Elephant Management and Owners Association, 122–125.

Slotow, R. & G. van Dyk 2002. *Ranging of older male elephants introduced to an existing small population without older males: Pilanesberg National Park.* Proceedings of a workshop on elephant research, Elephant Management and Owners Association, 16–21.

Slotow, R., T. Burke, & D. Balfour 2002. *Introduction of Kruger bulls to control young male elephant killing rhino at Hluhluwe-Umfolozi Park.* Proceedings of a workshop on elephant research, Elephant Management and Owners Association, 76–83.

REPRODUCTIVE CONTROL OF ELEPHANTS

Lead author: Henk Bertschinger
Author: Audrey Delsink
Contributing authors: JJ van Altena, Jay Kirkpatrick, Hanno Killian,
Andre Ganswindt, Rob Slotow, and Guy Castley

INTRODUCTION

CHAPTER 6 deals specifically with fertility control as a possible means of population management of free-ranging African elephants. Because methods that are described here for elephants function by preventing cows from conceiving, fertility control cannot immediately reduce the population. This will only happen once mortality rates exceed birth rates. Considering, however, that elephants given the necessary resources can double their numbers every 15 years, fertility control may have an important role to play in population management.

The first part of the chapter is devoted to the reproductive physiology of elephants in order to provide the reader with information and understanding which relate to fertility control. This is followed by examples of contraceptive methods that have been used in mammals, and a description of past and ongoing research specifically carried out in elephants. Finally guidelines for a contraception programme are provided, followed by a list of key research issues and gaps in our knowledge of elephants pertaining to reproduction and fertility control.

In this chapter we will also attempt to answer the following questions in regard to reproductive control of African elephants:

- Do antibodies to the porcine zona pellucida (pZP) proteins recognise elephant zona pellucida (eZP) proteins or is the vaccine likely to work in African elephant cows?
- Is it possible to implement a contraceptive programme using the pZP vaccine?
- Is it practical to implement such a programme?
- What contraceptive efficacy can one expect?
- Is the method safe, reversible and ethical?

- What effect does the implementation have on the behaviour of a population?
- What are the effects of contraception on behaviour?
- What are the proximate and ultimate effects of contraception?
- Given the current technology, what population sizes can be tackled?
- What are the costs involved?
- Are there alternatives to pZP for contraception of elephants?
- What developments are in the pipeline that could facilitate implementation?

ASPECTS OF ELEPHANT REPRODUCTION THAT RELATE TO REPRODUCTIVE CONTROL

Social organisation

Elephants live in female-dominated herds comprising an old female referred to as the matriarch together with her mature daughters and their offspring, including sexually immature male calves (Owen-Smith, 1988). Female elephants remain in their natal herds their whole lives; male elephants leave their natal groups at approximately 12–14 years or when they reach sexual maturity (Poole, 1996a & b). These young bulls are often driven out of their family groups by cows that bully and 'chivvy' them (Douglas-Hamilton & Douglas-Hamilton, 1975). These newly independent bulls may leave their families only to join up with another family for a few years, or go off to 'bull areas' and join up with other bulls to form bachelor herds, or they may stay in female areas moving from family to family (Poole, 1996b). These courses to sexual maturity result in mature males that live alone (13–60 per cent, Owen-Smith, 1988) or in small bachelor herds characterised by temporary associations (Owen-Smith, 1988).

The mating system of elephants can be considered as promiscuous, because males and females will mate with more than one individual during a given oestrus, but females usually have only one offspring per pregnancy (Rasmussen & Schulte, 1998). However, because only one male can be the sire, the mating system can also be described as sequential polygyny (Hollister-Smith, 2005).

Elephant communication is very complex and five main sensory receptor systems are used to communicate with other elephants and with their environment. These are tactile, visual, vibrational, auditory, and chemical receptor systems (Schulte *et al.*, 2007). In recent years chemical signalling has received a lot of attention, as it appears to play an important role in elephant societies. Chemosignals (also referred to as 'honest' signals because they cannot be faked) are released in urine, temporal gland fluid, vaginal mucus, from the toe glands in the feet and from a number of other sites. Chemosignals reflect the physiological status (age, sex, reproductive and metabolic condition) of the sender, and the response of the receiver also depends on receiver status (Schulte *et al.*, 2007). Importantly, chemical signals can be used for environmental enrichment of captive elephants as well as in the resolution of human-elephant conflict. Two examples of the latter are the use of musth chemosignals to deter wild Asian elephants from raiding crops (Rasmussen & Riddle, 2004), and conditioned aversion of African elephants with chilli peppers planted in between edible crops, also to prevent crop raiding (Parker & Osborn, 2006).

Reproductive physiology of African elephant cows

Puberty

In the wild, given adequate nutrition and social structure, a cow will reach puberty at about 10–12 years old. Laws (1969) found that culled elephant cows in five different parks were essentially ready to ovulate at the age of 10–11 years. However, the age at which first ovulation took place was dependent on population density (table 1). The parks studied – Mokomasi Game Reserve (MK), Tsavo National Park (TNP), Murchison Falls National Park North (MFPN), Murchison Falls National Park South (MFPS) and Budongo Central Forest Reserve (BCFR) – had elephant population densities of <3, 3, 2.8, 5.9–10.3 and 6–7 per sq mile (2.6 km^2), respectively. The mean age at first ovulation was 11, 12.5, 14, 18, and 20 years respectively. He attributes these differences to density-dependent physiological, social, and nutritional stresses.

The effects of density on age of ovulation, intercalving interval and incidence of anoestrus (see Gestation, intercalving interval, and lactation) must be taken into account when considering methods of population control. Methods that reduce population without affecting fertility will inevitably increase reproductive and thus population growth rate. In the Luangwa Valley females reached maturity (age at first ovulation) at 14 years of age (Hanks, 1972). First ovulation typically occurs between 11 and 14 years, while the earliest recorded age at first conception is 7 years (Owen-Smith, 1988). In Samburu National Park, the first oestrus was recorded in an 8-year-old primiparus cow. Faecal progestin metabolite studies showed that she in fact fell pregnant during this oestrus (Wittemyer *et al.*, 2007). The age of cows at first parturition ranges from 9 to 18 years (Owen-Smith, 1988). On the other hand, in translocated populations, births have been observed in cows as young as 9, indicating the onset of puberty as early as 7–8 years, a trend commonly observed in relocated populations (Delsink *et al.*, 2006; Slotow, pers. comm. 2006).

Oestrous cycle length, length of oestrus

Oestrus persists for 2–6 days, with a minimum of 2 and a maximum of 10 days, during which time a female may be mounted by several males (Western & Lindsay, 1984; Moss, 1983). The oestrous cycle lasts between 15 and 16 weeks with an 8–11 week luteal phase and a 4–6 week follicular phase (figure 1). It is considered that cows are at their most fertile during the late follicular to early luteal phase. Oestrous females may be observed in any month of the year, although oestrous frequency is highest during and following the wet season (Poole, 1987; Poole, 1989b; Brown *et al.*, 2004).

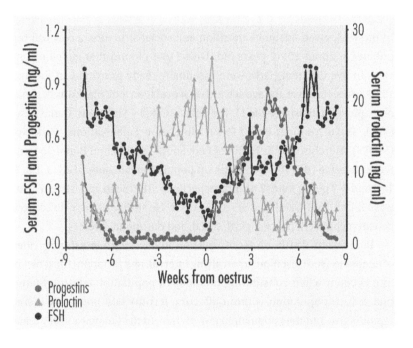

Figure 1: Mean profiles of serum progestins, prolactin, and FSH throughout the oestrous cycle in reproductively normal African elephants (*n* = 7 females; 15 cycles). Week 0 designates oestrus. The follicular phase is considered the period between successive luteal phases (week 6 to week 0) (Brown *et al.*, 2004)

Females remain reproductively active throughout the adult period and do not appear to display reproductive senescence (Lawley, 1994). Nevertheless, in the Luangwa Valley reproductive rate started to decrease after 40 years of age: 5 of 30 cows between 50 and 60 years old were neither pregnant nor lactating. This shows that a high percentage of animals are still reproductively active (Hanks, 1972). According to Owen-Smith (1988), fertility declines rapidly after 50 years of age. During one of the last elephant culls in the Kruger in 1995 a cow aged 55–60 had 11 placental scars, indicating that reproductive senescence was unlikely in this individual (Whyte, unpublished data).

Behaviours associated with oestrus

While this section has no direct bearing on the assessment process, it provides important background information which could be of importance in regard to future studies. One of the criticisms of contraception is that it may affect reproductive and social behaviour of breeding herds. In order to establish possible effects a sound knowledge of normal reproductive behaviour is essential.

Oestrous females exhibit conspicuous behaviour by calling loudly and frequently and by producing urine with particular olfactory components (Poole & Moss, 1989). Prior to the ovulatory phase the female uses low-frequency acoustic signals with a range of up to 8 km to attract males to her (Leong *et al.*, 2003). Five categories of oestrous behaviour were classified (Moss, 1983). The first sign of oestrous behaviour is that of 'wariness'. The female is noticeably alert and wary of males, carrying her head high and directing her gaze towards other elephants. When approached by males, the female avoids their approaches and attempts to test her reproductive status (Moss, 1983). During the ovulatory phase, bull elephants increase the frequency of genital inspections, flehmen and trunk contact towards the receptive cow between 1 and 9 days before ovulation, with most inspections occurring just a few days beforehand (Ortolani *et al.*, 2005). In Asian elephants the increase in male attention is thought to be due to the release of (z)-7-dodecene1-yl acetate in the urine during the pre-ovulatory phase. While this chemical or pheromone has not yet been detected in African elephants, it is likely that a similar compound exists, producing the same results (Rasmussen & Schulte, 1998). Male elephants detect this pheromone using the vomeronasal organ (a highly specialised and sensitive scent organ situated in the roof of the mouth) and when the concentrations are at their greatest, mating occurs after the sequence of oestrous behaviours have played out (Moss, 1983; Bagley *et al.*, 2006). This is the period just prior to ovulation. Females will often urinate on their tails and then raise them, presumably to help the spread of the olfactory cue (Freeman, 2005). The wariness phase gives rise to the 'oestrous walk', where the cow will exhibit a distinct 'walk' back and forth in which her back is arched, tail raised and head held at an angle (Moss, 1983). It is assumed that this is a visual and chemical cue to attract males (Freeman, 2005).

Increased vocalisations also occur when males arrive. It is thought that this is to further attract other males, allowing for female selection of the healthiest male to enable the optimum survival rate of her offspring (Vidya & Sukumar, 2005). The oestrous walk may develop into a 'chase', where both animals run. The male appears to be trying to catch the female. The female travels in a wide arc away from her group and may be separated from them for a number of hours. If the male is able to touch her with his trunk, the female will stop running, upon which he will attempt to mount her (Moss, 1983).

During oestrus, older experienced females show preferences for males in older age classes and preferably those in musth (Moss, 1983). There is a hierarchical dominance order among bulls of a particular range which is so strong that although as many as eight bulls have been seen escorting an oestrous cow, only the highest-ranking bull will mate or be accepted by the cow (Hall-Martin, 1987). In addition, females may solicit male-male competition by drawing attention to themselves in the early days of oestrus to have a wider selection pool (Moss, 1983). In the Amboseli study the female did not mate with each male in relation to his courting efforts, nor did she mate with males in proportion to the numbers in each male size/age class; furthermore, the male that she mated with was successful partly because of her own behaviour, letting herself be caught by or going into consortship with him (Moss, 1983).

Mountings occur after the 'chase' or a brief 'walk' with the oestrous female. Once the male is able to touch the female with his trunk and she has stopped, the male then places his trunk lengthways on her back, resting his head and tusks on her rump, rearing up on his back legs. The female remains standing still or may back closer to the male to facilitate intromission (Moss, 1983). The male stays mounted for about 45 seconds, while intromission lasts approximately 40 seconds. Temporary consortship occurs between the individual male and an individual female, characterised by close proximity, affinitive behaviour, and attempts to maintain exclusive copulatory behaviour (Moss, 1983). Consort behaviour may be displayed even when no oestrous walks or chases have been observed. However, the female approaches or follows the particular male if he moves away, even if other bulls are present in the area. Similarly, the bull follows the cow and remains in close proximity if she moves away (Moss, 1983).

When the oocyte is released (ovulation), the sperm will still be viable and capable of fertilisation. The ovulation fossa (follicular cavity) left behind is filled by the growing corpus luteum which secretes progesterone. Besides numerous other functions, progesterone is responsible for behavioural quiescence which lasts approximately 10 weeks.

There is also evidence to suggest that olfactory signals released during oestrus have an effect on other females, resulting in the synchronisation of ovulation in a herd. Presumably this would result in calves being born at a similar time, enabling the herd to create 'nurseries' which would increase individual calf survival (Archie *et al.*, 2006). The exact pheromones used are currently not known (Freeman, 2005; Rasmussen & Schulte, 1998).

Gestation, intercalving interval and lactation

Intercalving interval (ICI) is one of the major factors that affect the rate of population growth. It responds to a number of variables: most importantly resource availability and population density. The relationship between population density and intercalving interval was clearly shown by Laws (1969) in a study of cull material from five East African parks (table 1). The relationship between resource availability and reproductive endocrinology was also shown by Wittemyer *et al.* (2007) who studied a group of wild elephants in Samburu and Buffalo Springs National Parks. They compared Normalised Differential Vegetation Index to faecal progestogens metabolite concentrations during the oestrous cycle and in pregnancy. Faecal progestogens were significantly higher in pregnant and non-pregnant cows during the wet than dry seasons (figure 2). The relationship between resource and reproduction was thus clearly shown. Manipulating the ICI by means of reproductive control will therefore influence the rate at which a population grows. In several studies of East African populations, the ICI ranged from 2.9 to 9.1 years (Laws *et al.*, 1975). In Amboseli, ICI was 4.9 years during the period 1972–1980 (Moss, 1983), 5.6 years during dry years, and 3.5 years over a sequence of wet years (Owen-Smith, 1988). Variations in ICI occur according to region; between 3.3 and 5.5 years in Lake Manyara in Tanzania, Luangwa Valley in Zambia, Kruger National Park (Kruger) in South Africa, Gonarezhou and Hwange in Zimbabwe (Hanks, 1972; Owen-Smith, 1988).

Gestation lasts 22 months in African elephants, which accounts for approximately 50 per cent of the intercalving period. This means that for a period of up to two years after calving cows do not show an oestrous cycle. An example of the faecal profile of an acyclic (unknown cause) African elephant cow is shown in figure 3 (Brown *et al.*, 2004).

	Mokomasi Game Reserve	Tsavo National Park	Murchison Falls National Park North	Murchison Falls National Park South	Budongo Central Forest Reserve
Elephants per mile^2	<3	3	2.8	5.9 and 10.3	6–7
Age at first ovulation	11	12.5	14	18	20
Intercalving interval (yrs)	4–5	6–7	6–7	8–9	no data
% anoestrous cows ≥25 years and not lactating	no data	7.9	8.6	28.6	no data

Table 1: Influence of population density on reproductive variables of five East African parks (adapted from Laws, 1969)

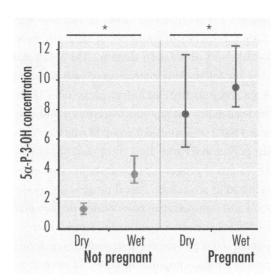

Figure 2: The median faecal 5α-pregnane-3-ol-20-one (5α-P-3-OH) concentrations (not pregnant, light grey circles; pregnant, dark grey circles) and inter-quartile ranges (error bars) of all females are presented in relation to reproductive state and season. The asterisks represent statistically significant differences between categorised individually paired median values (Wittemyer *et al.*, 2007)

Under normal conditions fertile cows would seem to lactate permanently (from one calving to the next). According to Whyte (2001), on data obtained from culling operations in the Kruger National Park (KNP) from 1989 to 1992, 344 lactating cows and 350 calves less than one, one, two, and three years old were found in the same populations, which had an average ICI of 54.4 months.

The gestation period of 22 months added to the age of the oldest group of calves (36 months) provides a lactation period of about 58 months, indicating that African elephant cows lactate from the birth of one calf to the next. The duration of lactation recorded in the Tsavo and Murchison districts of East Africa was 4–5 and 7–8 years, respectively (Laws, 1969) but in the same paper 40.7 per cent and 73.2 per cent of pregnant cows ≥25 years of age in Tsavo and Murchison North, and Murchison South were not lactating, respectively. It seems therefore that as intercalving interval increases, the percentage of lactating cows decreases.

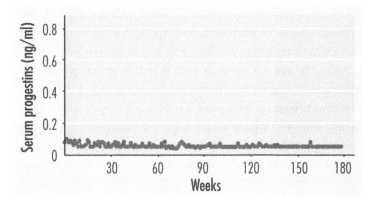

Figure 3: Serum progestin profile in a non-cycling African elephant female (Brown *et al.*, 2004)

Reproductive physiology of African elephant bulls

The social ranking of bulls and reproductive behaviour, which is driven by androgens of testicular origin, are the two most important factors that determine the likelihood of a bull being able to mate with an oestrous cow under natural conditions. A 30-year study using faecal DNA microsatellites to determine paternity of calves born in Amboseli showed that bulls 45 years and older sired 50 per cent of all calves. Furthermore, it showed that 75 per cent of progeny were sired by bulls that were in musth at the time of mating (Hollister-Smith, 2005). While physical stature and strength are known to affect ranking, there is another factor that explains why older bulls are more likely to be afforded the chance to breed. Recently Rasmussen & Riddle (2002) found distinct differences in the pheromone content of temporal gland secretions of Asian elephant bulls up to and over 35 years of age. The pheromone content of the younger bulls scares off cows in oestrus, whereas in older bulls it becomes highly attractive. Granted the work was carried out in Asian bulls, but the reproductive physiology of the males of both species has been found to be very similar and is hardly likely to differ in this respect (Schulte *et al.*, 2007).

The above highlights the more important factors associated with the natural selection of sires. Human interventions – particularly hunting, which targets trophy animals, but also translocation or culling of dominant bulls – potentially impact on the selection process and may affect the quality of progeny. As a result of hunting and poaching, tusk sizes have decreased in Africa. The complexity of the selection process also makes the bull a less attractive target for implementing reproductive control. If older bulls are targeted, younger, less dominant bulls will contribute more offspring than would normally be the case. In small, translocated populations with only one or two bulls, it may be an option. A watchful eye will, however, have to be kept on young bulls that have been introduced with their natal herds.

The following pages provide more insight into the reproductive behaviour of elephant bulls and link endocrinology and behaviour. This is valuable information upon which further studies related to various behaviours such as aggression, habitat degradation and reproductive control can be planned.

Puberty

In males, spermatogenesis begins between 7 and 15 years, but full sperm production is not reached until 10–17 years (Owen-Smith, 1988). In the Luangwa Valley spermatogenesis started at the mean age of 15 years, when the combined weight of the testes was 650–700 g (Hanks, 1972). According to Johnson & Buss (1967), however, the start of puberty ranged from 3 to 14 years of age. This was based on the histological appearance of the testes, whereas puberty is usually defined as the age at which sperm appear in the ejaculate. Male elephants leave their natal families at the onset of puberty, at about 14 years of age, and henceforth live in a highly dynamic world of changing sexual state, rank, associations, and behaviour, most of the time alone or in small groups of males (so-called bachelor groups) in specific bull areas (Moss, 1983; Poole, 1994; Lee, 1997). At this time, adolescent bulls produce fertile sperm and are physically able to mate successfully (Hall-Martin, 1987), giving a bull a potential reproductive lifespan of over 40 years. The age of dispersal and reproductive capability, however, may be affected by human activities (e.g. hunting, poaching and habitat loss). For example, the killing of older males may permit younger males to mate and may select for earlier dispersal by males (Sukumar, 1989; 1994; Owens & Owens, 1997). Although reproductive capability is already achieved during the teenage years, recent findings suggest that elephant bulls in the wild alternate between sexually active and inactive periods only from the age of 20 years onwards (Rasmussen, 2005). In captivity, however, the fact that elephant bulls are sometimes housed singly (http://www.elephant.se/elephant_database.php), could be the reason why reproductive activity is documented for captive bulls as young as seven years of age (Brown *et al.*, 2007).

Male reproductive behaviour and musth

Two alternative reproductive tactics are documented for sexually active bulls in the wild: (a) the sexually active non-musth tactic, a non-competitive tactic seen in less dominant, often younger males which is associated with low, prolonged investment, and (b) the musth tactic, a competitive tactic seen in dominant, often older individuals, associated with short periods of high investment (Rasmussen, 2005). Although recent findings underlined the importance of musth to male reproductive success, paternity analyses also revealed that approximately 20–25 per cent of the reproduction can be attributed to sexually active non-musth bulls (Hollister-Smith *et al.*, 2007; Rasmussen, 2005).

The phenomenon of musth was first described in adult Asian bulls in captivity (e.g. Eggeling, 1901). First studies on African elephants suggested that musth did not occur in this species (Perry, 1953; Sikes, 1971), but it has since also been shown to exist within the genus *Loxodonta* (Poole & Moss, 1981; Poole, 1987). The physical, behavioural and physiological changes associated with musth appear sporadically in adolescent Asian elephants from approximately 10–20 years of age, and periodically appear in all Asian bulls after the age of 30 (Eisenberg et al., 1971; Jainudeen et al., 1972a). In free-ranging African elephants, first signs of musth occurred at about 25 years of age. In the absence of older dominant bulls, however, musth may occur at an earlier age. As in the Asian elephants, older African bulls show longer and more predictable periods of musth (Poole, 1982; 1987; 1994). The long time span between the age when the first sporadic musth signs occur and the onset of relatively regular prolonged periods of musth (around 35–40 years of age) suggests that the optimal reproductive tactic (either sexually active non-musth or musth) may vary depending on ecological and/or demographic conditions during a period covering more than 10–15 years of life of a bull (Rasmussen, 2005). Although musth in general seems to be more predictable in older animals, the intensity and duration of musth is also variable and asynchronous between those bulls, and even the character of musth within such an individual can vary from year to year (Cooper *et al.*, 1990; Poole, 1987; 1994), indicating that the appearance of musth may vary depending on local conditions.

Male elephants in musth leave their normal home ranges, travel long distances and spend significantly less time feeding and resting in order to locate and associate with oestrous females (Hall-Martin, 1987; Poole, 1989b; 1994). A receptive female will preferably mate with the most dominant bull in the area, which is usually a bull in musth, because musth bulls show aggression which overrides normal social male hierarchies (Hall-Martin, 1987). During aggressive interactions, bulls in musth are invariably the winners irrespective of body size, the factor which normally determines dominance rank between males in non-musth condition (Poole, 1989a). Apparently, bulls in musth are more likely to be involved in fights, and several musth males have been killed by stronger bulls in musth (Hall-Martin, 1987; Poole, 1994). Furthermore, it is known that the presence of a dominant musth bull can suppress the physical and behavioural changes associated with musth in lower-ranking bulls (Poole, 1982).

Elephants in musth have a characteristic posture, which is particularly noticeable when they move. The head is carried well above rather than below the shoulder blades and held at such an angle that the chin looks tucked in. The ears are tense and carried high and spread (Poole, 1987; Kahl & Armstrong, 2002). Bulls in musth also repeatedly call at very low frequencies (infrasound) that travel over long distances without attenuation. African elephants in musth emit a distinct set of calls with frequencies as low as 14 Hz and sound pressure levels up to 108 decibels (Poole, 1987; 1994; Poole *et al.*, 1988). Males in musth call significantly more often when they are alone and apparently searching for female groups, than when they are in association with females (Poole, 1987). Their rumbles are often preceded or followed by listening behaviour, suggesting that they are either answering a call or calling and waiting for a response (Poole *et al.*, 1988), whereby similarly ranked musth males actively avoid one another (Poole, 1989a). Since males in musth criss-cross the range in order to locate and associate with oestrous females (Hall-Martin, 1987), it seems plausible that they may call and/or listen for other musth males' calls to avoid unexpected meetings with another equally aggressive and high-ranking bull (Poole *et al.*, 1988). Musth bulls emit further specific signals which notify other male and female elephants of their status. These musth-related signals are mostly characterised by the continuous discharge of urine in a series of discrete drops (urine dribbling) with the penis retained in sheath. This urine has a typical strong odour, especially when associated with a greenish discoloration of the penis and sheath (figure 4) (Poole & Moss, 1981; Poole, 1982; 1987; Hall-Martin, 1987). A further visual signal for a bull in musth is the copious secretion from and enlargement of the temporal glands (Jainudeen *et al.*, 1972a; Poole & Moss, 1981; Hall-Martin, 1987; Poole, 1987; Rasmussen and Schulte, 1998), unique paired modified apocrine sweat glands located in the temporal fossas. Watery secretion from the same glands is often also interpreted as a sign of musth. Elephants of all ages produce a watery secretion from the temporal glands in response to excitement (Bertschinger, unpublished data). Riddle *et al.* (2000) also described secretion of a watery liquid from the ears of captive and wild male and female African elephants of various ages. The fluid was noted when animals became excited and appeared in the form of squirts or slow dribbles. The compounds found in the secretions were also found in temporal gland secretions. Watery aural secretions have also been observed in captive elephant bulls and cows in South Africa (Bertschinger, unpublished data).

Figure 4: A – Bull showing all physical signs of musth (temporal gland swelling and secretion, urine dribbling and wet legs). B – greenish coloured sheath (photo: Bertschinger, Etosha National Park, Namibia, April 2007)

There is extensive anecdotal information from Asia that good nutrition and body condition are necessary for the successful expression of musth in elephant bulls, and that musth bulls in poor condition usually drop out of musth (Poole, 1989a, Rasmussen & Perrin, 1999). Although it seems that the musth-related weight and condition loss is largely attributed to the increased restlessness and reduced feeding activities of African elephant bulls during musth (Poole, 1982; 1989a), it has been shown that captive Asian elephants lose weight during musth even when they are chained and given normal rations of food (Poole, 1989a). Since musth is also associated with elevated androgen levels, weight loss may also be related to the increase in metabolic rate that is associated with high androgen levels (Poole, 1989a). In this respect, it could be demonstrated for an Asian elephant that musth-related changes in serum testosterone and triglyceride concentrations followed similar patterns: that lipase activity was significantly elevated immediately before and after musth and that urinary, especially albumin-like, protein concentrations increased during musth (Rasmussen & Perrin, 1999). However, the precise physiological links between body condition and musth are not clear as yet and due to the limited information available, additional data, particularly for the African elephant, are needed.

Endocrine profiles of elephant bulls

Physiologically, musth in both genera is particularly characterised by a periodic increase in androgen levels (e.g. Jainudeen *et al.*, 1972b; Poole *et al.*, 1984; Rasmussen *et al.*, 1996, Ganswindt *et al.*, 2002; 2005a). But it is still unknown which role the adrenal gland might play in this context, and if at all, whether increases in gonadal androgens precede the rise of androgens of adrenal origin, or vice versa. Recent findings also indicate that thyroid hormones might play a role in testicular steriodogenic activity (Brown *et al.*, 2007), but this possible relationship is far from clear and needs further investigations. Apart from the periodic increase in androgen levels, it has also been suggested that musth-related physical and behavioural changes are associated with elevated glucocorticoid levels, which would result from increased adrenal activity. As mentioned above, musth is known to be associated with increased restlessness and reduced feeding activities, often leading to a progressive loss of condition (Poole, 1989a). This has led to the hypothesis that musth represents a form of physiological stress, and a recent invasive study showed a modest positive correlation between testosterone and cortisol concentrations in captive bulls exhibiting musth (Brown *et al.*, 2007). Contrary to this result, Ganswindt and co-workers found no clues for an elevation in glucocorticoid output during musth or any other state of sexual activity in captive and free-ranging animals (Ganswindt *et al.*, 2003; 2005a), and additionally provide evidence for a suppressing effect of the musth condition on adrenal endocrine function (figure 5; Ganswindt *et al.*, 2005b). Further research is therefore necessary to determine whether characteristic conditions associated with musth represent a form of physiological stress, and which role the hypothalamic-pituitary-adrenal axis plays. Apart from the open questions regarding hormone involvement, little is also known about the time course and time-related occurrence of musth triggers influencing the onset and duration of musth. However, it could be recently demonstrated for captive African elephants, that temporal gland secretion and urine dribbling were typically first recorded after the elevation in androgens, indicating that these physical musth-related signs are downstream effects of increased androgen concentrations. In this respect, the results of the study show that temporal gland secretion responds earlier and to lower androgen levels than urine dribbling, which manifests itself later and requires a higher level of androgen stimulation (figure 5) (Ganswindt *et al.*, 2005b).

Nevertheless, more information about what regulates musth, and what the mechanisms are underlying the associated physiological and behavioural changes, is necessary and would not only be of scientific interest, but also useful in the development of new approaches to deal with the acute management problems of elephants in the wild.

BASIC METHODS OF CONTRACEPTION FOR WILDLIFE

If it is achievable, the ideal contraceptive must be efficacious, allow remote delivery, be reversible, produce no deleterious short- or long-term health effects, should cause no changes to social behaviours and group integrity, must not pass through the food chain, should be safe during pregnancy, and finally, be affordable.

Potential contraceptive methods available for animals

The following methods, broadly speaking, can be used for contraception in animals:

- surgical (gonadectomy, vasectomy, and salpingectomy)
- hormonal (oral contraceptives, depot-injections, or slow-release implants)
- immunocontraception.

Potential target tissues or reproductive processes that lend themselves to reproductive control were summarised by Asa (2005). Table 2 summarises methods that have been used for contraception in animals and how each method compares to the properties of an ideal contraceptive agent.

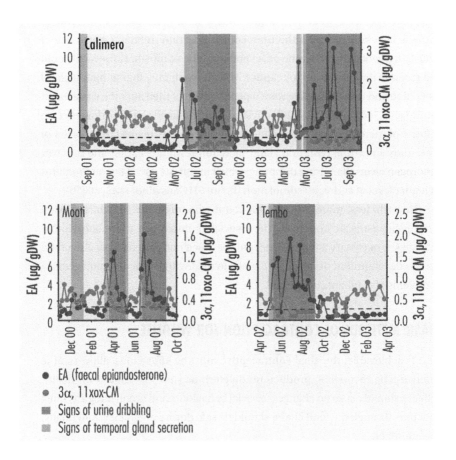

- EA (faecal epiandosterone)
- 3α, 11xox-CM
- ▨ Signs of urine dribbling
- ▨ Signs of temporal gland secretion

Figure 5: Profiles of faecal epiandrosterone (EA) and 3α, 11oxo-cortisol metabolites (3α, 11oxo-CM) immunoreactivity throughout a period of 11–26 months in three captive adult male African elephants (Calimero, Mooti, and Tembo). A range of months of 11–26 months has been given which addresses all three bulls. Dashed lines indicate the threshold for elevated epiandrosterone and decreased 3α, 11oxo-CM levels (Ganswindt et al., 2005b)

Surgical methods

Surgical methods to control reproduction in domestic species have been in use for many years. In males and females there are two options. The first is gonadectomy, which is irreversible and affects reproductive and probably territorial behaviour. The alternative is tying off the fallopian tubes in the female and vasectomy in the male, both of which leave reproductive and associated behaviours largely intact. Although vasectomies have been performed on African elephants (see *Surgical sterilisation of elephant bulls*) surgical methods are not considered practical for large numbers of wildlife. In addition they are invasive, expensive, and in elephants irreversible for practical purposes, as microsurgical techniques are needed to reverse the process.

Hormonal methods

Hormonal methods that have been used to contracept animals are oral or depot-type progestogens, oestrogens and androgens and GnRH super-agonists in the form of implants or depot-injections. Progestins in the form of long-acting implants were used extensively for contraception of large carnivores (figure 6). They are extremely effective but due to a number of serious side effects have largely gone out of use (Munson *et al.*, 2002; Munson *et al.*, 2005). The other possibility to consider, and this also applies to oestrogen implants, is that progestins may get into the food chain and affect reproductive performance of other species. A large number of bird, insect, and indirectly, reptile species could be exposed to such steroids through elephant faecal material should one consider such methods for elephants. Steroid implants are impractical to use since they require immobilisation of the target animal, and are too expensive.

Early attempts at contraception in wild horses relied primarily on steroids. It has long been known that exogenous androgens can exert a down-regulation of endogenous androgens and sperm production in the stallion (Blanchard, 1984). This fundamental biological strategy led to the first attempts at actual contraception aimed at free-ranging wild horses. Trials with domestic pony stallions demonstrated that six repeated monthly injections of testosterone propionate (TP) or 17-α-ethinyl oestradiol, 3-cyclopentyl ether (quinestrol), at doses of 1.7 g per 100 kg resulted in significant degrees of oligospermia and decreased motility (Turner & Kirkpatrick 1982). Practicality for application in the field, however, was limited because of the need for repeated treatments. Thus, testosterone propionate was microencapsulated in a polymer of D, L-lactide (mTP) (Southern Research Institute, Birmingham AL), permitting a sustained release after intramuscular injection for up to six months. On contact with intercellular water, the lactide coating erodes and releases the active steroid inside. The lactide coating is converted to carbon dioxide and lactic acid.

Property	Surgical methods		Steroids		GnRH agonists		Immunocontraception	
	Gonadect	Vasec/FallT	Oral	Implants	Injectable	Implants	pZP	GnRH
Contraceptive efficacy ♀	100%	100%	100%[b]	100%	≤100%	≤100%	70–100%	70–100%
Contraceptive efficacy ♂	100%	100%[a]	Poor	Poor	≤100%[c]	≤100%[c]	No	70–100%
Remote delivery	No	No	Only captive	No	Yes	No[c]	Yes	Yes
Reversible	No	No[d]	Yes	Yes	Yes	Yes	Yes	Yes
Social behaviours or organisation of groups or herds	Affects behaviour	Some in cats	Affects behaviour	Affects behaviour	No data	♀ anoestrus ♂ aggression decr.	♀ continue to cycle	♀ anoestrus ♂ aggression decr.
Deleterious short- or long-term health effects	Obesity	None	Carnivores some serious	Carnivores some serious	None	None	Local swelling	Local swelling
Contraceptive passes through the food chain	No	No	Possible	Yes	No	No	No	No
Safe to use during pregnancy	n/a	n/a	No	No	Yes/no	Yes/no	Yes	Yes
Production and/or application costs	Expensive	Expensive	Expensive	Medium	Medium	Medium	Medium	Medium

[a] Males can remain fertile for a number of weeks
[b] Only if given daily
[c] Does not work in male ungulates; remote delivery system being developed (Herbert & Vogelnest, 2007)
[d] Requires microsurgery – not feasible under field conditions and especially in megaherbivores
Gonadect = gonadectomy
Vasec = vasectomy
FallT = tying off fallopian tubes
n/a = not applicable

Table 2: Properties of an ideal contraceptive agent and evaluation of different options summarised (Bertschinger, unpublished)

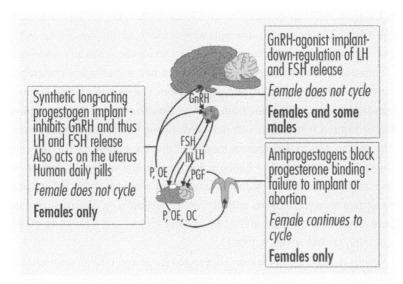

Synthetic long-acting progestogen implant - inhibits GnRH and thus LH and FSH release Also acts on the uterus Human daily pills *Female does not cycle* **Females only**

P, OE PGF

P, OE, OC

GnRH

FSH IN LH

GnRH-agonist implant-down-regulation of LH and FSH release *Female does not cycle* **Females and some males**

Antiprogestagens block progesterone binding - failure to implant or abortion *Female continues to cycle* **Females only**

Figure 6: Sites at which hormonal contraceptives act (Bertschinger, pers. comm.)

Experimental and control stallions in Idaho were immobilised from a helicopter and given mTP in the hip. Stallion libido and quantitative aspects of sexual behaviour, based on elimination marking behaviour (Turner *et al.*, 1981) were unaffected and breeding took place, but there was an 83 per cent reduction in foal production compared with mares bred by control stallions (Kirkpatrick *et al.*, 1982). Concerns for the safety of stallions and the dangers and high costs associated with helicopter use and immobilising drugs led to a second field trial in which the mTP was delivered remotely, to wild harem stallions on Assateague Island National Seashore, MD, (ASIS) with barbless darts, from the ground and without immobilisation. The pharmacological success of the mTP was evident, with a 28.9 per cent fertility rate for the mares accompanying the treated stallions and a 45 per cent fertility rate among the mares accompanying untreated stallions. Unfortunately, the logistics of delivering 3.0 g microcapsules in four separate doses to each stallion was daunting and impractical for routine use.

Logistical difficulties in treating stallions with steroids, concerns about steroid-related pathologies, and a general concern that the treatment of wild stallions would have serious genetic consequences to the gene flow in free-ranging herds turned the focus of contraception in horses to the mare. Based on experience with persistent corpora lutea (Stabenfeldt *et al.*, 1974) and data which indicated that plasma progesterone concentrations in excess of

0.5–1.0 ng.ml^{-1} inhibited ovulation in mares (Squires *et al.*, 1974; Noden *et al.*, 1978; Palmer & Jousett 1975), attempts were made to administer contraceptive doses of progestins to wild horses. Captive wild mares in Nevada were each implanted with silastic rods containing various doses of the synthetic oestrogen ethinyl oestradiol (EE$_2$) or EE$_2$ plus progesterone (Eagle *et al.*, 1992). Animals pregnant at the time of implantation delivered healthy foals, and contraceptive efficacy ranged from 88 to 100 per cent through two breeding seasons. Endocrine studies of these mares suggested that contraception was affected by blocking ovulation and/or implantation. In a similar study, intraperitoneal implants of EE$_2$ alone also resulted in contraceptive efficacy of 75–100 per cent through two breeding seasons, and rates of EE$_2$ decline in the plasma suggested a contraceptive life of 16, 26, and 48–60 months, for 1.5 g, 3.0 g, and 8.0 g of EE$_2$, respectively (Plotka & Vevea, 1990).

Results achieved with oestradiol, progesterone and ethinyl oestradiol in mares brings to focus advantages and disadvantages of natural versus synthetic steroids for contraceptive purposes in the horse. Steroids native to the mare, such as oestradiol and progesterone, are required in impractically large doses due to their rapid enzymatic degradation *in vivo*. The use of some long-acting synthetic steroids such as ethinyl oestradiol may delay metabolic degradation and permit more sustained contraception.

Because of the difficulties with delivering large masses of microencapsulated steroids, dangers associated with capture and restraint of horses, surgical procedures associated with intraperitoneal implants, concern over long-term effects of steroid contraception, and passage of synthetic steroids through the food chain, attention turned to immunocontraception.

Despite the problems experienced with steroids in horse trials, in 1996 a contraceptive trial was carried out on 10 elephant cows in KNP using slow-release oestradiol silicone implants (Compudose). Each cow received five implants providing a daily 17ß-oestradiol dose of 300 µg/animal (Goritz *et al.*, 1999). The cows were non-pregnant at the time and had small calves at foot. The results were reported by Bartlett (1997), Butler (1998) and Whyte & Grobler (1998). The implants were effective as a contraceptive and lasted at least 12 months when the uteri of all cows still showed signs of oestrogenisation. Reversibility of the method was inconclusive as some cows were still barren when the collars were removed. The major side effect of the implants is that the cows showed an almost permanent state of oestrus that lasted at least 12 months. The negative effects associated with this were constant presence of bulls, increased harassment of cows by bulls, and separation of calves from their mothers – two of the ten calves went missing and were presumed dead. On account of the side effects no further cows were made available for oestrogen-implant contraception. As in horses, the practicality and expenses related to administration of oestradiol implants in free-ranging elephants also renders the method a non-starter.

A newer and safer hormonal approach

GnRH super-agonist implants or depot formulations have replaced progestin implants in wild carnivores and to a large extent in non-human primates. Deslorelin marketed as a slow-release implant (Suprelorin®, Peptech Animal Health, Sydney) lasting 6-12 months and up to 24 months and leuprolide depot injection (Lupron Depot®, TAP Pharmaceuticals) lasting 1, 3 or 4 months have been the most commonly used products. They down-regulate the release of both LH and FSH and as such have the potential to be used in both sexes (figure 6). Suprelorin® has been used successfully in cheetahs (both sexes), lionesses, leopards (both sexes) and baboon and monkey species (Bertschinger et al., 2001; 2002; 2004b; 2006; 2007; 2008). Although well suited for the above species, the application in herds of female animals would be impractical since administration of implants requires immobilisation. Besides, the results with both products in ungulates have been highly variable and hardly reliable (Patton et al., 2005). The possibility for remote delivery of Suprelorin® implants is, however, being researched in kangaroos and preliminary results look promising (Herbert, 2007).

IMMUNOCONTRACEPTION

GnRH vaccine

The GnRH vaccine consists of one or more molecules of GnRH conjugated to a protein molecule to render the GnRH component antigenic. This is combined with an adjuvant and, when injected, antibodies to endogenous GnRH are formed. These antibodies neutralise GnRH released from the hypothalamus, thus down-regulating the release of the gonadotrophic hormones FSH and LH (figure 7). This is effective in both male and female animals. The process is reversible and if no further boosters are administered, the antibody titres will fall until insufficient to neutralise endogenous GnRH. Vaccinated females cease to show an oestrous cycle, while in males testosterone release is inhibited (meaning the vaccine also suppresses testosterone-related aggression) and eventually also spermatogenesis.

GnRH immunocontraception was originally developed for immunocastration of cattle (Hoskinson *et al.*, 1990). One of the main reasons for further development of the GnRH vaccine, however, was to immunocastrate male piglets as an alternative to surgical castration and so control the problem of boar taint in pork (D'Occhio, 1993; Oonk *et al.*, 1998; Dunshea *et al.*, 2001; Zeng *et al.*, 2001). The vaccine has also been used to control fertility of male feral pigs (Killian *et al.*, 2006), stallions (Dowsett *et al.*, 1996; Turkstra *et al.*, 2005; Burger *et al.*, 2006), rams (Janett *et al.*, 2003), bison bulls (Miller *et al.*, 2004), white-tailed deer (Becker *et al.*, 1999; Curtis *et al.*, 2001), and others (Ferro *et al.*, 2004).

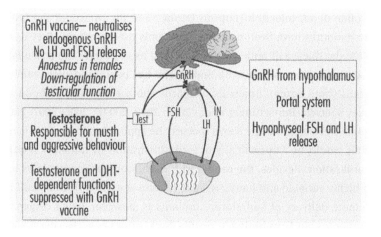

Figure 7: Endocrine control of testicular function in the male and site of action of anti-GnRH antibodies in males and females (Bertschinger *et al.*, 2004b)

Goodloe *et al.* (1997) immunised 29 wild mares on Cumberland Island National Seashore, GA, with GnRH conjugated to keyhole limpet haemocyanin (KLH). The vaccine was freeze-dried, sequestered in a solid biodegradable 0.25 calibre bullet and administered by an air-powered gun (Ballistivet, Inc., White Bear Lake, MN). After imbedding in the muscle of the target mare, the compressed compound forming the biobullet degrades over 24 hours, releasing the antigen. A total of 25 treated mares survived until the next foaling season and 17 (68 per cent) produced foals, which was not significantly different from control foaling rates.

A more recent attempt to immunise wild horses against GnRH was carried out in Nevada (Killian *et al.*, 2004). Mares received either 1 800 or 2 800 µg GnRH vaccine (National Wildlife Research Center, Fort Collins, CO) with Adjuvac adjuvant, which is a dilution of a commercial Johne's Disease vaccine (Fort

Dodge, Ames, IA). Following a single breeding season, none of 18 mares, in both treatment groups, were pregnant on the basis of ultrasound evaluations. All GnRH-treated mares had low concentrations of serum oestrogen and progesterone, which is consistent with the predicted actions of a GnRH vaccine. The largest study performed in mares so far made use of a GnRH vaccine developed for pigs (Improvac®, RnRF-protein conjugate, Pfizer Animal Health, Sandton, South Africa) (Botha *et al.*, 2008). Fifty-five mares were given a primary followed by a booster vaccination 35 days later, each containing 400 µg RnRF-protein conjugate. On Day 35 after the primary vaccination only 8 of 55 (14.5 per cent) experimental mares showed evidence of ovarian activity on clinical examination and by Day 70 all mares were quiescent. Baseline progesterone concentrations were attained by Day 42 and persisted until the end of the observation period on day 175. The application in elephants poses an exciting prospect. The vaccine is remotely deliverable, and the dose is small in terms of volume (3 ml), cheap and freely available. If the GnRH results in elephants follow the same pattern as pZP in elephants followed the pZP results in horses, it may provide an alternative method to pZP immunocontraception. One advantage that it may hold over pZP immunocontraception is that it induces anoestrus (see *Possible future developments*).

Porcine zona pellucida vaccine

The origins of porcine zona pellucida (pZP) contraceptive vaccine can be traced back to the work of Sacco & Shivers (1973) and Shivers *et al.* (1972), who demonstrated that the antibodies produced against the proteins of the porcine zona pellucida could inhibit sperm binding (figure 8). Shortly thereafter, it was discovered that spontaneous antibodies against the zona caused infertility in humans (Shivers & Dunbar, 1977) and that porcine zona proteins could block human fertilisation (Sacco, 1977). At about the same time, it was shown that these antibodies against native pZP were tissue-specific and did not cross-react with other tissues or hormones (Palm *et al.*, 1979). Collectively, these discoveries led to a surge of non-human primate research with native pZP, with an eye towards human contraception (Gulyas *et al.*, 1983; Sacco *et al.*, 1986, 1987).

Ultimately, interest in pZP for human contraception waned, for four major reasons. First, the immune systems of the target subjects were variable in their response to pZP and contraceptive efficacy was correspondingly variable. Second, the time to reversal of contraceptive effects was also quite variable and most pharmaceutical companies feared a wave of litigation by users. Third, it was discovered that in some species, such as baboons, rabbits, guinea pigs and dogs, but certainly not all species, ovarian abnormalities developed in response to the vaccine (Wood *et al.*, 1981, Mahi-Brown *et al.*, 1985, Dunbar *et al.*, 1989; Lee & Dunbar 1992). Finally, the inability to produce a synthetic or recombinant form of the native pZP, because of the difficulty in glycosylating the protein backbone, meant that large markets could never be serviced (Dunbar *et al.*, 1984).

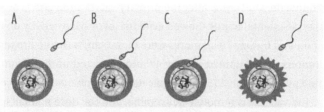

A When the egg (oocyte) is ovulated into the Fallopian tube it is surrounded by a capsular layer known as the zona pellucida capsule
B Before fertilisation can take place the sperm binds to one of thousands of receptor sites on one of the zona proteins. The sperm then undergoes the so-called acrosome reaction
C Only once the sperm has undergone the acrosome reaction can it penetrate the zP-capsule and then a single sperm fertilises the egg
D The antibodies formed in response to the pZP vaccine recognise and cover all sperm receptors on the ovulated egg. The binding of sperm is blocked, as is fertilisation and thus pregnancy

Figure 8: The proposed mechanism of pZP immunocontraception (Bertschinger *et al.*, 2004)

Liu *et al.* (1989) sparked new interest in pZP when they demonstrated that domestic mares could be rendered infertile with native pZP injections. Over the next 19 years the pZP vaccine was applied to wild horse herds in the US with a high degree of success with regard to both efficacy and safety (Kirkpatrick *et al.*, 1990, 1991, 1992, 1995a; Kirkpatrick & Turner 2002, 2003, 2007) as well as ability to achieve population effects (Turner & Kirkpatrick 2002; Kirkpatrick & Turner 2007). The work with wild horses rapidly led to the application of pZP to other species as well, including white-tailed deer (McShea *et al.*, 1997; Naugle *et al.*, 2002; Rutberg *et al.*, 2004), wapiti (Shideler *et al.*, 2002), and many species of captive exotics, in zoos (Kirkpatrick *et al.*, 1995b, 1996; Deigert *et al.*, 2003; Frank *et al.*, 2005).

Collectively, these applications of pZP to various wildlife species and their results conformed well to the hypothetical characteristics of the ideal wildlife contraceptive (Kirkpatrick & Turner, 1991). The vaccine resulted in an overall efficacy of near 90 per cent; it could be delivered remotely by means of small darts; the contraceptive effects were reversible; there were no changes in social behaviours or organisation among treated animals; no deleterious long-term health effects resulted from its use; the vaccine was protein in nature and therefore did not pass through the food chain; it was safe to administer to pregnant animals, and it could be produced and applied at relatively low costs.

RESEARCH ON CONTRACEPTION OF ELEPHANTS IN THE KRUGER NATIONAL PARK, 1995–2000

In 1995 the scientists associated with the pZP vaccine approached the Kruger National Park to propose a project that would test the safety and efficacy of pZP contraception on African elephants.

Since the pZP vaccine is made by purifying zona pellucida proteins derived from the ovaries of pigs, the team wanted to establish the degree of homology between porcine and elephant zona pellucida proteins (eZP) (Fayrer-Hosken *et al.*, 1999). To this end, an immunohistochemical study was conducted using elephant ovarian tissue obtained during the last Kruger culls in 1995 and rabbit-anti-pZP antibodies. After sections had been exposed to the rabbit antibodies they were rinsed and then treated with immuno-gold-labelled goat anti-rabbit antibodies rendering the antibody complex visible. The results showed distinct immuno-gold staining of the zona pellucida capsules of the elephant oocytes in the histological sections. Following this work three tractable elephant cows in zoos in North America were vaccinated with pZP in order to determine the vaccination regimen. All three cows developed antibody titres similar to those of horses that had been successfully immunocontracepted with pZP vaccine (Fayrer-Hosken *et al.*, 1997; 1999).

A field trial began in Kruger in late 1996 with the treatment of 21 female elephants and 20 control animals. The purpose of this and the second field trial with 10 cows was simply to test the contraceptive potential of the pZP vaccine. The initial primer inoculation (600 µg) was given by hand, after immobilisation, the animals were collared for later identification, and all subsequent booster inoculations were given remotely, by dart, from a helicopter. Each animal received two booster inoculations the first year, six weeks apart. A year later, 44 per cent of the treated animals and 89 per cent of controls were pregnant; this difference was statistically significant. One of the treated elephants had a 22-month foetus, indicating that the vaccine had no effects on the health of the pregnancy (refer to later trials).

The next experiment changed the timing of the booster inoculations, from six weeks apart to two and four weeks apart, and a year later only two were pregnant (20 per cent). Four of the seven cows that were non-pregnant during the first trial received a single booster inoculation. A year later ultrasound examinations revealed that none of the four pZP-boosted animals was pregnant; all were cycling and had normal looking reproductive tracts. The three that were not treated were all pregnant (Fayrer-Hosken *et al.*, 2000). The collective results of these experiments with the Kruger elephants were corroborations of previous work with wild horses, deer and a variety of captive exotic species, and suggested that pZP immunocontraception of African elephants was safe, effective and had reasonable efficacy as a population management tool.

MAKALALI CONTRACEPTION PROJECT

The Greater Makalali Private Game Reserve (GMPGR) is situated in the Limpopo Province, South Africa. Its contraception project was started in May 2000, and is the longest running same-population study (n = 7 years) on elephant immunocontraception to date.

Efficacy

Initially 18 target animals were vaccinated with 600 µg of pZP + 0.5 ml of Freund's Modified Adjuvant (Sigma Chemical Co., St Louis, MO) and two booster vaccinations of pZP (600 µg) emulsified in Freund's Incomplete Adjuvant (Sigma Chemical Co., St Louis, MO) each two to three weeks apart (Delsink *et al.*, 2002, 2006). By 2007, a further five target animals had been vaccinated under the same regime, totalling 23 vaccinated animals in the Makalali population. Reproductive control was demonstrated in all 23 targeted females who passed the population's average intercalving interval of 56 months (Delsink *et al.*, 2006). A 0 per cent growth rate has been maintained within this target group since August 2002.

Effects on population growth

Delsink's detailed population history (Delsink *et al.*, 2002; Delsink, 2006) allowed for the estimated rate of increase (excluding mortalities and introductions) to be determined for the population, on an individual elephant basis, using the average inter-calving interval (56 months) for the period 1994–2002. (While the programme was initiated in 2000, the first two years were not influenced by pZP, thus they were included in the intercalving rate calculation). The estimated population size for GMPGR for 2010 totals 108 animals (Delsink *et al.*, 2006) (figure 9). Contraception had a significant effect on the population's growth, as the difference between the estimated and observed population size was significantly different over time (Delsink *et al.*, 2006). The contraceptive effect over the period 2003-2010 produces an average population growth decline of 6.5 per cent, assuming all the original target animals remain on the programme and there are no further introductions or mortalities. This estimation includes the addition of eight calves from the current pre-pubertal cows that will only be contracepted after they have given birth to their first calves. The average population growth rate (excluding introductions and mortalities) from 1996 to June 2000 was 8.9 per cent, similar to the range of the model projected 6.2-8.9 per cent average annual population growth rates following introduction (Mackey *et al.*, 2006). Thus, the contraceptive will effectively reduce the population growth rate by 70 per cent for the period 2003 through 2010 (Delsink *et al.*, 2006).

Local and systemic side effects of the vaccine

Because many immunogens used for contraceptives have low antigenicity, their efficacy is dependent on concurrent delivery of potent adjuvants to stimulate an adequate immune response (Munson *et al.*, 2005). The optimal efficacy of immunocontraceptives such as pZP occurs when Freund's adjuvant is used, but Freund's adjuvant incites a marked granulomatous reaction (Munson *et al.*, 2005).

Of the 18 cows vaccinated and boosted between May and June 2000, 16 displayed local swellings – most likely abscesses (Bengis, 1993) – within 3 months after the initial vaccination (Delsink *et al.*, 2002, Delsink 2006). These swellings ranged from 20 to 100 mm in diameter and were all eventually resorbed. Only three cows developed swellings of approximately 100–120 mm in diameter after the primary vaccination in 2000. While these were still present though markedly reduced in size by the fourth annual vaccinations in 2004, they have now been completely resorbed (Delsink, 2006; Delsink, pers. obs. 2007). The reactions were not associated with lameness or with other visible ill health effects on the cows.

During subsequent annual vaccinations, Freund's Incomplete Adjuvant (less aggressive than Freund's Modified Adjuvant) was used, and no swellings greater than 50 mm were formed. The affected animals never displayed signs of irritation or discomfort as a result of the swellings, nor did they suppurate or need any treatment. In fact, the swellings that formed at the dart site were similar in appearance to those produced by thorns or other penetrating objects (Delsink, 2006).

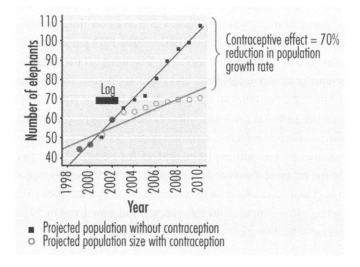

Figure 9: The effect of contraception on population size at Makalali Conservancy. The black bar above the curves indicates the lag effect before contraception as a result of elephants already pregnant prior to darting. See Delsink *et al.* (2006) for details, and text for statistical test

Behavioural effects

Effects of administration procedure on behaviour

The greatest impact of pZP implementation was on the interaction between the vaccination team and the herds during ground darting (Delsink *et al.*, 2007a). 'Avoidance' behaviour was clearly evident in the presence of and towards the darting team and was manifested by the herds either running away or remaining mobile away from the darting team (Delsink *et al.*, 2007a). Darting from the ground appeared to cause less stress than from the air, judging by the flight response and movement patterns of the elephants, but it was conducted over much longer periods ($n = 25$ days)(Delsink, 2006). This avoidance behaviour only continued until approximately two weeks after the last dart had been fired.

Conversely, vaccinations administered from the helicopter resulted in far greater perceivable stress in the animals. This perception was based on observations of bunching of the herd and flight patterns. The duration of disturbance, however, was much shorter than when darted from the ground, and the animals resumed normal movement patterns and appeared to be settled within a day of vaccinations (Delsink, 2006; Delsink *et al.*, 2007a). Aerial vaccinations were far more productive, with a darting time of about 30 seconds per female (Delsink *et al.*, 2007a). Thus, the average completion of aerial vaccinations within the Makalali population is approximately 30 mins. Furthermore, the helicopter vaccinations did not compromise further monitoring initiatives, as the elephants appeared to be reacting to the helicopter, rather than to being darted (Delsink *et al.*, 2007a).

Effects on ranging patterns of clan and individual herds

Although there was avoidance behaviour, the cumulative effect of vaccine administration did not cause an overall change in the treated population's core and total ranges over a five-year period for either individual herds or the whole clan (Delsink, 2006; Delsink *et al.*, 2007b). Thus, for reserves that are largely eco-tourism driven, the implementation of a pZP programme will have little effect on game-drive and safari activities (Delsink, 2006; Delsink *et al.*, 2007b). The shorter vaccine administration demonstrated by the helicopter darting appeared to have a more consistent effect on the herds, with the least shift in core range during and after darting (Delsink, 2006; Delsink *et al.*, 2007b).

Effects on reproductive behaviour (includes effects of non-conception)

The Makalali study demonstrated that although there was an increase in the number of musth and oestrous events over the years of treatment, this change was not significant over all treatment years for all herds (Delsink, 2006; Delsink *et al.*, 2007b).

Reproductive control is achieved in the third year of treatment. This means that all females that were pregnant would have given birth and after subsequent treatments, would have been contracepted, while those that were not pregnant would be contracepted after the initial vaccination series. Therefore, there was an expectant increase in oestrus observations in the third year. However, there was no significant difference in oestrus occurrence among the years of treatment, although the greatest frequency was observed in Year 3 – the year in which reproductive control was achieved (Delsink, 2006). Therefore, there was a higher incidence of oestrus in this year, as all treated females were cycling in the same year. Calf survival was not affected by increased cycling and, in fact, no calves were lost during the entire observation period. This is similar to experience with pZP-contracepted wild horses (Kirkpatrick & Turner, 2003).

Musth was most frequent in the oldest, dominant bulls, aged 25+ years (Delsink, 2006). Musth bulls dominated consort and matings in all instances, even in the presence of other bulls. Even in the absence of musth, dominant bulls still dominated all consort displays and matings, demonstrating that cow mate selection remained intact (Poole, 1996a). Thus, the treatments did not affect bull hierarchy or cow selection (Estes, 1991; Moss, 1983; Delsink, 2006; Delsink *et al.*, 2007b).

Bull association coupled to increased frequency of oestrus

Under the pZP treatment, the target animal displays normal oestrous cycles, cycling every 15–16 weeks, because although copulation still occurs, conception does not (Whyte, 2001). Therefore, under the pZP contraceptive, the frequency of mating and its accompanying disturbances is assumed to be far more frequent (Delsink, 2006). Thus, with an increased frequency of oestrus, there is the potential for change in the frequency of association of both sexually active musth and sexually active non-musth bulls with breeding herds as both sets of males compete for oestrous females (Poole, 1982; Delsink, 2006). In fact, bull association with herds decreased over the years, probably an effect of aging in this relatively young population. The decrease in herd-bull association further illustrated that the pZP implementation did not affect Makalali's bull hierarchy; that is, there were far more non-musth sexually active bulls than musth sexually active bulls and even in the absence of musth, mating and consort behaviour was highest in the three dominant musth bulls. Furthermore, the non-musth sexually active bulls did not increase their associations with the herds (Delsink, 2006; Delsink *et al.*, 2007b).

Herd fission/fusion

A change in association pattern and the proportion of time spent alone between herds was observed after the pZP implementation (Delsink, 2006; Delsink *et al.*, 2007b). The Makalali herds tended to associate less over the initial period of the study, but the integrity of the group remained strong throughout the study duration. These differences are attributed to the change in association pattern between herds – the natural formation of a bond group when the family unit becomes too large and splits along family lines (Moss, 1983) – and not to anti-fertility treatment. Furthermore, this is the general pattern of group formation in relocated herds (Slotow, pers. comm. 2006). This study demonstrates that the pZP implementation did not cause herd fragmentation and did not cause herds to become more isolated or alter the matriarchal group size – management concerns raised by Whyte (2001). Furthermore, behaviour between cows and their calves at foot was recorded, and cows were never separated from their calves (Delsink, 2006; Delsink *et al.*, 2007b).

The results from the Makalali study demonstrate that there were no aberrant or unusual behaviours with the medium-term and sustained use of pZP on bull and cow societies. As demonstrated in the Kruger trials (Fayrer-Hosken et al., 2000), the results of the pZP-treated cows were as expected from cows whose behavioural patterns were not affected by the treatment (Whyte, 2001); that is, there is no evidence to suggest that the pZP has any adverse effects on the behaviour of either the treated cows, their matriarchal groups or bulls (Delsink, 2006).

Continued management using contraception including reversibility

The original hypotheses of the programme have been successfully demonstrated (Delsink, 2006; Delsink et al., 2006; 2007a & b).While reversibility was demonstrated in the Kruger trials (after two successive years of treatment), the Makalali study aims to demonstrate reversibility in cows treated in the medium-long term. In 2005, five cows were removed from the programme to test reversibility; three cows were treated for five years, one for four years and one for three years, respectively (Delsink, pers. obs. 2007). All the cows have calved except for one who has never conceived. One of the three-year treated cows was accidentally vaccinated in 2006. However, all the other cows have been off treatment for two years (June 2007). Reversibility in 100 per cent of wild horse mares treated for one, two or three consecutive years has been demonstrated, while 68 per cent of those vaccinated for four consecutive years returned to fertility (Kirkpatrick and Turner [2002]). No mares treated for over seven years had returned to fertility at the time of publication, but they had started ovulating again (Kirkpatrick and Turner [2002]). A six-year study on white-tailed deer demonstrated that the treated does did return to fertility, though they became pregnant later in the breeding season than would normally occur (Miller et al., 1999).

OTHER GAME RESERVES WHERE PZP HAS BEEN USED

Including Makalali, immunocontraception has been applied in 10 discrete populations (table 3) (Bertschinger *et al.*, 2007). The Kapama cows belong to a captive population and were treated with a slow-release formulation of pZP. The other populations are wild and during Year 1 all cows of reproductive age (n = 103) received a primary (400 µg with 0.5 ml Freund's modified adjuvant) and two boosters (200 µg with 0.5 ml Freund's incomplete) followed by an annual booster of vaccine. Vaccines were administered remotely from the ground or helicopter with drop-out darts. The initial results of the populations vaccinated from 2000 to 2005 are shown in table 4.

The results of this ongoing elephant contraception project align with previous findings of the programme in Makalali (Delsink *et al.*, 2006). During Year 1 there is no effect of immunocontraception on calving percentage (28.8 per cent, table 4). There may be an effect in Year 2 (22.5 per cent), but in Year 3 the calving percentage was reduced to 15.1 per cent. Although the data from most of the populations are incomplete at this stage, from Year 4 and onwards no calves were born to treated cows. The lag time before the effects of the vaccine are visible is due to the long gestation period of 22 months in African elephants. Taking the gestation period into account it would appear that some cows are not immediately rendered infertile after the first booster vaccination. Eight of 52 cows contracepted during Year 1 produced calves during Year 3 (>24 months).

Occasional swellings were the only side effects seen following administration of the vaccine. It should also be noted that abscesses are common following darting of elephants with immobilising agents (Bengis, 1993). The cause of the abscesses during immobilisation procedures is more than likely related to contamination of elephant skin with large numbers of bacteria originating in stagnant water, dust, and sand. The dart needle merely provides a mechanical means of carrying bacteria into and below the skin. Big differences were noted in the incidence of swellings (presumed to be abscesses) at the primary vaccination site between Makalali and Thornybush, despite the same adjuvant having been used. Only 4 of 19 cows were affected at Thornybush, compared to 16 of 18 cows at Makalali. The incidence of abscesses can also be vastly different between reserves following immobilisation. According to Bengis (pers. comm.) the incidence in the South African Lowveld area is much higher than in Etosha. The latter is arid and elephants are much less likely to submerge themselves in contaminated water. This may indicate that the swellings noted after primary vaccination were bacterial abscesses rather than granulomas resulting from the adjuvant.

Reserve & inception	Method of darting	Population detail	Year							
			−1	1	2	3	4	5	6	7
Makalali Jun 2000	Yrs 1–3 Ground Yrs 4–7 Helicopter	Total population	45	47	52	60	64	64	67	69
		Cows contracepted	0	18	18+2	20+3	23	23	17+4	21
		Calves born	2	5	8	4	0	(3)	(2)	(1)
Mabula May 2002	Helicopter	Total population	9	11	11	11	11	11		
		Cows contracepted	0	4	4	4	4	2		
		Calves born	2	1	0	0	0	0		
Thaba Tholo Aug 2004	Helicopter	Total population	27	28	29	29				
		Cows contracepted	0	8	8	8				
		Calves born	?	1	0	1				
Shambala Jun 2004	Helicopter	Total population		10	11	12				
		Cows contracepted		4[a]	4	4				
		Calves born		1	1	0				
Phinda Jul 2004	Ground	Total population	71	77	83	90				
		Cows contracepted	0	19	19	18+3				
		Calves born	6	6	7	8				
Thornybush May 2005	Helicopter	Total population		35	38					
		Cows contracepted		19	19					
		Calves born		4[b]	4					
Welgevonden Sept 2006	Helicopter	Total population		117	129					
		Cows contracepted		35	35					
		Calves born		12	5					
Kaingo Oct 2005	Ground	Total population		9	11					
		Cows contracepted		4[a]	4					
		Calves born		2	0					
Karongwe May 2007	Helicopter	Total population	16	16						
		Cows contracepted	0	4[c]						
		Calves born	0	Too early						
Tembe E P May 2007	Helicopter	Total population		250						
		Cows contracepted	0	75						
		Calves born		Too early						

[a] One calf died after a rupture of the umbilicus

[b] These four cows only received a primary vaccination during Year 1

[c] These four cows were vaccinated once with a so-called 'one-shot' vaccine

Table 3: Summary of the elephants contracepted with pZP vaccine in 10 game reserves and the response in terms of calving data and total populations from Years 1–7 of the programme (Bertschinger et al., 2007)

	Year[a]						
	1	2	3	4	5	6	7
Number of reserves	9	8	5	2	1	1	1
Cows contracepted during Year 1	111[b]	3 111	53	22	18	18	18
Calves born	32	25	8	0	0	0	0
Calving %	28.8	22.5	15.1	0	0	0	0

[a] Year contraception programme was introduced
[b] Excludes the 75 cows from Tembe Elephant Park and Karongwe Private Game Reserve

Table 4: Percentage of cows vaccinated in Year 1 calving by year (Bertschinger *et al.*, 2007)

Excluding the largest reserve (Welgevonden; cow $n = 35$; total $= 117$), average helicopter darting time was 30 seconds per cow. Average flying time during Year 1 in Welgevonden, which is mountainous and heavily wooded in parts, was 6.9 minutes per cow.

The data available to date show that the pZP vaccine is effective in reducing the birth rate of free-ranging elephant cows. Birth rate starts to drop during the third year to reach zero during the fourth year. The safety of the vaccine has once again been demonstrated in target animals, the only side effect being occasional temporary lumps at the darting site.

The data available so far show that 64 of 111 cows vaccinated during Year 1 produced normal, healthy calves. An additional calf died a few days after birth due to a ruptured umbilicus that is unlikely to have been caused by the pZP vaccine. Short-term reversibility after one to two years of pZP vaccination of elephant cows was demonstrated in the Kruger (Fayrer-Hosken *et al.*, 2000). However medium- to long-term reversibility still needs to be demonstrated. To prove reversibility we have now withdrawn vaccination from five cows in Makalali (three five-year treated cows, one four-year treated cow and one three-year treated cow) and two at Mabula (four-year treated). In three reserves (Makalali, Phinda, and Thornybush), however, we have seen a total of 58 oestruses in a total population of 56 immunised cows. Because these data relied on behavioural observations, a number of oestruses may have been missed. The data reveal that the ovaries of these cows seen in oestrus must be functional and that there has been little or no immuno-destruction of follicular tissue. Currently there is one trial under way at Thornybush to monitor ovarian cycles more objectively than just by means of behavioural observations. Faecal progesterone metabolites will need to be used to achieve this objective

(Wittemyer *et al.*, 2007). This study will be extended to include faecal monitoring of eight cows at Makalali (Bertschinger *et al.*, 2007).

IMPLEMENTATION/METHODOLOGY

Identification of the population

In the Makalali study, it was essential that elephants could be identified individually as each elephant had to receive her primary and subsequent boosters timeously (Delsink *et al.*, 2002; Delsink, 2006). Furthermore, to ensure complete vaccine discharge, the darts were retrieved after each vaccination in the years 2000–2003. Thus, should incomplete vaccine delivery have been recorded, the target animal could be revaccinated (Delsink *et al.*, 2007).

The Makalali population was identified according to the methods described by Moss (1996), Poole (1996a & b), and Whitehouse and Kerley (2002), where animals were recognised by their individual characteristics including sex, age, unique ear pattern comprising nicks, tears, and holes, and the size and shape of tusks. When no distinguishing ear or tusk features were visible, ear venation patterns were completed according to the methods described by Whitehouse (2001). Other distinguishing features such as growths, lumps, scars and tail hairs were also recorded as identification criteria. The animals were sexed following the method of Moss (1996) and Hanks (1972), where head shape was observed (the profile of the head is rounded in males, steeply angled in females) when genitalia were obscured.

The elephants were aged according to accepted parameters. See box 3, Chapter 2.

The same detailed individual histories exist for the Mabula, Shambala, Thaba Tholo, Karongwe and Phinda elephants. However, no extensive individual data existed for the Thornybush, Welgevonden, or Tembe elephant populations. Individual elephants in the Thornybush population were only identified after the primary vaccination was administered, using the methods described above (Delsink, pers. obs. 2005). Welgevonden elephants were individually identified during the course of Year 1 of contraception implementation. Thus, these latter populations were managed not on individual-based elephant identifications, but rather on a broader age and sex class classification – adult and sub-adult females were identified based on size, grouping compositions and head shape from the air (Delsink, pers. obs. 2006; Van Altena, pers. comm. 2007).

Darting method

At Makalali, ground darting was the original source of delivery to facilitate dart retrieval, to test vaccine efficacy and for minimal impact on the herds (Delsink, 2006). However, during the collaring procedure in 2003, 17 individuals were vaccinated from the air. This greatly improved the team's productivity, and since 2004, all vaccinations have been conducted from the air (Delsink *et al.*, 2007). In 2003, 2004 and 2005, time spent in the field decreased while the average number of darts fired per day increased (figure 10). At Makalali, the cows have been vaccinated within 30–60 minutes since 2004 (Delsink, pers. obs. 2007).

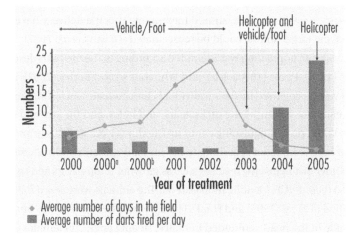

Figure 10: Efficiency in field implementation over successive treatment years 2000–2005, 2000, 2000a and 2000b reflect the primary vaccination, 1st and 2nd boosters in Year 2000, respectively. Vehicle/foot/helicopter refers to the source of vaccination delivery (Delsink *et al.*, 2007)

The experiences in the nine other reserves showed that vaccination is possible under a variety of conditions. In Phinda, which contains large areas of sand forest, ground darting was used for the first three years. The average daily rate of movement for five GPS-collared cows one week before, during the week-long contraception period and one week after contraception in Year 3 was 0.232, 0.760 and 0.250 km.h^{-1}, respectively (Burke, pers. comm.). This showed once again that elephants settle down very soon after the darting experience but not as rapidly as following helicopter darting. In Mabula, Thaba Tholo, Shambala, Thornybush, and Karongwe (all vaccinated from the helicopter), the flying time per cow was approximately the same as for 2004 in Makalali once the booster vaccinations commenced. Terrain and habitat had an influence on helicopter darting time. Average flying time during Year 1 in Welgevonden, which is mountainous and heavily wooded in parts, was 6.9 minutes per cow. In Tembe, the largest population vaccinated so far, helicopter darting was quick, provided the cows were in the open swamp area. Most, however, were darted in the sand forest, which has high tree canopies. Average darting time including total ferrying time of 90 minutes was 2.5 minutes per cow.

Implementation strategies for different conditions and population sizes

In small confined populations of up to 100 elephants the ideal starting point is to identify each individual animal on the property, concentrating especially on the females of breeding age. In accordance with the elephant management plan the number of cows to be immunised for contraception can be determined. Prepubertal cows can be either allowed to conceive and have their first calf or be immunised at the outset. Over time the response of the population is monitored and the contraception programme adapted according to the needs of the management plan and ongoing habitat response (figure 15). Darting is preferably carried out from a helicopter as the disturbance is brief and return to pre-darting patterns quick in comparison to ground darting. The use of GPS/radio collars placed strategically on a cow in each herd will decrease darting time but will add considerably to the cost of the programme. In time costs are likely to be reduced and collar battery-life increased. Where the population is small (4–8 cows), which allows darting to be carried out in one, maximum two days, ground darting may be considered.

Frequency of darting with regard to the primary vaccination and subsequent boosters can also be varied. The primary vaccination sensitises the animal to the antigen (in this case pZP or even GnRH), and B-lymphocytes respond by producing humoral antibodies. Some of these B-lymphocytes (also known as memory cells) remain present for years and when the animal is boosted they divide to form more active B-lymphocytes with a corresponding rise in antibodies. It stands to reason therefore that cows given a single vaccination during Year 1 followed by annual boosters will eventually develop sufficient immunity to prevent fertilisation from taking place – the process will, however, take longer (we speculate 4–6 years). This strategy could be useful to sensitise prepubertal cows but just as importantly could be employed to sensitise a translocated population where contraception is planned in the future. It would be sensible to apply this to all elephant cows being translocated to fenced smaller reserves. The cost would be minimal as the vaccination could be carried out when the cows are immobilised for translocation.

Research with pZP immunocontraception in horses has shown that time taken to reversal is approximately equal to the number of years a particular mare has been vaccinated for. If this holds true for elephant cows it may be possible to lengthen intervals between boosters after the initial three or four years. Currently we do not have data to prove such a possibility.

Returning to population size, the picture is somewhat different for larger populations where identification of individuals becomes increasingly difficult and time consuming as the population increases. Here the mass-darting approach would have to be applied. The largest population done to date was in the Tembe Elephant Park where 75 cows were immunised. Marker-darts were employed in this successful exercise which allowed the team to identify cows already darted. The same approach could be used for even larger populations, but as population size increases the percentage of cows vaccinated must decrease. Despite this it may be possible to achieve a considerable effect on the growth rate of large populations. According to Page (pers. comm.), models have shown that effective contraception of 60 per cent of cows will halve population growth rate over a period of 15 years.

COSTS

There is a concern that contraceptive implementation may be cost prohibitive. The Makalali study (table 5) has demonstrated that the highest costs incurred during contraception implementation are based on the helicopter costs, or more specifically, the costs of ferrying the helicopter to the site. Total annual contraceptive costs per elephant doubled from 2005 to 2006 simply because of the costs of ferrying the helicopter from Johannesburg and back during 2006 (4 hours).

At Makalali, implementation costs ranged from ZAR332 to ZAR1 170 per animal including the darts, vaccine and helicopter costs. Veterinary fees vary between ZAR2 500 and ZAR3 800 per day, and helicopter rates are approximately ZAR3 800 per hour (J Bassi, BassAir, pers. comm., 2005). As such, veterinary fees can be incorporated with vaccine and dart costs on an individual elephant basis.

As can be seen from the Thornybush contraception programme, initial costs of vaccination during the first year tend to be a lot higher than during the following years. During Year 1 the average all-inclusive cost was ZAR1 639 per vaccination per cow, whereas in Years 2 and 3 it was only ZAR645 per vaccination per cow. The main reason for this was that primary vaccination procedure was also used to identify the breeding units from the helicopter. This increased helicopter flying time considerably. With the subsequent two rounds the duration decreased. Another important factor is the experience of the helicopter/spotting/darting team, which improves with each round.

	2000[3]	2001	2002	2003	2004	2005	2006
Dan-Inject® @ ZAR178 per dart[1] (Number of darts in brackets)	5 518 (31)	2 136 (12)	2 136 (12)	712 (4)	356 (2)	0	0
Pneu-Darts® @ ZAR85 per dart (Number of darts in brackets)	0	0	0	1 445 (17)	1 785 (21)	1 955 (23)	2 394 (25)
Vaccine @ US$20 per dose (2000–2002)[2] (Number of doses in brackets)	11 160 (62)	4 500 (25)	4 500 (25)	0	0	0	0
Vaccine @ ZAR100 per dose (2003–2004) (Number of doses in brackets)	0	0	0	2 300 (23)	2 300 (23)	2 300 (23)	5 273
Helicopter	0	0	0	7 524[4]	7 980[5]	10 819[6]	18 058[7]
Number of elephants vaccinated	18	20	23	23	23	23	22
Average cost of vaccinations per elephant	927	332	289	520	540	655	1 170

[1] Dan-Inject darts were used at least twice each during the 2000–2002 vaccinations, thus, the number of actual darts fired is halved in order to determine actual cost of darts for these years. Four and two Dan-Inject darts were used in 2003 and 2004 respectively.

[2] Vaccine was obtained from the USA at an exchange rate of ZAR9:US$1.

[3] The year 2000 costs were high because each animal received a primary and two booster vaccinations. A total of 62 darts were fired, including 11 revaccinations due to unsuccessful darts.

[4] Helicopter time for 2003 included 5.5 hours of flying time including ferry to Makalali. Flying time was high because it included vaccinations and the collaring and biopsy procedure of 4 and 1 elephants respectively. Therefore, costs for 2003 vaccinations are calculated at the rate per hour for 2003 (ZAR3 300) and on the number of hours flown as per 2004, i.e. 2 hours in total.

[5] Helicopter time for 2004 included 1 hour of flying time and 1 hour of ferry costs.

[6] Helicopter time for 2005 included 1 hour of flying time and 2 hours of ferry costs.

[7] Helicopter time for 2006 included 1 hour of flying time and 4 hours of ferry costs.

The above costs do not include veterinary fees or salaries for the project darts man and elephant monitor.

Table 5: Costs incurred during the Makalali vaccinations 2000–2006. Amounts are given in ZAR (amended from Delsink *et al.*, 2007a)

SURGICAL STERILISATION IN ELEPHANTS

Methods

A safe, reliable and efficient technique for surgical castration of African and Indian (*Elephas maximus*) elephants has been developed (Foerner *et al.*, 1994). Although the work of Foerner *et al.* was pioneering, recent advances in elephant surgery, surgical equipment and anaesthesia, have made it feasible to consider the use of less invasive laparoscopic techniques that would avoid technical problems associated with this original procedure (Stetter *et al.*, 2005).

Surgical sterilisation of elephant cows

Laparoscopic surgery provides a direct view of internal organs and this allows for tissue manipulation via a minimally invasive procedure utilising relatively small incisions (Stetter *et al.*, 2005). In 2004, the first-ever laparoscopic sterilisations were performed on two free-ranging elephant cows at Phinda Game Reserve, South Africa (Delsink, 2006). The elephants were positioned in lateral recumbency for hand-assisted laparoscopic surgery (Stetter *et al.*, 2005). Once the surgery was complete on one side, the animals were rolled to the other side for the same procedure (Stetter *et al.*, 2005). This was the first reported abdominal surgery in free-ranging African elephants and has been considered a significant milestone. Whilst the surgery was successful, it was lengthy and the first female only returned to her herd two days after surgery (H. Genis, pers. comm. 2004). It was noted through post-operative monitoring that the surgical sites healed without complication and the treated cows showed no adverse social or behavioural issues after the sterilisation (Stetter *et al.*, 2005).

Surgical sterilisation of elephant bulls

In February 2005 Dr Mark Stetter and his team attempted the first laparoscopic vasectomies on two bulls in the Mabalingwe Game Reserve. To facilitate the procedure the bulls were suspended by straps from a crane (figure 11). In the first bull only one side could be completed, while the second bull died during the procedure. Another four bulls were operated on using the same technique in the GMPGR (Delsink, 2006). The procedure was altogether unsuccessful in one bull while in the other three the vasa deferentia could only be identified and ligated on one side. The duration of the whole procedure was approximately 5 hours per bull.

Figure 11: Suspension of elephant bull for vasectomy. Disney World Public Affairs (http://www.wdwpublicaffairs.com/PhotoAlbum)

In 2006, the team returned and successfully performed bilateral vasectomies on four of four bulls at Welgevonden Game Reserve. Procedure time was 3–4 hours per bull. In 2007, five more bulls were successfully vasectomised at Songimvelo, Mpumalanga Parks Board. The full procedure time varied from 2.5 to 3 hours. Greater efficiency was due to overcoming anatomical obstacles related to patient size, modifications in equipment, and the advancement of surgical techniques (Stetter *et al.*, 2006).

Possible applications, disadvantages and costs of bull vasectomy

Unlike castrations, vasectomies do not remove the testes and thus the treated animals should enter into musth, breed and maintain their social status (Stetter *et al.*, 2006). Recent improvements with this technique have greatly reduced the anaesthetic and surgery times and have paved the way for several animals to have surgery on a given day (Stetter, pers. comm. 2007). The potential to sterilise 10–40 breeding bulls in a small to medium-sized elephant population would provide for a significant reduction in birth rates approximately 22 months after completion of all surgeries. A further argument is that males need only be sterilised once, whereas using methods such as immunocontraception, females currently need to be treated three times in the first year and then annually, to prevent births (Bokhout *et al.*, 2005).

The limitation of vasectomy as a method for population control is that it is only suitable for smaller populations. This is dictated by cost, the invasiveness of the procedure, and the fact that it is, for all practical purposes, irreversible in elephants. The costs are summarised in table 6 (Bokhout *et al.*, 2005). In our view, the costs appear to be underestimated as they do not take into account equipment such as the crane and fieldwork to determine dominance status and locate bull areas. The average cost according Bokhout *et al.* (2005) is US$ 2 300 per bull (equivalent to ± ZAR16 100). Cows mate with sexually mature bulls during the middle of their oestrous period although they may mate with younger bulls during early and late oestrus (Moss & Poole, 1983; Poole, 1996a). This means that a high percentage of bulls would need to be treated to ensure that the females are not bred (Garrott *et al.*, 1992). Elephant populations are complex and bulls continually separate from their natal herds. These will also be potential breeders and especially with the increased frequency of heats in the population, young bulls may be afforded the chance to breed. Another disadvantage is that targeting the dominant bulls in the population will affect the natural genetic selection.

It is probable that similar levels of contraception can be achieved using a GnRH vaccine (see *The future of elephant contraception; GnRH vaccine*), which is reversible and cheaper to use.

Procedure	Cost per elephant (US$)
Capture and anaesthesia[a]	1 000
Equipment costs[b]	300
Surgery team	1 000
Total	2 300

[a] (Hofmeyr, 2003)
[b] Endoscopic instruments are calculated at US$30 000–60 000. If complete depreciation is assumed after 200 bull operations, cost per elephant = US$150–300.

Table 6: Estimated costs of laparoscopic bull vasectomy (Bokhout *et al.*, 2005)

THE FUTURE OF ELEPHANT CONTRACEPTION

Recently two papers were published that review or partly question the use of immunocontraception as a means of controlling population growth in African elephants. The first, by Perdock *et al.* (2007), is a fairly comprehensive review of the topic but with the shortcoming that the most recently quoted paper referring to elephants is from 2003. The other is by Kerley and Shrader (2007), with the rather sensational title 'Elephant contraception: silver bullet or a potentially bitter pill'. Both papers quite correctly list side effects on behaviour resulting from an increased incidence of oestrus, increased presence of bulls and a lack of calves within a breeding herd, with its possible ramifications.

While short-term studies have revealed no detectable behavioural changes (Delsink *et al.*, 2004b; 2006; 2007b) in populations that have been treated with pZP vaccines, extensive studies must be carried out to investigate possible long-term effects in this species, with its highly complex social structure. Once again, the same applies to any other form of population control. Kerley and Shrader (2007) state that our understanding of contraception is now at the stage that culling was at 30 years ago. This is only partly true. First, contraceptive trials began more than 10 years ago. Second, very little work has been carried out to study short-, medium- or long-term effects of culling on the behaviour of remaining elephants. The indications are that they may be considerable. Some other points made by the same authors with regard to behaviour are highly speculative. These are the loss of allomothering, increased stress levels and depression caused by the presence of fewer calves, and kidnapping of calves.

Elephants are not the first gregarious species to be exposed to contraception. The best example is probably the human, where contraception has been practised for hundreds of years, particularly stringently in so-called developed nations during the last 30 or so years. Another example is the domestic dog, which is an out-and-out pack animal. Have we been concerned about the effects of contraception in these two species? The answer is not really, despite the fact that there certainly are behavioural effects. Perhaps the most controversial point made by Kerley and Shrader (2007) is the possibility of injuries to cows as a result of increased mating attempts. There are no reports in the literature of injury to African or Asian elephant cows as a result of mounting. If one observes a bull in the process of mounting the stance of the bull on his back legs is very steep, meaning that he takes most of his weight on his hind legs placing very little on the cow. Elephants are quite nimble and their ability to balance on their hind legs only is plain to see when they reach for something high up in a tree requiring this stance. Besides, in domestic species like cattle and horses, where the weight difference between male and female can be considerable, service injuries are very rare despite the fact that these species are much less nimble.

The use of a GnRH vaccine, which has distinct possibilities, will of course cause anoestrus. Other points made by Perdock *et al.* (2007) are as follows:

- **The possibility of introducing new diseases with the pZP vaccine**. The vaccine is prepared from the ovaries of healthy pigs at abattoirs that are subjected to meat inspection. The manufacture of the vaccine involves washing of the ovarian material with large volumes of buffer fluid – 1000-fold larger than volumes used to free oocytes or embryos of virus particles. The zona ghosts are finally subjected to heat (65°C for

30 minutes) and aliquots are cultured for the presence of bacteria. All aliquots tested for use have been sterile. If a few viral particles were to survive the whole process they would not constitute an infective dose, and the viruses that may be found are specific pig viruses. Repeated injections of the same virus would have to be undertaken to possibly adapt a virus to a new host. pZP vaccine has been used on more than 80 different species all over the world, in many cases on captive populations, without the appearance of a new disease. In any case, elephants could be exposed to similar diseases in the wild where they come into contact with warthogs and bush pigs, both of which may carry the same diseases as domestic pigs.

- **Selection of cows that are immunocompromised as breeders, as they do not respond to the vaccine**. As with most free-ranging African mammals, natural selection of elephants that are resistant to or develop immunities to a range of infectious agents is rigorous. Immunocompromised animals will not survive under African conditions.

- **Ovarian damage may result, meaning that contraceptive effects could be permanent**. As the authors mention, short-term studies have demonstrated reversibility of pZP immunocontraception (Fayrer-Hosken *et al.*, 2000). Medium- to longer-term studies are under way to test reversibility after five years of vaccination and we expect that time taken to reversal will be approximately equal to the number of years a cow has been subjected to vaccination. We do, however, already have evidence that ovarian function in terms of the occurrence of normal oestrous cycles appears to be normal, with 58 oestruses observed in 56 cows that are seen intermittently (once to twice a week) (Bertschinger *et al.*, 2007). The time taken to reversal may, however, be a considerable advantage for the implementation of immunocontraception. It could mean that the interval between boosters can be lengthened to two or even three years after the first four or so years of annual vaccination. This would improve practicality and reduce cost.

The cost of contraception (at approximately ZAR1 000 per cow) and vaccination is considerably less than indicated in the papers above. A number of the points raised in the two papers and from other quarters presume sustained use of the vaccine on 100 per cent of the population. This does not have to be the case and will certainly depend on the management plan.

One-shot vaccines

One-shot vaccines are vaccine formulations that with a single administration would provide sufficient stimulus to the immune system to render the animal infertile for a year or more without any boosters. Such formulations either provide a slow continuous release of vaccine and adjuvant over time (e.g. liposomal system) or release at intervals (e.g. lactide-glycolide copolymer pellets, Turner *et al.*, 2002). The major advantages of one-shot vaccines are that they would be more practical and cheaper to administer, and provide less disturbance to the population being contracepted. The improved practicality would mean that much larger populations could be tackled with greater average efficacy. The lactide-glycolide copolymer pellets have been tested extensively in horses. The one-year formulation reduced fertility rates to 10.7 per cent in 266 hand-injected mares and to 25 per cent in 114 dart-injected mares. The fertility rate after the two-year formulation varied from 5.2 to 31.6 per cent in a total of 96 mares (Turner *et al.*, 2002). It should be remembered that annual boosters following the one-shot vaccine will further increase contraceptive efficacy.

Trial in captive elephants

A trial was performed on three captive elephant cows using the traditional vaccine (fluid) for primary vaccination and three different formulations of lactide-glycolide pellets that release after 1, 3 and 12 months respectively. The cows were hand-injected at two different sites – one for the fluid vaccine and one for the pellets. They were bled two months later (one month post-release of the one-month pellets) for antibody titre determination and the results compared to six cows one month after their first traditional booster vaccination (table 7). The titres of all three one-shot-vaccinated cows were higher than the cows treated with the traditional method (Van Rossum, 2006; Turner *et al.*, 2008). The trial is ongoing.

Trial in free-ranging elephants

This trial began in May 2007, when four cows were vaccinated from the helicopter. They will be captured at 3–6 month intervals during 2007–2008 to assess antibody titres and contraceptive effect (Bertschinger, pers. comm.).

Vaccine and cow	Anti-pZP antibody titre at 1:270 dilution (absorbency)
Traditional vaccine with boosters	
Cow 1	0.246
Cow 2	0.918
Cow 3	0.915
Cow 4	0.969
Cow 5	0.970
Cow 6	0.354
One-shot pellet vaccine	
Setombi	1.138
Bubi	1.431
Nandi	1.058

Table 7: Antibody titres of elephant cows vaccinated with either the traditional method or the one-shot pellet vaccine

Possible future developments

The one-shot pellet vaccine needs to be tested more extensively on free-ranging elephants to determine efficacy, duration of contraceptive effect and reversibility. An additional change to the lactide-glycolide copolymer pellets will extend the effect of the one-shot vaccine from two to three years after administration of a single dart (Turner *et al.*, 2008). Melodie Bates (MSc student 2006–2008, Faculty of Veterinary Science, University of Pretoria) is conducting a project in Thornybush Private Reserve to more objectively assess oestrous cycles and stress response of contracepted cows. She will monitor faecal progestins (cycle) and glucocorticoids (stress) to achieve this. Indirectly, if one can show that contracepted cows have ovarian cyclic activity, it will show that they have normal ovarian function. Evidence of this in at least four reserves is available through less reliable behavioural observations.

GnRH vaccine

Experience in elephant bulls

In 2003 a trial was initiated to test the efficacy of a GnRH vaccine (GnRH-tandem-dimer-ovalbumin conjugate, Pepscan Systems, Lelystad, The

Netherlands; Oonk et al., 1998) to control aggressive behaviour and musth in captive and free-ranging bulls (De Nys, 2005). Initially five captive bulls were vaccinated. Behaviour and faecal epiandrosterone were monitored in all bulls before the primary vaccination (Stage 1), after the primary, first and second booster vaccinations (Stages 2, 3 and 4), and two and four months after the second booster (Stages 6 and 7). Prior to vaccination two bulls were aggressive while the three others were not. During Stage 1 the behaviour of aggressive and non-aggressive bulls corresponded with faecal epiandrosterone concentrations of the two groups (figure 12). The vaccine produced encouraging results, with the two aggressive bulls showing a behavioural improvement and all bulls remaining non-aggressive during the remainder of the six-month observation period. The effect of the vaccine on one of the aggressive bulls (Thembo) is shown in figure 13. This bull is now 24 years old, vaccinated at six-monthly intervals, is non-aggressive and has yet to come into musth (Bertschinger et al., 2004c).

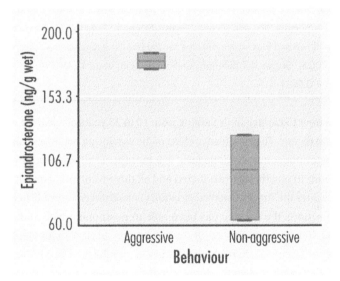

Figure 12: Epiandrosterone levels of non-musth aggressive and non-aggressive bulls during Stage 1 (De Nys, 2005). Two-sample t-test showed a significant difference between the two groups (t = 3.483, dF = 3, p = 0.04)

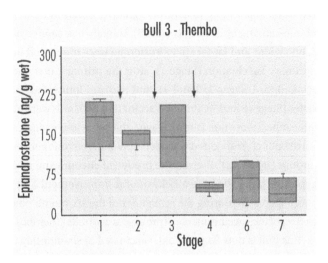

Figure 13: Faecal epiandrosterone concentrations of bull Thembo before (Stage 1) and after the primary, first and second booster (Stages 2, 3 and 4) GnRH vaccinations (arrows), and two and four months after the second booster (Stages 6 and 7, respectively) (De Nys, 2005). Stages 4-7 differed significantly from Stage 1 (AVOVA; dF = 5, F = 11.029, p < 0.001)

There are now 15 captive bulls varying from 10 to 23 years of age on the GnRH vaccination regime. The suppressive effect of the vaccine on behaviour and faecal epiandrosterone concentrations lasts 6–9 months. Furthermore, three free-ranging bulls in musth were vaccinated and all three went out of musth a week to 10 days after the first booster vaccination (Bertschinger, unpublished data).

Furthermore, the vaccine has been able to postpone musth and subdue aggressive behaviour in one adult Asian and one adult African elephant in Bowman Zoo (Canada) (Bertschinger & Korver, unpublished data). The African elephant died later unrelated to the vaccination, and on post mortem had small testes. This may well indicate that the vaccine not only suppressed testosterone production but also spermatogenesis. GnRH vaccines are known to be effective in down-regulating spermatogenesis and thus decreasing testicular size in a number of species such as cattle (Hoskinson *et al.*, 1990), pigs (Dunshea *et al.*, 2001; Zeng *et al.*, 2001; Killian *et al.*, 2006), stallions (Dowsett *et al.*, 1996; Turkstra *et al.*, 2005; Burger *et al.*, 2006), rams (Janett *et al.*, 2003) and bison bulls (Miller *et al.*, 2004). Testis size is largely determined by the diameter of individual seminiferous tubules and a decrease in spermatogenesis reduces the diameter of each tubule and thus the size of the testis. In the Asian bull musth

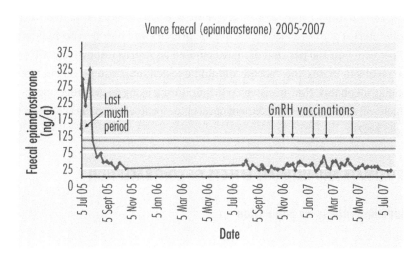

Figure 14: Faecal epiandrosterone concentrations (ng.g⁻¹) of a 35-year-old Asian elephant bull before and after repeated GnRH vaccinations (arrows) (Bertschinger & Korver, unpublished data)

has been postponed by at least 15 months as a result of GnRH vaccination (figure 14).

Possible application in elephant cows

One of the criticisms of GnRH vaccines is that they cause anoestrus (lack of cyclic activity), and in herd animals this could be regarded as an undesirable effect. From a behavioural point of view, anoestrus in horses may interfere with interactions between mares and the herd stallion and affect herd integrity. On the other hand, mares are seasonal breeders, cycling from spring to autumn with a gestation period of 11 months. Ten to 14 days post-foaling (foal heat) they often reconceive, body condition allowing. This means that normally they will cycle only once a year. African elephant cows commence cycling about two years after calving and probably fall pregnant during the first post-calving oestrus (Brown *et al.*, 2004). So which is better; anoestrus with GnRH vaccine or continuous cycling following pZP immunocontraception?

Only four elephant cows have been treated with GnRH vaccine, and as yet it is too early for conclusive results. In domestic mares the vaccine has been applied by various people, mostly with success (Garza *et al.*, 1986; Dalin *et al.*, 2002; Imboden *et al.*, 2004; Elhay *et al.*, 2007). A trial with probably the largest single group of mares to be treated with a GnRH vaccine was recently carried out in South Africa (Botha *et al.*, 2008). Fifty-five mares were vaccinated twice at an interval of five weeks. Following the primary and booster vaccinations 85 per cent and 100 per cent of mares entered anoestrus respectively. Given the results in mares, the vaccine should be tested more extensively in free-ranging elephants. Non-invasive cycle monitoring using faecal progestogens, which will allow early establishment of vaccine efficacy, is possible (Wittemyer *et al.*, 2007).

PROXIMATE AND ULTIMATE EFFECTS OF CONTRACEPTION

Modelling effects on population

Contraception rate

The contraception rate affects the growth of elephant populations and consequently the density of elephants on the landscape. The key elements of the modelling, using the Addo Elephant National Park (Addo) as an example, reveal that as the contraception rate is increased the population growth rate declines such that under a 100 per cent contraception regime the resultant declining growth rate will result in population extinction in the medium term (50–100 years) (Castley *et al.*, 2007). This assumes that the contraceptive treatment is 100 per cent effective in preventing pregnancy. A decline in the number of individuals in populations is expected only at contraception rates above 77 per cent of all breeding females in the population. Despite implementation of such contraceptive regimes the elephant density will remain above recommended densities in the short to medium term (table 8). This is park dependent, but is certainly the case in Addo, and similar responses may be found for other confined populations that have high growth rates (Slotow *et al.*, 2005). Increasing levels of contraceptive treatment in the population will also result in an overall aging effect on the population, with a higher number of individuals being represented in older age classes over time (Mackey *et al.*, 2006). This is likely to have social, behavioural and possibly ecological implications (Kerley & Shrader 2007).

Effects on sex ratio, age structure

The effects of contraceptive treatments on population demographics need to be carefully considered before implementing any contraception strategy as a means to control population growth. It may be necessary to highlight that the objectives to (a) control elephant population growth, and (b) maintain healthy viable elephant populations may be mutually exclusive but certain measures can be implemented to adopt an adaptive management approach that considers both of these objectives.

Scenario	Growth rate (%)	Population size after 15 years (2020)	Elephant density (recommended max. is 0.5 km⁻²)
Control	5.07	766	3.11
25%	3.78	643	2.58
50%	2.45	529	2.12
60%	1.84	484	1.94
75%	0.82	416	1.67
80%	0.44	393	1.67
100%	−1.29	303	1.22

Table 8: Elephant population growth rates for Addo over a 15-year period to illustrate the possible effects of variable rates of continuous contraceptive treatments (assumes 100 per cent contraceptive efficacy). Starting population size was 354 elephants in 2004 in all cases. The size of the elephant camp in 2020 is expected to be 249 km^2 after further consolidation (Castley *et al.*, 2007)

As stated previously, the elevation in contraception rates results in an aging population. Furthermore, this also results in the population slowly becoming dominated by females, given that the mortality rates for males (using the Addo population as our model) are higher than for females. Although the trends observed in the Addo population may not be transferable to other elephant populations in general, the model can be applied to those cases where demographic parameters are available to determine specific population responses in these situations. Notwithstanding any lack of generalisation of the detected trends in Addo, the model is still able to highlight potential areas of concern for protected area managers in relation to elephant management requirements. The model does not consider the possible impacts of changing herd dynamics on future mortality, as the design is too simple to cater for dynamic variability in population parameters.

Nonetheless under all contraceptive scenarios the populations become skewed towards females as well as older individuals. At higher rates of contraception the proportional representation of individuals within certain age classes is significantly different (Chi-square tests) to that expected from growth within a population that is not subjected to contraception (i.e. a control model).

Scenario	Growth rate	Density @50 yr	Mean density	Age class (mean % over 50 years) – ♂/♀											
				0 ♀	0 ♂	1–9	1–9	10–19	10–19	20–29	20–29	30–44	30–44	45–59	45–50
Control	5.14	17.01	6.36	3.27	3.13	21.73	20.55	13.99	12.89	7.87	5.98	5.79	2.90	2.05	0.03
Contiguous															
25	3.84	9.13	4.27	2.73	2.73	18.73	18.73	14.25	12.31	9.12	6.31	7.47	3.44	3.52	0
50	2.49	4.75	2.94	2.17	2.17	15.81	15.81	13.87	11.93	10.22	7.04	9.49	4.37	6.59	0
75	0.65	1.91	1.9	1.35	1.35	11.29	11.29	12.25	10.53	10.98	7.53	11.75	5.39	15.92	0
77	0.46	1.74	1.83	1.25	1.25	10.63	10.63	11.93	10.25	10.99	7.54	11.98	5.50	17.72	0
100	−3.04	0.3	1.13	0	0	0	0	0	0	0	0	0	0	100	0
First calf															
25	3.97	9.71	4.44	2.84	2.84	18.98	19.98	14.25	12.32	9.02	6.24	7.29	3.36	3.32	0
50	2.78	5.56	3.17	2.30	2.30	16.43	16.43	13.98	12.05	10	6.89	9.15	4.20	5.77	0
75	1.12	2.41	2.13	1.53	1.53	12.28	12.28	12.62	10.86	10.77	7.40	12.10	5.50	12.72	0
77	0.95	2.22	2.05	0.68	0.68	7.82	7.82	11.54	9.88	12.50	8.56	15.98	7.25	17.01	0
100	−2.15	0.47	1.3	0.03	0.03	0.56	0.56	1.28	1.08	3.94	2.69	18.10	7.80	63.90	0
90% Efficacy															
25	3.96	9.66	4.43	2.83	2.83	18.97	18.97	14.26	12.32	9.02	6.24	7.30	3.36	3.34	0
50	2.79	5.49	3.18	2.31	2.31	16.49	16.49	14.00	12.08	10	6.89	9.04	4.16	5.73	0
75	1.27	2.6	2.13	1.61	1.61	12.83	12.83	12.97	11.16	10.91	7.49	11.21	5.15	11.81	0
77	1.12	2.42	2.11	1.55	1.55	12.45	12.45	12.81	11.03	10.95	7.52	11.38	5.23	12.68	0
90	0.02	1.4	1.69	1.07	1.07	9.50	9.50	11.26	9.67	10.84	7.42	12.13	5.57	21.65	0
100	−1.11	0.79	1.41	0.62	0.62	6.29	6.29	8.70	7.46	9.39	6.42	11.19	5.13	37.67	0

Scenario	Growth rate	Density @50 yr	Mean density	Age class (mean % over 50 years) – ♂/♀											
				0♀	0♂	1–9	1–9	10–19	10–19	20–29	20–29	30–44	30–44	45–59	45–50
Stepwise															
100/80/77	−0.54	1.06	1.52	0	0	7.70	7.70	10.60	9.62	10.92	7.61	12.27	5.75	28.03	0
100/80/60	0.09	1.45	1.68	0	0	9.46	9.46	12.31	10.79	11.86	8.24	11.79	5.52	20.41	0
100/77/60	0.19	1.53	1.72	0	0	9.80	9.80	12.43	10.89	11.82	8.22	11.91	5.58	19.38	0
100/75/50	0.58	1.85	1.84	0	0	10.90	10.90	13.17	11.55	11.98	8.34	11.56	5.41	16.01	0
100/72/60	0.36	1.66	1.78	0	0	10.33	10.33	12.61	11.04	11.74	8.18	12.07	5.66	17.83	0
100/75	−0.17	1.27	1.67	1.46	1.46	4.85	4.85	11.38	9.83	10.26	6.87	10.54	4.86	33.47	0
100/50	1.02	2.3	2.08	2.58	2.58	7.28	7.28	14.53	12.59	11.26	7.55	11.18	5.15	17.76	0
Staggered															
100/75–25 (repeat)	0.4	1.7	1.83	2.53	2.53	3.74	3.74	19.44	17.03	10.14	6.52	11.31	5.47	17.45	0

Table 9: Modelled population response (growth rate, density, age/sex structure) by an African elephant population in the Addo Elephant National Park to a range of contraceptive scenarios

Furthermore, under a 100 per cent contraceptive scenario there is sequential extinction of all age classes for both male and female cohorts, such that after 50 years the entire population comprises females aged 45–59 (table 9). The sex- and age-specific responses under a 100 per cent contraception scenario are indicative of the possible implications of adopting such a management approach. Although managers may not have the local extinction of a population as an objective, the use of a 100 per cent contraceptive regime may be argued to maximise the reduction in population growth. It is clear from the model developed at Addo that this should not be considered without a complete appreciation for the possible consequences of such action.

At this stage Castley's model has not looked specifically at the breeding unit structure and impacts on the herd dynamics. Again, it would be feasible in Addo (and potentially the Makalali population, and other areas where the entire population history is known) to determine the possible effects of contracepting certain aged females and only certain percentages within family groups.

Adaptive management of elephants through modelling

The scenarios that have been discussed above present data based on standard contraceptive regimes that are simple in that they are continuous regimens of a single treatment option. Given that the need to reduce population growth and density would result in significant changes to population structure, Castley *et al.* (2007) investigated the possibility of achieving the dual objectives as stated previously by manipulating the frequency and amount of contraceptive treatments administered. Their model is able to incorporate any number of variable contraceptive scenarios, but the ones tested were based around potential to administer a single-dose, long-lasting vaccine (three years), and hence adopted a three-year cycle in the various models. The various stepwise and staggered models produce a diversity of management options to choose from in order to achieve the desired objectives, and the best model would need to be considered for the circumstances of a specific park.

Despite some models producing similar growth rates, the resultant changes to the population structure were quite variable. The model is also rather deterministic in the sense that it follows a standard series of cycles and does not therefore build any stochasticity into the modelling scenarios. However, it still highlights that it may be possible to manipulate elephant populations through an adaptive experimental approach to see how populations respond while maximising reductions in population growth at the same time. As with any management alternatives, though, there are some trade-offs to be made.

Building variability into the model, such that the population maintains a healthy structure, reveals that it may not be possible to simultaneously reduce the growth rate sufficiently over the long term. As a result it may be necessary to combine long-term contraception strategies with more intensive population reduction strategies such as culling. The benefit of the modelling reveals, however, that contraception may be effective in extending the period between consecutive culls.

Genetic diversity

Monitoring and research on the effect of contraception on the genetic diversity of contracepted elephant populations is required.

Habitat biodiversity effects and integrated management options

The Addo model, as well as common sense, tells us that contraception cannot have an immediate effect on a population. Assuming a 100 per cent efficacy, which is probably achievable in populations of up to 1 000 elephants, no more births will occur three years after implementation of a contraceptive programme. Population decline is then dependent on mortality rate, which in turn is dependent on age structure of the population and environmental factors like rainfall and disease. Contraception should be seen as a tool that can be used to prevent rather than cure overpopulation problems with elephants. On the other hand, where an overabundance of elephants is already present, whether perceived or real, contraception can significantly curb continued population growth rate. This is clearly visible in table 9 above, where using the Addo model even a 60 per cent efficacy restricts the total population to 484 instead of 776 over 15 years, having started with 354 elephants. By the same token, contraception will not prevent an already existing problem of habitat modification. Used as a preventive measure, it will restrict population growth and so also habitat impact. Used where an overabundance is already present, it will reduce further modification by an ever-growing population. Contraception can be combined with any of the three other management options – culling, translocation and creation of additional habitat, for example, the creation of transfrontier parks. One of the effects of each of these three management options is an increased reproductive rate as a result of density decrease (data or modelling needed). An effective way to combat the response is the use of contraception.

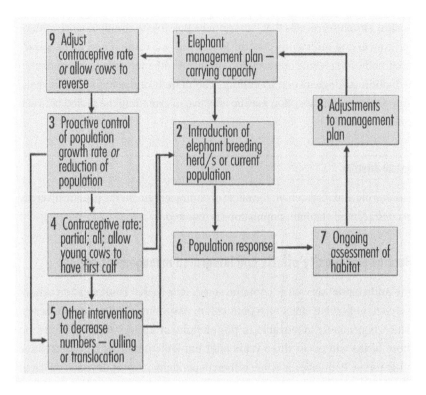

Figure 15: Adaptive management plan for the implementation and ongoing use of contraception using pZP vaccine for population control of elephants

GUIDELINES FOR IMPLEMENTATION OF A CONTRACEPTION PROGRAMME

Figure 15 suggests an adaptive management plan required for the implementation and ongoing use of pZP vaccine for elephant population control.

- Step 1 is central to the programme and is the elephant management plan for the reserve. It determines the approximate carrying capacity of the reserve. Step 1 also dictates the level of intervention necessary in the opinion of the managers.
- In Step 2 elephants are either introduced, or the reserve already has elephants that need managing, or more elephants are added.

- In Step 3 the contraceptive rate is determined, including decisions with regard to individual age groups such as young cows. Details of implementation will be determined by population size and habitat conditions.
- Step 4 is an additional step that may be required to reduce elephant population according to the requirements of the elephant management plan, bearing in mind the conditions laid down in the National Norms and Standards for Elephants in South Africa
- The population response is monitored (Step 5) bearing in mind that contracepted cows may continue to calve for up to three years following the primary vaccination.
- Parallel to Step 5 there will be an ongoing monitoring of the habitat (Step 6).
- In Step 7, according to the population response and habitat condition, adjustments to the management plan are made.
- In Step 8 the contraceptive rate is adjusted to suit reserve needs. Examples are: increasing the contraceptive rate as a result of habitat deterioration; allowing individual cows to reverse; adding young cows that have calved for the first time to the contraception programme.

KEY RESEARCH ISSUES AND GAPS IN THE KNOWLEDGE

pZP vaccine

- Vaccine production:
  - improve production and safety of native vaccine
  - develop glycosylated synthetic vaccine.
- In vitro tests:
  - develop in vitro tests which will assist in getting quicker answers than the live elephant model.
- Reversibility:
  - medium- to long-term reversibility with return to fertility
  - ovarian function assessed by faecal steroid assays
  - ovarian histology
  - T-cell response in immunised cows.
- Development and testing of one-shot/delayed release formulations in elephants. Test responses with antibody titres and contraceptive efficacy.

GnRH vaccine

- Test vaccine in population of cows – consider comparative trial with pZP.
- Long-acting formulations.
- Reversibility.
- Vaccinate bulls to test affects on spermatogenesis.
- Vaccinate adult bulls to investigate changes in chemical signalling.

Behavioural effects

- Measure stress by means of faecal steroids and behavioural changes in response to:
  - administration
  - repeated oestrous cycles
  - presence of bulls
  - lack of calves.
- Behavioural effects of:
  - cycling/anoestrus and presence of bulls
  - lack of calves on various aspects of family unit behaviour, integrity and movement.
- Need to apply fertility control to some large test populations to test the effects on population dynamics/behaviour.

Gaps in of knowledge of elephant reproduction and related behaviour

- Female reproductive physiology – what happens to infertile cows? How many cows cycle in a herd? How soon do they cycle after calving? How long does lactation anoestrus last and how can chemical signals be used to control movements and mate selection and many more?
- Male reproductive physiology – what are the triggers for musth? Why does a bull go out of musth? Can chemical signals be used to control their reproduction and movements? Are all adult bulls in fact potentially fertile?

REFERENCES

Archie, E.A., C.J. Moss & S.C. Alberts 2006. The ties that bind: Genetic relatedness predicts the fission and fusion of social groups in wild African elephants. *Proceedings of the Royal Society B: Biological Sciences* 273, 513-522.

Asa, C.S. 2005. A primer of reproductive processes: potential target tissues or processes for contraceptive intervention. In: C.S. Asa & I.J. Porton (eds) *Wildlife contraception.* The John Hopkins University Press, Baltimore, 30-52.

Bagley, K.R., T.E. Goodwin, L.E.L. Rasmussen & B.A. Schulte 2006. Male African elephants, *Loxodonta africana,* can distinguish oestrous status via urinary signals. *Animal Behaviour* 71, 1439-1445.

Bartlett, E. 1997. Jumbo birth control drives bull elephants wild. *New Scientist* 154, 5.

Becker, S.E., W.J. Enright & L.S. Katz 1999. Active immunization against gonadotropin-releasing hormone in female white-tailed deer. *Zoo Biology* 18, 385-396.

Bengis, R. 1993. Care of the African elephant *Loxodonta africana* in captivity. In: A.A. McKenzie (ed.) *The capture and care manual.* Wildlife Decision Support Services cc and The South African Veterinary Foundation, Lynnwood Ridge and Menlo Park, 506-511.

Bertschinger, H.J., C.S. Asa, P.P. Calle, J.A. Long, K. Bauman, K. Dematte, W. Jöchle & T.E. Trigg 2001. Control of reproduction and sex related behaviour in exotic carnivores with the GnRH analogue deslorelin: preliminary observations. *Journal of Reproduction and Fertility, Supplement* 57, 275-283.

Bertschinger, H.J., T.E. Trigg, W. Jöchle & A. Human 2002. Induction of contraception in some African wild carnivores by down-regulation of LH and FSH secretion using the GnRH analogue deslorelin. *Reproduction, Supplement* 60, 41-52.

Bertschinger, H.J., J.F. Kirkpatrick, R.A. Fayrer-Hosken, D. Grobler & J.J. van Altena 2004a. Immunocontraception of African elephants using porcine zona pellucida vaccine. In: B. Colenbrander, J. de Gooijer, R. Paling, S. Stout, T. Stout & T. Allen (eds) *Proceedings of an Expert Consultation on the Control of Wild Elephant Populations,* Utrecht University, 45-47.

Bertschinger, H.J., L.J. Venter & A. Human 2004b. Treatment of aggressive behaviour and contraception of some primate species and red pandas using deslorelin implants. *Advances in Eyhology* 38, 24.

Bertschinger, H.J., A.K. Delsink, J.F. Kirkpatrick, D. Grobler, J.J. van Altena, A. Human, B. Colenbrander & J. Turkstra 2004c. The use of pZP and GnRH vaccines for contraception and control of behaviour in African elephants. In: B. Colenbrander, J. de Gooijer, R. Paling, S. Stout, T. Stout & T. Allen (eds) *Proceedings of an expert consultation on the control of wild elephant populations*, Utrecht University, 69–72.

Bertschinger H.J., M. Jago, J.O. Nöthling & A. Human 2006. Repeated use of the GnRH analogue deslorelin to down-regulate reproduction in male cheetahs (*Acinonyx jubatus*). *Theriogenology* 66, 1762–1767.

Bertschinger, H.J., A. Delsink, J.J. van Altena, D. Grobler, J.F. Kirkpatrick, M. Bates & T. Burke 2007. Effects of porcine zona pellucida immunocontraception on annual calving percentages of six discrete African Elephant populations. 6th International Conference on Fertility Control for Wildlife. York, UK, 3–5 September, 48.

Bertschinger, H.J., M.A. de Barros Vaz Guimarães, T.E. Trigg & A. Human 2008. The use of deslorelin implants for the long-term contraception of lionesses and tigers *Wildlife Research,* 35: 525–530.

Blanchard, T.L. 1984. Some effects of anabolic steroids – especially on stallions. *The Compendium* 7, 1–8.

Bokhout, B., M. Nabuurs & M. de Jong 2005. Vasectomy of older bulls to manage elephant overpopulation in Africa. *Pachyderm* 39, 97–103.

Botha, A.E., M.L. Schulman, H.J. Bertschinger, A.J. Guthrie, C.H. Annandale & S.B. Hughes 2008. The use of a GnRH vaccine to suppress mare ovarian activity in a large group of mares under field conditions. *Wildlife Research*, 35: 548–5540.

Brown, J.L., S.L. Walker & T. Moeller 2004. Comparative endocrinology of cycling and non-cycling Asian (*Elephas maximus*) and African (*Loxodonta africana*) elephants. *General and Comparative Endocrinology* 136, 360–370.

Brown, J.L., M. Somerville, H.S. Riddle, M. Keele, C.K. Duer & W. Freeman 2007. Comparative endocrinology of testicular, andrenal and thyroid function in captive Asian and African elephant bulls. *General and Comparative Endocrinology* 151, 153–162.

Burger, D., F. Janett, M. Vidament, R. Stump, G. Fortier, I. Imboden & R. Thun 2006. Immunocastration against GnRH in adult stallions: effects on semen characteristics, behaviour and shedding of equine arteritis virus. *Animal Reproduction Science* 94, 107–111.

Butler, B. 1998. Elephants: trimming the herd. *BioScience* 48(2), 76–81.

Castley, J.G., G.I.H. Kerley & M.H. Knight 2007 (in prep.). Elephant population dynamics at different contraception levels using the Addo Elephant National Park as a model.

Cooper, K.A., J.D. Harder, D.H. Clawson, D.L. Fredrick, G.A. Lodge, H.C. Peachey, T.J. Spellmire & D.P. Winstel 1990. Serum testosterone and musth in captive male African and Asian elephants. *Zoo Biology* 9, 297–306.

Curtis, P.D., R.L. Pooler, M.E. Richmond, L.A. Miller, G.F. Marrfeld & F.W. Quimby 2001. Comparative effects of GnRH and porcine zona pellucida (pZP) immunocontraceptive vaccines for controlling reproduction in white-tailed deer (*Odocoileus virginianus*). *Reproduction* 60, 131–141.

Dalin, A.M., Ø. Andresen & L. Malmgren 2002. Immunization against GnRH in mature mares: antibody titres, ovarian function, hormonal levels and oestrus behaviour. *Journal of Veterinary Medicine* 49, 125–131.

Deigert, F.A., A. Duncan, R. Lyda, K. Frank & J. F. Kirkpatrick 2003. Immunocontraception of captive exotic species. III. Fallow deer (*Cervus dama*). *Zoo Biology* 22, 261–268.

Delsink, A.K., J.J. van Altena, J.F. Kirkpatrick, D. Grobler & R. Fayrer-Hosken 2002. Field applications of immunocontraception in African Elephants (*Loxodonta africana*). *Reproduction* 60, 117–124.

Delsink, A., H.J. Bertschinger, J.F. Kirkpatrick, H. De Nys, D. Grobler & J.J. van Altena 2004a. Contraception of African elephant cows in two private conservancies using porcine zona pellucida vaccine, and the control of aggressive behaviour in elephant bulls with GnRH vaccine. In: B. Colenbrander, J. de Gooijer, R. Paling, S. Stout, T. Stout & T. Allen (eds) *Proceedings of an expert consultation on the control of wild elephant populations*, Utrecht University, 69–72.

Delsink, A.K., H.J. Bertschinger, J.F. Kirkpatrick, D. Grobler, J.J. van Altena & R. Slotow 2004b. The preliminary behavioural and population dynamic response of African elephants to immunocontraception. In: J.H.A. de Gooijer & R.W. Paling (eds) *Proceedings of the 15th Symposium on Tropical Animal Health and Reproduction: Management of Elephant Reproduction*, Faculty of Veterinary Medicine, University of Utrecht, The Netherlands, 19–22.

Delsink, A.K. 2006. The costs and consequences of immunocontraception implementation in African elephants at Makalali Conservancy, South Africa. M.Sc. thesis, University of Kwa-Zulu Natal, Durban.

Delsink, A.K., J.J. van Altena, D. Grobler, H. Bertschinger, J.F. Kirkpatrick & R. Slotow 2006. Regulation of a small, discrete African elephant population through immunocontraception in the Makalali Conservancy, Limpopo, South Africa. *South African Journal of Science* 102, 403–405.

Delsink, A.K., J.J. van Altena, D. Grobler, H. Bertschinger, J.F. Kirkpatrick & R. Slotow 2007a. Implementing immunocontraception in free-ranging African elephants at Makalali Conservancy. *Journal of the South African Veterinary Association* 78(1), 25–30.

Delsink, A.K., J.F. Kirkpatrick, J.J. van Altena, D. Grobler, H. Bertschinger & R. Slotow 2007b. Lack of social and behavioural consequences of immunocontraception in African elephants. 6th International Conference on Fertility Control for Wildlife. York, UK, 3–5 September, 31.

De Nys, H.M. 2005. Control of testosterone secretion, musth and aggressive behaviour in African elephant (*Loxodonta africana*) bulls using a GnRH vaccine. MSc thesis, University of Pretoria.

D'Occhio, M.J. 1993. Immunological suppression of reproductive functions in male and female mammals. *Animal Reproduction Science* 33, 345–372.

Douglas-Hamilton, I. & O. Douglas-Hamilton 1975. *Among the elephants.* Collins, London.

Dowsett, K.F., L.M. Knott, U. Tshewang, A.E. Jackson, D.A. Bodero & T.E. Trigg 1996. Suppression of testicular function using two dose rates of reversible water soluble gonadotropin-releasing hormone (GnRH) vaccine in colts. *Australian Veterinary Journal* 74, 228–235.

Dunbar B. S., V. Lee, S. Prasad, D. Schwahn, E. Schwoebel, S. Skinner & B. Wilkins 1984. The mammalian zona pellucida: its biochemistry, immunochemistry, molecular biology, and developmental expression. *Reproduction, Fertility and Development* 6, 331–347.

Dunbar B.S., C. Lo & V. Stevens 1989. Effect of immunization with purified porcine zona pellucida proteins on ovarian function in baboons. *Fertility and Sterility* 52, 311–318.

Dunshea, F.R., C. Colantoni, K. Howard, I. Mc Cauley, P. Jackson, K.A. Long, S. Lopaticki, E.A. Nugent, J.A. Simons, J. Walker & D.P. Hennessy 2001. Vaccination of boars with GnRH vaccine (Improvac) eliminates boar taint and increases growth performance. *Journal of Animal Science* 79, 2525–2535.

Eagle, T.C., E.D. Plotka, R.A. Garrott, D.B. Siniff & J.R. Tester 1992. Efficacy of chemical contraception in feral mares. *Wildlife Society Bulletin* 20, 211–216.

Eggeling, H. 1901. Über die Schläfendrüse des Elephanten. Biol. *Centralblatt* 21, 443–453.

Eisenberg, J.F., G.M. McKay & M.R. Jainudeen 1971. Reproductive behaviour of the Asiatic elephant (*Elephas maximus*). *Behavior* 38, 193–225.

Elhay, M., A. Newbold, A. Britton, P. Turley, K. Dowsett & J. Walker 2007. Suppression of behavioural and physiological oestrus in the mare by vaccination against GnRH. *Australian Veterinary Journal* 85(1/2), 39–45.

Estes, R.D. 1991. *The behaviour guide to African mammals*. Russel Friedman Books CC, South Africa.

Fayrer-Hosken, R.A., P. Brooks, H.J. Bertschinger, J.F. Kirkpatrick, J.W. Turner & I.K.M. Liu 1997. Management of African elephant populations by immunocontraception. *Wildlife Society Bulletin* 25, 18–21.

Fayrer-Hosken, R.A., P. Brooks, H.J. Bertschinger, J.F. Kirkpatrick, D. Grobler, N. Lamberski, G. Honneyman & T. Ulrich 1999. Contraceptive potential of the porcine zona pellucida vaccine in the African elephant (*Loxodonta africana*). *Theriogenology* 52, 835–846.

Fayrer-Hosken, R. A., D. Grobler, J.J. Van Altena, J.F. Kirkpatrick & H. Bertschinger 2000. Immunocontraception of African elephants. *Nature* 407, 149.

Ferro, V.A., M.A.H. Khan, D. McAdam, A. Colston, E. Aughey, A.B. Mullen, M.M.Waterston & M.J.A. Harvey 2004. Efficacy of an anti-fertility vaccine based on mammalian gonadotrophic releasing hormone (GnRH-I) – a histological comparison in male animals. *Veterinary Immunology and Immunopathology* 101, 73–86.

Foerner, J.J., R.I. Houck & J.H. Olsen 1994. Surgical castration of the elephant (*Elephas maximus and Loxodonta africana*). *Journal of Zoo Wildlife Medicine* 25, 355.

Frank, K.M., R.O. Lyda & J.F. Kirkpatrick. 2005. Immunocontraception of captive exotic species. IV. Species differences in response to the porcine zona pellucida vaccine and the timing of booster inoculations. *Zoo Biology* 24, 349–358.

Freeman, E. 2005. Investigation of behavioural and socio-environmental factors associated with reproductive acyclicity in African elephants (*Loxodonta africana*). Ph.D. thesis, Department of Environmental Science and Policy, George Mason University, Fairfax, VA.

Ganswindt, A., M. Heistermann, S. Borragan & J.K. Hodges 2002: Assessment of testicular endocrine function in captive African elephants by measurement of urinary and fecal androgens. *Zoo Biology* 21, 27–36.

Ganswindt, A., R. Palme, M. Heistermann, S. Borragan & J.K. Hodges 2003. Non-invasive assessment of adrenocortical function in male African elephant (*Loxodonta africana*) and its relation to musth. *General and Comparative Endocrinology* 134, 156–166.

Ganswindt, A., H.B. Rasmussen, M. Heistermann & J.K. Hodges 2005a. The sexually active states of free-ranging male African elephants (*Loxodonta africana*): defining musth and non-musth using endocrinology, physical signals, and behaviour. *Hormones and Behaviour* 47(1), 83–91.

Ganswindt, A., M. Heistermann & J.K. Hodges. 2005b. Physical, physiological and behavioural correlates of musth in captive African elephants (*Loxodonta africana*). *Physiological and Biochemical Zoology* 78(4), 505–514.

Garrott, R.A., D.B. Siniff, J.R. Tester, T.C. Eagle & E.D. Plotka 1992. A comparison of contraceptive technologies for feral horse management. *Wildlife Society Bulletin* 20, 318–326.

Garza, F., D.L. Thompson, D.D. French, J.J. Wiest, R.L. St George, K.B. Ashley, *et al.* 1986. Active immunization of intact mares against gonadotropin-releasing hormone: differential effects on secretion of luteinizing hormone and follicle-stimulating hormone. *Biology of Reproduction* 35, 347–352.

Goodloe, R.B., R.J. Warren & D.C. Sharp 1997. Sterilization of feral and captive horses: a preliminary report. In: P.N. Cohn, E.D. Plotka & U.S. Seal (eds) *Contraception in Wildlife*. Edwin Mellon Press, Lewiston, NY, 229–246.

Goritz, F., T.B. Hildebrandt, R. Hermes, S. Quandt, D. Grobler, K. Jewgenow, M. Rohleder, H.H.D. Meyer & H. Hof 1999. Results of hormonal contraception program in free-ranging African elephants. In: *Verhandlungsbericht des 39 Internationalen Symposiums uber die Erkrankungen der Zoo- und Wildtiere*, Institut fur Zoo- und Wildtierforschung, Berlin, 39–40.

Gulyas, B.J., R.B.L. Gwatkin & L.C. Yuan 1983. Active immunization of cynomogolus monkeys (*Macacca fascicularis*) with porcine zona pellucida. *Gamete Research* 4, 299–307.

Hall-Martin, A.J. 1987. Role of musth in the reproductive strategy of the African elephant (*Loxodonta africana*). *South African Journal of Science* 83, 616–620.

Hanks, J. 1972. Reproduction of elephant, *Loxodonta africana*, in the Luangwai Valley, Zambia. *Journal of Reproduction and Fertility* 30, 13–26.

Herbert, C.A., L. Vogelnest 2007. Catching kangaroos on the hop: development of a remote delivery contraceptive for marsupials. *Proceedings of 6th International Conference on Fertility Control for Wildlife*, 3–5 September 2007, York, UK, 15.

Hollister-Smith, J.A. 2005. Reproductive behaviour in male African elephants (*Loxodonta africana)* and the role of musth: A genetic and experimental analysis. Ph.D. Thesis, Duke University, USA.

Hollister-Smith, J.A., J.H. Poole & E.A. Archie 2007. Age, musth and paternity success in wild male African elephants, *Loxodonta africana. Animal Behaviour* 74, 287–296.

Hoskinson, R.M., R.D. Rigby, P.E. Mattner, V.L. Huynh, M.J. D'Occhio, A. Neish, T.E. Trigg, B.A. Moss, M.J. Lindsey, G.D. Coleman & C.L. Schwartzkoff 1990. Vaxstrate: an anti-reproductive vaccine for cattle. *Australian Journal of Biotechnology* 4, 166–170.

Imboden, I., F. Janett, D. Burger, M. Hässig & R. Thun 2004. Influence of immunization against GnRH on cycling activity and estrous behaviour in the mare. *Theriogenology* 66(8), 1866–1875.

Jainudeen, M.R., G.M. McKay & J.F. Eisenberg 1972a. Observations on musth in the domesticated Asiatic elephant (*Elephas maximus*). *Mammalia* 36, 247–261.

Jainudeen, M.R., C.B. Katongole & R.V. Short 1972b. Plasma testosterone levels in relation to musth and sexual activity in the male Asiatic elephant, *Elephas maximus*. *Journal of Reproduction and Fertility* 29, 99–103.

Janett, F., U. Lanker, M. Jörg, M. Hässig & R. Thun 2003. Castration of male lambs by immunisation against GnRH. *Schweizer Archiv fürTierheilkunde* 145, 291–299.

Johnson, O.W. & I.O. Buss 1967. The testis of the African elephant (*Loxodonta africana*). II. Development, puberty and weight. *Journal of Reproduction and Fertility* 13, 23–30.

Kahl, M.P. and D.B. Armstrong 2002. Visual displays of wild African elephants during musth. *Mammalia* 66, 159–171.

Kerley, G.I.H. & A.M. Shrader 2007. Elephant contraception: silver bullet or a potentially bitter pill? *South African Journal of Science* 103, 181–182.

Killian, G., L.A. Miller, N.K. Diehl, J. Rhyan & D. Thain 2004. Evaluation of three contraceptive approaches for population control of wild horses. In: R.M. Tirron & W.P. Gorenzel (eds) *Proceedings of the 21st Vertebrate Pest Conference,* University of California, Davis, 263–268.

Killian, G., L. Miller, J. Rhyan & H. Doten 2006. Immunocontraception of Florida feral swine with a single-dose GnRH vaccine. *American Journal of Reproductive Immunology* 55, 378 384.

Kirkpatrick, J.F., A. Perkins & J.W. Turner 1982. Reversible fertility control in feral horses. *Journal of Equine Veterinary Science* 2, 114–118.

Kirkpatrick, J.F. & J.W. Turner 1991. Reversible fertility control in non-domestic animals. *Journal of Zoo and Wildlife Medicine* 22, 392–408.

Kirkpatrick, J.F. & A. Turner 2002. Reversibility of action and safety during pregnancy of immunizing against porcine zona pellucida in wild mares (*Equus caballus*). *Reproduction* (Suppl. 60), 197–202.

Kirkpatrick J.F. & A. Turner 2003. Absence of effects from immunocontraception on seasonal birth patterns and foal survival among barrier island horses. *Journal of Applied Animal Welfare Science* 6, 301-308.

Kirkpatrick, J.F. & A. Turner 2007. Immunocontraception and increased longevity in equids. *Zoo Biology* 25, 1-8.

Kirkpatrick, J.F., I.K.M. Liu & J. Turner 1990. Remotely delivered immuno-contraception in feral horses. *Wildlife Society Bulletin* 18, 326-330.

Kirkpatrick, J.F., I.K.M. Liu, J. Turner & M. Bernoco 1991. Antigen recognition in mares previously immunized with porcine zonae pellucidae. *Journal of Reproduction and Fertility* (Suppl. 44), 321-325.

Kirkpatrick, J.F., I.K.M. Liu, J.W. Turner, R. Naugle & R. Keiper 1992. Long-term effects of porcine zonae pellucidae immunocontraception on ovarian function of feral horses (*Equus caballus*). *Journal of Reproduction and Fertility* 94, 437-444.

Kirkpatrick, J.F., R. Naugle, I.K.M. Liu, M. Bernoco & J.W. Turner 1995a. Effects of seven consecutive years of porcine zona pellucida contraception on ovarian function in feral mares. *Biology of Reproduction Monograph Series 1, Equine Reproduction VI*, 411-418.

Kirkpatrick, J.F., W. Zimmermann, L. Kolter, I.K.M. Liu & J.W. Turner 1995b. Immunocontraception of captive exotic species. I. Przewalski's horse (*Equus caballus*) and banteng (*Bos javanacus*). *Zoo Biology* 14, 403-413.

Kirkpatrick, J. F., P.P. Calle, P. Kalk, I.K.M. Liu, M. Bernoco & J.W. Turner 1996. Immunocontraception of captive exotic species. II. Formosan sika deer (*Cervus nippon taiouanus*), Axis deer (*Cervus axis*), Himalayan tahr (*Hemitragus jemlahicus*), Roosevelt elk (*Cervus elaphus roosevelti*), Munjac deer (*Muntiacus reevesi*), and Sambar deer (*Cervus unicolor*). *Journal of Zoo and Wildlife Medicine* 27, 482-495.

Lawley, L. 1994. *The world of elephants*. Michael Friedman-Fairfax Publishing, New York.

Laws, R.M. 1969. Aspects of reproduction in the African elephant, *Loxodonta africana*. *Journal of Reproduction and Fertility* Supplement 6, 193-217.

Laws, R.M., I.S.C. Parker & R.C.B. Johnstone 1975. *Elephants and their habitats*. Clarendon Press, Oxford.

Lee, V. & B.S. Dunbar 1992. Immunization of guinea pigs results in polycystic ovaries. *Biology of Reproduction* (Suppl. 46), 131 (abstract).

Lee, P. 1997. Reproduction. In: G. Rogers & S. Watkinson (eds) *The illustrated encyclopedia of elephants*. Salamander Books Ltd., London, 64-77.

Liu, I.K.M., M. Bernoco, M. Feldman 1989. Contraception in mares heteroimmunized with pig zonae pellucidae. *Journal of Reproduction and Fertility* 85, 19–29.

Leong, K.M., A. Ortolani, L.H. Graham & A. Savage 2003. The use of low-frequency vocalizations in African elephant (*Loxodonta africana*) reproductive strategies. *Hormones and Behavior* 43, 433–43.

Liu, I.K.M., M. Bernoco, M. Feldman 1989. Contraception in mares heteroimmunized with pig zonae pellucidae. *Journal of Reproduction and Fertility* 85, 19–29.

Mackey, R.L.; B.R. Page, D. Duffy & R. Slotow 2006. Modelling elephant population growth in small, fenced, South African reserves. *South African Journal of Wildlife Research* 36, 33–43.

Mahi-Brown, C.A., R. Yanagimachi, J. Hoffman & T.T. Huang 1985. Fertility control in the bitch by active immunization with porcine zonae pellucidae: use of different adjuvants and patterns of estradiol and progesterone levels in the estrous bitch. *Biology of Reproduction* 32, 671–722.

McShea W.J., S.L. Monfort, S. Hakim, J.F. Kirkpatrick, I.K.M. Liu, J.W. Turner, L. Chassy & L. Munson 1997. Immunocontraceptive efficacy and the impact of contraception on the reproductive behaviors of white-tailed deer. *Journal of Wildlife Management* 61, 560–569.

Miller, L.A., B.E Johns & J. Killian 1999. Long-term effects of PZP immunization on reproduction in white-tailed deer. *Vaccine* 18, 568–574.

Miller, L.A., J.C. Rhyan & M. Drew 2004. Contraception of bison by GnRH vaccine: a possible means of decreasing transmission of brucellosis in bison. *Journal of Wildlife Diseases* 40(4), 725–730.

Moss, C.J. 1983. Oestrous behaviour and female choice in the African elephant. *Behavior* 86, 167–196.

Moss, C. J. & Poole, J.H. 1983. Relationships and social structure of African elephants. In: R.A. Hinde, (ed), *Primate Social Relations: an Integrated Approach*. Blackwell Scientific Publications, Oxford.

Moss, C. 1996. Getting to know a population. In: K. Kangwana (ed.) *Studying elephants*. African Wildlife Foundation, Kenya, 58–74.

Munson, L., I.A. Gardener, R.J. Mason, L.M. Chassy & U.S. Seal 2002. Endometrial hyperplasia and mineralization in zoo felids treated with melengestrol acetate contraceptives. *Veterinary Pathology* 39, 419–427.

Munson, L., A. Moresco & P.P. Calle 2005. Adverse effects of contraceptives. In: C.S. Asa & I.J. Porton (eds) *Wildlife contraception*, The John Hopkins University Press, Baltimore, 66–82.

Naugle, R., A.T. Rutberg, H.B. Underwood, J.W. Turner & I.K.M. Liu 2002. Field testing of immunocontraception on white-tailed deer (*Odocoileus virginianus*) on Fire Island National Seashore, U.S.A. *Reproduction* (Suppl. 60), 143–153.

Noden, P. A., W.D. Oxender & H.D. Hafs 1978. Early changes in serum progesterone, estradiol, and LH during prostaglandin F2α-induced luteolysis in mares. *Journal of Animal Science* 47, 666-761.

Oonk, H.B., J.A. Turkstra, W.M. Schaaper, J.H. Erkens, M.H. Schuitemaker-de Weerd, A. van Nes, J.H. Verheijden & R.H. Meloen 1998. New GnRH-like peptide construct to optimize efficient immunocastration of male pigs by immunoneutralization of GnRH. *Vaccine* 16, 1074–1082.

Ortolani, A., K. Leong, L. Graham & A. Savage 2005. Behavioral indices of estrus in a group of captive African elephants (*Loxodonta africana*). *Zoo Biology* 24, 311–329.

Owens, M. & D. Owens. 1997. Can time heal Zambia's elephants? *International Wildlife* 27, 28–35.

Owen-Smith, N. 1988. Megaherbivores. *The influence of very large body size on ecology*. Cambridge University Press, Cambridge.

Palm, V.S., A.G. Sacco, F.N. Snyder & M.G. Subramanian 1979. Tissue specificity of porcine zona pellucida antigen(s) tested by radioimmunoassay. *Biology of Reproduction* 21, 709-713.

Parker, G.E. & F.V. Osborn 2006.Growing chilli as a means of reducing human-wildlife conflict in Zimbabwe. Oryx, in press.

Patton, M.L., W. Jöchle & L.M. Penfold 2005. Contraception of ungulates. In: C.S. Asa & I.J. Porton (eds) *Wildlife Contraception*. The John Hopkins University Press, Baltimore, 149-167.

Perdock, A.A., W.F. de Boer, T.A.E. Stout 2007. Prospects for managing African elephant population growth by immunocontraception: a review. *Pachyderm* 42, 97–107.

Perry, J. S. 1953. The reproduction of the African elephants, *Loxodonta africana*. *Philosophical Transactions of the Royal Society B* 237, 93–149.

Poole, J.H. & C.J. Moss 1981. Musth in the African elephant. *Nature* 292, 830–831.

Poole, J.H. 1982. Musth and male-male competition in the African elephant. Ph.D. thesis, Cambridge University, UK.

Poole, J.H., L.H. Kasman, E.C. Ramsay & B.L. Lasley. 1984. Musth and urinary testosterone concentrations in the African elephant (*Loxodonta africana*). *Journal of Reproduction and Fertility* 70, 255–260.

Poole, J.H. 1987. Rutting behavior in African elephants: The phenomenon of musth. *Behavior* 102, 283–316.

Poole, J.H., K. Payne, W.R. Langbauer Jr. & C.J. Moss 1988. The social contexts of some very low frequency calls of African elephants. *Behavioural Ecology and Sociobiology* 22, 385–392.

Poole, H.H. & C.J. Moss 1989. Elephant mate searching: group dynamics and vocal and olfactory communication. In: *The Biology of Large African Mammals in Their Environment.* P.A. Jewell and G.M.O. Maloiy (eds). Clarendon Press, Oxford, UK: pp. 111–125.

Poole, J.H. 1989a: Announcing intent: the aggressive state of musth in African elephants. *Animal Behaviour* 37, 140–152.

Poole, J.H. 1989b. Mate guarding, reproductive success and female choice in African elephants. *Animal Behaviour* 37, 842–849.

Poole, J.H. 1994. Sex differences in the behavior of African elephants. In: R.V. Short & E. Balaban (eds) *The differences between the sexes.* Cambridge University Press, Cambridge, 331–346.

Poole, J. 1996a. The African Elephant. In: K. Kangwana (ed.) *Studying elephants.* African Wildlife Foundation, Kenya, 1–8.

Poole, J. 1996b. *Coming of age with elephants.* Hodder & Stoughton, London.

Palmer, E. & B. Jousett 1975. Urinary oestrogen and plasma progesterone levels in non-pregnant mares. *Journal of Reproduction and Fertility* (Suppl. 23), 213–221.

Plotka, E.D. & D.N. Vevea. 1990. Serum ethinylestradiol (EE_2) concentrations in feral mares following hormonal contraception with homogenous implants. *Biology of Reproduction* 42 (Suppl. 1), 43.

Rasmussen, H.B. 2005. Reproductive tactics of male African savannah elephants (*Loxodonta africana*). Ph.D. thesis, Oxford University, UK.

Rasmussen, L.E.L., A.J. Hall-Martin & D.L. Hess 1996. Chemical profiles of male African elephants, *Loxodonta africana*: Physiological and ecological implications. *Journal of Mammalogy* 77, 422–439.

Rasmussen, L.E.L. & B.A. Schulte 1998. Chemical signals in the reproduction of Asian (*Elephas maximus*) and African (*Loxodonta africana*) elephants. *Animal Reproduction Science* 53, 19–34.

Rasmussen, L.E.L. & T.E. Perrin 1999. Physiological correlates of musth: lipid metabolites and chemical composition of exudates. Physiology and Behavior 67, 539–549.

Rasmussen, L.E.L. & S.W. Riddle 2002. Meliferous matures to malodorous in musth. *Nature* 415, 975–976.

Rasmussen, L.E.L. & S.W. Riddle 2004. Development and initial testing of pheromone-enhanced mechanical devices for deterring crop raiding elephants: a positive conservation step. *Journal of the Elephant Management Association* 15, 30–37.

Riddle, H.S., S.W. Riddle, L.E.L. Rasmussen, T.E. Goodwin 2000. First disclosure and preliminary investigation of a liquid released from the ears of African elephants. *Zoo Biology* 19, 475–480.

Rutberg, A.T., R.E. Naugle, L.A. Thiele & I.K.M. Liu 2004. Effects of immunocontraception on a suburban population of white-tailed deer *Odocoileus virginianus*. *Biological Conservation* 116, 243–250.

Sacco, A.G. 1977. Antigenic cross-reactivity between human and pig zona pellucida. *Biology of Reproduction* 16, 164–173.

Sacco, A.G. & C.A. Shivers 1973. Effects of reproductive tissue-specific antisera on rabbit eggs. *Biology of Reproduction* 8, 481–490.

Sacco, A.G., D.L. Pierce, M.G. Subramanian, E.C. Yurewicz & W.R. Dukelow 1987. Ovaries remain functional in squirrel monkeys (*Saimiri sciureus*) immunized with porcine zona pellucida 55,000 macromolecule. *Biology of Reproduction* 36, 481–490.

Sacco, A.G., M.G. Subramanian, E.C. Yurewicz, F.J. DeMayo & W.R. Dukelow 1986. Heteroimmunization of squirrel monkeys (*Saimiri sciureus*) with a purified porcine zonae antigen (PPZA): immune response and biological activity of antiserum. *Fertility and Sterility* 39, 350–358.

Schulte, B.A., E.W. Freeman, T.E. Goodwin, J. Hollister-Smith & L.E.L. Rasmussen 2007. Honest signalling through chemicals by elephants with applications for care and conservation. *Applied Animal Behavior Science* 102, 344–363.

Shideler, S.E., M.A. Stoops, N.A. Gee, J.A. Howell & B.L. Lasley 2002. Use of porcine zona pellucida (pZP) vaccine as a contraceptive agent in free-ranging Tule elk (*Cervus elaphus nannodes*). *Reproduction* (Suppl. 60), 169–176.

Shivers, C.A., S.B. Dudkiewiez, L.E. Franklin & E.F. Russell 1972. Inhibition of sperm-egg interaction by specific antibody. *Science* 178, 1211–1213.

Shivers, C.A. & B.S. Dunbar 1977. Autoantibodies to zona pellucida: A possible cause for infertility in women. *Science* 197, 1182–1184.

Sikes, S.K. 1971. *The natural history of the African elephant*, Weidenfeld & Nicolson, London.

Slotow, R., M.E. Garai, B. Reilly, B. Page & D. Carr 2005. Population dynamics of elephants re-introduced to small fenced reserves in South Africa. *South African Journal of Wildlife Research* 35, 1–10.

Squires, E.L., B.C. Wentworth & O.J. Ginther 1974. Progesterone concentration in blood of mares during the estrous cycle, pregnancy and after hysterectomy. *Journal of Animal Science* 39, 759–767.

Stabenfeldt, G.H., J.P. Hughes, J.W. Evans & D.P. Neely 1974. Spontaneous prolongation of luteal activity in the mare. *Equine Veterinary Journal* 6, 158–163.

Stetter, M., D. Grobler, J.R. Zuba *et al.*, 2005. Laparascopic reproductive sterilization as a method of population control in free-ranging African elephants (*Loxodonta africana*). Proceedings AAZV, AAWV, AZA Nutrition Advisory Group, 199–200.

Stetter, M., D. Hendrickson, J. Zuba et al. 2006. Laparoscopic vasectomy as a potential population control method in free ranging African elephants (*Loxodonta africana*). *Proceedings International Elephant Conservation and Research Symposium*, 177

Sukumar, R. 1989. *The Asian Elephant: ecology and management*. Cambridge University Press, Cambridge.

Sukumar, R. 1994. *Elephant days and nights: Ten years with the Indian Elephant*. Oxford University Press, Delhi.

Turkstra, J.A., F.J.U.M. Meer, J. van der Knaap, P.J.M. Rottier, K.J. Teerds, B. Colenbrander & G.H. Meloen 2005. Effects of GnRH immunization in sexually mature pony stallions. *Animal Reproduction Science* 86, 247–259.

Turner, J.W., A. Perkins & J.F. Kirkpatrick 1981. Elimination marking behavior in feral horses. *Canadian Journal of Zoology* 59, 1561–1566.

Turner, J.W. & J.F. Kirkpatrick. 1982. Androgens, behaviour and fertility control in feral stallions. *Journal of Reproduction and Fertility* (Suppl. 32), 79–87.

Turner, A. & J.F. Kirkpatrick. 2002. Effects of immunocontraception on population, longevity and body condition in wild mares (*Equus caballus*). *Reproduction* (Suppl. 60), 187–195.

Turner, J.W., I.K.M., Liu, D.R. Flanagan, K.S Bynum & A.T. Rutberg 2002. Porcine zona pellucida (pZP) immunocontraception of wild horses (*Equus caballus*) in Nevada: a 10 year study. *Reproduction* Supplement 60, 177–186.

Turner, J.W., A.T. Rutberg, R.E. Naugle, M.A. Kaur, D.R. Flanagan, H.J. Bertschinger & I.K.M. Liu 2008. Controlled-release components of pZP contraceptive vaccine extend duration of infertility. *Wildlife Research*, 25: 555–562.

Van Rossum, R.J.W. 2006. pZP-immunocontraception in the African elephant (*Loxodonta africana*). Excellence track masters thesis, Utrecht University.

Vidya, T.N.C. & R. Sukumar 2005. Social organization of the Asian elephant (*Elephas maximus*) in southern India inferred from microsatellite DNA. *Journal of Ethology* 23, 205–210.

Western, D. & W.K. Lindsay 1984. Seasonal herd dynamics of a savanna elephant population. *African Journal of Ecology* 22, 229–244.

Whitehouse, A.M. 2001. *The Addo elephants: conservation biology of a small, closed population.* Ph.D. thesis, University of Port Elizabeth, Port Elizabeth.

Whitehouse, A.M. & G.I.H. Kerley 2002. Retrospective assessment of long-tern conservation management of elephants in Addo Elephant National Park, South Africa. *Oryx* 36, 243–248.

Whyte, I. 2001. *Conservation management of the Kruger National Park elephant population.* Ph.D. thesis, University of Pretoria, Pretoria.

Whyte, I.J. & D.G. Grobler 1998. Elephant contraception in the Kruger National Park. *Pachyderm* 25, 45–52.

Wittemyer, G., A. Ganswindt & K. Hodges 2007. The impact of ecological variability on the reproductive endocrinology of wild female African elephants. *Hormones and Behavior* 51, 346–354.

Wood, D.M., C. Liu & B.S. Dunbar 1981. Effect of alloimmunization and heteroimmunization with porcine zonae pellucidae on fertility in rabbits. *Biology of Reproduction* 25, 439–450.

Zeng, X.Y., J.A. Turkstra, D.F.M. Wiel, D.Z. van de Guo, X.Y. Liu, R.H. Meloen, W.M.M. Schaaper, F.Q. Chen, H.B. Oonk & X. Zhang 2001. Active immunization against gonadotropin-releasing hormone in Chinese male pigs. *Reproduction in Domestic Animals* 36, 101–105.

CONTROLLING THE DISTRIBUTION OF ELEPHANTS

7

Lead author: CC (Rina) Grant
Authors: Roy Bengis, Dave Balfour, and Mike Peel
Contributing authors: Warwick Davies-Mostert, Hanno Killian, Rob Little,
Izak Smit, Marion Garaï, Michelle Henley, Brandon Anthony, and Peter Hartley
Contributors to the fencing table: Meiring Prinsloo, Ian Bester, John Adendorf,
Paul Havemann, Bill Howells, Duncan MacFadyen, and Tim Parker

> The general impression left on my mind was that, with civilization closing in on all sides, ultimately something must be done to segregate the game areas from those used for farming; otherwise sooner or later some excuse for liquidation of the wild animals will be found ... North of the Letaba River the country West of the Park consists mainly of native locations and areas. Here the Park itself might be fenced off.
>
> Of course, a suitable fence over 200 miles long would be a most expensive undertaking, and its upkeep considerable. It would have to traverse all kinds of country, including stony hill ranges, and dense bush, but to my mind one of the chief difficulties would lie in the wide sand rivers running from west to east, and subject to annual heavy floods, which would carry away any kind of fence, and on their subsidence leave the way open for animals to pass freely up and down the river bed.
>
> J Stevenson-Hamilton, 23 January 1946, *Annual Report of Warden, Kruger National Park – 1945* (National Parks Board of Trustees, 1946, pp. 11–12)

INTRODUCTION

THE CONTAINMENT of elephants is an important aspect of their management when and where control of their movements is required. Physical barriers such as fences are passive control measures (Cumming & Jones, 2005) and are often seen as the most effective approach to containing elephants. Fences are not the only way to influence the distribution of elephants, however. Several other options are discussed in this chapter, including deterrents, water manipulation and behavioural manipulation. There are several reasons for the containment of wildlife, and particularly elephants. One is animal disease control (Freitag-Ronaldson & Foxcroft, 2003) – to protect livestock from wildlife-associated diseases, and also to protect wildlife from diseases of domestic species. Containment is a second important reason for fencing – to protect neighbouring communities and infrastructure from damage (especially by elephants and predators). Furthermore, by fencing a property, ownership of the species present is established and animals are somewhat protected from illegal hunting (see detailed discussion of this issue in Chapter 11).

PURPOSE OF FENCING

The containment of wildlife

Many small wildlife areas in South Africa are distributed amongst farms and villages with people, domestic stock and crops. This often leads to conflict between humans and elephants (Chapter 4). Fences allow people and elephants to share a landscape without the problems associated with this conflict (Hoare, 2001) (Chapter 4). To achieve this fences have to be upgraded to be able to contain the wildlife and elephant when elephant are included in a wildlife area (Chapter 11). Relatively small conservation areas located within agricultural areas require very efficient and sturdy fencing to avoid conflict. The legal requirements stipulated for such fences are described in various acts, for example the Animal Diseases Act 35 of 1984 and the National Environmental Management: Biodiversity (NEMBA) Act 10 of 2004 (Chapter 11).

Only in southern Africa, and South Africa in particular, does fencing play a large role in the wildlife and conservation industry (South African Savannas Network, 2001). In most other parts of Africa the national parks and game reserves have never been fenced, and yet seek to maintain and support wildlife populations. In addition, many of these conservation areas also seasonally support pastoralists. These communities had to adapt to the activities of their wild neighbours, and many types of localised (village level) physical barriers and deterrents (thorn bomas and ditches), as well as noise and smell, have been used to protect crops and livestock.

The wildlife industry in southern Africa has greatly expanded since the early 1980s (Smith & Wilson, 2002; South African Savannas Network, 2001). Much of this expansion took place in the middle of existing agricultural areas, or close to community settlements. Furthermore, while most of the remaining large wildlife used to be conserved in the larger national and provincial parks, smaller private reserves and game farms are playing an increasingly important role in the conservation of individual species and in ecotourism related to the presence of these species.

It is the responsibility of the landowner or manager of the particular conservation area, whether state- or privately-owned, to ensure that the animals they keep in the conservation areas do not interfere with neighbouring communities' livelihoods, including damage to their property or crops. The landowner has a legal obligation to all adjacent owners for damage that escaped animals can cause, as well as public liability in case of death or injuries or damage to property in the event of the animals breaking through the perimeter fence (Chapter 11).

Disease control

Elephants can be the major cause of fence breakages that allow the mingling of wildlife and livestock populations. Thus, although elephants do not carry these diseases, they are instrumental in their spread.

Diseases that can be transmitted from wildlife to domestic stock

Certain indigenous animal diseases carried and maintained by wild animals can be highly infectious to livestock and constitute a threat to the livestock industry. In southern Africa, the use of fencing (and other disease control measures such as proclamation of animal disease control zones, and permit requirements) to strictly control the movement of wildlife and livestock has enabled access to beef and other livestock markets in Europe and elsewhere in the developed world. Directly contagious diseases such as rinderpest, foot-and-mouth disease (FMD) and malignant catarrhal fever as well as diseases transmitted by flightless vectors such as African swine fever and corridor disease (theileriosis) can be effectively managed by barrier fencing (Bengis *et al.*, 2002). In contrast, barrier fences are ineffectual when dealing with diseases transmitted by winged vectors, such as trypanosomiasis, African horse sickness, bluetongue and Rift Valley fever.

FMD, rinderpest and African swine fever have the potential for very rapid spread, and are listed by the Organisation International Epizooties (OIE = World Organisation for Animal Health) as important animal health threats, because these diseases may have serious local, national and international animal health implications. These diseases not only cause local losses during outbreaks, but due to their epidemic character, they can become international in nature with serious socio-economic consequences.

In southern Africa, buffalo constitute the greatest risk in disease transfer to domestic livestock. They carry several diseases that affect livestock, including FMD, corridor disease, bovine tuberculosis and brucellosis. The Animal Diseases Act (35 of 1984) highlights specific responsibilities of owners or managers of properties with buffalo, including effective containment.

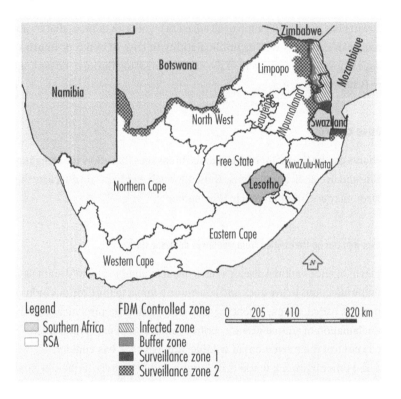

Figure 1: A map indicating the foot-and-mouth disease control areas in South Africa. Elephant-caused fence breakages on the boundary of these areas have serious consequences for disease control

In South Africa, FMD only occurs in the lowveld buffalo population (figure 1) of Mpumalanga and Limpopo provinces. This highly contagious 'trade sensitive' disease is therefore controlled by law (Standing Regulations of the Animal Diseases Act 35 of 1984) and was one of the major reasons for the erection of the animal disease control fence on the western and southern boundaries of Kruger by the Department of Agriculture in 1961–1963. At that time, the fence was constructed to contain cloven-hooved ungulates, including buffalo. Elephants (at that time) were present in relatively low numbers (population

Box 1: Diseases that can be transmitted from domestic stock to wildlife

Certain animal diseases can also spread from domestic animals to wildlife and constitute a threat to conservation efforts. A current example is bovine tuberculosis (BTB), which is considered an alien infection, and which entered the Kruger ecosystem relatively recently (about 1960). Indications are that it entered the Kruger across the southern boundary, from infected domestic cattle herds on two farms bordering the Crocodile River, just north of Hectorspruit. From there, the infection spread amongst the southern buffalo herds in the 1980s, and then progressed through the central district buffalo population in the 1990s, finally reaching the northernmost buffalo herds in the Levubu/Limpopo drainage in 2005. To date, spillover infection from buffalo has been documented in other sympatric species such as lion, leopard, kudu, warthog baboon, hyaena, cheetah, bushbuck, honey badger and genet (Keet *et al.*, 1996; Bengis *et al.*, 2001; Keet *et al.*, 2001). Although buffalo appear to be the main maintenance host of BTB in this ecosystem (De Vos *et al.*, 2001), recent indications are that kudu and warthog may also act as long-term maintenance hosts, and lions may act as short- to medium-term maintenance hosts.

There are also several viral infections that can spread from domestic stock to wildlife, including rinderpest, rabies, and canine distemper (Anderson, 1995). Historically, rinderpest, which is an alien viral infection, was introduced from Asia to the Horn of Africa with a shipment of cattle in 1888. This disease then rapidly spread westwards and southwards and killed millions of cattle and untold numbers of cloven hoofed wildlife in Africa. Many of the current distribution anomalies of certain African ungulates may have resulted from this pandemic. This disease eventually dissipated in 1902, and in more recent years, it has sporadically re-occurred in equatorial and eastern Africa.

Canine distemper, a disease of domestic dogs, is a threat to free-ranging carnivores, particularly small populations of endangered and susceptible species. In addition, canine rabies remains an ever-present threat to social wild carnivores and kudu.

estimate around 3 000), especially in the southern and central districts, with minimal pressure on the fences. The fences that were erected were 1.8 m high, consisting of 10 strands of barbed wire with no electrification (that technology did not yet exist), and were found to be adequate to prevent the movement of most ungulates. After the erection of these fences, the number of outbreaks of FMD in neighbouring cattle fell progressively, and not a single outbreak was detected in livestock adjacent to Kruger during the period 1983–1999.

Since the 1994 moratorium on lethal elephant population management in Kruger, the total elephant population has almost doubled (figure 2), and pressure on the Kruger fences has increased significantly. As an inferred result, during the period 2000–2006, five outbreaks of FMD have occurred in cattle adjacent to Kruger, four of which could be directly linked to buffalo escaping through fence breaks. This in spite of the fact that the fences had been upgraded to a 2.4 m, 20 strand fence electrified at 5 levels. However, many of these breaks probably occurred where electrification was not functioning properly. This can be attributed to poor quality of the fence workmanship and poor maintenance. Theft and vandalism have also played a role in providing opportunities for animals to escape from KNP. In addition to solar panels, fence wire, batteries, chargers, fencing standards, and droppers have also been stolen from the fence, rendering it ineffective.

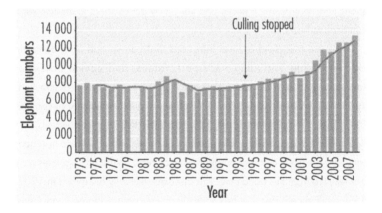

Figure 2: Elephant population numbers in the Kruger National Park between 1972 and 2007. The line represents the three point moving average to show the trend in population increase

Protection of livestock and crops

With increasing densities of elephants and depletion of natural foods in conservation areas (Smith & Kasiki, 2000), especially during dry seasons, the pressure for elephants to break out and look for more nutritious food sources increases (Naughton-Treves, 1998; O'Connell-Rodwell *et al.*, 2000). Most of the fence breaks are caused by single bulls that are brazen and strong enough to break the fence. Often conflicts with expanding human habitation displace elephants which then become dependent on crop-raiding to survive in resource-poor habitats (Tchamba, 1995). Cultivated crops are the perfect attractant for elephants; they are often highly nutritious (grains), and/or taste good (fruits and vegetables). The result is that elephants become crop raiders (Wasilwa, 2003) (see also figure 3).

Figure 3: Maize, inter-cropped with pumpkin and beans, cultivated adjacent to KNP fence near Altein village. Note elephant path leading from KNP (foreground) towards crop (photo courtesy of Brandon Anthony)

Taylor (1994) reported that fences can decrease the incidence of crop-raiding. In Negande (Zimbabwe), crop-raiding incidents dropped by 65 per cent after the erection of an elephant-proof fence but rose again by 42 per cent the following season, indicating that under specific circumstances, fences are not very effective in reducing crop losses. A small circular fence erected around irrigated crops was also successful in avoiding crop loss. However, in spite of agreeing to the project, villagers were reluctant to maintain the fence after the first success. The net economic benefit of the erection of elephant-proof fences is questionable. The main benefit may be that fewer animals are killed because of causing damage.

Thus farming of crops and livestock in areas which contain free-ranging elephants and lions (fence breaks by elephants facilitates the escape of lions) results in increased human-wildlife conflict. In arid environments, communal agricultural activity is concentrated along riparian zones. These zones are also favoured by elephants.

Elephant habitat expansion corridors will increase the human contact interface. In most situations, such corridors will need to be fenced. Where elephants have learned to avoid contact with humans, fencing corridors may pose an unnecessary expense. Douglas-Hamilton *et al.* (2005) found that elephants crossed corridors at significantly faster travelling speeds and during the cover of darkness to avoid conflict with humans. These results indicate that elephants are aware of danger within a space and time context. Consequently, where corridors are short enough to enable overnight travel from food and water sources within protected areas and where disturbance mechanisms are present that would prevent elephants from lingering along corridors, movement across corridors will likely occur with minimal incidents of conflict with people. However, the value of unfenced corridors to other animals is still not understood and requires further investigation.

In smaller protected areas that have elephants (e.g. Addo), more substantial and robust fences are needed because the rate of contact of elephants with the fence increases as the length of the fence decreases. This type of fence does not have to be electrified to be effective if the animals are trained to respect the fence (Anderson, 1994). Simple electric fences with only three strands and a voltage of 5.5 kV have been successful in controlling damage-causing animals in Mwea District, Kenya. This required very active community involvement and a full-time fence attendant, paid by a development agency (Omondi *et al.*, 2004).

CONSEQUENCES OF FENCE BREAKAGES

Over and above the negative consequences that elephant breakouts may have due to crop-raiding, or creating conduits for large carnivores or disease-carrying wildlife to exit protected areas (as discussed above), there are several other consequences.

Loss of animals

Animals that escape from conservation areas, including those considered dangerous, have to be returned to the conservation area, or destroyed where they are. In the case of elephants, the costs can be high (Lubow, 1996). The capture and transport of elephants needs specialised equipment, helicopters and vets experienced in elephant capture to be a success (Nelson *et al.*, 2003). If the elephants are close enough to the conservation area, they can be chased back (Hoare, 2001); a helicopter is usually necessary for this to succeed. The more common option is to destroy the animal/s (SANParks, 2005).

Several reserves which have elephant also have expensive, rare or endangered species including rhino, roan, sable, and tsessebe. These species have usually been introduced at great expense to the reserves, and although these species would seldom cross fences themselves, they can escape through fences damaged by elephants.

DOMESTIC STOCK ENTERING WILDLIFE AREAS

Economic impacts

Domestic stock entering wildlife areas, especially those aimed at tourism, can have a negative effect on the product on offer. Studies done in the Zambezi valley named wild animal species roaming free, indigenous plant species and lack of people as important factors in the perception of the tourist of an area to be wild. Pollution, litter, vehicles, noise, and the presence of domestic animals are factors that negatively influence tourists' perceptions of wilderness (Wynn, 2003) (see also figure 4 below).

OTHER USES OF FENCES IN CONSERVATION AREAS

Protection of vegetation

In Addo, exclosure fencing has been used effectively to protect endemic plants from utilisation by elephants. Five botanical reserves were identified within the Park which would represent 91 per cent of the Park's special plant species in less than 8 per cent of its area (Lombard *et al.*, 2001). Mature plants within such enclosures can then act as valuable seed banks to populate surrounding areas (Western & Muitumo, 2004).

Figure 4: Three head of cattle approximately 30 km within KNP, east of Hlomela village (October 2004). It was later discovered that these were part of a stolen herd that was being taken through KNP to Mozambique. Lions killed one of these animals, and the remaining two were killed by KNP rangers to control the threat of disease transfer (photo courtesy of Brandon Anthony)

Understanding system function

Enclosures have been very useful in studies of the effect of browsers and grazers on selected areas in Kruger. This information is essential for management decisions such as avoiding mistakenly controlling elephant populations to address impact concerns that they are not responsible for. Differences between areas inside and outside the enclosure help to understand the effect of elephant on the vegetation (Trollope *et al.*, 1998).

Enclosures are also useful to develop an understanding of the time needed for different plant types to recover after heavy use by elephant and other browsers (African Elephant Specialist Group Meeting, 1993). In Addo, such enclosures have contributed substantially to our understanding of how the thicket vegetation responds to elephant use (Kerley & Landman, 2006) (Chapter 3).

Protection of individual trees

Individual large trees can be physically protected from elephants. In East Africa and in the Associated Private Nature Reserves (APNR) on the western boundary of Kruger, 13 mm mesh wire netting wrapped around the trunk of mature tree stems has prevented such trees from being extensively bark stripped by elephants (Gordon, 2003; Henley & Henley, 2007) (figure 5). Heavy wire netting was more efficient in protecting trees against debarking and required less maintenance but was also more visible than 13 mm mesh wire at distances further than 5 m from the protected tree. Wire netting techniques did not protect trees from being uprooted or broken. Results from these studies indicate that the absolute use or avoidance of protected trees may not be as important as the degree to which the wire-netting reduces bark-stripping and consequently increases the survival rate of trees that are susceptible to bark-stripping by elephants.

Figure 5: Wire netting can be used to protect large trees from ring barking. It does not stop trees from being pushed over or broken (photos from Mapungubwe National Park)

Protection of infrastructure and people

Sturdy fences have been specifically designed to protect infrastructure such as water tanks, pipelines, windmills, dams, weirs, and buildings from elephants. In addition, tourist facilities, aircraft, and landing fields need barrier protection.

In the Mwea region of Kenya, an electric fence was erected to separate people and elephants. Before fence construction, an average of three people were killed yearly by elephants. Since the fence was completed, no elephant-related deaths have been reported (Omondi et al., 2004).

EFFICACY OF FENCES

To contain elephants

The long-term existence of small wildlife areas will probably depend on the efficacy of barriers to prevent animals escaping. Well-maintained fencing, especially electric fencing, appears to be the most effective barrier to restrict movement for most of the larger wildlife species (Nelson *et al.*, 2007). Elephants are capable of going through the most sophisticated barriers, including fences that are highly electrified, although this is often associated with a break in the electric current (figure 6). Elephants in particular are difficult to restrict as a result of their large size and the ease with which they can break fences, which make them the most important fence-breaking species (SANParks, 2005) (also see figure 7). Their home ranges are large, and migration and movement patterns often extend not only beyond park or reserve boundaries, but national boundaries as well (Craig, 1997) (Chapter 2).

Elephants most often cross fences because of the availability of water and food in adjacent areas (Buss, 1961). Studies on crop-raiding by elephants at Kibale Forest National Park, Uganda, showed that crop-raiding occurred throughout the year with peaks in dry seasons when crop availability was high. Bananas and maize were the main crops raided. Monthly crop-raiding incidences were not influenced by forage quality but by ripening of maize. Crop availability seems to be a more important driver of elephant breakages in forest habitats, whereas in savanna habitats large seasonal fluctuations in forage quality have a greater influence on temporal patterns of crop-raiding (Chiyo *et al.*, 2005). Osborn (2004) also found that the point at which the quality of the available forage declined below the quality of crop species corresponded to the movement of bull elephants out of a protected area and into fields. However there are differences in the behaviour of bulls and cows towards fences – bulls tend to be more inclined to break fences than females (Sakumar & Gadgil, 1988). Breakages in the Kruger fence in Limpopo Province in South Africa are illustrated in figure 7 and seem to also coincide with periods when forage may be scarce in the park.

Figure 6: Male elephant returning to KNP over border fence (photo courtesy of Peter Scott)

There seems to be a spatial and temporal correlation between elephant densities and the number of fence breaks. The elephant population of Kruger has almost doubled since 1995 when culling stopped. Using the incomplete reports available, Anthony (2006) recorded 386 incidents of damage-causing animals in the area between the Shingwedzi and Klein Letaba rivers between October 1998 and October 2004 (figure 7). Elephants caused 55 of these incidents and eight reports indicate that elephants were destroyed. The most common problem animals were buffaloes (137), lions (72), elephants (55), hippopotamuses (33) and crocodiles (18). It is important to note that many of the problem buffaloes, lions and even hippos probably exited through elephant fence breaks.

Standard electric fences work well to protect small areas for experimental purposes or to protect infrastructure. The maintenance of the fence is essential (see box 2). Breakages are relatively rare and breakages that did occur into these enclosures were due to failure of the electric fencing.

Box 2: The maintenance of fences

Fences need to be permanently maintained to restrict elephant movement effectively. Once elephants realise that they can cross a barrier they will be more inclined to repeat the effort. Thus the maintenance of fences must be financially and technologically within the capacities of the people maintaining them, if they are to be long-term solutions (Kangwana, 1995). Studies in Laikipia, Kenya, confirmed this statement and found that there was no clear relationship between the effectiveness of fences and their design and construction. Some simple fences worked, some high-tech fences (including high-voltage electric fencing) did not. Fences built to keep elephants and people apart may only be efficient if their construction follows a particular process which imbues a clear sense of common ownership and responsibility (Dublin et al., 1997). This aspect is very important, as some communities rather remove parts of the fencing material to use around their homesteads, while others may cut the fence to gain entry into the wildlife areas. Nevertheless, in South Africa and in some parts of Zimbabwe, fencing is used fairly effectively to contain elephants within protected areas.

The integrity of the Kruger Park western boundary fence is regularly compromised by certain human activities. These include:

- sabotage of the electrification by illegal transmigrants from neighbouring countries
- theft of electrical components, especially solar panels and batteries
- theft of structural components and material for own use or for sale.

Thus in areas where there are significant human pressures on the fence, and in areas where cable (Eskom) power is unavailable, electrification is not a good option because electric fences are easily sabotaged and solar panels and batteries have a high theft potential. In such situations, a more robust structure made of cables and 'I' beams that is elephant resistant but people friendly is a better option.

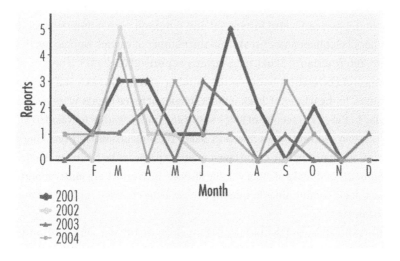

Figure 7: Reports of elephant breakages of fences between Kruger and the Limpopo province from January 2001 to October 2004 (Anthony, 2006)

Disease control

Between 1983 and 1999, the elephant density in Kruger was relatively low (about 0.4 elephant.km^{-2}) and in that period no outbreaks of FMD were detected in livestock adjacent to the park. Fence-breaking bulls and problem peripheral herds were frequently targeted as part of problem animal and border control management. Therefore during this period, elephant fence-breaking activities were sporadic and rapidly dealt with.

However with the increasing elephant density (0.46–0.62 elephant.km^{-2} between 2000 and 2006), five major FMD outbreaks occurred in the adjacent livestock populations. Four of these outbreaks (Bushbuckridge 2001, Masisi 2003, Mopani 2004, and Thulamela 2006) could be linked directly to buffalo exiting Kruger through fence breaks.

The Bushbuckridge outbreak cost the tax payer ZAR20 million, the Masisi outbreak cost ZAR4 million, and the Mopani outbreak cost ZAR90 million to control. Mass vaccination in and around the outbreak as well as road blocks and the erection of additional cordons and barrier fences were necessary to avoid the further spread of the disease. Further costs of such outbreaks include indirect costs to farmers due to movement restrictions on agricultural products. Additional financial losses would have been incurred if the outbreak had not been contained within the declared FMD control area, as a result of trade barriers and millions of rand (ZAR) lost in export earnings.

There has been a striking spatial and temporal correlation between the number of elephant fence breaks and the number of vagrant buffalo incidents (State Veterinarian – Skukuza, quarterly reports 2005–2007). There is also a temporal correlation between the number of fence breaks and elephant densities. In the winter of 2005, up to 35 elephant fence breaks were recorded per day in the 12 km section of fence stretching from Sawutini to Naladzi (State Veterinarian – second quarterly report 2005). These elephants were breaking out to drink and bathe in one of the few remaining pools in the Klein Letaba River.

Sporadic outbreaks of other wildlife diseases in livestock are under-reported, because local communities frequently consume the carcasses, and no diagnosis can be made.

Corridor disease (theileriosis), with close to 100 per cent mortality of infected cattle, was also sporadically reported in areas where buffaloes crossed fences broken by elephants and dropped infected ticks (Skukuza, Nelspruit & Mkhuhlu State Veterinary Reports, 2006; 2007; 2008).

To give an idea of the potential scale of African swine fever outbreaks, one that was well documented in southern Mozambique in 1997 resulted in the deaths of an estimated 180 000 pigs (Penrith pers. com.).

CONSEQUENCES OF RESTRICTION OF MOVEMENT BY FENCES

In the African context restriction of elephant movement is generally a result of human encroachment or habitat change (Hoare & Du Toit, 1999). In South Africa, movement is mostly restricted by fencing which has been erected with the express intent of restricting the animals to a certain area. Contrary to the situation in open landscapes, where animals are not restricted and can select from all available resources and habitats, fences restrict direct access to other resources. Some of these may be key resources, such as water, as in the case of the elephants in Tembe Elephant Park, which no longer have access to the Pongola River. Apart from the fact that these restrictions may have significant effects on the elephant population dynamics (Illius & O'Connor, 2000), the ecology of the animals may be affected (Van Aarde & Jackson, 2007). The relative importance of how the different resources change with climatic and seasonal changes and the long-term effect of fencing in this regard is not well understood and requires further targeted research (Owen-Smith et al., 2006).

The 'overabundance' of elephants has often been attributed to fences restricting elephants to confined areas (Gillson & Lindsay, 2003; Van Aarde & Jackson, 2007). The argument is that by restricting movement the natural regulators of elephant populations are weakened and this results in excessive impact and homogenisation of the local biodiversity, particularly the vegetation (Owen-Smith *et al.*, 2006).

The mechanisms underlying this hypothesis are not well understood, but may be linked to elephants being very adaptable in their ability to eat poor-quality food (Owen-Smith, 1988). Thus even when confronted with limited choice in quality and quantity of forage, they can continue to increase in numbers. There seems to be general agreement that fencing elephants into small areas will have a greater negative effect on the natural system heterogeneity than in larger areas, possibly because larger areas have an inherently wider range of different habitats (Owen-Smith *et al.*, 2006). Another argument is that the range of habitats which elephants normally have access to includes areas that serve as dispersal sinks (*sensu* Dias, 1996). By preventing animals from moving into these areas, the remaining areas are exposed to continuous high impacts, leading to loss of habitat variability (Van Aarde & Jackson, 2007).

A large build-up of elephant numbers in small, fenced areas is often followed by a decline in woodland cover due to a combination of tree destruction by elephants and often also by the interaction of the effects of fire (Jachmann & Bell, 1984). In a number of parks this has led to the temporary disappearance of large areas of *Acacia* (Mwalyosi, 1990) and *Commiphora* woodland (Leuthold, 1996), and in some cases local extinction of tree species including baobab (*Adansonia digitata*), which is highly favoured by elephants. Because of the fences the elephants were not capable of responding through migration to these radical changes in the food supply, and thus had a more severe effect than they would have had in an unfenced system.

Prior to the erection of a veterinary fence on the western boundary of Kruger in the 1960s, there was evidence of an east–west seasonal migration of herbivores (figure 8) (Whyte, 1985). With the initial erection of the fence, many animals were killed, such as giraffe, wildebeest, zebra, and kudu (Whyte & Joubert, 1988; Albertson, 1998). In Botswana the disease control veterinary fences also prevented vital wildlife movements, fragmented populations, separated young animals from herds, and caused the death of animals that got stuck in the fence (Albertson, 1998).

Fences do not only affect the migration routes of animals between resource areas, but also affect other ecological factors such as fire. Increased grazing pressure due to the confinement of animals led to reduction in the frequency of hot fires, and this commonly precipitated bush thickening (Peel, 2005). Wildlife-based tourist operations in the region are adversely affected by such bush encroachment because the dense woody layer reduces game visibility.

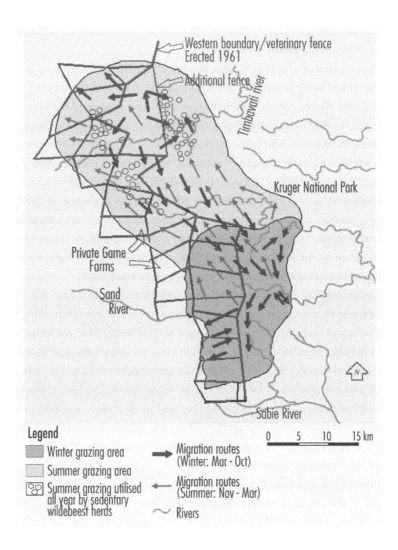

Figure 8: Hypothesised animal migration routes prior to the erection of the foot-and-mouth fence (Whyte, 1985)

The erection of the veterinary fence between the Kruger and private land to the west also led to the provision of water in previously seasonally waterless areas of both Kruger and the private reserves. Water shortages in such a confined area with inadequate surface water may increase fence breakages, conflict with humans (especially around water sources) and risk of disease spread. Artificially provided water sources will counter this effect, but alter the spatial and temporal foraging and trampling patterns of both elephants and other water-dependent animals (Chamaillé-Jammes *et al.*, 2007a; Smit *et al.*, 2007a). This may ultimately influence the vegetation (e.g. Thrash, 2000; Brits *et al.*, 2002), soil (e.g. Thrash, 1997) and nutrient patterns (e.g. Tolsma *et al.*, 1987) on multiple scales (multiple piosphere effect). Additional permanent water sources have also been blamed for influencing predator/prey relationships (Harrington *et al.*, 1999; McLoughlin & Owen-Smith, 2003; Mills and Funston, 2003), creating unnaturally high herbivore numbers with consequent population crashes during droughts (Walker *et al.*, 1987), compromising system resilience (Grant *et al.*, 2002), and degrading the quality of the herbaceous layer (Parker & Witkowski, 1999). The effects of fencing and water provision are thought to be reflected in the change in the status of impala. Impala did not occur west of 31°30'E in the 1800s (Kirby, 1896) and were in fact not found west of the Orpen Gate until the 1920s (Porter, 1970). They are now the most prolific herbivore in the lowveld. Both elephant and impala are strong competitors, have a great impact on areas they inhabit and are ultimately able to change the habitat to suit their requirements by maintaining the forage in an actively growing and palatable state. They can also switch easily from their preferred grazing to browsing when the quality or quantity of the grazing drops too low for their maintenance (Collinson & Goodman, 1982). According to Collinson & Goodman (1982), weak competitors such as roan, sable, and tsessebe cannot compete with species such as elephant and impala and are only successful within intensive breeding camps such as found at Selati Game Reserve.

Table 1 summarises the situation on two adjacent protected areas, both less than 15 km² in extent. The annual vegetation survey indicated that both areas were under nutritional stress due to drought conditions and high stocking densities. This was confirmed by the annual aerial game count and it was recommended to remove some game from both properties. Only reserve 2 implemented the suggested game control.

Subsequent aerial game counts showed large-scale mortalities on reserve 1 compared to a few mortalities on reserve 2. The latter example serves to illustrate the need for hands-on management, particularly in small fenced areas.

A further consequence of fencing is that depending on the timing of the fence erection, it may split a population of elephants, as in the case of the Tembe Elephant population, which was split between South Africa (Tembe Elephant Park) and Mozambique (Maputo Special Elephant Reserve).

Fences also separate local communities from resources such as water and medicinal plants, and this leads to people cutting the fences to obtain these resources, which then acts as an entry point for damage-causing animals.

Animal biomass (kg.km⁻²)		Impala		Wildebeest		Economic value at gate (ZAR)
Pre-drought	Post-drought	mortality (%)	removal (*n*)	mortality (%)	removal (*n*)	
Reserve 1 5 499	2 881	81	0	93	0	−343 000 (mortality)
Reserve 2 4 607	3 347	<5	35	35	28	−33 000 (mortality) +59 500 (live removal)

Table 1: Case study illustrating the ecological and economic effect of fencing and water provision on the ecology of areas of small size when animals are removed or not, according to predictions of available forage (Peel, 2006)

FENCES AND ELEPHANT WELFARE

From a welfare perspective, the needs of an elephant population could be satisfied in an enclosed area as small as 150 km². Elephants do not immediately increase their ranges when boundary fences are removed (Druce *et al.*, 2007) from areas of this size.

Additionally, work done by Space For Elephants Foundation indicated that the summer peak in animals breaking out of conservation areas coincided with the rainy season when cloud formations are consistently low. This allowed easier communication and was associated with the attraction of the abundance of suitable forage and marula fruit. Furthermore, elephant attempts to escape seemed to be due to confrontational stress, and they often tried to return to the area from whence they came (http://www.space4elephants.org).

TECHNICAL SPECIFICATIONS FOR FENCES AND THEIR MAINTENANCE

Given the present state of technology, well-constructed electric fences can act as a powerful deterrent to elephant entry and trespass (Hoare, 1992). A typical electrified game fence is illustrated in figure 9. The different types of fences and their efficacy are summarised in table 2.

ENSURING EFFICIENCY OF FENCES

Long-term success using fences to contain elephants is dependent on meticulous routine maintenance and the use of solid, durable material that is well anchored. Electric fencing technology is simple and definitely deters elephants, but has to be continuously maintained to be efficient (Hoare, 2003). Fencing is very expensive as a management tool, especially in view of the damage and the direct costs involved in fixing and/or replacing fences that have been destroyed by elephants (WWF, 1998; Hoare, 1995).

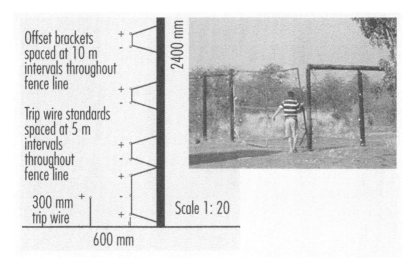

Figure 9: Diagram of electric wires for elephant-proof fence with an example of such a fence in Mapungubwe National Park

To ensure that fences are effective against elephants requires that:

- sufficient trained staff and transport must be available to ensure that fences are patrolled every day on a rotational system to effect fence repairs
- responsibilities for maintenance and costs associated are defined clearly and appropriately budgeted for
- neighbouring communities agree about the importance of fences and do not remove fencing material for their private use
- human interference is avoided by using cabling instead of wire as it is less sought after
- there is a reliable supply of electricity with sufficient power to deliver the required voltage

Fence type	Height (m)	No. electrified strands	No. & type of other strands	Type of straining post	Density of straining post	Erection cost (ZAR. km⁻⁴)	Maintenance cost (incl. staff cost)	Breakages by elephant	Comments
Legal requirements for elephant containment	2.4	3	20	NS*					
Recommend Elephant Managers & Owners Association	2.4	4	17 min			34 000	25 000 per month	Very few	Sacrificial fence at river crossings
Stout mechanical fence	2.4	0	4 x 12 mm cable 6 x straying wire 3 x barbed wire 1.9 m high diamond mesh	30 kg.m post 5 m apart planted 1 m deep		190 000	5 000 000 per year		Works fairly well if maintained. Mainly bulls that try to cross. Skilled and dedicated personnel needed to maintain. Used in Kruger.
Electrified Armstrong fence	2.4	6	6 cables	Railway tracks 2 m deep every 10 m		150 000	Very low, if any	0	Costly to construct, very little if no maintenance, but electrified part has a very high maintenance cost
Twenty-one-strand fence	2.4	5	21 steel wires		10 m intervals	51 666	90 000 per year	4 per year	Cost exclude grading costs and warden's salary
Nineteen-strand fence reserve	2.4	4 Top at shoulder height & bottom 1 m from ground	19			62 568.74	300 000 per year	None	

Fence type	Height (m)	No. electrified strands	No. & type of other strands	Type of straining post	Density of straining post	Erection cost (ZAR. km⁻¹)	Maintenance cost (incl. staff cost)	Breakages by elephant	Comments
Standard 17-strand game fence with wire netting	2.4	5	17 and 1.2 m mesh			46 000	Approx. 5 000 per month	Approx. 3 per year	Maintenance costs include a person to patrol all fences daily on motorbike
Six-strand fence	2	3	6	NS	NS	29 000			
Three-strand fence	2	3	3	NS	NS	29 000	Broken less than 6 strand press		
Chili paste fence poles painted	2					820	Needs repainting after each rainfall event		
Wire mesh around trees	1.25					50 per tree			
Trip alarms around individual farms						560			Decline in conflicts around homestead

Note: NS = not specified

Table 2: Specifications, erection and maintenance costs (2007) for different types of fences. All electrified strands must have a minimum voltage of 6 000 V and must have sufficient energisers to supply power to maintain this voltage over a distance of 8 km

- vegetation around fences is removed to avoid shorts in the electric current and damage by fire; this can be achieved by physically clearing the area or the judicious use of herbicides
- fences are checked after fires, flash floods, and lightning
- gates at access points are securely closed
- there may be a strategic opening of boreholes during the dry season to reduce fence breaks in areas where elephant movements are associated with accessible water outside the fenced area.

ALTERNATIVE METHODS OF MANAGING ELEPHANT DISTRIBUTION

Drinking water manipulation as a management tool for elephant distribution

Elephant distribution is often associated with the distribution of surface water and rivers (Stokke & Du Toit, 2000; Redfern *et al.*, 2003; Chamaillé-Jammes *et al.*, 2007a; Smit *et al.*, 2007a & b). It has been shown that the addition of surface water to areas with limited natural water availability can increase the density of elephants (Cumming, 1981) and expand their spatial distribution (Chamaillé-Jammes *et al.*, 2007a). Surface water manipulation has therefore been proposed as a 'non-intrusive and natural' management tool with which to alter elephant distribution patterns (Owen-Smith, 1996; Chamaillé-Jammes *et al.*, 2007a & b). However, considering the mobility of elephants (e.g. Viljoen & Bothma, 1990; Verlinden & Gavor, 1998), it is arguable how effective surface water manipulation will be to manipulate elephant distribution in areas like the Kruger National Park, where water is usually widely available (South African National Parks, 2005; Redfern *et al.*, 2005; Owen-Smith *et al.*, 2006; Smit *et al.*, 2007c). Depending on the availability of natural water, artificial waterholes may not influence large-scale elephant distribution patterns as much as they affect the local-scale activity patterns (i.e. piosphere effect). In Kruger, for example, the landscape-scale dry season distribution of mixed herds and breeding herds is more closely linked to the river system (figure 10) than to the artificial waterhole network, which tends to be more preferred by bull groups (Smit *et al.*, 2007a & b). Considering this, together with the ability of elephants to move between the (usually widespread) ephemeral and permanent water sources, it is debatable how effectively the density and distribution patterns of the elephants could be manipulated under normal conditions by means of water provision in Kruger (Redfern *et al.*, 2005; Smit *et al.*, 2007c); this is an area that requires further research.

The effect of the provision of artificial water on elephant distribution is much more pronounced in an arid system, as can be seen in the Addo Elephant National Park. In Addo the impact on the endemic subtropical thicket has been very extensive around the artificial waterholes where elephant tend to concentrate, while areas far from the waterpoints have been substantially less used (Knight *et al.*, 2002). Other studies have also indicated that the distribution and subsequent use of vegetation by elephants is higher in closer vicinity to water (e.g. Ben-Shahar, 1983; Thrash et al., 1991; Nelleman *et al.*, 2002). If water is artificially provided, it should preferably be restricted to areas close to localities where natural sources occur, minimising spatial alterations to grazing patterns (Pienaar *et al.*, 1997). Thus, a uniform distribution of water by the addition of artificial water sources may homogenise the natural variability in impact brought about by the uneven natural water availability. This is not desirable for biodiversity conservation (Owen-Smith, 1996; Knight *et al.*, 2002).

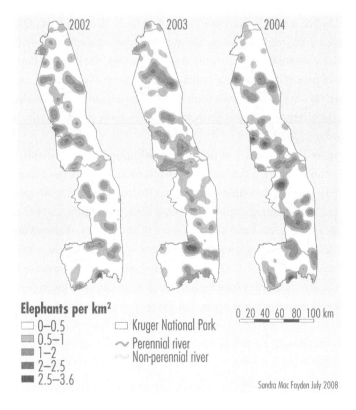

Elephants per km²
- 0–0.5
- 0.5–1
- 1–2
- 2–2.5
- 2.5–3.6

- Kruger National Park
- ~ Perennial river
- ~ Non-perennial river

0 20 40 60 80 100 km

Sandra Mac Fayden July 2008

Figure 10: Distribution and density patterns of elephants in Kruger. Note the concentration along the larger perennial and seasonal rivers (courtesy of Sandra MacFayden (Grant, 2005))

The effectiveness of surface water manipulation as a management tool for elephant distribution will depend, *inter alia*, on (1) natural surface water availability, (2) forage quality, (3) local densities, and (4) size and objectives of the confined area – that is, whether objectives are defined by biodiversity or sustainablility (Peel *et al.,* 1999). Surface water manipulation will be most effective as a management tool in large systems with very limited natural water distribution. In such systems the distribution patterns may be substantially influenced by water provision (Jackson & Erasmus, 2005). In small, enclosed areas with adequate natural water, artificial water provision can be expected to have a relatively small and localised effect, since any water provided will effectively be within walking distance for elephants.

Disturbance as a way to deter elephants

Disturbance methods may be used to deter elephants, but elephants soon become habituated (Hoare, 1995; O'Connell-Rodwell *et al.*, 2000; Osborn & Rasmussen, 1995), especially if the same animals are regularly involved (Hoare, 1999). These methods require trained personnel and they can be dangerous because of proximity to the elephants. They are generally cheap to apply and have been shown to have at least some effect. They are not fatal to the elephants and the involvement of the authorities provides some public relations value (Nelson *et al.,* 2007).

Villagers in Sumatra use powerful flashlights to deter elephants, in combination with noise, fire, and explosives, while fireworks and flares have been used in Zimbabwe with initial success (Hoare, 2001). Firing weapons over the heads of crop-raiding elephants to chase them from fields has been used in Zimbabwe (Hoare, 2001) and Niassa Reserve in Mozambique (Macadona, pers. comm.). In Niassa, it is used successfully in combination with electric fences.

O'Connell-Rodwell *et al.* (2000) experimented with trip alarms in villages in East Caprivi, Namibia. They found shorter wires around individual farms to be effective in the short term, but there was no impact on the overall number of conflict incidents reported in a year as elephants initially moved into neighbouring farms before becoming habituated. Each alarm cost US$78, less than the average elephant crop-damage claim. Between 1993 and 1995 an estimated US$1 800 was saved.

Massive disturbance (e.g. people, vehicles and/or helicopters) to drive elephants away from a conflict area has been tried with some immediate, although short-term, success in Zimbabwe (Hoare, 2001).

CHANGING BEHAVIOUR AS A MANAGEMENT TOOL

Elephants are intelligent animals capable of learning, and these attributes may be used to influence their distribution. This is currently a very active area of behaviour and ecosystem management research (Provenza *et al.*, 2003; Provenza & Villalba, 2006; Davis & Stamps, 2004; Provenza, 2003; Provenza, 2007).

This research is based on the fundamental understanding that all animals choose their behaviour based on the consequences they experience: positive consequences increase and negative consequences decrease the likelihood of behaviours recurring. Consequences involve two general behavioural systems in animals – skin-defence systems evolved under the threat of predation and gut-defence systems evolved under the threat of toxins in foods (Garcia *et al.*, 1985). These two systems form the basis for changing food and habitat selection behaviours in animals. Changing food/habitat selection behaviours requires making the food/habitat an animal is currently using less desirable (stick) relative to other foods/habitats (carrots).

Strategic hunting

Hunting can have significant and lasting impacts on the movement and distribution of game animals (Conner, 2002; Vieira *et al.*, 2003). As an example of this approach: elk are hunted in locations where they are not wanted, such as the former feeding areas, and they are not hunted in areas where they can stay. For instance, prior to 1986, both bull and cow elk migrated to lower elevations on the eastern portion of a ranch in Utah, USA. In mid-October in 1986, 100 hunters were allowed access to the ranch to hunt cow elk; they harvested 86 cows in one morning. For the past 20 years since that date cow elk have not migrated to lower elevations until snow pushes them down later in November or December. Bull elk, which have not been hunted in the lower elevations of the ranch, have continued to migrate to lower elevations in mid-October. One of the most striking examples of this involves a population of moose in central Norway that migrates from low-lying summer areas to high-elevation winter areas, contrary to the general pattern of migration (Andersen, 1991). Archaeological evidence shows their migratory behaviour follows a traditional pattern unchanged since 5000 BP despite deterioration in the quality of their winter range. Incongruously, there are no physical barriers preventing the moose using better habitat. Rather, the barriers are cultural, and they began

5000 BP when humans hunted (pit trapped) the moose. Humans no longer pit trap the moose and the behaviours are held in place by 'culture'.

In making such major changes in management, from field experience it is assessed that a minimum of three years typically are required to change the behaviours of long-lived social animals. The first year is the most difficult, as none of the adults has any experience with the new system. The second year is better because all those involved have a year of experience with the new system and the animals that were unable to adjust to the new system have been weaned. By the third year, all of the adults have two years of experience with the new system and young animals born into the new system are becoming members of the herd. In behaviour-based management, people become agents of change over time in animal cultures. Social organisation leads to culture, the knowledge and habits acquired by ancestors and passed from one generation to the next about how to survive in an environment (De Waal, 2001). A culture develops when learned practices contribute to the group's success in solving problems. Cultures evolve as individuals in groups discover new ways of behaving, as with finding new foods or habitats and better ways to use foods and habitats (Skinner, 1981).

Similarly, extended families with matriarchal leadership may provide a means for changing elephant behaviour. Efforts could be focused on individual families, and given the importance of the matriarch in behaviour of the family, specific efforts might be directed at the matriarch of each family. It may be best to test how to train elephants using a variety of techniques with a small number of families. Long-term mother-daughter associations should lead to the learning behaviour being transferred thus limiting the time needed to train the animals to avoid certain areas (Douglas-Hamilton, 1973; Moss & Poole, 1983).

Repellents

The use of chili extracts has shown particular promise not only because *Capsicum*-based products are non-toxic and environmentally friendly, but specifically because elephant's advanced olfactory and memory capabilities make them suitable for aversion conditioning (Osborn & Rasmussen, 1995; Osborn, 1997). Numerous evaluations with chili extracts have been completed, particularly in Zimbabwe where the objective was to protect crops belonging to rural populations that adjoin nature reserves or where elephants have caused extensive damage to crops (Osborn & Parker, 2002, 2003). These evaluations have been directed mainly at a practical and cost-effective means of applying *Capsicum* oleoresin in different forms like sprays and treated ropes which are strung around crops. Research has shown the effectiveness of chili extracts as a spray, when administered upwind of elephants and compared to traditional methods of trying to deter elephants during crop-raiding. When traditional measures are utilised, there is normally an aggressive reaction from elephants, whereas in the case of aerial spraying of *Capsicum* oleoresin, the response by the elephants was more rapid and resulted in prompt withdrawal from the crops without aggression (Osborn, 2002).

Other ways to protect crops or particular specimens of vulnerable trees include the placement of bee hives in strategic trees as elephants are sensitive to the sound and sting of bees (Karidozo & Osborn 2005; King *et al.*, 2007; Vollrath & Douglas-Hamilton, 2005a & b). Using bees as a selective repellent offers the added benefit that as a deterrent, bees could pay for themselves through the sale of honey (Vollrath & Douglas-Hamilton, 2005a; King *et al.*, 2007).

Buffer crops

Unpalatable crops such as tea, spiny plants such as sisal, timber plantations, and *Opuntia* barriers have all been tried but none have deterred elephants (Hoare, 2003). The cactus species *Opuntia dillenii* was used as a barrier in some parts of Laikipia and Narok, Kenya. Its potential to spread as a weed, however, is a major limitation. Another species, Mauritius thorn (*Caesalpinia decapetala*), has also been tried in Transmara, albeit with little success (Omondi *et al.*, 2004).

Moats and ditches

Ditches and moats have been tried in the past in Laikipia, Mt Kenya and Aberdares. However, due to lack of proper maintenance, they have not been successful in containing the elephants in protected areas. This method may be ideal only for small-scale sites of 3 or 4 km² and is not recommended for high rainfall areas as they may cause considerable soil erosion (Omondi *et al.*, 2004).

Stone walls

This method can only be considered where stones are available on site and the size of the area to be fenced is not extensive. Stone walls are not effective for containing elephants, as they soon learn to remove the rocks (Omondi *et al.*, 2004).

Sonic barriers

The use of sonic barriers may prove effective in deterring elephants from entering demarcated areas. High-frequency sound devices have already proved effective in preventing motorists from colliding with wildlife. In Australia vehicles are fitted with devices that provide a safety sound zone of 400 m and 50 m either side of the vehicle (http://www.shuroo.com/). As humans cannot hear the sound emitted by these sonic barriers and as they are not visible, such techniques may prove to be effective and aesthetically appealing when controlling elephant movements in particular areas.

EFFECTS OF FENCE REMOVAL OR THE LACK OF FENCING

Elephants typically disperse at rates of 7–10 km per year after the removal of a fence. Hence the 20 000 km^2 of Kruger was colonised within 50 years due to migration from Mozambique and the establishment of breeding herds in the Kruger National Park (Porter, 1970), after starting off with very few elephants in the early 1900s (Kirby, 1896).

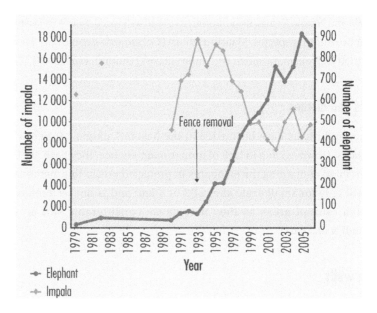

Figure 11: There was a steady increase in the elephant population in the Sabi Sand Wildtuin after the removal of the fence between Kruger and the private reserves in 1993

Elephants may readily move into new, unexplored areas, as can be seen by the increase in the elephant population in the private reserves next to Kruger after removal of part of the western boundary fence in 1993 (Peel & Grant, Chapter 8 in Grant, 2005) (figure 11).

The most recent addition to the Associated Private Nature Reserves is the Balule Nature Reserve, which had a low elephant density. Numbers in this area have increased from zero in the 1990s to almost 500 in 2006 (Peel, 2006). Even though it may still be too early to note the re-establishment of migration paths after the removal of the fence between Kruger and Sabi Sand Wildtuin it does appear that there is some seasonal movement in and out of areas such as the Sabi Sand (15 years). Satellite-collared animals are followed over time and movement between Kruger and Sabi Sand is already apparent in certain groups in both summer and winter (figure 12).

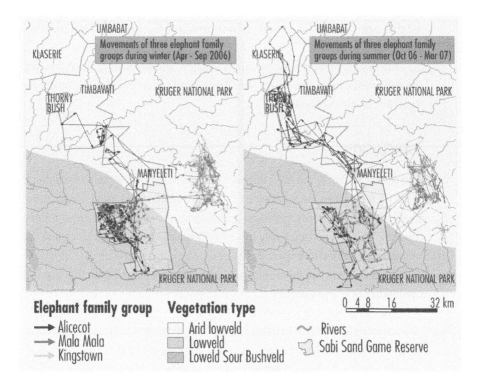

Figure 12: Seasonal movement of three elephant families between Sabi Sand Game Reserve and Kruger

During August 2004, the boundary fences between Phinda Private Game Reserve and two neighbouring reserves were removed. Initially family groups only moved into the new area at night and spent minimal time there, while older bulls spent longer periods of time, regardless of time of day. One year after the fence removal, most of the elephants had only expanded their home ranges slightly into the new area (Druce *et al.*, 2007). Similarly, elephants that were introduced into Marakele National Park in 1996 took a few years to move to the adjacent Marakele Pty Ltd after the fence was removed in 2001 (Bezuidenhout, 2004).

LEGAL OBLIGATIONS FOR FENCING

In any area where wildlife may be carriers of foot-and-mouth disease, the Animals Diseases Act (Act 35 of 1984) requires that the animals are separated from domestic stock. Any damage-causing animal that can be clearly identified by marking, collars, branding, microchip, etc. must be monitored and cases of damage need to be investigated thoroughly using these identification techniques as proof.

The quality of fences for wildlife is legally stipulated and defined for each type of animal to be contained. See Chapter 11 for a further discussion on the legal implications of fencing.

CONCLUSION

Fences are probably the most efficient barriers to restrict elephant movement. Electric fences can work very well if they are maintained at all times. These fences have to be sturdy and durable as elephants will tend to re-cross a fence once they have been previously successful. Fences are more efficient when the animals are trained to respect them.

Other barriers can be of some use, and may be cheaper than fencing, but maintenance is also essential.

'Teaching' animals to avoid certain areas is an option worth investigating. Disturbance in the form of noise or even local culling/hunting can be used to teach the animals to avoid certain areas. If this could be done successfully it may be possible to protect sensitive areas, at least to a certain extent, without fencing or other barriers.

RESEARCH GAPS

- Examine further ways of controlling elephant movement, e.g. learned behaviour or barriers.

- Understanding the factors, in particular water distribution, that determine distribution and density of elephant in enclosed areas.
- Examine effective techniques for fence line monitoring to enable fence-breaking individuals to be identified.

REFERENCES

African Elephant Specialist Group Meeting. 1993. Working group discussion Three Elephant - Habitat Working Group. *Pachyderm* 17, 9–16.

Albertson, A. 1998. Northern Botswana veterinary fences: critical ecological impacts. Unpublished manuscript.

Anderson, E.C. 1995. Morbillivirus infections in wildlife (in relation to their population biology and disease control in domestic animals). *Veterinary Microbiology* 44, 319–332.

Andersen, R. 1991. Habitat deterioration and the migratory behaviour of moose (*Alces alces* L.) in Norway. *Journal of Applied Ecology* 28, 102–108.

Anderson, J. L. 1994. The introduction of elephant into medium-sized conservation areas. *Pachyderm* 18, 33–38.

Anthony, B. P. 2006. A view from the other side of the fence: Tsonga communities and the Kruger National Park, South Africa. Ph.D. Environmental Sciences and Policy, Central European University.

Bengis, R.G., D.F. Keet, A.L. Michel & N.P.J. Kriek 2001. Tuberculosis caused by *Mycobacterium bovis*, in a kudu (*Tragelaphus strepiceros*) from a commercial game farm in the Malelane area of Mpumalanga Province, South Africa. *Onderstepoort Journal of Veterinary Research* 68, 239–241.

Bengis, R.G., R.A. Kock & J. Fischer 2002. Infectious animal diseases: the wildlife /livestock interface. *Scientific and Technical Review of the International Office of Epizootics* 21, 53–65.

Ben-Shahar, R. 1983. Patterns of elephant damage to vegetation in northern Botswana. *Biological Conservation* 65, 249–256.

Bezuidenhout, H. 2004. Internal report on the impact of elephants on the vegetation of the Zwarthoek section, Marakele National Park. Arid Ecosystems Research Unit, Conservation Services.

Brits, J., M.W. van Rooyen & N. van Rooyen 2002. Ecological impact of large herbivores on the woody vegetation at selected watering points on the eastern basaltic soils in the Kruger National Park. *African Journal of Ecology* 40, 53–60.

Buss, I.O. 1961. Some observations on food habits and behaviour of the African elephant. *Journal of Wildlife Management* 25, 130–149.

Chamaillé-Jammes, S., M. Valeix & H. Fritz 2007(a). Managing heterogeneity in elephant distribution: interactions between elephant population density and surface-water availability. *Journal of Applied Ecology* 44, 625–633.

Chamaillé-Jammes, S., M. Valeix, & H. Fritz 2007b. Elephant management: why can't we throw the babies out with the artificial bathwater? *Diversity and Distributions* 13, 663–665.

Chiyo, P. I., E.O.P. Cochrane, L. Naughton & G.I. Basuta 2005. *Temporal patterns of crop-raiding by elephants: a response to changes in forage quality or crop availability?* African Journal of Ecology 43, 48–55.

Collinson, R. F. H. & P.S. Goodman 1982. An assessment of range condition and large herbivore carrying capacity of the Pilanesberg Game Reserve, with guidelines and recommendations for management. *Inkwe* 11–54.

Conner, M.M. 2002. *Movements of mule deer and elk in response to human disturbance: a literature review.* Colorado Division of Wildlife, Mammals Research. Fort Collins, USA.

Craig, C. 1997. *The ELESMAP Project Report, Namibia.* Namibia Nature Foundation.

Cumming, D.H. 1981. The management of elephants and other large mammals in Zimbabwe. In: P. Jewell, S. Holt & D. Hart (eds). *Problems in management of locally abundant wild mammals.* Academic Press, New York, 91–118.

Cumming, D. H. & B. Jones 2005. *Elephants in southern Africa: management issues and options.* WWF, WWF–SAPRO Occasional Paper Number 11. WWF, Harare.

Davis, J.M. & J.A. Stamps 2004. The effect of natal experience on habitat preferences. *TREE* 19, 411–416.

De Vos, V., R.G. Bengis, N.P. Kriek, A. Michel, D.F. Keet, J.P. Raath & H.F. Huchzermeyer 2001. The epidemiology of tuberculosis in free-ranging African buffalo (*Syncerus caffer*) in the Kruger National Park, South Africa. *Onderstepoort Journal of Veterinary Research* 68, 19–30.

De Waal, F. 2001. *The ape and the sushi master: cultural reflections of a primatologist.* Basic Books, New York.

Dias, P.C. 1996. Sources and sinks in population biology. *Trends in Ecology and Evolution* 11, 326–330.

Douglas-Hamilton, I. 1973. On the ecology and behaviour of the Lake Manyara elephants. *East African Wildlife Journal* 11, 401–403.

Douglas-Hamilton, I., T. Krink & F. Vollrath 2005. Movement and corridors of African elephants in relation to protected areas. *Naturwissenschaften* 92, 158–163.

Druce, H. C., K. Pretorius & R. Slotow 2007. The response of an elephant population to conservation area expansion: Phinda Private Game Reserve, South Africa. *Biological Conservation* (in press).

Dublin, H.T., O.M. McShane & J. Newby 1997. Conserving *Africa's elephants. Current issues & priorities for action*. WWF International, Gland, Switzerland.

Freitag-Ronaldson, S. & L.C. Foxcroft 2003. Anthropogenic influences at the ecosystem level. In: J.T. du Toit, K.H. Rogers and H.C. Biggs (eds). *The Kruger experience: ecology and management of savanna heterogeneity*. Island Press, Washington, 391–421.

Garcia, J., P.A. Lasiter, F. Bermudez-Rattoni & D.A. Deems 1985. A general theory of aversion learning. In: N.S. Braveman & P. Bronstein (eds) *Experimental assessments and clinical applications of conditioned food aversions*. New York Academy of Science, New York, 8–21.

Gillson, L. & K. Lindsay 2003. Ivory and ecology – changing perspectives on elephant management and the international trade in ivory. *Environmental Science & Policy* 6, 411–419.

Gordon, C.H. 2003. The impact of elephants on the riverine woody vegetation of Samburu National Reserve, Kenya. Unpublished report for Save the Elephants.

Grant, C. C. 2005. Elephant effects on biodiversity: an assessment of current knowledge and understanding as a basis for elephant management in SANParks: A compilation of contributions by the Scientific community for SANParks. *Internal Report* 3/2005. Scientific Services, Skukuza.

Grant, C. C., T. Davidson, P.J. Funston & D.J. Pienaar 2002. Challenges faced in the conservation of rare antelope: a case study on the northern basalt plains of the Kruger National Park. *Koedoe* 45, 45–66.

Harrington, R., N. Owen-Smith, P.C. Viljoen, D.R. Mason & P.J. Funston 1999. Establishing the causes of the roan antelope decline in the Kruger National Park, South Africa. *Biological Conservation* 90, 69 78.

Henley, M.D. & S.R. Henley 2007. Population dynamics and elephant movements within the Associated Private Nature Reserves (APNR) adjoining the Kruger National Park. Unpublished May progress report to the Associated Private Nature Reserves.

Hoare, R.E. 1992. Present and future use of fencing in the management of larger African mammals. *Environmental Conservation* 19, 160–164.

Hoare, R.E. 1995. Options for the control of elephants in conflict with people. *Pachyderm* 19: 54–63.

Hoare, R. 1999. Determinants of human-elephant conflict in a land-use mosaic. *Journal of Applied Ecology* 36, 689–700.

Hoare, R. 2003. Fencing and other barriers against problem elephants. AfESG website HEC section.

Hoare, R.E. & J.T. du Toit 1999. Coexistence between people and elephants in African savannas. *Conservation Biology* 13, 633–639.

Hoare, R.E. 2001. *A decision support system for managing human-elephant conflict situations in Africa.* IUCN African Elephant Specialist Group Report.

Illius, A. & T.G. O'Connor 2000. Resource heterogeneity and ungulate population dynamics. *Oikos* 89, 283–294.

Jachmann, H. & R.H.V. Bell 1984. Why do elephants destroy woodland? *Pachyderm* 3, 9-10.

Jackson, T. & D.G. Erasmus 2005. Assessment of seasonal home-range use by elephants in the Great Limpopo Transfrontier Park. University of Pretoria, Pretoria.

Kangwana, K. 1995. Human-elephant conflict: the challenge ahead. *Pachyderm* 19, 9–14.

Karidozo, M. & F.V. Osborn 2005. Can bees deter elephants from raiding crops? An experiment in the communal lands of Zimbabwe. *Pachyderm* 39, 26–32.

Keet, D.F., N.P.J. Kriek, R.G. Bengis & A.L Michel 2001. Tuberculosis in kudu (*Tragelaphus strepsiceros*) in the Kruger National Park. *Onderstepoort Journal of Veterinary Research* 68, 225–230.

Keet, D.F., N.P.J. Kriek, M-L. Penrith, A. Michel & H. Huchzermeyer 1996. Tuberculosis in buffaloes (*Syncerus caffer*) in the Kruger National Park: spread of disease to other species. *Onderstepoort Journal of Veterinary Research*, 239–244.

Kerley, G.I.H. & M. Landman 2006. The impact of elephant on biodiversity in the Eastern Cape Subtropical thickets. *South African Journal of Science* 102, 1–8.

King, L.E., I. Douglas-Hamilton & F. Vollrath 2007. African elephants run from the sound of disturbed bees. *Current Biology* 17, 832–833.

Kirby F.V. 1896. *In haunts of wild game.* Blackwell & Sons, Edinburgh.

Knight, M., G. Castley, L. Moolman, & J. Adendorff 2002. Elephant management in Addo Elephant National Park. In: G. Kerley, S. Wilson & A. Massey (eds). *Elephant conservation and management in the Eastern Cape – workshop proceedings.* Terrestrial Ecology Research Unit University of Port Elizabeth, Port Elizabeth, 32–40.

Leuthold, W. 1996. Recovery of woody vegetation in Tsavo National Park, Kenya, 1970-94. *African Journal of Ecology* 34, 101-112.

Lombard, A.T., C.F. Johnson, R.M. Cowling & R.L Pressey 2001. Protecting plants from elephants: botanical reserve scenarios within the Addo Elephant National Park, South Africa. *Biological Conservation* 102, 191-203.

Lubow, B.C. 1996. Optimal translocation strategies for enhancing stochastic metapopulation viability. *Ecological Applications* 6, 1268-1280.

McLoughhlin, C.A. and N Owen-Smith 2003. Viability of a Diminishing Roan Antelope Population: Predation and Threat. *Animal Conservation,* 6231-6236.

Mills, M.G.L. & P.J. Funston 2003. Large carnivores and savanna heterogeneity. In: J.T. du Toit, K.H. Rogers & H.C. Biggs (eds) *The Kruger experience: Ecology and management of savanna heterogeneity.* Island Press, Washington DC., 370-388.

Moss, C.J. & J.H. Poole 1983. Relationships and social structure of African elephants. R.A. Hinde (ed.). *Primate social relationships: an integrated approach.* Sinauer, Sutherland, MA, 315-325.

Mwalyosi, R.B.B. 1990. The dynamics ecology of Acacia tortilis woodland in Lake Manyara National Park, Tanzania. *African Journal of Ecology* 28, 189-199.

Naughton-Treves, L. 1998. Predicting patterns of crop damage by wildlife around Kibale National Park, Uganda. *Conservation Biology* 12, 156-158.

Nelleman, C., R.M. Stein & L.P. Rutina 2002. Links between terrain characteristics and forage patterns of elephants (*Loxodonta africana*) in northern Botswana. *Journal of Tropical Studies* 18, 835-844.

Nelson, A., P. Bidwell & C. Sillero-Zubiri 2007. *A review of human-elephant conflict management strategies.* Wildlife Conservation Research Unit, Oxford University, People and Wildlife Initiative.

Nelson, A., P. Bidwell & C. Sillero-Zubiri 2007. *A review of human-elephant conflict management strategies.* People and Wildlife Initiative, Wildlife Conservation Research Unit, Oxford University, Oxford.

O'Connell-Rodwell, C.E., T. Rodwell, M. Rice & L.A Hart 2000. Living with the modern conservation paradigm: can agricultural communities co-exist with elephants? A five-year case study in East Caprivi, Namibia. *Biological Conservation* 93, 381-391.

Omondi, P., E. Bitok, & J. Kagiri 2004. Managing human-elephant conflicts: the Kenyan experience. *Pachyderm* 36, 80-86.

Osborn, F.V. 1997. The ecology and deterrence of crop-raiding elephants: final technical report. Unpublished report to USFWS.

Osborn, F.V. 2002. Capsicum oleoresin as an elephant repellent: field trials in the communal lands of Zimbabwe. *Journal of Wildlife Management* 66, 674–677.

Osborn, F.V. 2004. Seasonal variation of feeding patterns and food selection by crop-raiding elephants in Zimbabwe. *African Journal of Ecology* 423, 22–27.

Osborn, F.V. & G.E. Parker 2002. Community-based methods to reduce crop loss to elephants: experiments in the communal lands of Zimbabwe. *Pachyderm* 33: 32–38.

Osborn, F.V. & G.E. Parker 2003. Linking two elephant refuges with a corridor in the communal lands of Zimbabwe. *African Journal of Ecology* 41: 68–74.

Osborn, F.V. & L.E.L.Rasmussen 1995. Evidence for the effectiveness of an oleo-resin capsicum aerosol as a repellent against wild elephants in Zimbabwe. *Pachyderm* 20, 55–64.

Owen-Smith, R.N. 1988. *Megaherbivores: the influence of very large body size on ecology.* Cambridge Studies in Ecology. Cambridge University Press, Cambridge.

Owen-Smith, N. 1996. Ecological guidelines for waterpoints in extensive protected areas. *South African Journal of Wildlife Research* 26, 107–112.

Owen-Smith, N., G.I.H. Kerley, B. Page, R. Slotow & R.J. van Aarde 2006. A scientific perspective on the management of elephants in the Kruger National Park and elsewhere. *South African Journal of Science* 102, 389–394.

Parker, A.H. & E.T.E. Witkofski. 1999. Long-term impacts of abundant perennial water provision for game on herbaceous vegetation in a semi-arid African savanna woodland. *Journal of Arid Environments* 41, 309–321.

Peel, M.J.S. 2005. Towards a predictive understanding of savanna vegetation dynamics in the eastern Lowveld of South Africa: with implications for effective management. Ph.D. thesis, University of KwaZulu-Natal.

Peel, M. 2006. Ecological monitoring: Association of Private Nature Reserves. Unpublished landowner report.

Peel, M.J.S., H.C. Biggs & P.J.K. Zacharias 1999. The evolving use of indices currently based on animal number and type in semi-arid heterogeneous landscapes and complex systems. *African Journal of Range and Forage Science* 15, 117–127.

Pienaar, D.J., H.C. Biggs, A. Deacon, W. Gertenbach, S. Joubert, F. Nel, L. van Rooyen & F. Venter 1997. A revised water-distribution policy for biodiversity maintenance in the Kruger National Park. Internal report. South African National Parks, Skukuza.

Porter, R.N. 1970. An ecological reconnaissance of the TPNR Private Nature Reserve. Unpublished report. Timbavati Private Nature Reserve, Hoedspruit.

Provenza, F.D. 2003. Foraging behavior: managing to survive in a world of change. Utah State University, Logan.

Provenza, F.D. 2007. What does it mean to be locally adapted and who cares anyway? *Journal of Animal Science* (accepted).

Provenza, F.D. & J.J. Villalba 2006. Foraging in domestic vertebrates: linking the internal and external milieu. In: V.L. Bels (ed.). *Feeding in domestic vertebrates: from structure to function*. CABI Publications, Oxfordshire, UK, 210–240.

Provenza, F.D., J.J. Villalba, L.E. Dziba, S.B. Atwood & R.E. Banner 2003. Linking herbivore experience, varied diets, and plant biochemical diversity. *Small Ruminant Research* 49, 257–274.

Redfern, J.V., R. Grant, H. Biggs & W.M. Getz 2003. Surface-water constraints on herbivore foraging in the Kruger National Park, South Africa. *Ecology (Durh.)* 84, 2092–2107.

Redfern, J.V., C.C. Grant, A. Gaylard & W.M. Getz 2005. Surface water availability and the management of herbivore distributions in an African savanna ecosystem. *Journal of Arid Environments* 63, 406–424.

Sakumar, R. & M. Gadgil. 1988. Male-female differences in foraging on crops by Asian elephants. *Animal Behaviour* 36, 1233–1235.

Skinner, B.F. 1981. Selection by consequences. *Science* 213, 501–504.

Skukuza, Nelspruit & Mkhuhlu State Annunal Reports, 2006, 2007, and Quaterly Reports, 2008.

Smit, I.P.J., C.C. Grant, & B.J.Devereux 2007(a). Do artificial waterholes influence the way herbivores use the landscape? Herbivore distribution patterns around rivers and artificial surface water sources in a large African savanna park. *Biological Conservation* 136, 85–99.

Smit, I.P.J., C.C. Grant & I.J. Whyte 2007(b). Landscape-scale sexual segregation in the dry season distribution and resource utilisation of elephants in Kruger National Park, South Africa. *Diversity and Distributions* 13, 225–236.

Smit, I.P.J., C.C. Grant & I.J. Whyte 2007(c). Elephants and water provision: what are the management links? *Diversity and Distributions* 13, 666–669.

Smith, N. & S.L. Wilson 2002. Changing land use trends in the thicket biome: pastoralism to game farming. Unpublished report No. 38 to the Terrestrial Ecology Research Unit, University of Port Elizabeth, Port Elizabeth

Smith, R.J. & S.M. Kasiki. 2000. *A spatial analysis of human-elephant conflict in the Tsavo-ecosystem, Kenya.* A report to the African Elephant Specialist Group, Human-Elephant Conflict Task Force, of IUCN. Gland, Switzerland.

South African National Parks (SANParks). 2005. Report on the Elephant Management Strategy: Report to the Minister: Environmental Affairs and Tourism on Developing Elephant Management Plans for National Parks with Recommendations on the process to be followed. Accessible at: http://www.sanparks.org/events/elephants/strategy_19-09-2005.pdf.

South African Savannas Network 2001. The status of southern Africa's savannas. Report to United Nations Environment Programme, University of London, London.

Stokke, S. & J.T. du Toit 2000. Sex and size related differences in the dry season feeding patterns of elephants in Chobe National Park, Botswana. *Ecography* 23, 70–80.

Taylor, R.D. 1994. Elephant management in Nyaminyami District, Zimbabwe: turning a liability into an asset. *Pachyderm* 18, 17–29.

Tchamba, M.N. 1995. The problem elephants of Kaele: a challenge for elephant conservation in northern Cameroon. *Pachyderm* 19, 26–31.

Thrash, I. 1997. Infiltration rate of soil around drinking troughs in the Kruger National Park, South Africa. *Journal of arid environments* 35, 617–625.

Thrash, I. 2000. Determinants of the extent of indigenous large herbivore impact on herbaceous vegetation at watering points in the north-eastern lowveld. South Africa. *Journal of Arid Environments* 44, 71–72.

Thrash, I., P.J. Nel, G.K. Theron & J.D.P. Bothma 1991. The impact of the provision of water for game on the woody vegetation around a dam in the Kruger National Park. *Koedoe* 34, 131–148.

Tolsma, D. J., W.H.O. Ernst & R.A. Verwey 1987. Nutrients in soil and vegetation around two artificial waterpoints in eastern Botswana. *Journal of Applied Ecology* 24, 991–1000.

Trollope, W.S.W., L.A. Trollope, H.C. Biggs, D.J. Pienaar & A.L.F. Potgieter 1998. Long term changes in the woody vegetation of the Kruger National Park, with special reference to the effects of elephants and fire. *Koedoe* 41, 103–112.

Van Aarde, R.J. & T.P. Jackson 2007. Megaparks for metapopulations: addressing the causes of locally high elephant numbers in southern Africa. *Biological Conservation* 134, 289–297.

Verlinden, A. & I.K.N. Gavor 1998. Satellite tracking of elephants in northern Botswana. *African Journal of Ecology* 36, 105–116.

Vieira, M.E., M.M. Conner, G.C. White & D.J. Freddy 2003. Effects of archery hunter numbers and opening dates on elk movement. *Journal of Wildlife Management* 67, 717–728.

Viljoen, P.J. & J.D.P. Bothma 1990. Daily movements of desert-dwelling elephants in the northern Namib Desert. *South African Journal of Wildlife Research* 20, 69–72.

Vollrath, F. & I. Douglas-Hamilton 2005a. African bees to control African elephants. *Naturwissenschaften* 92, 508–511.

Vollrath, F. & I. Douglas-Hamilton. 2005b. Elephants buzz off! *Swara* 25, 20–21.

Walker, B. H., R. H. Emslie, N. Owen-Smith & R.J. Scholes. 1987. To cull or not to cull: lessons from a southern African Drought. *Journal of Applied Ecology* 24, 381–401.

Wasilwa, N.S. 2003. Human elephant conflict in Transmara district, Kenya. In: M.J. Walpole, G.G. Karanja, N.W. Sitati & N. Leader-Williams (eds). *Wildlife and people: conflict and conservation in Masai Mara, Kenya.* Wildlife and Development Series 14, International Institute for Environment and Development, London.

Western, D. & D. Muitumo 2004. Woodland loss and restoration in a savanna park: a 20-year experiment. *African Journal of Ecology* 42, 111–121.

Whyte, I.J. 1985. The present ecological status of the Blue Wildebeest (*Connochaetes taurinus taurinus*, Burchell, 1823) in the Central District of the Kruger National Park. MSc thesis, University of Natal, Pietermaritzburg.

Whyte, I.J. & S.C.J. Joubert 1988. Blue wildebeest population trends in the Kruger National Park and the effects of fencing. *South African Journal of Wildlife Research* 18, 78–87.

WWF. 1997. *Conserving Africa's elephants: Current issues and priorities for action.* In: H.T. Dublin, T.O. McShane & J. Newby. World Wide Fund for Nature International Report, Gland, Switzerland.

WWF 1998. *Wildlife electric fencing projects in communal areas of Zimbabwe – current efficacy and future role.* Price Waterhouse Coopers, Report for WWF Southern Africa Programme Office, Harare, Zimbabwe.

Wynn, S. 2003. Zambezi River: wilderness and tourism research into visitor perceptions about wilderness and its value. USDA Forest Service Proceedings RMRS-P-27 200.

8

LETHAL MANAGEMENT OF ELEPHANTS

Lead author: Rob Slotow
Authors: Ian Whyte and Markus Hofmeyr
Contributing authors: Graham HI Kerley, Tony Conway, and Robert J Scholes

INTRODUCTION

FOR THE purpose of this chapter, we define two broad circumstances under which elephants are killed for management purposes. The first, which we will term culling, is where a significant fraction of the elephant population are killed with the objective of reducing the population size or controlling its growth rate. The second is when specific individuals are killed to prevent them from causing further damage or threatening human lives (hereafter referred to as 'problem animal control') (DEAT, 2008). Decisions on the implementation of problem animal control are relatively uncontentious. When an individual poses a threat to human life, or persistently causes damage to infrastructure or agriculture, that identified individual is dealt with according to set decision-making norms and procedures, which may include lethal management (DEAT, 2008). Culling for population management is much more complex and is at the root of much of the elephant debate (Caughley, 1976).

Imposed population control may be necessary when natural mechanisms of population regulation are not operating, for whatever reason (Chapter 2). Besides controlling population numbers, manipulation of age-sex class composition may be necessary to correct historical effects, e.g. populations founded by young elephants only (Garaï *et al.*, 2004).

The purpose of this chapter is to provide a resource for decision-making around culling of elephants and problem animal control and to evaluate our current understanding, knowledge, and gaps regarding lethal management of elephants. Culling has been applied as a management tool for elephants since elephants have been managed, and we have a good understanding of certain aspects. However, increased accountability to the broader community has necessitated that all aspects are well considered.

The specific objectives of this chapter are to: (1) describe the history of culling, (2) briefly describe and evaluate the methods for culling, (3) describe the various management contexts and objectives for culling as a viable intervention, (4) highlight the constraints and consequences of culling, and (5) define gaps in our knowledge.

THE HISTORY OF CULLING

Zimbabwe and other southern African countries

An overview of culling in southern Africa is provided to place the South African experience in context. Although this assessment focuses on elephant management in South Africa, numerically, most culling that has taken place to date has occurred in Zimbabwe, where culling to control population numbers was first implemented in 1966. By August 1996, a cumulative total of almost 50 000 elephants had been culled in Zimbabwe (Martin *et al.*, 1996; see also Cumming & Slotow, 2003 cited in Cumming & Jones, 2005). The estimated elephant population in Zimbabwe in 1996 was 68 000 (Martin *et al.*, 1996).

Culling of family groups also occurred in the Luangwa valley in Zambia in the 1960s and 1970s (Cumming & Jones, 2005). Culling to reduce numbers also took place in Etosha, Namibia in 1983 and 1985 (Cumming & Jones, 2005). Culling to reduce population size has not been undertaken in Botswana or Mozambique. Poaching, however, has resulted in a major reduction in population size in Mozambique (and Zambia) since the 1960s (Cumming & Jones, 2005).

The first culling within national parks took place in Hwange in 1966 and 1967, and then in Mana Pools in 1968 and 1969 (Cumming, 1981). Culling was scaled up in 1971 with 1 300 elephants removed from Hwange, and 665 elephants from Gonarezhou (Cumming, 1981). Culling was initiated in Chizarira and Matusadona in 1972 (Cumming, 1981). Full details of removals in Zimbabwe up until 1979 are included in Cumming (1981); by 1979 18 216 animals had been removed. Details of culling after that time are provided in Martin *et al.* (1996).

Kruger National Park

The decision to cull elephants in Kruger is dealt with as a case history in Chapter 1, and is not repeated here. A policy of elephant population control by culling and live removals was implemented from 1965 until 1995.

A total of 14 629 elephants were culled between 1967 and 1997 in the Kruger (Whyte, 2001; table 1). The highest number of elephants culled in any year was 1 846 in 1970, the median number for years in which more than 100 were culled was 348. The median percentage of the existing population that was culled per year was 5.35 per cent.

Addo Elephant National Park

The attempt in 1919 to eradicate the population of elephants in the Addo bush near Port Elizabeth in the Eastern Cape was one of the earliest specific management attempts to cull elephants. This population of about 120 elephants represented what was then the largest elephant population in South Africa (Whitehouse, 2001), but these animals were in conflict with local farmers due to the elephants causing destruction of dams and other infrastructure (see Chapter 1 for details). Pressure by the local farmers led the then Cape Provincial administration to contract Major P.J. Pretorius to eradicate this population (Hoffman, 1993). Pretorius, who described the habitat as a 'hunter's hell' (Pretorius, 1947), managed to shoot about 120 animals and sold the meat and ivory. He also sold some specimens to the then South African Museum. Pretorius later petitioned the Administrator of the Cape to be released from his contract when an estimated 16 elephants remained, pleading the need for their conservation. This was granted in 1920, and the remaining animals became the basis for the now famous Addo elephants. After the establishment of the Addo Elephant National Park (Addo) in 1931, further culling of some problem animals occurred. This included one bull shot in self-defence in 1931, a second shot in 1932 to protect a windmill, a third in 1937 in retaliation for the death of a ranger (Whitehouse & Kerley, 2002) and a fourth bull in 1968 (the well-known Hapoor) who had escaped from the fenced park (Whitehouse, 2001).

The population was ultimately reduced to 11 animals. This strong bottleneck effect has been expressed at a genetic level, with the Addo elephants being genetically impoverished compared to the parent population (i.e. the pre-Pretorius Addo population) or the Kruger population (Whitehouse & Harley, 2001). The current social structure of the Addo population (seven family groups) reflects the presence of five cows in the founder population (Whitehouse, 2001). Finally, the culling of the three bulls in the 1930s left the Addo population without any mature breeding bulls for nine years (until a bull calf matured), hindering the population's recovery from low numbers (Whitehouse & Kerley, 2002).

Year	Census total	Culling quota	Total culled	Juveniles translocated	Family units translocated	Adult bulls translocated	Total removed after census
1966	No census	–	–	26	–	–	26
1967	6 586	650	355	–	–	–	355
1968	7 701	1 230	460	–	–	–	460
1969	8 312	1 408	1 160	–	–	–	1 160
1970	8 821	2 093	1 846	–	–	–	1 846
1971	7 916	889	602	–	–	–	602
1972	7 611	618	608	–	–	–	608
1973	7 965	738	732	–	–	–	732
1974	7 702	853	764	–	–	–	764
1975	7 408	601	567	–	–	–	567
1976	7 275	350	285	–	–	–	285
1977	7 715	663	544	26	–	–	570
1978	7 478	392	348	35	–	–	383
1979	–	380	322	48	–	–	370
1980	7 454	395	356	55	–	–	411
1981	7 343	71	16	0	–	–	16
1982	8 051	555	427	46	–	–	473
1983	8 678	2 229	1 290	66	–	–	1 356
1984	8 273	1 890	1 289	88	–	–	1 377
1985	6 887	369	268	101	–	–	369
1986	7 617	495	404	94	–	–	498
1987	6 898	305	245	59	–	–	304
1988	7 344	367	273	83	–	–	356
1989	7 468	367	281	85	–	–	366
1990	7 287	367	232	132	–	–	364
1991	7 470	367	218	140	–	–	358
1992	7 632	350	185	150	144	–	479
1993	7 834	577	308	74	8	–	390
1994	7 806	600	177	31	146	2	356
1995	8 064	0	44	0	83	0	127
1996	8 320	0	18	0	52	6	76
1997	8 371	0	5	0	12	34	51
1998	8 869	0	0	0	13	18	31
1999	9 152	0	0	0	0	12	12
2000	8 356	0	0	0	22	27	49
Total	–	20 169	14 629	1 339	458	72	16 520

Table 1: Annual elephant census totals and culling quotas in the Kruger National Park since the initiation of the census and culling programmes in 1966, and numbers removed from the population (from Whyte, 2007)

Tembe Elephant Park – KwaZulu-Natal

Tembe Elephant Park houses one of the original (non-introduced) populations in South Africa. The park was proclaimed in 1983, and the south, west and east boundaries were fenced to protect the local population from elephants. The northern boundary with Mozambique was fenced in 1989, until which time elephants could move freely in and out of the park.

A number of individuals had been injured by people in Mozambique, resulting in human-elephant conflict within the park. Nineteen elephants have been culled in Tembe to date, and four males were hunted in Tembe in 1996 (table 2).

Reserve	Problem males	Problem females	Hunted	Source
Marakele	2			Hofmeyr, SANParks
Ithala	10			Conway, EKZNW
Mkhuze	1	1		Conway, EKZNW
St. Lucia				Conway, EKZNW
Tembe	19[a]		4	Matthews, EKZNW
Hluhluwe-Umfolozi	10			Conway, EKZNW
Pilanesberg	?		24[b]	Nel, NWP&TB
Madikwe	?	2	22[b]	Nel, NWP&TB
Fish River Complex	1			Kerley, NMMU
Songimvelo [c]	?		7[b]	Steyn, MPB
Mthethomusa	?		3[b]	Steyn, MPB
Kwa Madwala			2	Steyn, MPB
Private reserves 2001 data [c]	8	1	9	Slotow, UKZN

[a] 3 animals were culled because of wounds received in Mozambique.

[b] Some of the hunted males were problem animals. A ? is inserted in that column to indicate that there were some problem animals on those reserves.

[c] Based on the database from a survey done by the Elephant and Owners Association in 2001 ($N = 56$ reserves with wild elephants), extracted by Slotow. Here we present the number of reserves which had killed elephants as problem animals or had hunted elephants.

Note: See text for details from Addo, which last culled an animal in 1968. Whether animals have been hunted or culled in Limpopo provincial reserves is unknown.

Table 2: Killing of problem animals and hunting in small reserves (data up to 2007 except for private reserves)

Small populations in South Africa

Elephants had been introduced to 58 small, fenced reserves (<1 000 km²) by 2001 (Garaï *et al.*, 2004; Slotow *et al.*, 2005), and the number of small reserves with elephants probably exceeded 80 in 2007 (see Chapter 6). Two reserves have to date instituted culling as a means of population control. In both cases a family group was culled (Steyn, pers. comm.; Van Altena, pers. comm.).

A number of these reserves have removed elephants alive. These capture and removal events, while not resulting in elephant deaths, present an opportunity to learn about the effects of population reduction on the remaining population. Elephant groups have been translocated from Madikwe Game Reserve (five groups), Hluhluwe-Umfolozi Park (three groups), Phinda Private Game Reserve (five groups), Weenen Biosphere (one group) and Magudu Game Reserve (one group). These elephants were relocated to other reserves to form new populations.

Hunting has taken place in a large number of small reserves (table 2), but the effects on the population have been comprehensively studied only in Pilanesberg National Park (Burke *et al.*, 2008). See the relevant section below for key conclusions.

Problem animal control

A large number of small reserves have experienced problems with elephants, to the extent that at least 15 reserves have shot problem animals (table 2) (also see Chapter 4). At least six small reserves (Tembe, Pilanesberg, Hluhluwe/Umfolozi Park, Madikwe, Phinda, Greater St. Lucia Wetland Park) have destroyed both male and female elephants that killed people (unpublished information provided to Slotow by reserve management staff). In all cases involving females, the particular individual female was killed rather than the whole group. Elephants have been hunted in at least 15 small reserves (table 2).

Controlling problem elephants in and around the Kruger National Park has been ongoing since large-scale culling of elephants was stopped in 1994. In 2006/2007 fewer than 20 elephants were destroyed in and around Kruger (data extracted from Kruger internal diaries and reports housed at Skukuza archives by M. Hofmeyer). All of these animals were killed because they were deemed problem animals, because of threat to outside communities when they broke out of the park, damage to property inside the park, or a threat to guests or staff in the park. Some animals were destroyed because they had sustained serious injuries such as snare wounds.

Elephant that are shot on communal land around Kruger are given to the communities to salvage the meat and other products. Tusks are removed for safekeeping by the provincial conservation agency. In Kruger, the carcasses are left to scavengers in the areas far away from the wildlife products plant

---

Box 1: Proposed management of problem elephants at KNP (extracted from SANParks standard operating procedure by Hofmeyr):

Inside KNP: When an elephant (usually it is an individual) causes problems in staff villages or enters tourism facilities, it must be chased out/away. If an individual displays aggressive behaviour over an extended period, then a decision is taken on whether to destroy the animal or to capture and translocate it.

Elephants leaving KNP: The decision on the actions to be taken against elephants that left the park will be the responsibility of the relevant Provincial Environmental Authority. The first option will be to chase herds back to the park, and if this cannot be achieved the next consideration should be capture (in the case of family groups) and lastly to destroy them. Individuals and bulls that cause problems will most probably have to be destroyed, as it is not viable to capture and translocate these animals. Translocation of elephants is an expensive exercise and the areas for translocation in South Africa are limited (Chapter Six). It is therefore not a viable option to deal with problem elephants in this manner; they also often become repeat offenders and subsequently break out of their new locality.

---

in Skukuza, otherwise the carcass is transported to the plant and as much is used as possible (skin and meat mainly). The meat is sold internally to staff or restaurants. Skins and tusks are securely stored at Skukuza. At the time of writing (2007) the wildlife products plant is not fully functional due to limited usage since 1994. It is estimated that it will cost in excess of R14 million in repairs and maintenance to restore the plant to the required standard (internal report – Kruger National Park). The economics and broader consequences of this are dealt with in Chapter 10.

CULLING METHODS

Shooting methods

In Zimbabwe elephants to be culled were located using a fixed-wing aircraft, and a ground crew of three to five marksmen (Thompson (2003) suggests only three should be used because of safety concerns) were directed to the target group (Bengis, 1996; Thompson, 2003; Cumming & Jones, 2005). Elephants were brain-shot with heavy-calibre weapons (FN 7.62 mm automatic rifles were used on sub-adult animals, Conway pers. obs.), with up to 50 animals being killed within two minutes (Thompson, 2003; Cumming & Jones, 2005). A helicopter was used on only one occasion in Gonarezhou, and was discarded because the sound of the blades disturbed the elephants and caused them to run (Thompson, 2003). Details of the culling procedures used in Zimbabwe are provided by Thompson (2003: 272). Bulls in the breeding herds were generally shot first, and then the older females, generally the matriarch, in order to anchor the group (Thompson, 2003). Large trophy bulls were generally spared in most of these culling operations. In East Africa a helicopter was used to herd the animals to the 'killing ground' and to pursue any escapees, but they were shot from the ground as per the Zimbabwe method (Bengis, 1996).

Culling of elephants in Kruger was always conducted from a helicopter that herded the elephants until all in the group had been targeted (Bengis, 1996). Initially, Kruger elephants were immobilised using Scoline-loaded darts fired from a helicopter (see below) followed up with lethal brain-shots from a ground team (Bengis, 1996). However, this technique was shown to be physiologically stressful and inhumane, and was discontinued (Bengis, 1996). In an attempt to reduce the time interval between motor paralysis ('going down') and the administration of the lethal brain-shot, the technique was modified as follows: as soon as the animal went down following darting with Scoline, it was brain-shot from the helicopter. Inevitable delays and the wider scattering of carcasses of the target group meant this was not ideal (Bengis, 1996).

After the discontinuation of the use of Scoline, all elephants culled in Kruger were brain-shot at close range from a helicopter using heavy calibre rifles (.375 or .458) to cull bulls, and a semi-automatic rifle firing .308 (7.62 mm) brass monolithic solid bullets for females and other age classes. The advantage of the latter rifle was low recoil, large magazine capacity and rapid repeat fire if necessary. The helicopter continuously circled the herd, keeping them within the confines of the identified 'killing zone'. Shooting of the matriarch first anchored the rest of the family group, allowing all to be quickly shot in a small area (Bengis, 1996). For details see box 2. This technique was also used briefly in Etosha, Namibia, during the early 1980s (Bengis, 1996).

In Kruger, the processing of the carcasses constrained the culling rate and thus the total number of elephants culled. Between 1968 and 1970 two separate culling operations were conducted in Kruger, one in the southern half which supplied carcasses to the permanent abattoir in Skukuza, and one in the north which supplied carcasses to a temporary facility on the Shingwedzi River near the western boundary. The latter facility collected the ivory, treated the skins and cooked the meat in large cast iron pots, for sale at a low price to communities neighbouring the park. Cooking of the meat before sale was necessary due to the uncertainty at the time regarding the possibility that elephant might be carriers of FMD. This has subsequently been found not to be the case. The temporary northern facility allowed increased culls in the years 1968–1970 (see table 1). In Hwange, Zimbabwe, up to 5 000 elephants were killed over the three-month winter period in the 1980s (Cumming & Jones, 2005). A target of 50 animals per day could be achieved if the animals were breeding herds, but even half that would be difficult to reach if bull groups were targeted (Thompson, 2003).

Chemical methods with specific reference to Scoline

For reasons of safety to operators and the public, the culling of elephants in Kruger was initially conducted using the drug Scoline (succinyldicholine chloride). This compound paralysed the animal, rendering it immobile and harmless once it was recumbent until it could be dispatched by means of a brain-shot. It was shown by Hattingh *et al.* (1984a; 1984b; 1990a; 1990b) that the use of Scoline for culling elephants was inhumane. These authors showed that in elephants the locomotory muscles are immobilised initially, rendering the animal recumbent yet totally aware of its surroundings. A while thereafter the diaphragm is affected, stopping respiration. The heart muscle continues to function for several minutes thereafter and the animal eventually dies of asphyxiation if it is not brain-shot. The use of Scoline was therefore discontinued and is not approved in the Norms and Standards (DEAT, 2008) or SANParks Standard Operating Procedures for either culling or euthanasia.

Financial assessment of culling

Culling, if humanely conducted, and with recovery of the animal products, can be a complicated process. It needs specialist equipment and personnel, and is distracting to management (Owen-Smith *et al.*, 2006). 'Culling of elephants whether on a small or large scale is expensive' (Martin *et al.*, 1996). The costs in KNP for culling approximately 800 elephants and processing them was calculated in year 2005 values as ZAR5 298 260 (table 3), about ZAR6 600 per elephant. The field costs alone (excluding salaries) were ZAR1 761 per elephant. The non-field costs excluded the commissioning of the abattoir facilities, and some of the processing costs such as canning (Grant, 2005).

Box 2: Proposed management of problem elephants at KNP (extracted from SANParks standard operating procedure for elephant culling by Hofmeyr):

Elephants are culled using rifles. The procedure differs slightly for bulls compared to breeding herds. For bulls, single animals or small groups (2–4) are selected. The helicopter slowly herds the animal(s) to a selected work site, which is chosen for its distance from the tourist road system, proximity to a patrol or firebreak road, heavy vehicle access, and terrain. At the selected site, the helicopter flies low and at slow airspeed, lining up the marksman over and behind the target, in a direct line with the animal's direction of movement. This is more easily accomplished with a 'right door off' helicopter for a right-handed marksman, and for a 'left door off' helicopter for a left-handed marksman. It is essential that the marksman is brought over the midline of the elephant, in line with its direction of movement. For opposite-handed marksmen, it may be necessary for the pilot to 'crab' the helicopter slightly to give the marksman the correct approach and aiming angle.

Elephants have a very smooth gait with little head movement and generally move in a straight line. As the helicopter flies above the elephant, approaching from behind, the marksman shoots the animal in the midline, aiming at the lower part of the back of the skull. Excellent landmarks for this aiming point are the two large longitudinal muscle masses, clearly visible on the back of the neck. As the helicopter overflies the elephant, the marksman needs to aim between these two muscle masses at a forward angle of between 30° and 45°. It is important to note that an elephant's brain lies very deep, and a shot at the skull (rather than neck) from behind, frequently passes over the top of the brain cavity. If the shot placement is too far back on the neck or at too steep an angle, as long as it is in the midline, it should strike the cervical spinal chord, which is also instantly effective. For lateral brain-shots from the side of an elephant, the aiming point should be at the angle between the ear-slit and the cheekbone.

The elephant should collapse instantly at the shot, frequently in sternal recumbency with tusks ploughed into the ground. If falling in lateral recumbency, the spasmodic jerking of one or both of the hind legs is frequently indicative of a good brain-shot.

If more than one bull elephant is to be culled, the second (or more) bull(s) is herded by the helicopter to the proximity of the first carcass, and then shot. A minimum calibre of .375 inch firing a monolithic solid bullet must be used in the culling of elephants. Only shots to the brain should be used and once the animal has collapsed a second shot (insurance shot) should be administered to make sure that it is dead. In the rare event of an elephant only being stunned by a shot close to the brain, this animal may regain consciousness and be dangerous.

In the case of elephant breeding herds, the helicopter separates a family group out of the breeding herd, and then slowly herds this group to the working site. The selection of family group depends on its size (relative to the day's quota) and direction of movement (towards the selected working site). At the working site, the matriarch is shot first, which anchors and confuses the rest of the group. The rest of the group can then be quickly dispatched, as they mill around or are herded around the fallen matriarch.

Shooting of elephants from the ground is usually done in the case of solitary injured or problem animals, in the absence of helicopter back-up. When firing frontal brain-shots in elephants, one should take cognisance of the fact that an elephant's brain is also located low down in the skull. Good landmarks for the brain level are the bilateral protuberances caused by the 'cheek bones' (zygomatic arches). When aiming from the frontal position, another important factor is whether the head is in the neutral position or lifted alert/aggression position. In the neutral position, the aiming point is on a level with the eye line, whereas when the head is in the lifted alert/aggression position, the aiming point is lower, approximately on the second or third crease of the trunk. From a lateral position, the angle where the ear slit meets the cheekbone makes a good landmark for the position of the brain. For an oblique rear or rear shot, the back of the neck or behind the ear, at ear slit level, are good aiming points.

Elephant products are potentially valuable. In KNP only the blood and intestines were left in the field, and the rest used (Whyte, 2001). The ivory is the most valuable single product, but if the hides are properly treated they represent an almost equivalent value. All meat was used either for biltong or for a canned meat product. Biltong and canned meat were sold to tourists while the cans were also issued to field staff as rations. Excess carcass fat was rendered and sold to the cosmetic industry, while all other parts of the carcass were made into carcass meal for sale to the agricultural industry. Taking into account all of the potential income excluding ivory, 800 elephants would generate about

ZAR10 976 000 per year. This provides a profit of ZAR5 677 740, or just over ZAR7 000 per year per elephant (excluding recapitalisation of the abattoir). The addition of ivory to the income would effectively double the profit. Note that these figures have not been externally audited, and should be regarded as approximate.

North West Parks and Tourism Board conducted a feasibility study for culling of elephants in Madikwe Game Reserve. Based on their investigations, there is no local market for the elephant products, and carcasses would have to be transported to a major centre for processing (Pieter Nel, NWP&TB, pers. comm.). Based on transport of carcasses to Gauteng, and on culling groups of 10 elephants, the costs per culling were calculated at about ZAR4 000 per elephant (table 4). If a suitable processing plant could be found closer, in Mafikeng or Rustenburg, the costs would be reduced to about ZAR3 578 per elephant. The potential income from culls was not calculated.

THE VARIOUS MANAGEMENT CONTEXTS FOR CULLING

Overpopulation and population control

Elephants, being large, have populations limited by bottom-up effects (*sensu* Sinclair, 2003). In other words, their population number is expected to be limited by some underlying resource, most likely food. Top-down control, i.e. from predators such as humans, has been proposed as a historical elephant number-limiting mechanism by some (Cumming, 2007). Bottom-up control of population size manifests through density-dependent effects, where as the population increases, the per capita population growth declines (Sinclair, 2003; see also Coulson *et al.*, 2004; Owen-Smith *et al.*, 2006). Such density dependence has been found for elephants in the Serengeti (Sinclair, 2003). There is some evidence for density dependence in Kruger National Park (Van Aarde *et al.*, 1999), but it is tenuous. Errors in the population estimates can lead to spurious regressions of population change against population density, and migration between subpopulations may further confuse the results. Nevertheless, Van Aarde *et al.* (1999) concluded that: 'The method has small flaws, but it is contended that these will not obscure the general results.'

While it is inevitable that elephant numbers within a confined area must eventually reach a limit (Owen-Smith *et al.*, 2006), and thus that the net population growth rate must decline to zero at some maximum density, it is not certain that elephants in all areas will show a smoothly declining population growth rate as a function of population density, commencing at moderate elephant densities. For example, there is no evidence yet for density dependence in elephants at Addo (Gough & Kerley, 2007). The issue of density dependence is explored in greater depth in Chapter 2.

Costs (based on an annual cull of 800 elephants in Kruger):	
Item	Per year costs (ZAR)
Average daily helicopter costs (1.5 hours per day @ 2 200 per hour = 3 300 per day)	440 000
Transport truck @ 62 200 per month	311 000
Trailer for transport truck @ 21 000 per month	105 000
Tractor (x2) costs @ 38 826 per month each	388 260
Trailer (x2) @ 620 per month each	6 200
Mobile crane @ 21 160 per month	105 800
Ground crew transport @ 10 600 per month	53 000
Staff salaries (if all staff are SANParks)	3 009 000
Operating costs: salt, spices, cleaning materials, PPE, etc.	100 000
Abattoir costs: water, electricity, etc.	420 000
Abattoir maintenance	360 000
Hidden costs not included: operating costs of processing (e.g. canning); commissioning and capital costs.	??*
Total costs (excluding hidden costs)	**5 298 260**
Income (based on an annual cull of 800 elephants in Kruger):	
Item	Per year income (ZAR)
Meat products @ 4–11 per kg ('average ZAR5.00')	1 200 000
Hides (average 200 kg per elephant: last sold at 60 per kg dry salted)	9 600 000
Front feet (220 for front feet per elephant)	176 000
Carcass meal (sold at 'break even' prices)	0
Total income	**10 976 000**
Potential profit (income – cost, excluding some processing and capital costs)	**5 677 740**

* Commissioning and capital costs for the meat processing plant estimated at ZAR14 000 000 (see text).

Note: Above calculations exclude the sale of ivory (average of 5 kg per tusk per elephant, sold at an average price of ZAR750 per kg) = ZAR6 000 000

Table 3: Estimated costs and income from culling 800 elephants per year in Kruger National Park (Grant, 2005: 315)

The absence of natural density-dependent population controls at moderate population densities (i.e. densities lower than those which result in dramatic habitat transformation and starvation of the elephants and other species) may necessitate management intervention if such scenarios are to be avoided. Culling by removal of the annual population increment results in population numbers remaining approximately constant from year to year. Culling in excess of the population growth rate is the only viable mechanism by which populations can be reduced in size in the short term. The 'short term' means up to 5–15 years, depending on whether the animals in the population are mostly old or young. Because of the high cost of capture and translocation, and the increasing scarcity of receiving habitat (Chapter 5), this is not a viable alternative to culling where a near-immediate reduction in the population size is the objective. Contraception is a potential alternative to culling if population size reduction *per se* is not required (Chapter 6).

Item	Units	Cost per unit (ZAR)	Units	Total costs (ZAR)
Veterinarian travel	1	4	710	2 840
Pilot's travel	1	4	710	2 840
Helicopter costs	1	4 500	3	13 500
Equipment and drugs	2	500	2	2 000
Park staff				Not included
Veterinarian	1	350	5	1 750
Assistants	2	200	5	2 000
Tractor and trailer	1	8	60	480
Heavy duty truck and trailer	2	15	355	10 650
Bakkie 1 ton		4	60	720
Bakkie 1 ton (Gauteng return)		4	710	2 840
Total cost for 10 elephants				39 620
Cost per elephant				3 962

Note: Total cost/elephant if carcasses are transported to Mafikeng/Rustenburg = ZAR 3 578

Table 4: Estimated costs of culling a group of 10 elephants at Madikwe Game Reserve, North West Province, with carcasses transported to Gauteng for processing (source: unpublished estimates provided by Pieter Nel, North West Parks and Tourism Board)

The key issue involves how to define 'overpopulation' and ensure management actions are triggered by ecological indicators rather than elephant numbers (see Owen-Smith *et al.*, 2006). Historically, elephant target densities were set arbitrarily (Van Aarde & Jackson, 2007, Chapter 1).

Decision-making around elephant numbers has moved away from a simple maximum density-based approach to one using indicators from the environment (there is a historical review in Chapter 1, and see current examples of such decision-making processes in Chapter 12 and Slotow *et al.* (2003)). The main reasons for abandoning a single, constant maximum elephant density (sometimes characterised as a 'carrying capacity') are that (1) carrying capacity is not a constant in environments with a highly varying climate (McLeod, 1997); (2) a coupled plant–herbivore system, especially one involving long-lived plants and animals, may not smoothly reach a maximum, but may oscillate or alter to a new state at high herbivore density; (3) herbivore numbers are probably constrained by the availability of spatially or temporally restricted key resources rather than general conditions; and (4) there is insufficient quantitative information relating to elephants at high densities to set density-based estimates with any reliability in most circumstances.

The use of ecological indicators to drive decisions requires them to be spatially explicit. For a reserve the size of Kruger such indicators will vary regionally, and as such management may be implemented only in specific areas – that is, the total population size may not be relevant. Even for small reserves, population density *per se* presents a weak indicator for management intervention when it is based on extrapolation from other situations because elephant effects are governed by the specific ecology of each reserve (see Chapter 3). It is therefore preferable that local ecological indicators be used to trigger management interventions.

Population control should have specific objectives, and depending on the objectives, different approaches can be used to achieve them. The longevity of elephants should always be considered. There may be effective short-term approaches that have long-term negative consequences.

Approach 1: Culling adult males

The culling of males is an extremely inefficient and usually ineffective way of reducing the population size or growth rate. The reason is that the breeding herds continue to produce young at a rate faster than adult males can be culled (even if nearly all males are culled – Martin *et al.*, 1996). Removal of adult males should therefore occur for specific objectives unrelated to population size or growth, such as preventing crop-raiding, fence-breaking or excessive damage to trees. The adult bull elephants are disproportionately linked to the pushing over of large trees (see Chapter 3, and Midgley *et al.*, 2005), but not necessarily to ringbarking, another important cause of tree death. Tree-pushing is widespread among bull elephants, therefore there is reasonable cause to believe that a reduction in bull density will lead to a reduction (but not elimination) of tree damage. On the other hand, breakouts are not necessarily dependent on the *number* of males, since some individuals have a greater propensity to break out, so specific individuals would have to be targeted to reduce breakouts, or fencing would have to be improved/maintained (Chapter 7). Male elephants are more valuable in terms of trophy hunting (see Chapter 10, and consumptive use below), because of the larger size of their tusks, but trophy hunting, by itself, is an ineffective means of population control.

Targeting of adult males would, therefore, be warranted when an objective is to reduce the probability of a particular action that is most commonly attributed to adult male elephants, rather than to reduce the population *per se*. In the first comprehensive aerial census in Kruger in 1967, adult bulls comprised 15 per cent of the total population (Whyte, 2001). In order to maintain this proportion, the prescribed culling quota for bulls was always 15 per cent of the total cull, with the balance from breeding herds (Whyte, 2001). Selection of bulls for any particular cull was a random process and was based on location, group composition and individual characteristics (large tuskers were excluded from culls) (Whyte, 2001). Usually only two or three bulls were culled per day (Whyte, 2001), due to the limited capacity of the processing plant (Whyte, pers. obs.). The bull quota in Kruger was frequently used to address the problem of break-out elephants and other problem elephants (which were mainly bulls), as well as to reduce the numbers of tree-breaking individuals in designated 'botanical reserves' in the KNP (Bengis, pers. comm.).

Removal of individuals or small groups of males is relatively simple, and they can be shot from the air (Whyte, 2001) or the ground (Burke *et al.*, 2008). Such removals do affect the local population, but for a relatively short time (Burke *et al.*, 2008), and the elevated stress that results in the remaining animals is deemed acceptable (Burke *et al.*, 2008; and in the opinion of the authors).

There are a number of concerns about targeting only adult males. Firstly, larger (older) adult males are the more spectacular tourism animals, and may be the most important contributors of genes to following generations (Martin *et al.*, 1996). Secondly, selective removal of adult males over an extended period will result in a compression in age of the adult male population in the future due to simple numerical effects (see Milner *et al.*, 2007 for review). Such a compressed age-group is relatively young, and may swamp the dominance hierarchy within the adult male population. Distorted male age hierarchies have been implicated in abnormal behaviour in these younger males, such as elevated aggression, killing people, and killing rhino (Bradshaw *et al.*, 2005; Slotow *et al.*, 2000; Slotow *et al.*, 2001; Slotow & van Dyk, 2001). Thirdly, older males have greater reproductive success, and longevity may reflect fitness (Hollister-Smith *et al.*, 2007). Because of the above, any significant manipulation of the adult male population needs to be carefully considered, and should be a truly random process across all age classes if skewing of the age structure and the associated problems are to be avoided. Martin *et al.* (1996) also point out concerns over selective removal of males.

Approach 2: Removal of entire family groups (and associated males)

The only feasible manner of culling to reduce population growth rate (and thus population size, in the long term) is by removal of females. Where the total population size must be significantly and rapidly reduced, in practice this is best achieved by culling entire family groups, along with their associated young males (Whyte, 2001; Cumming & Jones, 2005). It has the further advantage of leaving no traumatised family members, although nearby (and possibly related) family groups may still be disturbed. Because of the complex social system of elephants, removal of entire family groups is considered the most ethical approach to population reduction (DEAT 2008; Martin *et al.*, 1996). In Kruger, all members of the selected group were culled regardless of sex or age class, unless young animals were to be translocated. This practice was terminated in 1994 (Whyte, 2001). Usually a daily cull would have comprised about 15 animals, with carcass recovery to a centralised abattoir being the limiting factor (Whyte, 2001). In Zimbabwe, up to 50 animals were killed by a team per day (Thompson, 2003), and carcasses were processed *in situ*, with the meat and skins taken to a temporary base in the protected area for cleaning and drying.

The approach is not, in fact, completely age and sex neutral, because the proportion of juvenile to older individuals tends to increase in harvested populations (see Gordon *et al.*, 2004 and references therein). In a modelling exercise, even if herds are removed randomly, the average age of the matriarch leading the group may decrease (Mackey & Slotow, unpublished manuscript).

Some degree of disruption of the complex social network is inevitable (McComb *et al.*, 2001; Wittemyer *et al.*, 2005; Wittemyer & Getz, 2007) with culling as it is with live removals. The consequences include long-term stress to the population (Gobush *et al.*, 2007). It is possible to remove female groups from relatively small populations. Family groups have apparently been successfully culled from small reserves in two cases (Steyn, pers. comm.; Van Altena, pers. comm.), although the consequences were not studied in detail. Groups have been successfully removed live from Phinda on four occasions and Hluhluwe-Umfolozi Park on three occasions (Slotow pers. obs.). In all cases there were no major disruptions to the animals that remained behind (Phinda: H. Druce (Elephant Monitor), pers. comm.; K. Pretorius (Reserve Manager), pers. comm.; Slotow, pers. obs.; Hluhluwe-Umfolozi Park: T. Burke (Elephant Monitor), pers. comm.; Slotow, pers. obs.). Five groups (Hofmeyr, pers. obs.; Slotow, pers. obs.) have been removed from Madikwe, but the consequences were not systematically studied.

Approach 3: Selective removal within family groups

Natural mortality from droughts or predation would in the first instance be just-weaned calves (Moss, 2001; Dudley *et al.*, 2001; Leggatt, 2003; Woolley *et al.*, 2008). Simulation studies, based on southern African data, indicate that episodic droughts with about a five-year frequency that resulted in 100 per cent mortality of just-weaned calves, or about a eight-year frequency that resulted in 85 per cent mortality of infants and weaned calves (0–7 years old), would lead to a zero net population growth (Woolley *et al.*, 2008). Mimicking natural processes by selective removal of young elephants has not been attempted (Cumming & Jones, 2005), although it has been considered in a number of reserves (Goodman, pers. comm.).

The most efficient means of reducing future population growth is removal of young adult females (Van Aarde *et al.*, 1999; Woolley *et al.*, 2008), where efficiency is defined as minimising the total number of individuals that need to be culled to achieve a given reduction in the population growth rate. Population models indicate that removing an annual number of prepubertal females equivalent to just 2 per cent of an elephant population would stop its growth (Whyte *et al.*, 1998; Van Aarde *et al.*, 1999), compared to the 6 per cent that would need to be removed if an age-and-sex neutral approach were to be used.

Removal of selected young individuals (as opposed to other herd members) from a herd on a periodic basis would simulate 'natural mortality' of elephants from drought or predation, and could be relatively easily implemented from the ground using a rifle (possibly silenced) or a lethal dose of drugs. The stress caused to the remainder of the family is unknown. It could cause major gaps in age classes (Martin *et al.*, 1996), and there are ethical considerations (Chapter 9). This approach precludes natural selection from acting on particular individuals; it would disrupt more different herds than removal of entire herds, and like contraception, would not lead to a significant immediate reduction in elephant impacts.

The Norms and Standards precludes removal of individuals from a breeding herd: 'an elephant may not be culled if it is part of a cow-calf unit unless the entire cow-calf unit, including the matriarch and juvenile bulls, is culled' (DEAT, 2008), so this is no longer an option for management.

Synthesis of approaches

The relative effectiveness of the different approaches to solving particular management problems are summarised in table 5. Each approach has merits, depending on the objective. Note that culling is generally not the only potential management intervention to solve a particular problem. The process of evaluating the relative merits of various alternatives is dealt with in Chapter 12.

Elephant population age–sex structure

The age–sex structure of a population is important for two reasons (see also Milner *et al.*, 2007 for further discussion). Firstly, when specific age classes are missing from a population, behavioural abnormalities can occur (e.g. Slotow *et al.*, 2000). Secondly, future population growth is governed by the current population structure. A relatively young population displays a higher growth rate than an older population, and this effect persists for an extended period, while the population comes to a stable-state distribution (e.g. Mackey *et al.*, 2006; Mackey & Slotow, unpublished manuscript). What might be considered a normal age–sex structure is dealt with separately in Chapter 2.

The tendency of culled populations to 'overshoot' their natural limits once culling has been discontinued is termed 'eruptive growth'. An eruptive population temporarily exceeds its ecological carrying capacity – that is, the long-term limit imposed by its key limiting resource (Caughley, 1970) – causing a decline in that resource, and subsequently in the population itself (Caughley, 1970). Density dependence does not act to constrain an eruptive population as it would a population with a stable age-structure (Mackey & Slotow, unpublished manuscript). Eruptive growth is related to the population response inertia introduced by the age-structure distortions discussed above. There is no field-based evidence that eruptive growth does or does not occur in elephant populations, since this question has not been adequately addressed for any post-culling elephant population.

There is also no reason why elephants should behave differently from other herbivores, where eruptive growth has been widely observed (see for example Forsyth & Caley, 2006 for population models of several species showing eruptive growth). What is evident is that the formerly culled populations in both Kruger and Zimbabwe have grown at near-maximum rates since culling was suspended (Van Aarde & Jackson, 2007).

It must also be noted that the elephant population in Kruger is still in a growth phase from the deep reductions at the end of the nineteenth century, and thus the age structure was probably not a stable one during the period of culling. Although there is currently no compelling evidence available to assess this, the possibility of eruptive growth, either from the 'founding' effects of the Kruger population, or as a consequence of culling, cannot be eliminated. The relatively high population growth rates recorded in Kruger in the decade since culling ceased may simply be because this population is far from its resource limitation level (see Owen-Smith *et al.*, 2006 for a discussion of the ecological carrying capacity of elephants in Kruger).

Many of the problems associated with skewed age-sex structure have manifested in small populations. Many of the smaller reserves have a population structure clearly biased towards young adults (Slotow *et al.*, 2005), and do display abnormally high growth rates (Mackey *et al.*, 2006), which may in time show the characteristics of eruptive growth. Until 1993, the founder populations in all the elephant reintroduction areas consisted of young (<8 years old) elephants (Slotow *et al.*, 2005; Whyte, 2001). Since 1993, elephants have been moved as family groups including adult females, and since 1998 large adult males (>35years old) have been included as well (Slotow *et al.*, 2005; Whyte, 2001).

The best-known behavioural abnormality resulting from a skewed age-structure was the killing of white and black rhino by young male elephants in Pilanesberg, Hluhluwe-Umfolozi, and other reserves (Slotow *et al.*, 2000; Slotow & van Dyk 2001; Slotow *et al.*, 2001). These males had matured in the absence of an older male hierarchy, and were entering musth much earlier, and for much longer, than normal (Slotow *et al.*, 2000). This problem was corrected by the introduction of older males into the population at Pilanesberg in 1998 (Slotow *et al.*, 2000) and Hluhluwe-Umfolozi in 2000 (Slotow *et al.*, unpublished data). Subsequently, older males have been introduced to most populations founded by young elephants, and the problem of elephants killing rhino has been largely negated, although there are occasional mortalities from time to time in various reserves (Slotow, unpublished data).

Female elephants have been displaying an abnormal amount of aggression in circumstances where they would normally retreat (Slotow, unpublished data). This has resulted in human deaths in at least five small reserves in South Africa (Slotow, unpublished data). This behaviour may be due to female elephants maturing in the absence of an older experienced matriarch, and thus not learning how to behave in threatening circumstances. These females show abnormal responses when experimentally tested relative to females from a normally age-structured population (Shannon *et al.*, unpublished data). Alternatively, the behaviour may reflect long-term (>12 years) effects of elevated stress levels associated with capture and introduction into a new environment. Learnt abnormal behaviour related to social disruption will almost definitely require lethal control of those individuals responsible.

The problems displayed in smaller reserves may also be present in larger populations such as Kruger National Park, where the age-structure may have been somewhat affected by long-term culling despite the policy of age- and sex-specific neutral removals (see above), although there is little evidence to suggest that this is the case. In the past 10 years there have on average not been more than one incident of elephant–human conflict per year in the Kruger, with three incidents in 2007, and three elephants being culled after charging staff in the 12 months prior to October 2007 (data extracted from Kruger internal diaries and reports housed at Skukuza archives by M. Hofmeyr). No culling-related significance can be attributed to the recent apparent increase; it may simply be a result of increased numbers of elephants and tourists.

It is very difficult to define a 'normal' population age–sex structure, as the population structure varies somewhat under natural circumstances. However, the complete absence of a particular age–sex class (e.g. older male or female elephants) is definitely abnormal and results in problems. The preponderance of young breeding individuals, for example 10- to 20-year old females, will cause above-average growth rates until that age class stops breeding (i.e. only after >50 years; see discussion of population consequences below).

Problem animal control

Problem animal control occurs both inside and outside of reserves, although the focus tends to be on elephants that have broken out of reserves (DEAT, 2008). The traditional method of control is to shoot the culprits (Hoare, 1995). This represents symptomatic relief rather than a long-term solution to the problem (Hoare, 1995). Ongoing killing of problem animals on the periphery of protected areas may erode the quality of the remaining animals in terms of trophies (Hoare, 1995) and genetic diversity (Martin *et al.*, 1996). A number of alternatives to problem animal control are being investigated (e.g. better fencing, *Capsicum* repellents – Osborne & Rasmussen, 1996).

Male problem animals outside reserves tend to be alone or in small groups, and can be relatively easily controlled (Chapter 4). However, when the animal is still within the reserve, identification of the offending individual can be difficult because of the lag between reporting the incident and locating the elephant.

Such uncertainty has resulted in more than one animal being killed in order to ensure the culprit was killed (e.g. at Pilanesberg when three males were killed to ensure that the one that was killing rhino was eliminated – Slotow & Van Dyk, 2001). Alternatively, the culprit is not killed because the specific elephant could not be identified (e.g. at Hluhluwe-Umfolozi, from a group of four elephants that killed rhino, Slotow, pers. obs.).

Management objective	Random removal across population Effectiveness		Random removal of complete family groups Effectiveness		Selective removal of young individuals from family groups Effectiveness		Selective removal of males Effectiveness	
	Short	Long	Short	Long	Short	Long	Short	Long
Reduce impact on large trees	2	1	3	3	4	4	1	2
Reduce biomass removal	1	3	2	2	4	1	3	4
Problem animal control	2	2	2	2	4	4	1	1
Population reduction (for whatever reason)	2	3	2	2	3	1	4	4
Correcting eruptive growth	4	4	4	4	1	1	4	4
Correcting age–sex problems (too many males)	4	4	4	4	4	4	1	1
Correcting age–sex problems (population young)	4	4	4	4	3	2	4	4
Correcting age–sex problems (too many breeding females)	4	4	4	3	4	4	4	4

Rankings: rank 1 = most effective; rank 4 = least effective.
Short-term indicates a period up to 5 years. Long-term indicates a period >5 years.

Table 5: Relative effectiveness of different culling approaches in achieving specific management objectives

Problem animal behaviour has been observed to escalate within an individual over time, as the elephant becomes habituated to humans and their infrastructure (Slotow, pers. obs., from Pilanesberg, Hluhluwe-Umfolozi, Mkhuze). Individuals progress from pulling up pipes, to damaging cars, to killing people. Accordingly, one approach that is being used in populations where each individual can be identified is to create 'rap sheets' (records of 'bad' behaviour that can be attributed to identified individuals in the population) of problem animals, and to identify repeat offenders for subsequent removal (Slotow, unpublished information).

Not all problem animals are male, and particularly within some reserves, female elephants may be posing a greater risk than males. It is very difficult to remove a single female from the population, either because of difficulties of identification, or because of the stress placed on the remaining animals. This has been successfully done in Greater St. Lucia Wetland Park, where a number of female elephants were suspected of being responsible for an unprovoked attack on a vehicle, which resulted in the death of people. All the adult females in the group were sequentially immobilised and the culprit identified by metal scrape marks and chips on the tusks (Conway, pers. obs.).

Inside Kruger, problem elephants are rare, usually animals that have attacked people or consistently broken into rest camps (Whyte, pers. obs.). Culprits are sometimes identified by the tear patterns of ears or tusk shape or marked with paint ball guns. Where the individuals can be positively identified, they are usually brain-shot using heavy calibre rifles from a helicopter, or else shot by the local Ranger on foot from the ground (Whyte, pers. obs.).

When elephants break out of Kruger, or other protected areas, attempts are usually made to drive them back inside (Whyte, pers. obs.). However, once outside of a National Park, legislation dictates that they become the responsibility of the provincial conservation agencies (see Chapter 11). Often they are shot from a helicopter or from the ground by staff of these agencies (Whyte, pers. obs.).

The situation around Kruger (and potentially at Mapungubwe, Madikwe, Pongola and Tembe) is complex because the boundaries of the park stretch across international and provincial borders. The control and management of problem animals is covered by different legislation in each jurisdiction (Chapter 11). Currently, if an elephant leaves the Kruger boundaries, ranger staff from Kruger will respond only when specifically requested by the provincial authorities of Limpopo Province or Mpumalanga Province (Hofmeyr, pers. obs.). In Mpumalanga most problem animal cases are handled directly by Provincial staff. In Limpopo Province the ranger staff are frequently asked to help out with destroying elephants that are causing problems in community areas. The situation is more distinct with the international boundaries, and Kruger staff do not follow up elephants that leave the park into Mozambique or Zimbabwe. Only if there was a direct request, and in collaboration with the relevant authorities in those countries, would Kruger staff help with elephant problems. The human population is very sparse along both the Zimbabwe and Mozambique boundaries so such requests have not been made.

Smaller reserves in South Africa have well-maintained fences and there have been relatively few break-outs from such reserves (Slotow, pers. obs.). Within these reserves problem animal control will be primarily to protect humans, other species, and infrastructure on the reserve. There have been no break-outs of elephants in North-West Province, but there have been break-outs from at least eight small reserves in KwaZulu-Natal and three in Limpopo Provinces (Slotow, unpublished data). Ezemvelo KwaZulu-Natal Wildlife routinely either chased elephants back, or immobilised and transported them back into the reserve (Slotow, pers. obs.).

Disturbance culling to move elephants around the landscape

One possible application for culling is to reduce the density of elephants in a particular sub-area of the range. The effect could be direct (by removal of animals) or indirect (by making an area less attractive to elephants through the culling-related disturbance). Such disturbed elephants – that is, elephants with experience of a cull – may move out of the area, resulting in a lower localised density. The trade-off between risks and benefits of disturbance culling need to be carefully evaluated.

In Kruger, when culls were compartmentalised into sections of the park, regional population totals varied in response (Whyte *et al.*, 1999). The year after a cull in a region, population totals declined more than could be accounted for by culling alone, while a year later, regional population totals increased in excess of that possible by reproduction alone. Movements in and out of these regions in response to culls must have been responsible for the trends (Van Aarde *et al.*, 1999; Whyte *et al.*, 1998). These results imply that localised culling results in some movement out of the affected area by the remaining elephants. The regional culling boundaries in Kruger did not conform to the elephant clans' home range boundaries, and such movements could have been within the normal home range of those elephants (Whyte, 2001). Elephants may return to the culling area at a later stage, and an analysis based on home ranges rather than on arbitrary logistical boundaries might have yielded rather different results (Whyte, pers. obs.).

In Hwange, elephant numbers remained low in areas where culling had taken place for up to two to three years after a cull (Cumming, 1981), indicating that there may have been some localised disturbance effect. These culls were in areas towards the edge of the park. There is no indication that elephants moved out of the area, but the implication was that elephants did not move into the area, at least over the medium term. Elephants have home ranges which they will defend against non-clan members with certain antagonistic behaviours (Moss, 1988). Thus one would expect that where a decline (but not complete removal) of a 'clan' has been effected, a replacement of the population through immigration is unlikely. When the western boundary fence of the KNP was removed, there was little or no increase in the elephant populations of the Klaserie and Timbavati Private Nature Reserves, where established populations of elephants already existed. In the Sabi-Sand, however, only a small population was present (mainly bulls with a small translocated herd of young females and males). This area was quickly colonised once the fence was removed (Whyte, 2007).

Prior to the erection of the game fence in Sebungwe (Zimbabwe), massive culling in 1968 in the area to the south of the fence caused 250 animals to move north of the fence (Cumming, 1981).

In Kasungu (Malawi), poaching occurred along one side of the park, which acted as a sink (Bell, 1981). Most of those killed were males (Bell, 1981), and the implication is that additional males were continually moving into the area. A radio-collared male elephant that was near an animal killed in agricultural fields adjacent to a national park in Zimbabwe returned to the agricultural fields within five days (Hoare, 2001). This 'disturbance culling' was clearly not discouraging elephants from moving into the area of risk.

During the 1960s (particularly after 1966), attempts were made to keep elephants out of the Bunyoro Forest (Uganda) through disturbance culling ('chasing elephants out of young regeneration areas' or 'keeping elephants out of the forest' (Laws, 1974)). Although initially successful, the programme failed because elephants modified their behaviour by entering the forest at night to simply avoid the disturbers working during the day (Laws, 1974). 'As well as causing considerable disturbance, the control shooting over the years has been wasteful, uneconomic and grossly inhumane' (Laws, 1974).

It may be that elephants do not avoid areas of high risk, but rather change their behaviour in response to the risk. Elephants in Tarangire, Tanzania, adopted different day-time (high-risk) and night-time (low-risk) behaviour, foraging while walking slowly at night, and 'streaking' through dangerous communal farming areas during the day (Douglas-Hamilton *et al.*, 2005).

Hunting elephants

Consumptive use is normally a by-product of culling or problem animal control. Although sustainable use is part of the mission statement and governing legislation of most conservation agencies in South Africa, the use of elephants (or any other species) is not a primary management objective in national parks administered by SANParks. However, elephants have been hunted in the Makuleke region (a community-owned area within the greater Kruger boundary, and co-managed by SANParks) and Pilanesberg National Park, which is managed by North West Parks and Tourism Board (table 2). Hunting has also occurred in the Tembe Elephant Park (owned by the Tembe people, but managed by EKZN Wildlife) and in Mpumalanga Parks Board reserves. Hunting also occurs in some privately owned reserves (table 2).

Hunting of bull elephants is a straightforward process when conducted professionally by experienced personnel, and has only a minor effect on the elephants that remain behind (Burke *et al.*, 2008). The Norms and Standards (DEAT, 2008) prescribes the manner in which elephants may be hunted, and excludes females from being hunted unless vagrant. 'Green hunts' (in which a client pays to fire a non-lethal immobilising dart into the elephant) are prohibited in terms of the Threatened and Protected Species Regulations (2008).

SHORT- AND LONG-TERM CONSEQUENCES OF CULLING

Short-term effects on remaining elephants (stress)

Disturbance causes stress, and an important consideration prior to culling is an assessment of the impact that culling will have on the animals that remain behind. 'It is naive to believe that, if an entire herd is killed, the remainder of the population know nothing about the event' (Martin *et al.*, 1996). This must also apply to translocations, as the disturbance and impact on animals remaining will be almost identical. Herds may be able to signal their distress over considerable distances using infrasound. Radio-collared elephants in Sengwa, Zimbabwe, occasionally visited cull-sites shortly after the event (Martin *et al.*, 1996). Elephants in an area adjacent to Hwange were inferred to have responded to culling 150 km away by disappearing into the bush, and were later found at the opposite end of the reserve, bunched up (Moss, 1988).

Cumming & Jones (2005) indicate that culling does result in disturbance, and conclude that major interventions involving 'more than one team operating in a culling season may result in unacceptable levels of disturbance'. They do not provide the basis for this statement, but the implication of the quote is that such disturbance is unacceptable because of the stress imposed on the elephant population. After major culling operations, elephants in Hwange may spend less time at waterholes during the daytime, but their behaviour towards tourists appeared not to alter (Martin *et al.*, 1996). Further, Whyte (1993) initiated a study of the effects of culling on elephants in Kruger because of concerns based on reports from rangers that after culls elephants 'disappeared from the area'. This effect was not based only on the helicopter actions, because Whyte (1993) noted that when elephants were immobilised from helicopters they did not react in the same manner.

Whyte (1993) observed the behaviour of collared female elephants that were within 7 km of a culled group. Four out of the 10 females 'reacted to the cull by undertaking significant movements', including, for example, direct movement of 23, 25, and 30 km overnight or within two days of the cull (Whyte, 1993). Such movements even took the elephants outside of their previously determined home ranges (Whyte, 1993), but subsequent observations suggest that they had moved to another part of their home range (Whyte, pers. obs.). The response was not consistent among all females studied – the other six showed a weak or no response.

The above results suggest that there is substantial short-term stress on the elephants, but none of the studies quantified stress directly. The effect of hunting on stress in the elephants that remain behind was studied in Pilanesberg, using both behavioural and hormonal assays (Burke *et al.*, 2008). The effect of immobilisation of a single female from 12 different herds was also examined (Burke, 2005). There were physiological and behavioural stress effects lasting about four days from any hunting or immobilisation event. The effect was apparent in corticosterone levels for longer than from behavioural assays (Burke *et al.*, 2008). Although there was increased stress throughout the population during these disturbances, the levels of corticosterone were still much lower than those associated with natural stress events, for example during thunderstorms (Millspaugh *et al.*, 2007).

Long-term elephant behavioural consequences

It is uncertain how the observed short-term responses may translate into longer-term consequences (but see Bradshaw & Schore, 2007). Very little information on the longer-term consequences of culling is available. Although some reports were received from staff and tourists that elephants were aggressive in areas where elephants had been culled in Kruger, no quantified data exist to support or refute this (Whyte, 2001). The statement in Cumming & Jones (2005) about the disturbance of culling clearly implies short-term consequences, but it is not clear if there are longer-term consequences. Whyte (1993) concluded that Kruger elephants react more to helicopters during culling events than during translocation events, but the sample was small and the study was prematurely discontinued as culling was terminated in 1994. Elephants in Chizarira (Zimbabwe) now associate aircraft with danger, and flee (Martin *et al.*, 1996).

Gobush *et al.* (2007) represents the only study to assess the long-term consequences of poaching and culling. The elephants at Mikumi National Park (Tanzania) experienced heavy poaching prior to the ivory ban, but pressure has declined since then (Gobush *et al.*, 2007). These elephants still show elevated stress levels over a decade later, potentially from a loss of kinship in the socially complex matrilineal network (Gobush *et al.*, 2007; see also McComb *et al.*, 2001; Wittemyer *et al.*, 2005).

Culling results in a loss of cultural information and experience from the population, especially if older individuals are targeted (McComb *et al.*, 2001). It also results in trauma associated with the culling (Bradshaw *et al.*, 2005). 'Calves witnessing culls, and those raised by inexperienced mothers are high-risk candidates for later disorders, including an inability to later regulate stress-reactive aggressive states' (Bradshaw, 2005). The Norms and Standards (DEAT, 2008) does not allow for capture of juveniles during culls. This is considered inhumane: juveniles should not witness the culling of their families. There have been human deaths from young female matriarchs in five small reserves in South Africa (Slotow, unpublished data), and younger matriarchs respond differently and less appropriately than older females (both in Pilanesberg and Amboseli) to unfamiliar elephant calls or lion roars (Shannon *et al.*, unpublished data). The effects of human-caused disruptions on neuro-endocrinological development of elephants, and subsequent non-normative behaviour, has been recently reviewed by Bradshaw & Schore (2007).

We should not expect elephants to live a completely stress-free life, and they have evolved to deal with a certain degree of stress. Top predators, including humans, hunted elephants throughout their evolutionary history, but modern technology allows humans to impose a more intensive disturbance event, of a scale unlikely in the evolutionary history of elephants. This generates concern about long-term effects on elephants from both a welfare and human risk perspective. There is a need to differentiate between evolutionary and ecologically selected stress responses (i.e. to which an animal such as an elephant is adapted), and artificial disturbances (such as mass culling) to which elephants would not have had the opportunity to adapt (Bradshaw, pers. comm.). In terms of the latter, there may not be adequate coping mechanisms to diffuse or ameliorate stress (Bradshaw, pers. comm.), and such unnatural trauma may have fundamental consequences.

Long-term elephant population consequences

When culling is implemented in strict proportion to the existing population age-sex structure, as has been the general practice to date (Whyte, 2001; Cumming & Jones, 2005), the intention is to retain the current structure, albeit in lower numbers. The reduction in population size will depend on the number of animals removed. The problem occurs once culling stops. Once a population is released from culling, even if the culling is age-and-sex neutral, it enters a growth phase that inevitably leads to an age distribution somewhat skewed toward younger individuals (e.g. Red Deer – Coulson *et al.*, 2004; Pronghorn antelope – White *et al.*, 2007; see also Mackey & Slotow, submitted manuscript). There are a number of reasons for this. The reduction in density leaves the underlying resources on which the population depends intact, thereby increasing the amount of resources available to each individual (Caughley, 1983), which then grows and breeds at a maximal rate. Culling automatically boosts the potential population growth rate of the population (Caughley, 1983). This effect persists for at least one generation (all newly born animals age and die), after which a new stable age-structure will gradually establish itself. The consequences of culling Red Deer included demographic and spatial effects that persisted for 30 years – almost four generations – after culling halted (Coulson *et al.*, 2004). Although the long-term consequences of culling have only been modelled in elephants, given their very long intergeneration times, the possibility exists that such effects may at least in theory persist up to a century. They could potentially be mitigated by applying age-and-sex neutral culling. Eruptive growth is exacerbated if older individuals are selectively targeted.

Eruptive growth overshoot could be minimised by artificially 'aging' the population, for example by contraception or selective culling of younger age-sex classes (Mackey & Slotow, unpublished manuscript). If the age-class distortion is effectively countered, the mechanism of density dependence may prevent population overshoot (Mackey & Slotow, unpublished manuscript). There are ethical and practical concerns about culling specific age-sex classes from within breeding groups (Martin *et al.*, 1996, see section above on selective culling), and such an approach is specifically excluded in the Norms and Standards (DEAT, 2008).

GAPS IN OUR KNOWLEDGE

Key gaps in our knowledge that emerge from above are:

1. Incomplete information on the population structure and dynamics of important elephant populations as they approach high-density limits. Such information would allow robust predictions of the population consequences of culling or not culling.
2. A lack of systematic information regarding problem animals. Monitoring and assessment of incidents will allow the identification of patterns and underlying causes, which will provide the basis for any management interventions.
3. Uncertainty as to what indicators should be used to trigger management intervention to reduce population numbers, and what the critical thresholds are.
4. The consequences of selective removal of particular age or size classes from the population from demographic, behavioural, social, economic, and genetic perspectives, particularly in small reserves, and how to mitigate undesirable outcomes.
5. The long-term viability and practicality of low-density areas created by disturbance culling.
6. Scientific studies, preferably including both behavioural and hormonal assays, on the disturbance effects and consequences of culling, e.g. for elephant social networks.

CONCLUSIONS

The shooting of identified individual elephants is the main means to control problem animals in the short term, but avoidance of problem-creating situations is a better long-term strategy. Culling is the only realistic mechanism to reduce population size in the short term if this is necessary to achieve specific management objectives. The technical aspects of culling and problem animal control are well understood and it is feasible in South Africa to cull on whatever scale may be needed.

There are, however, a number of issues that have emerged from this assessment relating to the elephants that remain behind after culling.

Firstly, possible resultant elevated population growth rates resulting from skewed age and sex structures could create long-term problems requiring ongoing management. Secondly, the behavioural consequences are uncertain, but this assessment indicates that they are potentially substantial, raising welfare concerns.

Our current understanding of culling relates mostly to what has happened in Kruger and Zimbabwe. There may be a more immediate need for culling intervention on smaller spatial scales and in smaller populations, for which our understanding is less developed.

REFERENCES

Bengis, R.G. 1996. Elephant population control in African National Parks. *Pachyderm* 22, 83–86.

Bell, R.H.V. 1981. An outline of a management plan for Kasungu National Park, Malawi. In: P.A. Jewell & S. Holt (eds) *Problems in management of locally abundant wild mammals.* Academic Press, New York, 69–89.

Buechner, H.K., I.O. Buss & W.M. Longhurst 1963. Numbers and migration of elephants in Murchison Falls National Park, Uganda. *Journal of Wildlife Management* 27, 36–53.

Bradshaw, G.A., A.N. Schore, L.B. Brown, J.H. Poole & C.J. Moss 2005. Elephant breakdown. *Nature* 433, 807.

Bradshaw, G.A. & A.N. Schore 2007. How elephants are opening doors: Developmental neuroethology, attachment and social context. *Ethology* 113, 426–436.

Burke, T. 2005. The effect of human disturbance on elephant behaviour, movement dynamics and stress in a small reserve: Pilanesberg National Park. M.Sc. thesis, University of KwaZulu-Natal, Durban.

Burke, T., B. Page, G. van Dyk, J. Millspaugh & R. Slotow 2008. Risk and ethical concerns of hunting male elephant: behavioural and physiological assays of the remaining elephants. *PLOS One* 3 (6), e2417.doi: 10,137/journal0, 1371/journal.pone.00024 0002417.

Caughley, G. 1970. Eruption of herbivore populations, with emphasis on Himalayan Tar in New Zealand. *Ecology* 51, 53–72.

Caughley, G. 1976. The elephant problem – an alternative hypothesis. *East African Wildlife Journal* 14, 265–283.

Caughley, G. 1983. Dynamics of large mammals and their relevance to culling. In: R. Owen-Smith (ed.) *Management of large mammals in African conservation areas.* HAUM Educational Publishers, Pretoria.

Coulson, T, F. Guinness, J. Pemberton & T. Clutton-Brock 2004. The demographic consequences of releasing a population of red deer from culling. *Ecology* 85, 411–422.

Cumming, D.H.M. 1981. The management of elephant and other large mammals in Zimbabwe. In: P.A. Jewell & S. Holt *Problems in management of locally abundant wild mammals*. Academic Press, New York, 91–118.

Cumming, D.H.M. 2007. *Of elephants, predators and plants in protected areas: A case of classic trophic cascades?* (Abstract.) 21st Annual Meeting of the Society for Conservation Biology, Nelson Mandela Metropole, South Africa.

Cumming, D. & B. Jones 2005. *Elephants in southern Africa: management issues and options*. WWF-SARPO Occasional Paper Number 11, Harare.

DEAT 2008. National Norms and Standards for the management of elephants in South Africa. Department of Environment and Tourism. *Government Gazette*, 8 May 2008.

Douglas-Hamilton, I., T. Krink & F. Vollrath 2005. Movements and corridors of African elephants in relation to protected areas. *Naturwissenschaften* 92, 158–163.

Druce, H., K. Pretorius & R. Slotow The response of an elephant population to conservation area expansion: Phinda Private Game Reserve, South Africa, unpublished manuscript.

Dudley, J.P., G.C. Criag, D. St. C. Gibson, G. Haynes & J. Klimowicz 2001. Drought mortality of bush elephants in Hwange National Park, Zimbabwe. *African Journal of Ecology* 39, 187–194.

Forsyth, D.M. & P. Caley 2006. Testing the irruptive paradigm in large-herbivore dynamics. *Ecology* 87, 297–303.

Garaï, M.E., R. Slotow, B. Reilly & R.D. Carr. 2004. History and success of elephant re-introductions to small fenced reserves in South Africa. *Pachyderm* 37, 28–36.

Glover, J. 1963. The elephant problem at Tsavo. *East African Wildlife Journal* 1, 30–39.

Gobush, K.S., B.M. Mutayoba & S.K. Wasser 2007. *Long-term consequences of poaching on relatedness and physiological health of African elephants*. (Abstract.) 21st Annual Meeting of the Society for Conservation Biology, Nelson Mandela Metropole, South Africa.

Gough, K.F. & G.I.H. Kerley 2007. Demography and population dynamics of the elephants *Loxodonta africana* of Addo Elephant National Park, South Africa: Is there any evidence for density-dependent regulation? *Oryx* 40, 434–431.

Gordon, I.J., A.J. Hester, & M. Festa-Bianchet 2004. The management of wild large herbivores to meet economic, conservation and environmental objectives. *Journal of Applied Ecology* 41, 1021–1031.

Grant, C.C. (compiler) 2005. *Elephant effects on biodiversity: An assessment of current knowledge and understanding as a basis for elephant management in SANParks.* SANParks Scientific Report 03/2005.

Hattingh, J., P.G. Wright, V. de Vos, S.T. Cornelius, M. Silove, I.S. McNarin, M.F. Ganhao & G. Wolverson 1984a. Blood composition in culled elephants and buffalo. *South African Journal of Science* 80, 133–134.

Hattingh, J., P.G. Wright, V. De Vos, L. Levine, M.F. Ganhao, I.S. McNairn, A. Russel, C. Knox, S.T. Cornelius & J. Bar-Noy 1984b. Effects of etorphine and succinyldicholine in blood composition in elephant and buffalo. *South African Journal of Zoology* 19, 286–290.

Hattingh, J., N.I. Pitts, M.F. Ganhao & V. de Vos 1990a. The responses of elephant and buffalo to succinylmonocholine. *South African Journal of Science* 86, 546.

Hattingh, J., N.I. Pitts, M.F. Ganhao, D.G. Moyes, & V. De Vos 1990b. Blood constituent responses of animals culled with succinyldicholine and hexamethonium. *Journal of the South African Veterinary Association* 61(3), 117–118.

Hoare, R. 1995. Options for the control of elephants in conflict with people. *Pachyderm* 19, 54–63.

Hoare, R. 2001. Management implications of new research on problem elephants. *Pachyderm* 30, 44–48.

Hoffman, M.T. 1993. Major P. J. Pretorius and the decimation of the Addo elephant herd in 1919-1920: important reassessments. *Koedoe* 36, 23–44.

Hollister-Smith, J.A., J.H. Poole, E.A Archie, E.A. Vance, N.J. Georgiadis, C.J. Moss & S.C. Alberts 2007. Age, musth and paternity success in wild male African elephants, *Loxodonta africana. Animal Behaviour* 74, 297–296.

Laws, R.M. 1974. Behaviour, dynamics and management of elephant populations. In: V. Geist & F. Walther (eds) *The behaviour of ungulates in relation to management.* IUCN, Morges, Switzerland.

Leggatt, K. 2003. *The effect of drought and low rainfall on elephant populations.* Proceedings of the Norwegian Bonnic Programme Workshop, March 2003. Kasane, Botswana.

Mackey, R.L., B.R. Page, K.L. Duffy & R. Slotow 2006. Modelling elephant population growth in small, fenced, South African reserves. *South African Journal of Wildlife Research* 36, 33–43.

Mackey, R.L. & R. Slotow. Controlling irrupting populations: A theoretical comparison of contraception vs. culling using African elephant as case study. Unpublished manuscript.

Martin, R.B., G.C. Craig & V.R. Boot 1996. Elephant management in Zimbabwe. Department of National Parks and Wildlife Management. Harare, Zimbabwe, 158.

McComb, K., C. Moss, S.M. Durant, L. Baker & S. Sayialel 2001. Matriarchs as repositories of social knowledge in African elephants. *Science* 292, 491–494.

McLeod, S.R. 1997. Is the concept of carrying capacity useful in varying environments? *Oikos* 79, 529–542.

Midgley, J., D. Balfour & G.H.I. Kerley 2005 Why do elephants damage savanna trees? *South African Journal of Science* 101, 213–215.

Millspaugh, J.J., T. Burke, G. van Dyk, R. Slotow, B.E. Washburn & R. Woods 2007. Stress response of working African elephants to transportation and Safari Adventures. *Journal of Wildlife Management* 71, 1257–1260.

Milner, J.M., E.B. Nilsen & H.P. Andreassen 2007. Demographic side effects of selective hunting in ungulates and carnivores. *Conservation Biology* 21, 36–47.

Moss, C.J. 1988. *Elephant memories: Thirteen years in the life of an elephant family.* First edition. Fontana/Collins, Glasgow.

Moss, C.J. 2001. The demography of an African elephant (*Loxodonta africana*) population in Amboseli, Kenya. *Journal of Zoology, London* 255, 145–156.

Osbourn, F.V. & L.E.L Rasmussen 1996. Evidence for the effectiveness of an Oleo-Resin capsicum as a repellent against wild elephant in Zimbabwe. *Pachyderm* 21, 55–64.

Owen-Smith, N., G.I.H. Kerley, B. Page, R. Slotow & R.J. van Aarde 2006. A scientific perspective on the management of elephants in the Kruger National Park and elsewhere. *South African Journal of Science* 102, 389–394.

Pretorius, P.J. 1947. *Jungle man.* Harrap & Co, London.

SANParks 2006. Kruger National Park: Park Management Plan. Draft. SANParks, Skukuza.

Sinclair, A.R.E. 2003. Mammal population regulation, keystone processes and ecosystem dynamics. *Philosophical Transactions of the Royal Society of London* B358, 1729–1740.

Slotow, R, G. van Dyk, J. Poole, B. Page & A. Klocke 2000. Older bull elephants control young males. *Nature* 408, 425–426.

Slotow, R., D. Balfour & O. Howison 2001. Killing of black and white rhinoceros by African elephant in Hluhluwe-Umfolozi Park, South Africa. *Pachyderm* 31, 14–20.

Slotow, R. & G. van Dyk 2001. Role of delinquent young 'orphan' male elephants in high mortality of white rhinoceros in Pilanesberg National Park, South Africa. *Koedoe* 44, 85–94.

Slotow, R., B. Reilly, P. Owen & R. Kettles 2003. *A generic management plan for African elephant.* Elephant Managers and Owners Association, Vaalwater.

Slotow, R, M.E. Garaï, B. Reilly, B. Page & R.D. Carr 2005. Population ecology of elephants re-introduced to small fenced reserves in South Africa. *South African Journal of Wildlife Research* 35, 23–32.

Thompson, R. 2003. *A game warden's report: The state of wildlife in Africa at the start of the third millennium.* Magron Publishers, Hartbeespoort, South Africa.

Van Aarde, R., I.J. Whyte & S. Pimm 1999. Culling and the dynamics of the Kruger National Park elephant population. *Animal Conservation* 2, 287–294.

Van Aarde, R. J. & T.P. Jackson 2007. Megaparks for metapopulations: Addressing the causes of locally high elephant numbers in southern Africa. *Biological Conservation* 134, 289–297.

Van Wyk, P. & N. Fairall 1969. The influence of the African elephant on the vegetation of the Kruger National Park. *Koedoe* 12, 66–75.

White, P.J., J.E. Bruggeman & R.A. Garrot 2007. Irruptive population dynamics in Yellowstone Pronghorn. *Ecological Applications* 17, 1598–1606.

Whitehouse, A.M. 2001. The Addo Elephants: Conservation biology of a small, closed population. Ph.D. thesis, University of Port Elizabeth.

Whitehouse, A.M. & E.C. Harley 2001. Post-bottleneck genetic diversity of elephant populations in South Africa, revealed using microsatellite analysis. *Molecular Ecology* 10, 2139–2149.

Whitehouse, A.M. & G.I.H. Kerley 2002. Retrospective assessment of long-term conservation management: recovery from near extinction of the elephants in Addo Elephant National Park, South Africa. *Oryx* 36, 243–248.

Whyte, I.J. 1993. The movement patterns of elephant in the Kruger National Park in response to culling and environmental stimuli. *Pachyderm* 16, 72–80.

Whyte, I.J. 2001. Conservation management of the Kruger National Park elephant population. Ph.D.. thesis, University of Pretoria, Pretoria.

Whyte, I.J. 2004. Ecological basis of the new elephant management policy for Kruger National Park and expected outcomes. *Pachyderm* 36, 99–108.

Whyte, I.J. 2007. *Census results for elephant and buffalo in the Kruger National Park in 2006 and 2007.* Internal Scientific Report No. 06/07. South African National Parks, Skukuza.

Whyte, I.J., H.C. Biggs, A. Gaylard & L.E.O. Braack 1997. A new policy for the management of the Kruger National Park's elephant population. *Koedoe* 42 (1), 111–132.

Whyte, I.J., R.J. van Aarde & S.L. Pimm 1998. Managing the elephants of Kruger National Park. *Animal Conservation* 1, 77–83.

Whyte, I.J., R.J. van Aarde & S. Pimm 2003. Kruger National Park's elephant population: its size and consequences for ecosystem heterogeneity. In: J. du Toit, K.H. Rogers & H.C. Biggs (eds) *The Kruger experience – ecology and management of savanna heterogeneity.* Island Press, Washington, 332–348.

Wittemyer, G., I. Douglas-Hamilton & W. Getz 2005. The socio-ecology of elephants: Analysis of processes creating multi-tiered social structures. *Animal Behaviour* 69, 1357–1371.

Wittemyer, G. & W.M. Getz 2007. Hierarchical dominance structure and social organisation in African elephants, *Loxodonta africana. Animal Behaviour* 73: 671–681.

Woolley, L.-A., R.L. Mackey, B.R. Page & R. Slotow 2008. Modelling the effect of age-specific mortality on African elephant population dynamics. *Oryx* 42, 49–57.

ETHICAL CONSIDERATIONS IN ELEPHANT MANAGEMENT

Lead author: HPP (Hennie) Lötter
Authors: Michelle Henley, Saliem Fakir, and Michele Pickover
Contributing author: Mogobe Ramose

PURPOSE OF THE CHAPTER

THE FATE of the half a million or so free-ranging elephants in Africa depends on the choices people will make. What 'moral standing' do elephants deserve, and thus what constraints should we impose on our behaviour towards them? These are ethical questions. In general terms, ethics tells us what is good and bad behaviour, which human actions are right or wrong. Usually theories of ethics indicate a range of moral duties we owe to human beings; either generally or to those with whom we have specific relationships. In some cases our ethics also alerts us to duties we have towards non-human living beings or things. Thus, in our ethical theories we attempt to indicate to what extent we should restrain our actions so as to avoid negative impacts on other humans and living beings as well. We also consider what duties or actions we have to perform that will be beneficial and helpful to other people or species.

To assess the state of our knowledge about ethics and elephants is no easy affair. Different views on the moral standing of elephants and thus the obligations humans owe elephants, are not really a matter of scientific knowledge, although such knowledge might deeply influence our chosen ethics. At stake are human value choices that are developed through argument and discussion into ethical positions that suggest, prescribe, or legislate acceptable behaviour, and proscribe or discourage unacceptable treatment of elephants. The point of this assessment is thus to determine which ethical positions have been developed on various matters concerning the management of elephants and have been justified through reasoning. In open societies the diversity of views that arise about controversial moral issues generates intense debate. Since the early 1990s, world views that were once silent or repressed in South Africa have gained ascendancy and voice. These world views need careful consideration to determine from which to choose our ethical values.

This chapter portrays the different ethical views relevant to the management of elephants that are present in some or other form in the public domain.

In some cases ethical views can be found in detailed academic reports, in other cases ethical views will be reconstructed from other sources, like presentations at public meetings, official documents, and research reports. The emphasis will be on showing the strengths and weaknesses of each view. The aim is to make readers aware of the multiplicity of ethical issues involved in elephant management, as seen from a variety of ethical viewpoints. The complexity of the benefits and harm that accompany different management options will be clarified. The readers as citizens must make up their own minds on the ethical considerations they judge appropriate for the management and care for elephants.

The chapter is divided into five sections: (1) our human responsibility towards elephants, (2) the accountability of elephant custodians, (3) the moral standing of elephants, (4) ethical theories, and (5) ethics and management options.

HUMAN RESPONSIBILITY FOR ELEPHANTS

Why decisions about elephants cannot responsibly be avoided

Whatever choice wildlife managers make when determining how humans treat elephant populations, has consequences that cannot be ignored in any way. Consequences of decisions about the management of elephants affect the lives of thousands of animals, plants, all other living species, human visitors, and concerned supporters of conservation areas (see Mosugelo *et al.*, 2002, 235, 237, 238; Mapaure & Campbell, 2002, 216).

Refusing to take a decision on the issue of limiting elephant numbers regardless of the consequences for other living organisms implies taking sides; it implies a choice to let nature be; a preference to let matters develop without any human intervention. By doing nothing, wildlife managers are actually making a choice with observable consequences for which they ought to be held accountable to the same extent as for any other conscious, deliberate choice. For this reason a 'consequentialist' ethical approach – which gives priority to consequences of action and inaction – is appropriate.

Can decisions responsibly be avoided because of a lack of knowledge? In general, no, because although responsible decisions must be informed by the best available knowledge, humans mostly make decisions without the security of perfect knowledge. Characteristically, humans act in the light of available knowledge and that makes our actions fallible and revisable. What we have done today might be judged wrong tomorrow in the light of new knowledge and fresh insight. How well-meaning conservationists managed elephants yesterday is not good enough for today and tomorrow, as scientific understanding and ethical insight have progressed by leaps and bounds in the last three decades. Neither is our current state of knowledge a valid basis for making judgements on the ethical value of decisions taken in the past.

If responsibility requires us to make decisions in the light of currently available but incomplete knowledge, is precautionary action ethically acceptable? Precautionary action means that we act to prevent serious or irreversible harm from occurring despite large uncertainties about the likelihood of such harm actually taking place (and in some interpretations, largely *because* uncertainty about future outcomes is high). Precautionary action cannot be ethically acceptable if no evidence can be presented that points to such potential harm. However, precautionary action based on the best and latest knowledge available, that took account of all scientific sides to a debate, that investigated all possible concerns, and that weighed the possible consequences of inaction against those of the various forms of action that could be taken, must surely stand a good chance of being judged ethically responsible.

We have a much greater ethical responsibility towards elephants now

Why is our ethical responsibility towards elephants much greater today than a few centuries ago? One possible answer can be developed along the following lines. For many centuries humans were more prey than predator and could easily be threatened and harmed by elephants; our only defences were bow and arrows, spears, stones, fire, and holes dug in the ground. When humans had only primitive weapons, elephants might easily have had the upper hand. With the mastery of fire and the availability of metals, the power relation started shifting in our favour. Nevertheless, a more-or-less equal relationship existed for most of our shared history.

Our ever-increasing knowledge of the world and our sophisticated technologies have now made it possible for us to interfere with elephant lives on a scale unimaginable in the past. We can now control elephant lives through chemicals to immobilise them and practise birth control, kill them instantly with a range of weapons, restrict their movements by using elephant-proof fences, round them up and transport them with automobiles and helicopters, and track their movements by using satellites.

In the past 50 years humans have investigated elephants in depth. As a result our knowledge about the ecology, physiology, behaviour, social structure, communication, and mental characteristics of elephants has deepened our perception and understanding of elephants (see Douglas-Hamilton, 1975; Moss, 1988; 1992; and Payne, 1998). We therefore need a new ethics as our newly acquired knowledge about elephants requires a redefinition of our relationship with animals that are more like us than we previously realised.

Our new knowledge of elephants has increased our ability to exercise almost absolute power over them. Not only can we now effectively manage elephants in conservation areas according to standards we decide on, but we have also developed and refined various techniques of taming and training elephants. We are the dominant species on earth. The massive increase in the human population has reduced the land available to elephants to a fraction of their former range. Not only have we conquered their land, but we have also reduced their numbers through killing them. We thus need ethics that reflect our status as the most powerful species this planet has ever had.

Why ethical decisions about elephants are so controversial

Why do we have persistent moral disagreements about some issues? What is the nature of the moral dilemmas that human interaction with elephants generates? Why do ethical issues about the management of elephants lead to such strong disagreement?

In their seminal work on how to deal with moral conflict in what they define as 'deliberative democracy', Amy Gutmann and Dennis Thompson (1996) explain persistent moral disagreements as a result of four factors that occur in all human societies.

Incomplete understanding

A lack of in-depth knowledge or detailed understanding of the issue under investigation results in different people judging different ethical values applicable.

The debate on the culling of elephants is affected by our incomplete understanding of the dynamics of elephants in the African savanna ecosystem. What level of elephant impact on vegetation is 'normal'? What degree of vegetation change is acceptable in the savanna ecosystems? What are the effects of fires, other browsers and artificially provided water, compared to the impact of elephants? What kept elephant numbers stable throughout the greater part of their history? Our understanding of the role of elephants in ecosystems is fragmented and incomplete at best.

Moderate scarcity

The goods judged as valuable by living beings are those not available in sufficient quantities to ensure that every living being can comfortably get as much as they need or prefer. For this reason we will always debate appropriate ways of distributing valuable goods, dividing scarce resources, assigning precious opportunities, or of recognising merit, strength, and beauty. Land available to elephants has shrunk dramatically over the past century. In many parts of Africa elephants now compete for land with farmers and with other animals in fenced-off, or otherwise clearly demarcated conservation areas. Many managers of conservation areas report an overpopulation of elephants that apparently have too much of an impact on the quality of the habitat available to other living beings. Whether such reports are accurate is not the issue here, but the fact that many knowledgeable people observe the lack of sufficient resources for not only growing numbers of wildlife but also burgeoning human populations.

Limited generosity

Human beings are not renowned for being altruistic in nature. The interests of our selves or our group often make us partial. In the process we often deny others things, opportunities, and recognition they ought to be able to rightfully claim from us. Why should humans be generous to elephants and allow them large tracts of land where humans could have made productive livelihoods? Why not use elephants as resources to combat poverty and create jobs for the unemployed? Should some elephants not sacrifice their lives to ensure the long-term survival of their own species and others? Why should we not judge elephants as part of the natural resources of Africa that will help us provide a better life for every human involved? How can we justify safeguarding elephant lives and caring for elephant well-being if human lives are wasted through devastating poverty? Many would argue that our generosity towards higher mammalian species only goes as far as first taking care of human well-being allows us.

Incompatible values

The incompatibility between the directives of some of our ethical values forms the fourth factor that creates persistent moral disagreement. We are not always clear on how to specify a particular ethical value, nor are we confident about its exact range or scope. If we balance competing values differently or assign varying strengths to them we can get conflicting outcomes. The end result is disagreement about complex ethical issues, regardless of a possible consensus about the fundamental ethical values involved.

Incompatible values that give rise to moral dilemmas present some of the most difficult ethical issues to resolve. A moral dilemma occurs when one ethical value emphasises safeguarding certain interests, while other ethical values point in a different direction. The correct solution depends on which ethical perspective you adopt. Sometimes several prescribed actions appear acceptable, and in other cases none seem palatable. When this kind of conflict between ethical values with contrasting prescriptions occurs in the context of non-ideal conditions, moral disagreement appears almost insoluble.

The unfortunate characteristic of moral dilemmas is that they seem to require that we sacrifice one or more aspects of our ethical values and their supporting arguments and evidence. This loss appears unacceptable to people strongly committed to their set of ethical values. The clash between an individualist perspective, that values the life of every animal affected, and the ecosystem perspective, that is willing to sacrifice individual lives for the sake of the well-being of other elements within the complex interactive web of life, is a good example of such incompatible values.

We are responsible for the environment because of our impact

Humans have had an exponentially increasing impact on the earth's environment and its inhabitants. The acceleration of our exploitation of wildlife throughout the nineteenth and twentieth centuries, our increased occupation of land through our rapidly growing numbers, and our destruction of the environment through pollution, deforestation and global warming are major factors depriving wildlife of places where nature can function without significant human influence.

In South Africa, space for those species of wildlife that cannot easily co-exist with humans is no longer available outside conservation areas. Protected areas have become the sanctuaries of wildlife. They are artificial human constructions that represent small dots and islands in the cultivated and inhabited areas on the maps of Africa (see Chadwick, 1992, 40). Whether we can still speak of 'natural processes' in Africa's small areas of land available for conservation is a decisive factor in debates about elephants, but as yet not well enough understood.

Human impact also occurs inside conservation areas. They are heavily influenced by human settlements surrounding them, even in some of the supposedly most natural conservation areas without any fences, like Chobe

National Park in northern Botswana (see Cumming & Cumming, 2003, 566). For example, rivers running through conservation areas are used and polluted by humans where those rivers flow through their agricultural land or urban areas before these rivers enter conservation areas (Whyte, 2001, 9).

In a situation of massive human influence, letting 'nature take its course' does not imply no further action. Humans have already massively interfered with nature and must take responsibility for this interference. We thus ought to interfere responsibly to conserve as natural a state as possible for future generations. 'Letting nature take its course' in this situation implies doing research to address the problems created by humans. In such cases human intervention keeps nature on track. Malevolent human interference in nature has become so prevalent that humans must now interfere benevolently so as to 'let nature be' (see Lötter, 2005).

THE ACCOUNTABILITY OF CUSTODIANS TO STAKEHOLDERS

Protected areas exist and operate within the framework of a political system and its associated constitution and laws. Governments have agencies and bureaucracies charged with the management, development, and extension of such areas. To have conservation areas properly managed and protected, to increase the number of habitats, landscapes, and ecosystems to be preserved, and to ensure appropriate conservation policies, require political action to lobby, pressurise, and influence governmental policy makers. To do so successfully, conservation areas and game reserves must have some value for citizens (Regenstein, 1985, 132).

Governance and accountability

The South African government has formally accepted full responsibility as trustee to ensure the management and conservation of biodiversity in the laws enacted to deal with protected areas, biodiversity, and the sustainable use of natural resources. Wildlife scientists and managers, as well as operational and administrative managers and staff, are appointed to run these conservation areas under the guidance of national or provincial conservation governing bodies. These people are custodians entrusted to guard, protect, and maintain conservation areas according to goals formulated by national or provincial legislatures and embodied in laws and policies. Conservation areas as public property have been legally placed in their care as trustees to administer for the benefit of all citizens. As custodians and trustees they use their professional, scientifically informed judgement within the broad goals and purposes set by national and provincial governments on behalf of citizens. Within this framework they have a degree of discretion and independent judgement to do what is best for a particular conservation area. They are accountable to government and citizens through regular reports and feedback.

When a matter excites so much emotion and generates such controversy as elephant management, democratic theory and practice require that wildlife managers of public conservation areas demonstrate their accountability to the public for whatever decisions they take. In such cases they ought to consult thoroughly with all stakeholders, as has become accepted practice in modern constitutional democracies like South Africa (see Gould, 2002; Begg, 1995).

Not all protected areas are managed by public bodies. Private institutions, social organisations and individual citizens manage the majority of protected areas in South Africa, out of their own volition, for their own benefit, or as a civic duty. These civic bodies and private individuals also take responsibility on behalf of the public or in the name of public interest when they deal with the natural heritage and life-enabling biospheres of citizens. They are accountable to the general public just like state institutions.

In moral dilemmas generated by controversial aspects of elephant management, the decision makers take on a collective moral responsibility similar to individual moral agents in their ethical decision making. They must give a public account of how they discharge their moral duties in their custodial role. They have the responsibility to take all information available into account and to place the information in the public domain for inspection and discussion by interested parties. They must be transparent in their decision making so that everyone can follow the logic of their reasoning and the factual, scientific basis of their claims. They are accountable to their stakeholders and must be prepared to engage stakeholders in dialogue (see Gutmann & Thompson, 1996). This much is required of any person in public office in a constitutional democracy that is paid by public funds and makes decisions about contentious issues.

Categories of stakeholders

Stakeholders do not all have the same interests, nor do they have claims of equal value or weight. The categories of stakeholders and the weight of their interests must be carefully distinguished. For example, the interests of villagers whose lives, bodies, and livelihoods are threatened by elephants crossing the boundaries of conservation areas must be judged more urgent than the interests of people in distant locations.

In the global village, it can be argued that conservation areas do not only belong to the citizens of the country in which they lie. Most conservation areas have special significance as a result of their globally unique ecosystems with accompanying biodiversity. Such areas can thus be judged to be common property of all human inhabitants of our planet, a kind of global commons. Many people judge that what some humans do in the biosphere affects all other humans. Similarly, many people judge that the natural resources and wonders on Planet Earth ought to be held in trust for all citizens of the globe. The creation of world heritage sites by the United Nations captures this idea. The various international treaties relating to biodiversity conservation give effect to this notion of a global commons.

THE MORAL STANDING OF ELEPHANTS

There is no doubt that humans regard themselves as beings with moral standing – that is, as beings whose interests must be taken into account in any ethical decision making. In general, humans believe we owe it to one another to consider the ways our actions impact the well-being of other people. Thus we ascribe a moral obligation on ourselves to be aware of, and care about the possible benefit or harm our actions cause to other humans.

Do we extend this moral consideration to other living organisms? If so, what organisms do we include and to what extent do we take their interests into account? Thus, what level of moral standing do we assign them compared to the standing we believe we owe to members of our own species?

Humans and moral standing

Many people have a human-centred bias in the way they ascribe moral standing to other living organisms. This means we are biased toward our own species and use ourselves as benchmark in determining moral standing. We generally look at the characteristics such living organisms have in common with us, characteristics we find impressive. Somehow this bias makes sense, as we are the only beings, as far as we know, to make such judgements. The only thing we have as benchmark for moral standing is our own flawed attempts to ascribe moral standing to members of our own species. This starting point seems to be as good as we might get. The crucial question is whether we are *fair* in our judgement of other living organisms. We must be open to the inherent differences in other species and appropriately acknowledge their qualities and characteristics.

Elephants as agents

How can we responsibly determine the moral standing of elephants? Perhaps what follows is a way forward. Humans experience elephants as 'intriguing animals' (Bell-Leask, 2006). What are the characteristics of elephants that so fascinate us and lead so many of us to judge them as belonging to a superior class of animals deserving high moral standing, like dolphins, whales, dogs, chimpanzees, gorillas, and lions?

Elephant researchers have convincingly demonstrated that individual elephants are complex *agents,* sources of self-originating activities (Taylor, 2002, 89). The concept 'agent' at its basic level refers to something with the potential to exert power, produce an effect, cause an outcome, or influence its environment. This 'something' is a point of origin of one or more forces that can be activated under the right conditions to start a chain of events.

There is a continuous spectrum of agents of increasing complexity with higher degrees of agency. At one end are lifeless chemical agents, such as acid. At the other end of the spectrum we find human beings. Human agency can be seen in our ability to act intentionally, author events, produce effects, make things happen, bring about change, and cause consequences. Agency also manifests itself by our nature as centres of experience through which we process information about our world to become aware of its possibilities. Our agency shows in how we make decisions about appropriate courses of action in the light of values and goals we have set and appropriated for ourselves.

Obviously human agency has limits, as we cannot act to alter the movements of the stars, cannot effectively intervene in the course of terminal disease, nor bring to bear appropriate force on two individuals to make them fall in love. Nevertheless, our collective agency as humans on earth seems powerful enough to alter climates on our planet.

Elephant agency similarly manifests in various ways. They are important sources of activity within the African ecosystems, with functions often described as those of 'engineers' that stimulate, affect, and even create habitats for other living beings. They are centres of experience that observe the world through complex sensory organs. They store the information thus received in long-term memories that provide guidance about resources crucial for survival. Their experience of their world is filtered through complex brain processes that include a range of emotions and linguistic symbols. Their complex agency functions resemble ours to a significant degree (see Antonites, 2007).

Similarities between humans and elephants

There are many other characteristics of elephants that are similar to ours (see Douglas-Hamilton, 1975; Moss, 1988; 1992; Payne, 1998; Whyte 2001; 2002; Chadwick, 1992; Meredith, 2001; McComb *et al.,* 2002; Gröning, 1999; Hanks, 1979, and Larom, 2002). Elephants have senses similar to human beings: their eyesight might be worse than ours, but their sense of hearing and sense of smell are far better than what we possess. They can experience a range of emotions, of which their acute awareness of death and resultant mourning the loss of family and friends move us (see Moss, 1988; 1992).

Their lifespan roughly matches ours and their young need similarly many years of upbringing before they are judged to be adults. They have complex social behaviour and organisation. The playfulness of younger elephants in matriarchal herds, the joy of family groups at reunions, the stand-offs between bulls of all ages, the care and protection older females display towards the young, the 'discussion' between senior members in family herds about decisions, and the gentle but firm leadership of the matriarch are all forms of behaviour we can identify with. We are intrigued by their regionally unique languages with up to 80 different calls, commands, and other elements.

Societies capable of socially complex behaviour are societies (1) with unique individuals as members, (2) that are reasonably stable over the longer term, (3) that have individuals capable of acquiring social skills, and (4) that have experienced members that transfer acquired habits and knowledge to younger ones. Elephant society can clearly be described as socially complex, though less so than human societies (see De Waal *et al.*, 2003; Payne, 2003). Furthermore, their social bonds and their sense of death, and in general, the close resemblance between their lives and ours engender our sympathy and love for all those qualities that make ascriptions such as 'intelligent' and 'gentle giants' seem appropriate.

Differences between humans and elephants

Having pointed to the similarities between humans and elephants, we should not ignore the enormous differences between the two species, nor the fact that many other non-human species exhibit these features to varying degrees. One can argue that both the similarities as well as some of the differences between humans and elephants are reasons for the feelings of awe and appreciation we have for them. Note the important role of some differences in this case. If elephants were only similar to us, but had no significant differences, we would have treated them solely as beings of lesser qualities and worth than ourselves. The differences that matter in this case are dissimilar, distinct, and impressive qualities of elephants that we do not possess.

Some of the differences that add to our appreciation and valuation of elephants are their superior physical size and strength. Similarly, we value their acute sense of smell, we are amazed by their communicative abilities through infrasound, and we are thrilled by their stealth in moving silently and unobtrusively through thick bush despite their massive size. The fact that such huge animals are vegetarian adds to their allure as well.

Although elephants can destroy us through their enormous physical power in any one-on-one fight, humans are the dominating species that control so much of the lives of elephants. Perhaps the most important difference between the two species is the fact that elephants cannot call a meeting and discuss the challenges their feeding habits create for other species. They cannot come up with a plan to deal appropriately with such an issue, as far as we can see. We must do it for them, although we struggle to implement such plans effectively for our own species! Elephants do not have our highly sophisticated communication skills, including natural and symbolic languages. They do not have our amazing organisational capacities. Elephants cannot transform natural resources into useful products such as computers, like we can. Their impact on their environment is dwarfed by our impact. Our capacities for suffering and mourning the loss of our dead manifest in far more complex ways than similar capacities do in elephant society.

Elephants are also not capable of the full range of moral behaviour that would make them into moral agents on a par with humans. We are as yet not sure if, and to what extent, elephants have a moral sense like ours or follow moral rules (see Antonites, 2007). Yet, they are still important moral patients, beings to whom we owe considerable moral respect, although not to the same degree as to members of our own species.

Thus, in the light of these significant differences, the interests of elephants cannot have the same weight as those of humans, as our complexities in terms of features we define as relevant to moral standing far outstrip theirs. Elephants do not have equal moral standing with humans, as they do not match the intellectual, behavioural, or emotional complexities of our species that demand so much moral respect. Obviously this is spoken as a human being with a biased perspective! Moreover, we must also acknowledge that despite our self-assigned moral standing, humans have negative qualities that elephants cannot match. For example, their potential to impact negatively on our shared world is nowhere close to ours. Similarly, our repeatedly demonstrated capacity for maiming and killing of living beings of all species, our own included, far exceeds theirs. In an important sense they are by far a more peaceful species than we are.

The moral standing of elephants and other animals

Elephants are not the only animals with characteristics that we judge to be amazing, although for many people they are members of the small group of 'most special' non-human beings. Many animal species have characteristics that we value or admire, or qualities that make them unique, appreciable, and astounding. For example, we prize owls for eyesight in the dark, their sharp hearing, and their stealth flying. We are amazed by the navigational skills of pigeons and marine turtles. Dogs are highly valued animals for their acute sense of smell, their ability to be trained for specialist functions to assist the police, emergency services, and disabled people, their sensitivity to human emotions, and their companionship coupled with immense loyalty. We admire and fear lions for their regal demeanour, strength, ferociousness, and their hunting prowess. The differences in the complexity of mental life between humans and elephants are perhaps much more than the differences between elephants and owls, dogs, or lions. There seems to be no convincing reason why elephants should deserve a moral status equivalent to humans, as they are closer to other animals than to humans. With other higher species, however, they do deserve a special moral status, as they have some of the most complex sets of behaviour and intricate inner lives of all animals.

As humans we differentiate between living beings in terms of their moral standing, mostly based on the level of complexity they express in their consciousness, individual behaviour, social organisation, or physiology (see Antonites, 2007). Most people have no problems eating meat from cattle and sheep, but would shrink from having dogs or primates killed for human culinary purposes. Many people do not mind killing a rat that nests in their ceiling, but would find it far more difficult to kill a cat doing the same. Elephants definitely belong to the upper class of animals that we judge to have higher moral standing than the rest.

What are the implications of their moral standing for our behaviour towards them? Perhaps some of the ethical theories can guide us.

ETHICAL THEORIES

Many of the harshest critics of any human intervention in the lives of elephants, especially those causing suffering or death, are referred to as animal-rights activists or animal-welfare activists. Do animals really have rights that humans must respect at all times? If so, who has assigned them their rights and why should humans refrain from violating these rights? Or should we perhaps argue that all sentient beings have interests that humans ought to respect to the degree that those beings can experience welfare – that is, pleasure and satisfaction or pain and distress?

If the interests and rights of individual animals have to be taken into account, should these interests and rights get priority above the well-being of ecosystems and other species? To what extent can we use wildlife as a resource to fulfil our human needs and wants? The ethical theories discussed in this section aim to answer questions like these.

Singer's consequentialist individualism

Most animal welfare organisations have their intellectual roots in the environmental ethics of Peter Singer. He offers one possible justification for placing the interests of animals much higher on our human list of priorities than most people actually do. Singer makes the apparently controversial claim that humans have no special place in nature and cannot claim any superior position to any other animal in any process of ethical decision making. This strong claim is qualified by other aspects of his theory (Singer, 1985, 6). Singer counts all beings as morally relevant and able to experience pain and distress or enjoy things and have pleasure. His view acknowledges that taking a human life can be worse than killing a snake. The reasons are that humans have more complex and sophisticated experiences of pain and pleasure and humans have more complex mental lives that include pasts and futures (Singer, 1985, 9).

Singer's utilitarian ethics determine the correct action by calculating the amount of pleasure, happiness, and well-being generated by an action versus the pain, suffering, or ill-being it incurs. Thus, in the case of the human versus the snake, the more complex and sophisticated human experiences of pain and suffering far outweigh the painful experiences of the snake.

If Singer's intuitively plausible views are applied to the elephant problem, the interests of an individual elephant will outweigh the interests of most other individual animals belonging to species other than *Homo sapiens*. Elephants would have a moral standing lower than humans, but higher than most other animals. However, despite the moral standing of individual elephants and their species, Singer does not intend his utilitarian ethics to be applied in individualistic fashion. When a conservation area has an overpopulation of elephants that are altering the habitat of other species and themselves, a careful weighing of the interests of different forms of life has to be done. The issue is to determine the effect that the consequences of different decisions will have for all parties involved. The interests of all individual elephants, millions of other living beings, tourists, wildlife managers, and all other stakeholders must be weighed against one another.

It is doubtful whether Singer's ethics that treats animals as equals implies that human interference in nature is never justified. There are too many other animals that might lose their lives as a result of elephant impact and, in some cases, even whole species might be driven to extinction. Singer's view would definitely require some kind of intervention in favour of the multitudes of animals with threatened livelihoods. Some kind of management intervention, potentially including culling, would be justified if all interests are fairly added up.

Regan's deontological individualism

Many people and organisations are committed to the idea that animals have rights. Tom Regan (1984) is regarded as champion of the idea that animals have rights which all humans must respect (see also Cohen & Regan, 2001). Regan's stance rests on the idea that many living beings are similar to humans as they possess mental capacities and can experience their lives in terms of better or worse welfare. Such animals are subjects-of-a-life and they thus have inherent value. Therefore, animals must be treated respectfully as rights-holders that have the same moral status as humans. Respectful treatment implies that such beings may not be killed, their bodies may not be invaded or injured, and their choices may not be restricted nor their freedom limited. Regan strongly rejects all utilitarian positions, as such views cannot protect innocent individual animals from being sacrificed for the benefit of others whose interests count more (see Sagoff, 2002, 42). Regan emphatically rejects the killing of any rights-holder and strengthens his position by saying that killing is unacceptable regardless of the consequences for others.

When he discusses wildlife, Regan often states his view simply as 'let them be!' (Regan, 1984, 357, 361). He refuses to see wildlife as a natural resource available for human benefit and recommends that wildlife managers should aim to keep 'human predators out of their affairs' (Regan, 1984, 357). It is doubtful

Box 1: Limited rights for elephants?

Rights are generally understood as justified claims to specific things that a society guarantees its citizens for certain strongly defended reasons. Rights can only be assigned if the majority of citizens in a society have decided the claim is acceptable and that members have a duty to provide that thing to one another. If a society would decide that elephants deserve rights, what might the contents be of such legal protection?

A first possible right builds on the idea that humans should not lightly kill elephants: 'No human may kill an elephant unless in self-defence, or when an independent panel of appropriate experts find compelling reasons to do so.' The biggest harm we can do to elephants is to kill them. We thus first of all owe elephants the security of their lives that we cannot take away without good reasons. Elephants are subjects-of-a-life or agents similar to us, though of slightly lesser complexity, who make decisions and experience a wide range of emotions. They have consciousness like us and are deeply aware of death. They thus deserve similar protection of their lives to that which humans get.

The second possible right articulates the idea of liberty for elephants: 'No human may deprive an elephant of its liberty to live its own life in a fitting habitat without convincing justificatory reasons.' If we can ascribe free choice to elephants, then ethical treatment of elephants implies that we ought to give them liberty, as they have a clear and distinct inclination to live their lives in suitable habitat according to their lights. If elephants are agents with high levels of sentience, they have a compelling interest to live their lives in the light of their own best judgement of where to find food, water, shelter, and companionship.

What justifies this right? Most living beings exhibit a whole range of behavioural signs that they detest being held in captivity or resist being captured and held in human hands. Elephants are no different than any other living being that prefers (i) to settle the boundaries of its home range for itself in competition with other members of its species and (ii) determine its own life within that territory.

What does liberty for elephants imply? It means we must give them space and opportunity to live in near-natural conditions. We must also respect their autonomy to choose themselves how they want to live their lives, as elephants have done for millennia. Elephants present us with strong evidence that they are agents and we must respect their capacity for informed decision making. There is no doubt that elephants are competent to make their own choices and thus do not need anything more from us than to allow them to be, i.e. to live their social life on sufficiently large tracts of land with suitable habitat.

A third possible right outlines the importance of privacy for elephants: 'No human may intrude or interfere in an elephant's life without strong reasons to do so.' Privacy can be defined as a state or condition of limited access to a life, or zones and spheres of lives that are not to be invaded or violated. Privacy is important to allow living beings to act freely in the absence of scrutiny and interference. Elephants apparently do not have a need to limit humans' access through observation to any part of their lives. For example, birth, death, and sexual relations occur in public spaces visible to any living being close by. If they do not withdraw themselves into the cover of vegetation, we might assume they are not too bothered by our prying eyes. However, elephants do seem to need lots of personal space around them. There is no doubt that they insist on enough space whenever other animals or motor vehicles get too close to them.

To give elephants the privacy they require thus has an important implication for tourists and researchers, i.e. they must stay at a respectful distance from elephants. A respectful distance will be determined by elephants themselves, who can often be seen threatening either wild animals that violate their private space or motor vehicles that are driven too close to them for comfort.

The final right elephants might deserve goes as follows: 'Owners, managers, or keepers must give elephants appropriate care and compassion that will ensure both their well-being and that no harm or suffering from non-natural causes will befall the elephants they are custodians of.' This means we must protect their habitat and not beat them in abusive fashion. We do not have to interfere in their struggles within their ecosystems, unless some kind of prior human interference impacts negatively on the functioning of ecosystems.

Is the idea of limited rights for elephants far-fetched? Perhaps not. The norms and standards for elephant management proposed for legal enforcement by the South African government's Department of Environmental Affairs and Tourism embody many of these 'elephant rights'.

whether Regan's views on animal rights can be applied so simply to conservation dilemmas. He touches on such issues briefly, but does not highlight the full implications of his view that individual animals have rights that need almost absolute protection. Applied to the issue of controlling elephant numbers, one can usefully extend his views by taking a cue from his discussion of what is ethically acceptable when a rabid dog attacks you in your backyard (Regan, 1984, 296). Although he reiterates his position that animals can do no moral wrong, in this case the dog is a threat to our bodily integrity and maybe even our life. We can thus defend ourselves and harm the dog in the process (Regan, 1984, 296). What Regan does here is to weigh the rights of humans, whom the dog might violate, against the rights of the dog as aggressor that intends bodily harm to a fellow animal (the human). The rights of the victim thus trump the rights of the aggressor through legitimate self-defence.

Let us assume for the sake of argument that Regan's view on animal rights is generally accepted as true and correct. If individual elephants have rights, and so too thousands of other individual animals qualify as rights-holders, how are we going to solve the ensuing complex conflict of rights when elephants alter the habitat and thus endanger the livelihood of millions of other rights-holders? (see Cumming & Cumming, 2003, 561).

Animals cannot manage and administer their own rights under the best of circumstances, thus needing humans to assist them. If humans have to solve this problem in terms of animal rights, then we should interfere in this conflict of rights to life. Or could an animal-rights supporter be so callous and insensitive to say that millions of living beings can be allowed to die in the name of 'letting nature be', but not one animal may die as a result of benevolent human intervention to protect species and ecosystems? Perhaps management interventions with the explicit motive of removing excess numbers to protect the habitat for millions of living beings seem more in line with an animal-rights approach than merely letting nature be?

Holistic protection of ecosystems

Many people and organisations involved in conservation believe that it is the complex of ecosystems, landscapes, and diversity of life forms that must be preserved for posterity. This approach is championed in South Africa by SANParks (and by WWF and other large conservation NGOs), who interpret their mandate as custodians of South Africa's conservation areas in this light.

What is the goal of protected areas according to the holistic view of ecosystem conservation? According to this view, conservation in national parks should be comprehensive, with the goal to protect the full scope of biodiversity (see Holmes Rolston III, 2002, 38 and Whyte *et al.*, 1999, 113). The focus is on all aspects of life and its enabling conditions, thus including the biosphere, landscapes, ecosystems, species of all the different life forms, and individual organisms. The approach implies that all aspects of conservation areas should be protected so as to allow and enable nature to function, as far as possible, on its own without human interference or even without benevolent human intervention. The comprehensive, holistic focus on the well-being of greater systems is the strength of this approach, while its willingness to sacrifice individuals and groups for the sake of the overall health of ecosystems and landscapes is its downside.

According to this view, conservation areas should ideally have limited human presence and even less human interference, so as to allow natural ecological processes to function as they did for millennia. These places should be free from all forms of human domination and exploitation. Such conservation areas provide opportunities to establish different 'biocentric' or 'ecocentric' worlds where biodiversity flourishes and free animals pursue their interests as they see fit within their preferred habitats. Such 'worlds' can allow evolutionary processes to follow their ways. Eco-tourists should behave like visitors and guests who show deep respect for the 'citizens' of these 'worlds'. They should know and appreciate the fact that conservation areas are neither cattle ranches nor zoos. In these areas nature must follow its course and human interests must be subservient to the dictates of the wilderness. Eco-tourists in these 'worlds' can imagine themselves entering past worlds, worlds similar to the ones in which humans first evolved thousands of years ago and akin to those in which our early hunter-gatherer ancestors survived for millennia. In the same way that tourists respect items on display in museums and art galleries, we should foster respect for all elements within these natural museums and galleries of our evolutionary past.

The idea of ecocentric worlds implies holistic conservation with the aim to keep intact the enabling conditions and prerequisites for the effective functioning of the earth's biosphere. These ideas have gained new relevance in recent years. Global environmental challenges seemingly require major changes to our population growth, lifestyles, and use of natural resources if we want to preserve the global biosphere's life-enabling qualities. The elephant issue encapsulates these challenges that confront us with the history and consequences of our impact on elephant lives and habitat. This issue presents an opportunity to redefine our relationship with elephants and to rethink how we take care of them.

The holistic view about the conservation of nature's functioning through ensuring multiple continuing interactions within various ecosystems leads to ethical principles similar to the famous one articulated by Aldo Leopold (1981): 'A thing is right when it tends to preserve the integrity, stability and beauty of the biotic community and wrong when it tends otherwise.' This principle implies that elephants, or any other living beings for that matter, are expendable for the sake of the health and beauty of the larger wholes, like the biosphere or a specific ecosystem.

Note how an animal rights perspective believes the holistic view sacrifices individual animals for the sake of the larger whole. Pickover (2006) rejects the holistic view that holds that as 'long as the species is perceived to be sustained it does not matter what that might involve, or what the plight might be of individual animals or groups of animals.'

Respectful sustainable use in traditional African communities

The idea of sustainable use of natural resources has been widely discussed. Not only is the idea of sustainable use of natural resources part of the South African constitution and conservation legislation, it is also part of the policies of the Southern African Development Community and the IUCN. Large differences of opinion exist about the correct understanding of this idea. Instead of unpacking these debates, we let pre-colonial African communities serve as example of the sustainable use of the natural resources of the African savanna that forms the habitat of most of Africa's elephants.

No one really knows how big the impact of human hunting was on African wildlife, elephants included. What can be inferred is that the impact was sufficiently minimal that the wildlife persisted in the presence of humans for millennia, and thus the use of African fauna as a food source was sustainable most of the time. If not, we would not have had reports from early European explorers describing Africa as a place 'teeming with wildlife.'

Although contemporary academic theories that develop traditional African values about the environment into theories of environmental ethics are scarce in South Africa, sufficient clues exist that enable a partial reconstruction of such values. Perhaps the most important fact to consider is that African people lived alongside wildlife for centuries without hunting any species of wildlife into extinction that we know of, as has happened on other continents like Europe. Reports by early explorers and anthropologists point to lifestyles that made respectful and constrained use of wildlife for survival, trade, and adornment. Oral and literary reports speak of Africans with a deep love for nature that found expression in a comprehensive knowledge and profound understanding of African wildlife.

Credo Mutwa (1996, 11–26) explains some of the traditional African values that he encountered in different communities across southern Africa. Mutwa claims that these pre-colonial values often persist in some contemporary communities, albeit sometimes in fractured forms.

Mutwa believes pre-colonial Africans had a deep awareness of their dependence on nature. They saw themselves as part and parcel of nature, not as dominant conquerors. They had respect for animals and plants that can be seen in a host of regulations aimed at the protection of plants, animals, and water sources. At least some individuals had impressive knowledge and understanding of all the elements in ecosystems that are involved in intricate, intimate, interwoven interactions. They made use of natural resources *inter alia* through ethically regulated hunting for survival purposes.

The phenomenon of tribal totems – plants or animals that functioned as symbols of the identity of a tribal community – illustrates the ideas these communities implemented in their conservation practices. Mutwa argues that preservation of the totem animal not only required protection of the specific animals in question, but also its habitat, the animals that live in close association with it, and its predators. In this way the food sources of the totem species are safeguarded, its natural allies who assist in vigilant watchfulness continue to play their role, and their predators ensure the survival of only the fittest of the species.

Totemism is a crucial theme in an African philosophy of conservation (Ramose, 2007). In African culture, a totem animal or item, including the elephant (*tlou* (Sotho), *ndlovu* (Zulu), *Zhou* (Shona)) is an object that demands reverence and not mere respect. It is revered because it is deemed to be a special, mysterious, representative of the power of the gods (*badimo, madlozi, badzimu*). This quality confers upon it the aura of untouchability. The effect of this is the preservation and conservation of the totem animal. The *Batloung* clan among the Sotho-speaking peoples and the *BakaNdlovu* clan among the Nguni-speaking peoples are to promote and defend the preservation and conservation of the elephant because it is their totem animal.

According to Ramose (2007), totemism must be seen against the background of the African conception of community. It is crucial to note that relationships within this community extend very consciously beyond the sphere of human beings. If this were not so, then totemism would be meaningless. The traditional African community is a three-dimensional community comprising the living, the living dead ('ancestors') and the yet-to-be-born. The bonding of this community, which thus includes animals, rests upon (a) mutual care and concern, (b) solidarity through the preservation of the network of relationships, as encapsulated by the idea that 'I am related, therefore we are,' and (c) the imperative to strive after and maintain harmony in the prevailing relationships. Thus reverence to the totemic animal is not equal to, but akin to reverence to the living-dead whose power and influence over the lives of the living are overwhelming. It is this philosophic outlook which in pre-colonial Africa ensured the preservation and conservation of nature in general, and elephants in particular.

Many of these attitudes towards the environment can still be observed, for instance among the Maasai people in Kenya and in the revival of community-based conservation projects and sustainable use practices in Zimbabwe and Namibia. In contemporary South Africa contractual agreements between local communities and SANParks in the Kruger National Park (Makulekes in the north), the Richtersveld National Park (the local Nama people), and in the Kgalagadi Transfrontier Park (the Khomani San) similarly point to these values being reinvigorated and put to new use. One can also detect these values from the presentations made by traditional African community representatives at the Great Elephant Indaba organised by SANParks in 2004.

The American philosopher and economist David Schmidtz (1997) takes up some of these ideas selectively in his writings on environmental ethics. He argues persuasively in favour of the sustainable use of African wildlife through community-based conservation efforts. He bases his argument on his observations of, and extensive interviews with many people in southern Africa involved in such projects. His argument is that poor Africans will protect their natural environment and its inhabitants if they derive some value and enjoy some benefits from their efforts. He thinks that protection and sustainable use of the environment will be the rational thing to do if the impoverished communities can make a decent living from benefits that accrue from hunting and ecotourism. He consistently cautions that wildlife and natural landscapes will disappear if communities have no proper incentives to care for them. His view raises these questions again: if local communities can make livelihoods from African wildlife on land outside protected areas, why should they not? If resources can be sustainably harvested within protected areas, why should they not be?

Wildlife has been a major resource on the African continent for centuries. Can it be used as a sustainable resource for fighting desperate human poverty in African countries (Osborn & Parker, 2003, 73; Du Toit, 2002, 1403–1416)? If yes, what kind of sustainable use is acceptable? Will it be ethically acceptable to use some conservation areas not only for the purposes of ecotourism, but also for hunting, culling, harvesting excess wildlife, thus, in short, for any kind of commercial exploitation? Several projects in different African countries have shown the idea to be viable if managed carefully (Bonner, 2002, 320–329). The idea also makes sense, as many African savanna areas are by far more suitable for wildlife farming than for cattle ranching or cash crops. If implemented on a large scale, much more land will become available for African wildlife, as has happened in South Africa's explosive development of commercial conservation for the purposes of ecotourism and sport hunting (Bulte & Horan, 2003, 110).

Many wildlife enthusiasts immediately reject proposals for sustainable consumptive use of African wildlife, based on their view that killing animals is ethically unacceptable. They find the idea that conservation can obtain income through using natural resources, particularly where this involves 'harvesting' of wildlife (Hanks, 1979, 165), to be abhorrent. Indeed, southern African conservation agencies are unusual in the degree to which they are able and expected to 'pay their own way'. Whether harvesting takes place by means of culling excess animals or issuing hunting licenses, the whole idea of a conservation area conforming to the economic logic of cattle ranching seems repulsive to many (Ginsberg, 2002, 1185; Du Toit, 2002, 1403–1416). The reasons behind this feeling against utilisation are that in this case human interests stand paramount in determining the value of wildlife, with the implication that whatever humans do not find valuable, can be neglected, abandoned, or wasted. People against this kind of harvesting, or sustainable utilisation of wildlife resources, try to articulate an intrinsic value for conservation areas, assigning value to them that is independent of human concerns and interests.

Note how Michele Pickover (2006) describes the conception of elephants she believes inherent in sustainable use practices: 'intelligent and sentient beings who are capable of deep emotions and who, at the very least, deserve our respect and compassion, are being classified as goods and chattel'. She finds these practices objectionable, arguing that 'using animals as resources to serve human needs is wrong for some of the same reasons that slavery is wrong'. Her view does not mean that she has no compassion with the everyday struggle of poor people to survive. She acknowledges the 'need to focus on and foster other, more sustainable and humane forms of income generation', but she denies that poor communities can only benefit from conservation 'if the animals pay with their lives'. Her alternative is to design 'poverty alleviation programmes ... that avoid animal suffering and take into account respect for other species'.

Can the conflicting ethical theories be harmonised?

Most democratic societies experience reasonable moral pluralism, which means that over a range of issues, reasonable and morally mature adults make conflicting moral judgements on the same issue. Humans in democracies have learned to live with such moral differences about serious matters, such as abortion, by being tolerant towards one another and acknowledging that there are no universally applicable moral principles for solving some moral dilemmas (Willott & Schmidtz, 2002). Of course, there are certain fundamental moral values embodied in a society's conception of justice, such as the injunction not to kill fellow citizens. But even the detailed understanding and application of such absolute moral values do not necessarily rest on full consensus, as we can see in controversies about whether the right to life can be squared with the death penalty or abortion.

Let's take the proposal of the sustainable use of African wildlife through hunting and culling and its critics as an example of the possible resolution of ethical issues in a morally pluralist society. If we live in a human world where we have reasonable differences about serious moral issues (see Gutmann & Thompson, 1996), do those of us whose personal morality does not allow hunting, rejects eating the carcasses of wildlife, and disapproves of animals being killed for human purposes, have a right to prohibit these practices for those of a different opinion? (see Schmidtz, 1997, 327–329).

One must note that a vast majority of people accept the use of cattle, sheep, and pigs as nutrition for human beings or as religious sacrifices. Are there any particular reasons why these commercially used animals should have much less of a moral status than most species of African wildlife? This state of affairs implies that prohibiting commercial use of African wildlife as a sustainable natural resource for Africans to better their lives might be labelled as a case of cultural-ethical imperialism. Do rich, privileged environmental activists – who can afford a healthy vegetarian diet (or neatly packaged meat from a supermarket) – have the right to impose their cultural and personal ethical views about deeply controversial moral issues of hunting and eating meat on poor rural people with centuries-old traditions of sustainable use of wildlife? If these poor communities develop such a deep commitment to the value package African wildlife offers humans, and thus contribute substantially to enlarging areas available to wildlife, should they be refused the chance to do so?

If one takes the claims of people who have lived with African wildlife for centuries seriously, then the idea of a morally pluralist world opens the possibility for legitimate use of elephants through culling and hunting.

There might, of course, be good moral arguments that restrict or reject both these options. Wisely managed culling and hunting are two manifestations of sustainable use that reject the moral standing of higher mammals like elephants. Is that acceptable? Other forms of sustainable use like ecotourist activities, such as safaris, hiking, and camping, are forms of use that respect the moral standing of elephants.

Somehow we will have to learn to engage fellow citizens who have ethical viewpoints substantially different from ours in dialogue. We will have to learn how to deal respectfully with the moral differences between us and our fellow citizens through moral deliberation.

ETHICS AND MANAGEMENT OPTIONS

In an ideal world all humans would treat elephants in ways that appropriately acknowledge and respect their moral standing. Elephants would have enough land available to freely live their lives as they see fit and to migrate to other areas when they deem it appropriate. In such a world humans would have no reason to intervene in their lives. However, we do not live in such a world. As a result of the violent history between our species, the exponential growth in human population, and the resultant loss of elephant habitat, conservationists must explore various management options to create the best life possible for elephants within current constraints.

Translocation

Translocation is at best an experience that traumatises elephants in several ways. The trauma begins with a helicopter flying low over their heads and the elephants being darted. The older cows are darted first to ensure the matriarch goes down quickly. This practice confuses and disorientates the younger ones and they thus do not run off, but stay close to the matriarch. The anaesthetic takes several minutes to knock out an elephant. The elephants are aware of being drugged and that they are losing bodily functions and consciousness. When the elephants awake, they find themselves inside a cramped steel compartment, with humans both injecting them to keep them sedated and prodding them with electric shocks to move them into position.

The captured elephants travel for hours in a semi-sedated condition until they are offloaded in a strange place. Once there they are disoriented – their store of knowledge about the physical features, feeding areas, and waterholes of their home range has been disabled. They must start all over again, this time perhaps without their complete family and bond groups. In translocation operations, reliable and exact selection of a smallish herd is difficult. Some family members might have wandered off on their own, or might be socialising with another herd close by. Selecting a herd from a helicopter can fail to get it right and some close family members might consequently be permanently separated from the herd despite the best intentions of a capture team.

If one weighs and compares the costs and benefits produced by culling or translocating elephants, the limited trauma of translocation and possible separation of members from their herd are not as bad for elephants as to have their lives terminated through culling. For this reason translocation is ethically preferable to culling. To avoid wrong selection of elephants an ethologist with keen observation skills and deep understanding of elephant behaviour ought to work with the capture team.

Besides wrong selection of elephants for translocation, the wrong selection of habitat for the introduction of elephants can also be made. Humans with elephants in their care must ensure that the habitat is suited for elephants, in the sense of (1) having enough space for the normal size of an elephant home range in the relevant kind of habitat with adequate refuge areas, (2) offering adequate food and water sources through various climate cycles, and (3) providing habitats with suitable space and vegetation types to accommodate so-called 'bull-areas'. The habitat set aside for elephants must be appropriate to avoid an unnatural increase of conflict between sexually active bulls, thus giving them fitting 'social landscapes'.

Although the financial cost and required expertise might in some cases limit the use of the translocation option, a far more important factor almost precludes translocation as a serious alternative to culling. Human encroachment on elephant habitat has diminished the land available for elephant relocation. Only small pockets of land are available for the specialised needs of elephants.

Contraception

Contraception is clearly more ethical than culling, as no existing elephants are deliberately killed. Contraception merely prevents elephants being born and thus can be administered to slow down the birth rate to reach the desired population size over a longer period of time.

Contraception thus seems to be a promising alternative that might soon go a long way to satisfy opposition to culling. But note the words used: 'a promising alternative', and 'might soon'. We still have to wait for the outcome of long-term scientific studies with strongly confirmed evidence on the effects of vaccination on elephant physiology and social behaviour. The logistics and cost of vaccination are other complex issues that have not yet been sorted out. There is no ethical justification to use methods in an experimental stage and not yet adequately tested on large elephant populations. There are good reasons for caution when implementing new management strategies for elephant populations. Human understanding of the complexities of elephant life is not yet well enough advanced to be able to predict the outcomes of such management interventions. The consequences of these interventions may also take several years to become manifest, due in part to the longevity of elephants and the complexities of their social structure and their reproductive systems (Whitehouse & Kerley, 2002).

Contraception is not without ethical problems. This invasive method is a drastic human intervention in the bodies of elephants. The possibility that contraception can cause sterility over the longer term must be examined, as well as the effects on cows that normally come into oestrus and mate once every 5-9 years, now coming into oestrus every 15 weeks and mating without falling pregnant (Whyte, 2001, 164). The social effect of fewer calves on the size of herds might not be so problematic, as smaller herds (between 10 and 20) often have kin groups with whom they might rejoin if under stress. The more important issue is that young elephant cows might be denied the process of learning to become a mother. Young female elephants learn how to be mothers from their elders, a process called allomothering. If their own mothers and aunts won't have any calves for five years or more, they might not get the chance to serve their motherhood apprenticeship properly before they give birth for the first time.

Contraception will have to be developed and applied with ethologists with keen observation skills and deep understanding of elephant behaviour, and veterinarians who can monitor physiological impacts.

Culling

Culling is the deliberate killing of animals for the purpose of reducing the size of an animal population. Whilst the scientific jury is still out deliberating whether culling is absolutely necessary in some or all cases for the sake of conserving living organisms, landscapes, and natural processes, this section asks the question: if culling is recommended by scientists, should it be done?

Culling raises serious ethical issues (see Chadwick, 1992, 430-436; Payne, 1998, 213-224).

- Is it wrong to kill special mammals solely for the reason that there are too many?
- If we do have to kill elephants, which methods are the most humane?
- Does the practice of killing the matriarch first and then the others cause unnecessary, though very brief, suffering?
- What is the significance of elephants communicating their experience of culling through infrasound to other herds in a radius of approximately 10 kilometres? (see McComb *et al.*, 2002, 317–329; Larom, 2002, 133–136)?
- Will elephants that are aware of culling practices in or close to their home range become aggressive to humans and threaten tourists as a result?
- Is it ethical to require people to participate in culling and the removal and disposal of carcasses?

Some people do not accept that elephants have any moral standing and thus find no problem in advocating culling for population reduction. In terms of a strongly perceived moral obligation not to harm or destroy animals of exceptional psychological, social, behavioural, and physical complexity, other people argue that culling elephants can only be justified in situations as extreme as those used to justify killing humans in a just war (Lötter, 2006). As in a just war, where the interests of the state and the larger community of citizens override the well-being and safety of the individual, so the interests and well-being of a diverse network of ecosystems and the life forms they sustain can trump the interests of groups of individuals if those individuals threaten the continued well-being of the greater whole. So, according to this view, culling can only be ethically justified if a clear and convincing case can be made that it has a reasonable probability of solving an urgent problem *after all other options have convincingly been shown to have failed* (Lötter, 2006). Analogous to justifying a war in which fellow humans will be killed, culling can be justifiable only as an ethically flawed procedure to be employed under strict conditions. These conditions are as follows.

1. Culling can only be employed to deal with a serious and imminent threat to the continued existence of the rich diversities of the natural world. The intention must be to protect other living beings and their habitats from destruction or significant degradation. Elephants are too special to be killed for anything other than the most serious and weighty reasons (Whyte, 2002, 299).

When only the weightiest moral considerations can justify the killing of elephants, a decision to this effect must be grounded in the best possible information. Reasons for culling elephants must be firmly supported by the best available scientific information. One reason is that the behaviour and circumstances of these adaptable mammals vary quite dramatically. These variations between elephant populations in different geographic locations must be taken into account.

In terms of the preservation of the diverse individuals, species, landscapes, and ecosystems of the natural world, the impact of elephants appropriate for a conservation area ought to be set where that impact can still function to modify the habitat to set up spaces that provide living opportunities for other forms of life. However, the impact cannot cause long-term effectively irreversible degradation of the environment to the point that other populations are jeopardised (see Whyte *et al.*, 1999, 120). Thus, the interests of individual animals or an individual species are made subservient to the well-being of the larger whole.

The complexity of the judgement to determine how many elephants a particular conservation area can accommodate should not be underestimated. There are many variables to take account of and, seemingly, no general rules can be laid down for all climatic conditions and vegetation types. Aristotle's advice about the kind of judgement a virtuous person would make is apposite in a case where people deal with such variations. A virtuous person would not respond either too much or too little, but would respond at the right time, in the right amount, in the right way, and for the right reasons (Rosenstand, 2000, 350). Custodians of wilderness areas are required to make this kind of refined judgement that accurately fits the specific situation at hand.

2. Culling elephants is only ethically acceptable when all other less drastic options have been proven to be fruitless for solving the problem of overpopulation. Culling can never be the only option considered. All other options must be explored to determine if the killing can be avoided at acceptable cost to other interests. For this reason, wildlife managers must peruse all scientific information on all aspects of the elephant problem and be clear in their minds about the goals and purposes of their conservation area. Only if they have explored all other options diligently and urgently to no avail, can they seriously consider culling. If culling is chosen, it must be the only option left to

avoid a clearly defined and highly probable unacceptable outcome. Note that this stricture applies equally to many other forms of elephant management as well; all actions with large consequences must be carefully evaluated against their alternatives.

3. In the process of making a decision on culling, custodians of conservation areas and their scientific advisers must be just and fair in their judgements. They must be able to produce accurate, sufficient, and convincing evidence that the impact of elephants on the habitat of other species and their own has become destructive and excessive. Custodians, responsible for the natural world diversities in their care and accountable in democratic terms to concerned citizens everywhere, must sketch management alternatives, publicise their discussions and debates of the alternatives, and indicate the processes they followed to reach a decision.

4. The aim of culling must be to establish a 'just peace' – that is, a situation where elephants and all other living beings, individuals, and species, can prosper. If conservation managers choose culling they must ensure that they use just enough force to counter the threat, i.e. not one more elephant must be culled than is absolutely necessary. Thus, the number of elephants to be culled must be proportionate to the threat they pose. Only so many elephants must be killed as is necessary to achieve the objective. Our imperfect knowledge and the dearth of accurate foresight will make it difficult to judge correctly every time!

5. Are there elephants that should definitely not be killed, and some that should be killed regardless? In some cases there might be convincing arguments not to select certain elephants as part of a culling programme. One could argue a case that magnificent animals ought to be excluded from culling to be kept for tourist viewing – few people have had the privilege of observing huge tuskers since the ivory slaughter of the 1970s and 1980s in Africa. The case for not killing elephants in special relationships with humans needs almost no argument. For example, to kill elephants that are being studied by elephant researchers violates not only the lives of those elephants, but the emotional and psychological lives of the researchers as well. In addition, it seems pointless to wreck research projects and to waste precious intellectual and financial research investments.

Can one assume that elephants that escaped from protected areas and elephants that cause damage to human property or threaten human lives should automatically be killed? Not so. When deciding this matter, one should take into account human responsibility for fencing protected areas, the cost of returning elephants to the wild, and the efforts required for successful rehabilitation of elephants in appropriate areas.

6. As much as possible of the physical evidence of a cull must be removed from the location of the kill for the sake of the remaining elephants. Elephants are very aware of death and fascinated by the dead bodies of their kin. They show specific reactions when they encounter an elephant carcass or merely dry elephant bones. Some elephant researchers suggest that elephants can recognise the identity of the remains of an elephant if they were known to each other. Carcasses and other evidence must be removed as soon as possible so as not to confront the remaining elephants with the signs of the slaughter and so instil fear in them. It is unimaginable to leave the carcasses for scavengers, fully exposed to the particularly sharp senses of the remaining elephants. It would also be grim to set up Auschwitz-like structures where the carcasses can be burnt.

7. If culling is justified in a specific case, then the meat, hide, and ivory must be used for the benefit of conservation agencies and to support research that ultimately benefits elephants and other species. Such use can also result in projects to set up imaginative partnerships with a conservation area's poor neighbours.

Methods of killing and their impact

If elephants have to be killed, well-trained, professional teams should avoid prolonging any suffering by killing them as humanely as possible in as short a time as possible, and with the least possible disturbance. The killing methods must be as humane as current knowledge and technology allow. Issues that need careful attention are (1) how to reliably select a herd when all close family members are together and none has wandered off elsewhere, so as to avoid leaving some herd members behind on their own and deeply traumatised; (2) to know which animals to shoot first so that the herd does not run away in all directions and some escape the culling with terrible memories of the killings of family members, resulting in deep and long-term trauma; (3) to use only highly trained sharpshooters who almost never miss their target, so as to reduce the suffering of their last moments to a minimum; (4) to avoid using substances like Scoline that immobilise elephants so that they slowly suffocate to death while still being conscious; (5) to use a method of killing that is as instantaneous as possible so as not to prolong the suffering caused by dying.

Current wisdom suggests that if elephants are to be killed, the best option is to cull whole family herds or bachelor herds (Chapter 8). One important reason for killing herds is that young orphaned elephants cannot become proper elephants without the teaching and guidance from older elephants. Elephant adolescents need a hierarchy of seniority determined by age and strength to keep their levels of aggression within limits (Meredith, 2001, 198). Recent scientific research on the occurrence of post-traumatic stress disorder in elephants exposed to the trauma of seeing family members killed suggests the effects of these experiences are significant factors to explain such delinquent behaviour (see Bradshaw *et al.*, 2005, 807).

The technical culling option that would result in the fewest number of elephants being killed to achieve population stabilisation would be to kill young female calves. This precludes their future breeding potential, and it is argued that culling this age group is merely simulating what would happen to this vulnerable age group in the elephant population during a severe drought (Chapter 8). They would be among the first ones to die anyway. Although it might be true, selective culling is still a drastic *human* intervention through lethal means that will cause suffering to the mothers, siblings and the extended family of such youngsters, and will deprive matriarchal herds of their child care assistants. Traumatising elephant herds through human intervention known to them might also affect their behaviour towards humans.

How should we choose between these two options if they are the only ones available? Matters to take into account when considering culling a whole herd are as follows. No elephants will survive with deeply traumatic experiences that might induce behavioural changes that could take years to settle down, if ever. No elephant will remain that might develop a grudge against humans for killing its family members. However, such culling implies destroying a whole herd's genetic pool, which can diminish the genetic diversity of the larger population. Furthermore, the history of the herd will be wiped out, as embodied in decades of memory of acquired knowledge. The herd's unique set of behavioural traits and communicative skills will also be lost. Nevertheless, culling a whole herd is more an intervention in an ecosystem by taking out one 'unit' rather than an intervention in the social lives of elephants by taking out one or more of their family members.

To cull individual elephants raises serious concerns as well. Human intervention through selective culling may result in serious psychological trauma that could disrupt social behaviour for every herd targeted in this way.

To cull the same number of elephants as through the removal of a herd, many more herds will have to be traumatised through selective culling. But perhaps elephant herds are capable of dealing with the stress and trauma of individual deaths? One may doubt whether selective culling can be done without elephants detecting humans as the source responsible for the death of family members. For this reason elephant attitudes towards humans may deteriorate. Selective culling thus seems a much more direct and widespread human intervention in the personal and social lives of elephants that might have negative impact on their lives and their relationships with us, but this needs to be balanced against the smaller number of elephants, in total, that would need to be culled.

The remarks above are still speculative, as the impact and consequences of selective culling have not yet been adequately studied. This matter requires deeper reflection, as well as intense discussion between specialists and citizens.

Hunting

Is it a good idea to allow the hunting of elephants? Elephant hunting is allowed in six African countries (South Africa, Zimbabwe, Botswana, Namibia, Cameroon, and Tanzania) at approximately R70 000 per trophy animal (Owen, 2006, 83). The controversy about hunting, says Chadwick (1992, 121), is 'universally such a bitter, emotionally charged disagreement'. Many committed conservationists are opposed to hunting on moral grounds, while others find it perfectly acceptable.

The arguments in favour of hunting African fauna and flora are as follows (see Fakir, 2006). Some people claim that humans have evolved in such a way that an 'instinct for hunting' became hardwired in their brains. For this reason it is part of human nature, a kind of instinctual drive that produces the desire to hunt animals. Some hunters acknowledge that part of the thrill of hunting is to experience the power of killing, of pursuing, outwitting, and eventually taking the life of a prime specimen of a species. This does not mean that a hunter has no sympathy with the prey.

The supporters of hunting point to its valuable consequences. As a result of sport or commercial hunting large areas of land are now again used for conservation purposes instead of cattle farming. The income from sport hunting is substantial and leads to significant job creation. In many cases hunters have ethical codes to regulate hunting adventures. Ideas about 'fair chase' abound that require the hunter to use tracking and other skills to outwit the animal. In some cases the hunter is required to confront 'the natural fierceness of the animal, its threat to one's own life, and experiencing the fear of the hunt as the animal fears being preyed upon by the hunter' (Fakir, 2006). In this context the link between a particular conception of masculinity and hunting becomes clear as well. Learning to deal with danger, fear and the conflicting emotions accompanying deliberate killing is regarded as things that have educational value because 'it makes a man out of you' (see Fakir, 2006).

There are several arguments that oppose hunting. Whilst Fakir (2006) can see some role for hunting that has human survival or religious sacrifice as goal, he has little sympathy for sport hunting that serves the interests of a small minority of well-off people. He objects to the idea that the only value animals have is to be a 'pleasurable utility to serve the hedonistic needs of humans'. Besides the possibilities for abuse that have so often manifested in the past when governments allowed unregulated hunting, Fakir's biggest concern is whether hunting animals for pleasure and entertainment can in any way be squared with the regard we ought to have for animals, the moral standing we are obliged to assign them.

If elephants indeed have the moral standing we earlier in the chapter argued for, Fakir's ethical problem intensifies. If elephants in so many respects share the characteristics that give humans moral standing, can humans hunt them for fun, adventure, and the satisfaction of a presumed but unproven instinctual drive? The hunting issue thus becomes much more complex when hunting elephants is specifically considered. Hunting clearly has negative and harmful consequences on elephants, perhaps much more so than on any other species of African wildlife. One such negative effect is their hostile or nervous reaction to humans in response to being shot at.

Is it strange that hunters often describe regularly hunted elephants as 'very aggressive'? Not really, if we take into account recent research that suggests elephants too suffer from PTSD, almost just like humans. Bradshaw *et al.* (2005, 807) note that many wild African elephants display typical PTSD symptoms, like 'abnormal startle response, depression, unpredictable asocial behaviour, and hyperaggression'. They ascribe PTSD directly to elephant society being 'decimated by mass deaths and social breakdown from poaching, culls, and habitat loss'. Thus, killing elephants can have harmful consequences that persist for decades in elephant society, as 'trauma early in life has lasting psychophysiological effects on brain and behaviour' (Bradshaw *et al.*, 2005, 807). Whether controlled professional hunting has the same impact as poaching or culling is yet to be determined.

Taming and training

Although this Assessment explicitly does not deal with captive elephants, the capture of elephants from the wild has sometimes been mooted as an alternative to culling them. Note that the South African Elephant Norms and Standards do not permit this. However, a brief discussion of the relative ethical merits of this option is included here for completeness.

Many people have firmly believed in the past that the African elephant cannot be tamed and trained.

Claims that new training methods are used that apparently successfully tame and train African elephants require careful scrutiny. The use of such tamed and trained elephants in the tourism industry in South Africa has not yet been studied in depth by scientists. A few preliminary ethical remarks can be made in the light of the available information (see Van Wyk, 2006a & b; 2007).

If their training could be judged ethically justifiable, if they belong to a newly constituted bond group, if they have a daily option of returning to the wild, and if the elephants are not required to do demeaning, humiliating tricks, are there any counter-arguments against the taming and training of elephants? There are strong arguments available. Some people consider the taming of African elephants unnatural and thus unbecoming such wonderful animals. Elephants ought not to be used as mere objects for commercial exploitation and also not as instruments for human recreational and tourist purposes either. Others judge it immoral to separate young elephants between the ages of 8 and 11 years old from family herds to train them, as they are still in need of the contact and guidance of the older elephants in the herd.

Most people find such close encounters with elephants awe-inspiring, much like the close contact between humans and dolphins. If humans who have had such experiences as a result of interaction with captive elephants develop a deep appreciation for elephants and fight for their survival in the remaining areas of African wilderness, have these elephants not served their species well as ambassadors of good will, rather than having been killed? Even if one can answer yes to these questions, taming and training elephants should not be done if appropriate respect cannot be shown to them. Whether taming and training allows proper respect for elephants requires further investigation and debate.

CONCLUSION

Even before scientific information gives a clear and unambiguous picture of the nature and consequences of elephant overpopulation, the ongoing debate about the most ethical 'management plan' has to take place. We need to critically examine our moral values, assign them priorities, and choose which ones we are prepared to violate or ignore in our attempts to balance the competing claims they make and the contrary implications they suggest. Ethical matters that are not yet clear enough are the nature and current status of traditional African values on the environment and wildlife, the wisdom of selective culling of young elephants, the ethical acceptability of taming and training African elephants for the tourism industry, and the links between local action to prevent habitat degradation due to elephant impact on protected areas and global action to reverse human impact on the earth's biosphere.

The way we deal with elephants and one another in debating elephant issues betrays the quality of our humanity. Can we deal with deep moral conflict in ways that still show respect for one another and value one another's contributions to solve intractable moral problems? Can we continue the conversation regardless of our differences and still listen attentively to both the contents and justification of the viewpoints of our opponents? The deep emotions associated with the debate on elephants threaten to overwhelm the tolerance and critical reasoning we require for meaningful engagement through dialogue. Hopefully we can interrogate and engage our emotions fittingly so that we will always treat our opponents in debate respectfully as fellow human beings with dignity and equal worth.

Our humanity will also be tested in our interactions with elephants. Can we treat elephants appropriately as beings dependent on our benevolence for opportunities to live their lives according to their lights? Can we use our vast store of knowledge about nature, ecosystems, mammals, and elephants to fully respect elephants for what they are: beings so close to us and yet so impressively different? Through our astonishing cultural evolution we have become the most knowledgeable and powerful species ever to set foot on earth. Do we want to live up to our species name, *Homo sapiens,* in our interaction with elephants? Can we live as the 'wise beings', those who understand the most about all forms of life on earth? Can we then appropriately value elephant lives and accordingly act respectfully towards them?

Although the elephant issue is of minor consequence compared to, for instance, to the challenges of global climate change, both these matters offer us an opportunity to question and revise how we live decent human lives on this planet.

REFERENCES

Antonites, A. 2007. Elephant management – a philosophical ethical approach. Written submission to the Assessment of South African Elephant Management.

Begg, A. 1995. The great elephant debate: to cull or not to cull. *African Wildlife* 49(4), 6, 7, 9.

Bell-Leask, J. 2006. An elephantine ethical dilemma. Paper presented at the Elephant Symposium at the Annual Conference of the Ethics Society of South Africa (ESSA), 11–13 September 2006.

Bonner, R. 2002. At the hand of man: peril and hope for African wildlife. In: D. Schmidtz & E. Willott (eds) *Environmental ethics: what really matters, what really works.* Oxford University Press, New York, 306–319.

Bradshaw, I.G.A., A.N. Schore, J.L. Brown, J.H. Poole & C.J. Moss 2005. Elephant breakdown. Social trauma: early trauma and social disruption can affect the physiology, behaviour and culture of animals and humans over generations. *Nature,* 433, 807.

Bulte, E.H. &. Horan, R.D. 2003. Habitat conservation, wildlife extraction and agricultural expansion. *Journal of Environmental Economics and Management,* 45, 109–127.

Chadwick, D.H. 1992. *The fate of the elephant.* Sierra Club Books, San Francisco.

Cohen, C. & T. Regan. 2001. *The animal rights debate.* Rowman & Littlefield, Lanham.

Cumming, D.H.M. & G.S. Cumming 2003. Ungulate community structure and ecological processes: body size, hoof area and trampling in African savannas. *Oecologica* 134, 560–568.

De Waal, F.B.M. & P.L. Tyack 2003. *Animal social complexity: intelligence, culture, and individualized societies.* Harvard University Press, Cambridge, Mass.

Douglas-Hamilton, I. 1975. *Among the elephants.* Collins and Harvill Press, London.

Du Toit, J.T. 2002. Wildlife harvesting guidelines for community-based wildlife management: a southern African perspective. *Biodiversity and Conservation* 11, 1403–1416.

Fakir, S. 2006. Sentience and sensibility: considering the ethics of hunting. Manuscript submitted to the Assessment of South African Elephant Management.

Ginsberg, J. 2002. CITES at 30, or 40. *Conservation Biology* 16(5), 1184–1191.

Gould, C.C. 2002. Does stakeholder theory require democratic management? *Business & Professional Ethics Journal* 21(1), 3–20.

Gröning, K. with text by Saller, M. 1999. *Elephants: a cultural and natural history.* Könemann Verlagsgesellschaft.

Gutmann, A. & D. Thompson. 1996. *Democracy and disagreement.* Harvard University Press, Cambridge, Mass.

Hanks, J. 1979. *A struggle for survival: the elephant problem.* C. Struik Publishers, Cape Town.

Larom, D. 2002. Auditory communication, meteorology, and the *Umwelt. Journal of Comparative Psychology* 116(2), 133–136.

Leopold, A. 1981. *A sand county almanac.* Oxford University Press, Oxford.

Lötter, H.P.P. 2005. Should humans interfere in the lives of elephants? *Koers* 70(4), 775–813.

Lötter, H.P.P. 2006. The ethics of managing elephants. *Acta Academica* 38(1), 55–90.

Mapaure, I.N. & B.M. Campbell. 2002. Changes in miombo woodland cover in and around Sengwa wildlife research area, Zimbabwe, in relation to elephants and fire. *African Journal of Ecology* 40, 212–219.

McComb, K.D.R, L. Baker, C. Moss & C. Sayialel 2002. Long-distance communication of acoustic cues to social identity in African elephants. *Animal Behaviour* 65, 317–329.

Meredith, M. 2001. *Africa's elephant: a biography.* Hodder & Stoughton, London.

Moss, C. 1988. *Elephant memories: thirteen years in the life of an elephant family.* Elm Tree Books, London.

Moss, C. 1992. *Echo of the elephants: the story of an elephant family.* BBC Books, London.

Mosugelo, D.K., S.R. Moe, S. Ringrose & C. Nellemann. 2002. Vegetation changes during a 36-year period in northern Chobe National Park, Botswana. *African Journal of Ecology* 40, 232–240.

Mutwa, C.V. 1996. *Isilwane. The animal. Tales and fables of Africa.* Struik Publishers, Cape Town.

Osborn, F.V. & G.E. Parker. 2003. Towards an integrated approach for reducing the conflict between elephants and people: a review of the current research. *Oryx* 37(1), 80–84.

Owen, C. 2006. Trophy hunting: a sustainable option? *Africa Geographic* 14(3) (April), 83.

Payne, K. 1998. *Silent thunder: in the presence of elephants.* Penguin Books, New York.

Payne, K. 2003. Sources of social complexity in three elephant species. In: B.M. Frans de Waal & P.L. Tyack (eds) *Animal social complexity: intelligence, culture, and individualized societies.* Harvard University Press, Cambridge, Mass., 58–85.

Pickover, M. 2006. The reprehensible dictum of 'If it pays it stays.' An elephant perspective. Paper presented at the Elephant Symposium at the Annual Conference of the Ethics Society of South Africa (ESSA), 11–13 September 2006.

Ramose, M. 2007. Written submission on request to the Assessment of South African Elephant Management.

Regan, T. 1984. *The case for animal rights.* Routledge & Kegan Paul, London.

Regenstein, L. 1985. Animal rights, endangered species and human survival. In: P. Singer (ed.) *In defense of animals.* Basil Blackwell, New York, 118–132.

Rolston III, H. 2002. Values in and duties to the natural world. In: D. Schmidtz & E. Willott (eds). *Environmental ethics: what really matters, what really works.* Oxford University Press, New York, 33–38.

Rosenstand, N. 2000. *The moral of the story. An introduction to ethics.* Third edition. Mayfield Publishing Company, Mountain View, California.

Sagoff, M. 2002. Animal liberation and environmental ethics: bad marriage, quick divorce. In: D. Schmidtz & E. Willott (eds) *Environmental ethics: what really matters, what really works.* Oxford University Press, New York, 38–44.

Schmidtz, D. 1997. When preservationism doesn't preserve. *Environmental Values*, 6(3), 327–339.

Schmidtz, D. & E. Willott 2002. *Environmental ethics: what really matters, what really works.* Oxford University Press, New York.

Singer, P. 1985. Prologue: ethics and the new animal liberation. In: P. Singer (ed.) *In defense of animals.* Basil Blackwell, New York, 1–10.

Singer, P. (ed.). 1985. *In Defense of Animals.* Basil Blackwell, New York.

Taylor, P.W. 2002. The ethics of respect for nature. In: D. Schmidtz & E. Willott *Environmental ethics: what really matters, what really works.* Oxford University Press, New York, 83–95.

Van Wyk, S. 2006a. On the backs of giants. *Africa Geographic* 14(7) (August), 54–64.

Van Wyk, S. 2006b. Riding high. *Africa Geographic* 14(3) (April), 84.

Van Wyk, S. 2007. Wild rides. *Africa Geographic* 15(9) (October), 52.

Whitehouse, A.M. & G.H.I. Kerley 2002. Retrospective assessment of long-term conservation management of elephants in Addo Elephant National Park, South Africa. *Oryx* 36(3), 243–248.

Whyte, I.J. 2001. Conservation management of the Kruger National Park elephant population. Ph.D. thesis, Pretoria University.

Whyte, I.J. 2002. Headaches and heartaches: The elephant management dilemma. In: D. Schmidtz & E. Willott (eds) *Environmental ethics: what really matters, what really works.* Oxford University Press, New York, 293–305.

Whyte, I.J., H.C. Biggs, A. Gaylard & L.E.O. Braack 1999. A new policy for the management of the Kruger National Park's elephant population. *Koedoe* 42(1), 111.

Willott, E. & D. Schmidtz 2002. Why environmental ethics? In: D. Schmidtz & E. Willott (eds) *Environmental ethics: what really matters, what really works.* Oxford University Press, New York, xi–xxi.

10

THE ECONOMIC VALUE OF ELEPHANTS

Lead author: James Blignaut
Authors: Martin de Wit and Jon Barnes

INTRODUCTION

ELEPHANTS PLAY a huge role within any landscape where they occur. They are habitat engineers. As charismatic species they awaken emotions among people like few others. As keystone species, they contribute significantly to the integrity of ecosystems and must be very carefully managed. From an economic perspective, they are also value generators. In this broad context, we first consider the range of relevant economic values, using the Total Economic Value approach in a generic sense, and then apply this framework to identify the specific factors that determine the economic value of elephants in South Africa. Thereafter we summarise both regional (southern African) and international studies that consider the economic value of elephants. We conclude with an assessment of the state of knowledge on elephants' contribution to the economic value of elephant-containing ecosystems and the economy as a whole.

This assessment borrows heavily from studies concerning the economic value of elephants carried out in Botswana, Namibia, and Zimbabwe, since similar studies in South Africa could not be located. To date, published studies in South Africa focused either on the cost of the individual elephant management options – which is a subject treated in the relevant management chapters of this book – or else investigations of the value of tourism. The specific contribution of elephants to the value of tourism was not isolated in these studies.

BACKGROUND ON ECONOMIC VALUE

Adam Smith, the 'father of modern economics', distinguishes between two types of economic values: exchange values and use values. He clarifies as follows (quoted from reprint in Smith, 1997, 131):

> The word VALUE ... has two different meanings, and sometimes express the utility of some particular object, and sometimes the power of purchasing

other goods which the possession of that object conveys. The one may be called 'value in use'; the other, 'value in exchange'. The things which have the greatest value in use have frequently little or no value in exchange; and, on the contrary, those which have the greatest value in exchange have frequently little or no value in use.

He explains the distinction between exchange and use value by referring to the well-known water-diamond paradox. Nothing is more useful than water, yet it has almost no exchange value. In contrast, diamonds have relatively little *real* use, but have extremely high exchange values. Exchange values are easy to observe. They are the market values of a product, good, or service. Use values, however, are not observed. If care is not taken one could easily ignore these use values when making decisions. The economic valuation of ecosystem goods and services is an attempt to mitigate the impact of either the absence of markets or the wrong signals markets send by estimating the value of natural capital in terms of what these resources contribute to society. Some are opposed to the quantification of the value of natural resources (McCauley, 2006), but most of these antagonists are ignorant about the way economists distinguish between the environment's use value and exchange value. Ecological economists are fully aware of the fact that it might not always be possible, or even necessary or desirable, to estimate the use value of a resource – especially when dealing with so-called *critical* natural capital (Ekins *et al.,* 2003; Farley & Gaddis, 2007; Blignaut *et al.,* 2007). Yet, by estimating the values that are deemed appropriate, economists acknowledge the fact that environmental values exist and that they contribute meaningfully and significantly to social welfare.

Figure 1 provides a breakdown of the suite of environmental values by first distinguishing between the primary and secondary value of the environment. Primary values – values without an economic purpose – are also called intrinsic values and reflect the non-demand values of ecosystems. In some instances, primary values could also be considered as the value of life itself.

Economists do not place a monetary value on these, but often take cognisance of them in a qualitative sense. Ecosystems' secondary values, also called the Total Economic Value (TEV) of ecosystems, comprise direct, indirect, option, existence, and bequest values. See box 1 for a discussion as to the different components of TEV.

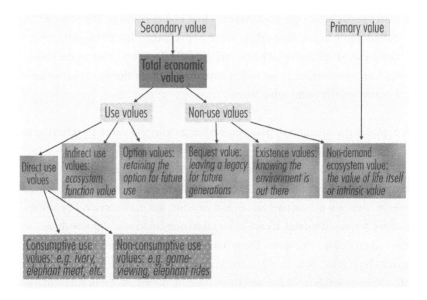

Figure 1: Values for the environment (adapted from Turner *et al.*, 1994)

In the next section we discuss this suite of values with specific reference to elephants.

FACTORS DETERMINING THE TOTAL ECONOMIC VALUE (TEV) OF ELEPHANTS

The TEV of elephants cannot be calculated by summing up all the animal's use and non-use values. There is conflict, even 'rivalry', among some of the categories. For example, the direct consumptive use of an elephant for its ivory excludes the possibility to enjoy any non-consumptive or non-use value from that individual animal. The direct consumptive use of the individual, however, does not – at least theoretically – exclude any non-consumptive or non-use value of the population as a whole. In some cases the direct consumptive use of a resource could have a negative impact on non-use values, depending on how people act and react to such direct use. This is due to the fact that non-use values are driven by perceptions and heavily influenced by specific contexts, which can change over time and in response to events. Neither are these values easily transferable from one setting to another.

The impact of elephants on their surroundings can also lead to a decline in the TEV of the return on the ecosystem in general. If not managed properly, elephants can lead to environmental degradation. Such degradation could

Box 1: Total Economic Value (TEV): A description

Direct use values are often exchange values since markets can exist for them. The estimation thereof is conceptually straightforward, but not necessarily easy. The fact that markets do (or can) exist does not imply that they are functioning well. Market imperfections such as legislations, trade-bans, and spatial and temporal differences between resources, can distort such a market and hence the market outcome. Direct use values can be sub-categorised as:

- consumptive use values (e.g., elephant meat, ivory, trophy hunting)
- non-consumptive use values (e.g., game-viewing, elephant rides, etc.).

Indirect use values correspond closely to the value of ecosystem functions (e.g., watershed protection, carbon sequestration, nutrient recycling). In the past these values tended to be use values but this is changing, with the advent of the carbon and water markets, and they are increasingly becoming exchange values. Biodiversity markets, however, are not well developed yet and the role an individual species, such as an elephant, plays within an ecosystem is also not isolated within this market. This is not to imply that this cannot change in future. Much discussion is under way to develop a biodiversity market of which both South Africa and all of southern Africa could be beneficiaries. Indirect use values are, however, not just positive. Individual species, such as an invasive alien plant, can have a negative impact on the social and economic value, and the ecological functioning of an ecosystem in general, and likewise the over-population of an endemic species such as an elephant can be globally negative.

Option value is an expression of an individual's preference not to make use of a resource today because he/she prefers to retain the option to use the resource in future and, therefore, is willing to pay for today's conservation to retain the option for any possible future use.

Bequest value is a measure of an individual's willingness-to-pay to ensure that an environmental resource is preserved for the benefit of his/her descendants. Bequest values are non-use values for the current generation, but a potential future use or non-use value for their descendants.

> Existence value measures the willingness to pay for the preservation of the environment that is not related to either current or optional use, thereby being the only true 'non-use' value. Existence values are based on the concept of the environment [or an individual species] being there. In some cases, bequest values are treated as part of existence values as it is often difficult to differentiate between the two on an empirical level.

lead to a loss in ecosystem function (indirect use value), which not only implies a loss in ecosystem productivity and resilience, but also the need for ecosystem restoration. The damage to field crops by elephants that escape from conservation areas and the ensuing challenges between humans and elephants are a direct cost to the affected human community. But this cost is not reflected in, for example, the value an international tourist derives from viewing elephants in the protected area where the damage-causing individual lives. This implies that space and context matter when considering economic valuation. Additionally, partial analyses may skew perception of the TEV. For example, should a study only focus on one aspect of the total economic value, say its non-consumptive use value, but not consider any other value – such as the loss of plausible consumptive use values or its nuisance value – this can lead to partial or even wrong conclusions. It is best to consider the suite of values as a package and, from an economic vantage point, optimise the suite of them rather than any one individual component. This implies the need for systems thinking and adaptive management, well informed by good data.

Lastly, two entrenched problems, inherent to all forms of economic valuation, are the issues of time and income difference. Studies have to make adequate provision for both the time preference of money – which usually depreciates over time – and the change in value of ecosystems goods and services – which often increases over time, should they become more scarce due to habitat loss. As for income differences, often communities adjacent to conservation areas are poor, while visitors to the park are affluent. These two constituencies tend to value and evaluate a resource such as elephants quite differently because of their different perspectives, and their different relationships with, or uses of, elephants. One has to consider and seek to either optimise the value of the system as a whole or to manage it sustainably and not just that of an individual value.

Most of the economic valuation studies of elephants done in the past focused on direct consumptive use value. Since 1989, when the African elephant

Box 2: Non-consumptive use values of elephants

Direct (non-consumptive) use: Within the tourism industry, elephants are important drawcards or attractions. The benefits of elephants within the ecosystem from a tourism perspective include direct income to households through employment, ownership, or equity in tourism-linked businesses, as well as foreign exchange earnings for the government, and government income through taxation of individual earnings, sales taxes and corporate taxes. It is, however, costly and a management-intensive exercise to host elephants. Elephant tourism options include either low numbers/high paying options (no self-drive; overnight lodges) or high numbers/low budget options (self-drive and camping or self-catering lodges). Elephant-related tourism expenditure is therefore a good indicator of people's willingness to pay for them.

Indirect use: Elephants are a keystone species in any biome where they occur and they play an important biological role in ecosystem functioning, ensuring the survival and continued evolution of many species. These values are generally not measured and can go two ways. One could value the indirect value of elephants either as an umbrella species, and therefore incorporating a range of other values in their value as well, or, individually by considering its role in the ecosystem. This could be positive, as an important habitat engineer, or negative, as a megaherbivore whose actions can lead to ecosystem degradation requiring restoration and intensive management. This is especially the case when population densities become too high.

Non-use values: There is an ongoing global concern for the continued existence of elephants. This concern is expressed mainly in the form of donations focusing on the protection of the elephant. In Kenya, for example, the elephant conservation industry is largely dependent on this form of money transfer for its continued survival. How sustainable and efficient it is, however, can and is being questioned (Norton-Griffith, 2007). Wildlife policies create the enabling environment for wildlife conservation, also for elephants, which, if designed appropriately, will be conducive to both conservation and the development of economic opportunities through markets. Market mechanisms can be developed to harness the non-use values of elephants in conjunction with their direct and indirect use values. (Based on Geach, 1997.)

was listed in Appendix I of the CITES list of endangered species (becoming effective in 1990), the direct consumptive use of elephants has been reduced dramatically and is effectively zero at present. Over time, however, it is likely to recover some of its importance thanks to the ongoing debate within CITES, especially between China, Japan and the other Far Eastern countries, on the one hand, and Europe and the United States on the other. The Far Eastern countries view the CITES trade ban as unnecessary and would like to see it annulled. By and large, the countries in southern Africa also support the removal of the trade ban, but for completely different reasons. They are concerned with the impact of their large and increasing populations of elephants on their habitat (see Chapter 3). Together, these countries form a lobby canvassing for the lifting of the ban, either in full or in part. Relaxation of the ban will lead to a new series of economic drivers influencing elephant conservation management. Such a change would also affect other, non-consumptive use factors, which determine the TEV of elephants, as is listed in box 2.

LITERATURE OVERVIEW

Southern Africa

Several studies estimating the economic value of elephants have been undertaken in Botswana, Namibia, and Zimbabwe. Nearly all of this work focused on direct use values associated with the elephant. Policy in all three countries is aimed at promoting generation of income and employment from wildlife, and research has thus been focused primarily on the value of elephant utilisation.

Prior to the Appendix I CITES listing of the African elephant, Child & Child (1986) and Child & White (1988) documented the financial values associated with elephant culling, which was being undertaken at that time in Zimbabwe to control the growing numbers of elephants in national parks. They showed that the culling programme, operated by a special unit within government, was profitable. Sales of ivory and dry, salted hides exceeded the costs of low-budget culling of matriarchal herds in the national parks. In addition, low-quality dried meat was provided cheaply to neighbouring communities in an attempt to engender local support for elephant conservation by offsetting the need for poaching for bush meat. The numbers culled varied between 800 and 1 500 per annum.

In 1989 the Botswana Department of Wildlife and National Parks undertook an analysis of the options for utilisation of its large and rapidly growing elephant population. At that time, the only use of elephants was non-consumptive, as part of the general wildlife viewing experience. Hunting was banned and culling had not been introduced. The Appendix II listing for elephants at the time would have allowed reintroduction of elephant hunting and the introduction of a culling programme. Soon after that, initiatives among the CITES parties were made to have elephants listed in Appendix I. This was enacted in 1990, effectively closing all trade among CITES parties in consumptive products for the species. Botswana, which was against the listing, undertook a study to compare the economic values of the options for use of its elephant resource. Barnes (1990) estimated and documented the contribution that use of elephants for wildlife-viewing tourism, trophy-hunting tourism, hunting by citizens, and culling, could make to Botswana's national economy. This was followed by analyses for 1990 and 1992 of the effects that the international policy environment had on these values (Barnes, 1992; 1996a). The studies involved detailed financial and economic, budget/cost-benefit models of wildlife viewing activities in elephant areas, trophy hunting, and elephant culling, as developed by Barnes (1998). These models were based on empirical evidence from users, including data from the elephant use activities in Zimbabwe. The proportions of value attributable specifically to elephants were estimated as representing 41 per cent of wildlife viewing value, and 37 per cent of trophy hunting value. The models provided measures of the private profitability for the investor, as well as the net contribution of the activity to the national income. The net present value of various combinations of this income over 15 years, taking into account policy and plans for development of utilisation in the wildlife sector, were estimated, as summarised in table 1 (see Barnes, 1996a and 1998 for the details on the research methods employed).

As indicated in table 1, among the list of options for elephant use in Botswana in 1989, the combination with the highest value is Scenario 6, which contained all possible uses except hunting by citizens. To a large extent, elephant-viewing tourism, trophy hunting, and elephant culling were complementary spatially, allowing the highest values to be generated. The introduction of trophy hunting and culling of elephants was assumed to have a moderate effect on the values of elephant viewing through disturbance. In 1990, after the Appendix I listing, trophy hunting under quota was still permitted, and the option of culling was still a possibility, with some products marketed domestically and to non-CITES parties. Since 1990, culling could therefore add very little to the economic use value of Botswana's elephants, implying that the CITES listing effectively reduced the use value of elephants by some 47 per cent, as represented by the decline in value from P293 million in 1989 to P155 million in 1990 (table 1).

Scenario (option)	15 year present value @ 6%[a] (Pula million: 1989)[b]	
Viewing only with no consumptive uses	108.9	108.9
Viewing with trophy hunting only	153.2	153.2
Viewing with hunting by citizens only	130.7	–
Viewing with culling only	248.7	110.5
Viewing, trophy hunting, hunting by citizens and culling	282.3	–
Viewing, trophy hunting and culling	293.5	155.3

[a] Cumulative contribution to gross national income by year 15, after discounting at 6% per annum and after partial shadow pricing

[b] In 1989 Pula 1.00 was equal to ZAR 1.32 and US$ 0.51; Pula inflation factor from 1989 to 2007 is 3.50

Table 1: Present values of increases in Botswana's gross national income over 15 years, attributable to options for elephant management (1989 and 1990 analyses) (source: Barnes, 1996a; 1998)

Expenditure category[c]	15 year net present value @ 6%[a] (P million, 1992)[b] Utilisation option			
	Viewing only (no consumptive use)	Viewing with trophy hunting only	Viewing with hunting by citizens only	Viewing with culling only
Base case (costs rising from P16 to P242 per square km over 15 years)	123.5	181.5	122.6	181.2
Slow increase (costs rising from P16 to P510 per square km over 15 years)	84.0	142.0	83.2	141.8
Medium increase (costs rising from P16 to P510 per square km in first 10 years)	−1.5	56.5	−2.3	56.3
Fast increase (costs rising from P16 to P510 per square km in first 5 years)	−20.0	37.8	−20.9	37.6

[a] Value added over 15 years to national income, net of government expenditures, after discounting at 6% and after shadow pricing (April, 1992)

[b] In 1992 Pula 1.00 was equal to ZAR1.34 and US$ 0.47; Pula inflation factor from 1989 to 2007 is 3.02

[c] Different patterns of increase to a stable maximum for government expenditure on elephant management over the northern range (49 000 square kilometres)

Table 2: Effect of different scenarios for government expenditure on elephant management on economic net present values of different options for elephant utilisation in Botswana (1992 analysis) (source: Barnes, 1996a; 1998)

A second analysis carried out two years later, in 1992, showed similar results (Barnes, 1996a). Culling was not able to generate additional national income due to the restrictions on the ivory market. Elephant trophy hunting could, however, increase the value added by between 36 per cent and 58 per cent, depending on how much it disturbed elephant viewing activities. At the same time a cost-benefit analysis was conducted (Barnes, 1996a), comparing predicted national income streams generated from different possible use options with predicted government expenditure streams for elephant conservation. Future net income streams with management costs increasing to P242 per km² over 15 years generated positive returns in national income for all options. When costs were increased to P510 per km² (i.e. US$246.km⁻², after taking inflation and exchange rate fluctuations into account), as might occur with a surge in poaching, the inclusion of elephant trophy hunting was an important factor in ensuring a positive return for investment in elephant conservation. Table 2 shows the results of this analysis.

Table 3 shows the breakdown of value in terms of potential contribution to national income for all the different elephant products when all uses were included under conditions prevailing in 1989, 1990 and 1992. The salient point is that the culling values, which would have amounted collectively to 40 per cent of the total elephant use value in 1989, were reduced to negligible levels after that. The analysis of Barnes (1996a; 1998) provided evidence of the negative impact of the Appendix I listing on the economic viability of elephant conservation in Botswana. Combating elephant poaching for ivory was the prime motivation for the Appendix I listing, but this eliminated all culling values. It is noteworthy that values attributable to ivory (ivory sales and ivory carving in table 3) made up only 42 per cent of the total value of culling which was lost with the listing. Southern African countries have been trying to re-establish ivory markets within the CITES framework, but even if this is successful, it is unlikely that the 1989 markets for other elephant culling products, such as hides, could be revived.

Culling as a use option appears to have irreversibly lost the economic viability it had in 1989. In addition, culling as an activity has increasingly faced opposition from an animal rights perspective (see Chapter 9). Recent elephant utilisation policy in Botswana has allowed for a combination of elephant viewing and elephant trophy hunting only, with culling retained as a possible option for management purposes only. Since loss of culling value has resulted from attempts to conserve elephants, an argument could be made for compensation through the capture and transfer to Botswana of international non-use values for elephants.

Work on the economics of consumptive tourism (i.e. recreational hunting) in Namibia and Botswana (Novelli *et al.*, 2006) has shown that trophy hunting occupies a spatial niche that is complementary to and does not oppose or displace wildlife viewing tourism. The inclusion of elephants in trophy hunting quotas adds significant value to trophy hunting tourism. In addition to the elephant trophy fees, income from daily hunter fees is enhanced by the inclusion of a high-value elephant in the hunting bag. Using data from a northern Botswana trophy hunting enterprise model (Turpie *et al.*, 2006), and comparing values from trophy hunting in Botswana, where elephants are important (ULG Northumbrian, 2001), and Namibia, where less valuable plains game species are important (Novelli *et al.*, 2006), it was possible to impute a proportion of hunting income to elephants. Based on these calculations we estimate that some 44 per cent of the income from an elephant-inclusive hunting experience in northern Botswana is attributable to elephants.

	Year of analysis		
	1989	1990	1992
Total present value[b] (Pula million, 1989)[c]	293.5	155.3	133.0
Use category (%)			
Tourism – viewing	44.2	70.1	71.3
Tourism – trophy hunting	16.4	26.0	26.5
Culling – raw ivory	8.7	2.3	–
Culling – ivory carving	7.9	–	–
Culling – fresh or dried meat [d]	0.8	1.2	0.8
Culling – meat processing [e]	11.6	–	0.3
Culling – dry salted hides	6.6	–	0.6
Culling – hide tanning	3.7	–	0.2
Culling – live sale (calves) [f]	0.2	0.4	0.3
Total	**100.0**	**100.0**	**100.0**

[a] Management option 6, which included viewing, trophy hunting and culling for each year of analysis
[b] Present values for June 1989 and October 1990, and net present value for April 1992; all at 1989 prices
[c] In 1989 Pula 1.00 was equal to ZAR1.32, and US 0.51; Pula inflation factor from 1989 to 2007 is 3.50
[d] Carcass value after field recovery and field dressing
[e] Including (in 1989) use of meat as feed in crocodile breeding and rearing for production of skins and meat, and (in 1992) production of carcass meal
[f] Sale of calves between six months and one year old

Table 3: Proportional contributions of different products to the economic present values of elephant uses[a] in Botswana in the 1989, 1990, and 1992 analyses (sources: Barnes, 1996a; 1998)

No such comparative studies for South Africa have been conducted, but the live sale of elephants and the occasional hunting thereof on private land are permitted and the values known. Table 4 provides an overview of the average prices and number of trades over the past three years for various categories of animals. The trade in the number of live animals is restricted since conservation areas have commonly reached their carrying capacities. Trades are therefore restricted to private game farms. Similarly, the number of animals available for hunting is restricted by the fact that only animals from private game farms are eligible. The price per elephant, whether as a live sale or for a hunt, is very high, but this is attributable to the restricted nature of the market. It is therefore not possible to derive a total market value for all elephants in South Africa from these numbers. This is also the case in South Africa's neighbouring countries. The trophy values in the neighbouring countries are much lower, though. In Zimbabwe, for example, the trophy fee, set by government, for an elephant was US$10 000 for 2006/2007.

This figure is lower than that of 2000/2001, which was US$15 000, due to a decline in the quality of the animals. In a recent government auction for individual hunts in the Zambezi Valley safari areas, where a private individual can buy an elephant hunt as part of an associated bag of species, elephant hunts were sold for between US$25 000–30 000 per animal. In Botswana (2007), elephant trophy fees are US$18 000 and in Tanzania an elephant hunt (including all fees) is estimated to be US$23 000. In Mozambique the trophy fee for an elephant is only US$5 000, but this low value could be a reflection of the lack of a CITES trophy quota for that country (Cumming, pers. comm.).

The parties at the 12th Conference of Parties (CoP) to CITES in 2002 agreed to a one-off sale of 30 tons of ivory originating from the Kruger National Park. The prospective buyers had to register with the CITES Secretariat, fulfilling various requirements as laid down by the Conference. Only Japan and China indicated an interest in buying the ivory. To date (November 2007) only Japan has been verified as an acceptable trading partner. China will most probably be verified as a trading partner during the Standing Committee meeting scheduled for July 2008. CITES approved of the trade taking place at the CITES Standing Committee meeting in the Netherlands in June 2007. A further one-off sale has been approved by the 13th CoP of CITES (June 2007), which includes legally obtained ivory stock from South Africa, registered with the CITES Secretariat by 31 January 2007. Before the sale can take place, the ivory must be verified by the CITES Secretariat to be eligible for sale within the CITES framework and agreement.

| Category | Live sales | | Hunts | | |
	Price per animal (ZAR)*	Number	Category**	Price per animal (ZAR)*	Number
Trained animals	575 000–1 100 000	_	15–20 kg	290 000	10
Juveniles	50 000–500 000	–	20–25 kg	325 000	7
Cows plus family	15 000	150	30–35 kg	430 000	2–3
Bulls: approx. 20 kg**	70 000	30	35+ kg	500 000	2
Bulls: approx. 30 kg**	100 000	20	–	–	–

* Numbers quoted in rand, but most trading takes place in US$ and an exchange rate of ZAR7.2 per US$ has been used
** Weight of tusks

Table 4: Average prices and number of elephants traded in South Africa per year over the period 2005–2007* (Grobler, pers. comm.)

Can people living in areas adjacent to and in elephant-containing ecosystems benefit in any way from the presence of the elephants? One mechanism through which elephants can benefit local communities is through community-based natural resource management (CBNRM) programmes. CBNRM programmes that aim to partially devolve property rights over wildlife to communities on communal land have been under development in nearly all southern African countries since the 1980s, and are well developed in Namibia, Zimbabwe, and Botswana. Wildlife use, involving elephants for both wildlife viewing and trophy hunting, is commonly associated with these programmes. CBNRM in Namibia (Libanda & Blignaut, 2007), and in Botswana, involve both non-consumptive and consumptive tourism, but in Zimbabwe's CAMPFIRE programme, over 80 per cent of income derives from trophy hunting, which in the 1990s was dominated by elephant values (Bond, 1994; 1999). This figure seems to have risen above 90 per cent in recent years (Muchapondwa, 2003).

Elephants are therefore quite important as generators of income both nationally and for local communities in Botswana, Namibia, and Zimbabwe. However, they also generate costs in the form of damage to crops and infrastructure wherever they occur outside of fenced conservation areas. Sutton (2001) and Sutton *et al.* (2004) conducted a detailed household survey to measure the costs and benefits of living with elephants in the Caprivi Region of Namibia. Sutton determined that in the agro-pastoral system, which predominates in this region, elephants generate fewer damage costs than other wildlife, and that livestock actually causes more crop damage than all wildlife put together. Nevertheless, elephants still manage to reduce crop yields significantly. Jones & Barnes (2007) used crop damage data in crop enterprise models to show that average crop losses due to elephants reduced net profits for small-scale crop growers by some 30 per cent. Crop damage varies spatially, and in areas where it is the highest (some two or three times the average), crop profits can be eliminated altogether. Barnes (2006) used a similar crop enterprise approach to estimate the value of crop losses due to elephants in the Okavango Delta area of Botswana. Here, damage levels were generally higher, and average small-scale, rain-fed crop production profits were reduced by some 75 per cent, and even entirely eliminated in some cases.

| | Elephant crop damage cost level | | |
	Basic damage cost	2 x damage cost	3 x damage cost
Trust profit	604 200	333 600	−155 900
Community net benefit	1 199 400	928 800	439 300
Gross output	2 578 300	2 578 300	2 578 300
Gross national income (GNI)	2 002 900	1 777 600	1 349 800
Net national income (NNI)	1 894 400	1 669 100	1 241 400

[a] In 2006 Pula 1.00 was equal to ZAR 1.14, and US$ 0.16; Pula inflation factor from 2006 to 2007 is 1.06

Table 5: Impact of elephant crop damage costs on the measures of private and economic viability for a model CBNRM community trust investment in the Okavango Delta, Botswana (Pula per annum, 2006)[a] (source: Barnes, 2006)

Of importance here is the degree to which elephant damage costs incurred by communities can be offset by the benefits they derive from use of elephants through CBNRM. Models of community investments in CBNRM, developed by Barnes *et al.* (2001; 2002) were used to compare the wildlife crop damage costs with the utilisation benefits incurred by communities in both of the Caprivi and Okavango delta study sites. Table 5 and figure 2 (derived from Barnes, 2006) show the results for a typical CBNRM investment in the Okavango delta. The impacts of various crop damage levels (based on average figures) on the profits made by the community trust, the community members as a group, and the contribution made by the investment to the gross and net national income, were measured. Generally, benefits outweighed costs for all measures. In the case of the community trust, losses were only incurred when damage costs of three times the average levels were sustained over time. Jones & Barnes' (2007) results for the Caprivi Strip, Namibia, also established that CBNRM benefits generally outweighed crop damage costs. Various policy options are available to address elephant and wildlife damage costs. These studies suggested that human-elephant conflicts could be internalised with CBNRM programmes. For a further discussion on the human–elephant link within a CBNRM context, please see Chapter 4.

While it appears that in southern Africa rural people at the community level can derive positive net benefits from wildlife, do they actually derive direct financial gains from it? Libanda & Blignaut (2007) found that in Namibia households do generally benefit significantly from CBNRM and that sufficient institutional mechanisms are in place to ensure broad-based support for the programme, as indicated by the rapid growth of the CBNRM programme from its inception in 1996, to the end of 2006, when it included 50 CBNRM areas and covered an area of 118 705 km^2. The area under CBNRM management comprises 15 per cent of the land surface of Namibia and is adding to the 16.5 per cent of the land surface area that is already formally protected. CBNRM areas already host 37 per cent of Namibia's rural population and a further 31 conservancies are in various stages of development, clearly indicating the widespread interest in, and support for, the programme.

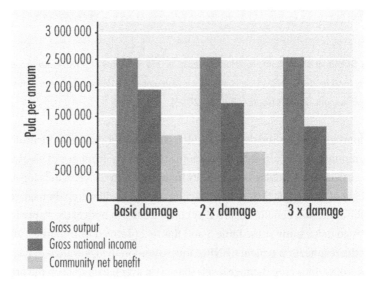

Figure 2: Impact of crop damage costs due to elephants on the economic gross output, the contribution to the gross national income, and the private community net benefits for a model CBNRM community trust investment in the Okavango Delta, Botswana (Pula per annum: 2006) (Barnes, 2006)

In contrast, this success of CBNRM is not unequivocally shared in Zimbabwe. Muchapondwa (2003) and Muchapondwa *et al.* (2003) conducted contingent valuation studies in Mudzi District, a CAMPFIRE district since 1992, where households' willingness to pay for the preservation of elephant was measured. Some 570 households, randomly selected from within two similar wards in Mudzi District were surveyed, and, along with the willingness to pay bids, variables such as household size and income, sex, age, and education of household head, distance from an elephant reserve, size of intruding elephant herds, existence of mitigation, support for government conservation, participation in agriculture, and labour spent on mitigation were tested. The studies found that 34 per cent of households were willing to pay for elephant preservation, with a median willingness to pay (WTP) of Z$300 or US$5.45. This was 3.87 per cent of median annual income. However, 62 per cent of households had a negative willingness to pay for elephant – they were willing to pay to have elephants removed from their area, with a median WTP of Z$98 or US$1.78. This was 1.27 per cent of median annual income.

The results indicated that the community as a whole had a net positive willingness to pay for elephant preservation, but that the majority of community members did not support elephant preservation. This suggested that any net benefits that the community might have derived from CBNRM must not have been reaching many households. Muchapondwa *et al.* (2003) recommended external transfers to households in Mudzi to increase incentives for elephant conservation. The willingness to pay values estimated by Muchapondwa *et al.* (2003) can be said to represent non-use values, namely, any or all of option, bequest, or existence values. In the CBNRM context, they are likely to be made up largely of option values. Apart from these findings on local non-use values, no other studies appear to have been carried out.

Other examples

While we have emphasised the studies estimating the economic value of elephants in southern Africa thus far, a large number of other, non-regional, studies have been conducted as well, a selection of which is summarised in table 6. It must be noted that values derived in these studies are not always comparable, either between themselves or with the studies listed above, since different methods and measures are used.

Using an open-ended stated preference technique, Vredin (1997) estimated the median Swedish household's willingness-to-pay (WTP) for the preservation of African elephants, which is an attempt to capture the non-use values of elephants. With a resulting median value of SEK100 (= US$14.92) per household for the year 1996, it was estimated that the aggregated WTP of the Swedish population for the preservation of the African fauna and flora (using the African elephant as indicator) is SEK383 million (=US$53.7 million). The main motives stated were: existence value (30 per cent of valid observations), care for future generations (28 per cent – bequest values) and own experiences (18 per cent – option values). This WTP is sensitive to changing income, as follows: a 1 per cent increase in income would lead to a 0.3 per cent increase in WTP (Hökby & Soderqvist, 2003). When taking this income elasticity into account as well as an average growth rate of 2.8 per cent, and changes in population since 1996, but with all other things being equal, aggregated WTP in 2006 has increased to SEK420 million (US$57 million). At average 2006 exchange rates, this amounts to US$14.73 per household per year. Currently, there are 470 000–690 000 African elephants in the wild (WWF, undated). Assuming 500 000 elephants and extrapolating to all 150 million European and US households (see Bulte *et al.*, 2006), this amounts to an indicative total WTP of US$2.2 billion per annum, or US$4 420 per elephant per annum. These numbers are, however, only indicative of the fact that the WTP for elephant conservation is potentially significant. They cannot be used in absolute terms since they are based on too many assumptions.

The estimated total gross tourism viewing value of elephants, in particular, was estimated at between US$25 and 30 million in Kenya in 1989 (Brown & Henry, 1989). This value was based on the travel costs of European and North American visitors and their stated purpose of travel. With an estimated 16 000 elephants in Kenya in 1989 (Ivory Trade Review Group, as quoted on the website http://www.american.edu/ted/elephant.htm), and using a low value of US$25 million per annum, that amounts to a mean WTP of US$1 562 per elephant in Kenya. Assuming declining travel costs and rising income over time this figure can be used as indicative for current values, but with low levels of confidence. Assuming that only three-quarters of Africa's elephants (375 000) are accessible to tourism this provides an indicative value of US$585 million or US$3.91 per European and US household per year. This is probably a low estimate, as up to 90 per cent of African elephants occur in southern and eastern Africa (Blanc *et al.*, 2007), both of which regions are readily accessible to international tourism. With low levels of confidence in these numbers – due to the fact that the studies on which they are based are dated and were carried out by various researchers in a variety of places using different methods – all of these discrepancies make comparisons difficult.

What has been valued	Valuation technique	Source	Values	Remarks
WTP of Swedes to preserve the population of African elephant	• Open Ended Contingent Valuation Method • Linear aggregation	Vredin (1997) Hokby & Soderqvist (2003)	1996; US$53.7 million for all Swedes Median: SEK 100 per household	1500 Swedish residents in age group 19–75
The cost of preventing a decline of elephants from severe commercial poaching for their ivory	Defensive Expenditure Method (Cost of protection)	Leader-Williams (1994)	1981: US$215 per km^2 (adjusted to 1994 values: $340 per km^2)	The relationship between spending and success in protecting elephants was significant but only explained 32% of the variance
Tourism value of elephants in Kenya	Travel costs	Brown & Henry (1989)	1989: US$25–30 million pa, $1 562 per elephant	Estimating consumer surplus from European and North American visitors
Conservation of 650 elephants in Amboseli NP	Marginal cost of PES scheme to conserve elephants	Van Kooten & Bulte (2000) Bulte *et al.* (2006)	US$10 per acre per year (US$2 470 per km^2) or US$175 per elephant per year; equal to an estimated minimum of US$0.60 per European and US household per year for all African elephants	Current estimates of the African elephant population amount to some 500 000 head. Assuming a minimum benchmark cost of $175 per elephant per year, the total benefits of elephant conservation should amount to $87.5 x 106 per year. Dividing by the number of households (150 x 106) this amounts to $0.60 per household per year (Bulte *et al.* 2006)
Value of ivory exports from Africa	Market price	Ivory Trade Review Group(as quoted http://www.american. edu/ted/elephant.htm) Cobb (1989)	1979: US$36.889 million 1987: US$19.18 million 1979–1987: >US$500 million	
Ivory value	Market price	Vredin (1995)	1987: US$2 734 per elephant	Estimated 1.3 million elephants killed for their tusks during 1970s and 1980s

What has been valued	Valuation technique	Source	Values	Remarks
Trophy value	Market price	Vredin (1995)	1989: US$2 366 per elephant	
Relocation of elephants from KNP to Shamwari Game Reserve	Market price	Wilderness Conservancy (http://www.wilderness conservancy.org/projects/ongoing.html)	US$2 850 per elephant	

Table 6: Valuation studies on African elephants (excluding studies from southern Africa)

Another way to value elephants is to estimate the minimum costs to sustain an elephant or elephant population. This would normally provide a measure of minimum value. The minimum cost to conserve elephants in Luangwa Valley, Zambia, during a time of intensive poaching was estimated at around US$215 per km^2 in 1981 values, and when adjusted for inflation amounts to US$340 per km^2 in 1994 terms (Leader-Williams, 1994). Using the same average 4.5 per cent annual increase in costs from 1981 to 2006 as used by Leader-Williams (1994), current cost levels are estimated at around US$600 per km^2. Assuming desired density of two elephants.km^{-2} in savanna habitat – which is high – this amounts to a cost for elephant conservation of US$300 per elephant or US$150 million per annum. In relation to the number of households in Europe and the US this amounts to US$1 burden per household per annum. These results should be interpreted with caution as only 32 per cent of conservation success could be explained by spending levels in the original study (Leader-Williams, 1994, 31). This implies that more spending, i.e. a bigger budget, is insufficient to assure elephant conservation, but institutional factors and management practices play a significant role as well.

When elephants cross protected area boundaries into adjacent human-inhabited areas, the costs of protection increase. In a study on the minimum cost of implementing a payments for ecosystem goods and services (PES) scheme in the Amboseli National Park of Kenya it was estimated that Maasai farmers needed compensation equal to US$10 per acre per year for roaming elephant populations in their croplands (Bulte *et al.*, 2006). For the 650 elephants of the Amboseli Park, this amounts to a compensation cost of US$175 per elephant. Assuming that this study is representative of all African farmers confronted with elephants (a very strict assumption) and that all of the 500 000 elephants in Africa can migrate across protected area boundaries (a clear worst-case situation), this amounts to a maximum of US$87.5 million per annum in compensation payments. For comparison, this amounts to a theoretical burden of US$0.60 per household per annum for all European and US households, which implies that if all these households pay US$0.60 per year, sufficient money could be collected to offset the damage caused by the elephants to crops.

Care should be taken interpreting this number since it is based only on one study consisting of 650 animals, but, indeed, it does indicate that the value from tourism (estimated above as US$3.91 per European and US household per year) is significantly more than the damage cost caused by elephants. This appears to create a unique opportunity for the implementation of a PES system.

The cost of translocation is also an indication of the socio-political WTP for the conservation of elephants. In South Africa, costs of up to US$2 850 per elephant were reported for translocation within the country (Wilderness Conservancy, no date); see also Chapter 5 for a detailed discussion. The total WTP for elephant relocation has not yet been estimated.

Vredin (1995) estimated the ivory value per elephant at US$2 734 (1987 prices). According to a recent report by CWI (2007), ivory prices for unworked pieces of ivory range from US$121 to US$900 (average US$390) per kilogram. Another recent release by CITES stated that the black market value of African ivory is approaching a high of US$700 per kilogram (CITES, 2007). It is well known that ivory per elephant is declining rapidly, and currently estimated at between 7 kg and 12 kg of ivory per African elephant (Van Kooten, 1995; Hunter *et al.*, 2004). Multiplying this by the price range of US$121–900 provides an estimate of US$850–US$6 300 per elephant. At an average price of US$390.kg^{-1} the current average value is estimated at around US$2 725 per elephant. Given the illegal nature of the ivory trade, it is very difficult to estimate the number of elephants involved. Nevertheless, Hunter *et al.* (2004) used one set of data, and careful extrapolation methods, to estimate that the ivory from between 4 862 and 12 249 African elephants is required annually to supply the unregulated markets in Africa. Although it is only a best guess at this stage, this would imply a market of between US$4.1 and US$77.2 million annually. This represents a theoretical burden of between US$0.03 and US$0.51 per European and US household. The trophy value of elephants was closely matched to the value of ivory and estimated at US$2 366 at 1989 prices (Verdin, 1995).

ASSESSING ELEPHANTS' CONTRIBUTION TO THE ECONOMIC VALUE OF ELEPHANT-CONTAINING ECOSYSTEMS

The suite of economic values of elephants are summarised in table 7. Though these values are by no means definitive and are often based on outdated data and various assumptions, using different valuation techniques, a clear picture appears. The consumptive benefits (e.g. ivory, trophy hunting) of the African elephant are much less than its non-consumptive (e.g. tourism) and non-use (e.g. existence, option, and bequest) values. The stated WTP for the preservation of the African elephant for just the Swedish population (US$57 million) is only 28 per cent less than the high-end estimate for the value of the total ivory market (US$77 million). If we hypothesise that this same WTP is shared by all European and American households – which are more or less, relatively speaking, on the same welfare level when compared to the average African household – then the high-end value of the ivory market is only 3.5 per cent of the potential Euro–North American WTP for the preservation of the African elephant. This analysis also points out that a compensation programme for both the direct damage costs of elephants to farmers and lost ivory income (a combined cost of US\$165 million per annum) is 7.5 per cent of the estimated WTP for preservation by European and American households. Such a voluntary conservation aid programme would also save an additional US\$150 million in protection costs. Obviously, there is little confidence in the absolute level of these numbers, or how much of this market could actually be realised, or what South Africa's portion of it could be, but they are sufficiently high to indicate that options for alternative scenarios exist when considering the potential scope for the creation of a market for the preservation of the African elephant.

Type value	Comparative value per US & EU household (US\$)[1]	Value per elephant (US\$)[2]	Total estimated value per annum (US\$)
Mainly existence, bequest and experience value	14.73	4 420	2.2 billion
Non-consumptive tourism value	3.91	1 562	585 million
Protection costs against poaching	1	300	159 million
Compensation costs to surrounding land owners	0.60	175	87.5 million
Offsetting consumptive value of ivory	0.03–0.51	2 730	4.1–77.2 million
Consumptive value of trophy hunting	n/a	2 360	n/a
Translocation costs	n/a	2 850	n/a
Trade in live elephants[3]	n/a	2 000–70 000	n/a
Hunting values[3]	n/a	40 000–70 000	n/a

[1] For comparison all values are expressed in terms of 150 million European and US households willing to pay, see Bulte *et al.*, 2006 for a similar approach

[2] Values adjusted to reflect 2006/07 estimates

[3] These values, from the South African studies, are inflated due to the restricted market

n/a Not available

Table 7: Summary of main economic values of African elephants

The formally measured and accounted-for direct consumptive use values of the African elephant are low, as is to be expected given the heavy impact of the CITES ban. As noted by Barnes and his colleagues, the realised TEV, excluding non-use values, of elephants has declined due to the CITES listing of elephants, probably by as much as 47 per cent. Although the non-consumptive, indirect, and non-use values of elephants are high (Vredin, 1997; table 7), the CITES listing has reduced the real cash flow to both nations and communities.

This is because there are currently few mechanisms to retrieve or capture the non-use values. What is required is measures to protect, compensate, translocate, and even consume elephants, in a sustainable fashion, and, concurrently, for local communities, the nation, and the elephants to derive direct, measurable, and tangible benefits from all such activities. Within the development of such a 'conservation, preservation and sustained use' market, and of institutions to support it, Far Eastern countries can likely play an important role, especially related to the direct 'consumption' of elephant tusks. Additionally, if communities do not directly benefit from the presence of elephants, whether through consumptive or non-consumptive use or a combination thereof, indications are that they will not support elephant conservation in future (see the example from Zimbabwe). If, however, they are integrated, and made part of the 'solution', then indications are that they would readily support conservation (see example from Namibia). The experiences of these countries offer South Africa excellent learning references.

What is also apparent is that an inclusive conservation package that allows for all the possible economic benefits to be realised would be easily offset by the sum of economic benefits that could be gained. The challenge remains to create an efficient institution that would be able to capture these gains – that is, the economic rent – and distribute this to the benefit of both landowners and elephants. Evidence from all the studies cited previously suggests that international willingness to pay for elephant conservation in African countries exists, which implies that South Africa has a range of options to choose from. Barnes *et al.* (2002) supports this view and maintains that much of the hitherto substantial international NGO and donor support for CBNRM is a form of non-use values. Additionally, contingent valuation studies among wildlife viewing tourists in Botswana and Namibia (Barnes, 1996b; Barnes *et al.*, 1999) revealed a significant willingness to pay for wildlife conservation. The tourists surveyed generally had trip consumer surpluses and were willing to pay more for their trips than they had paid, a view supported by South African studies as well (Turpie, 2003; Turpie & Joubert, 2001; Geach, 1997). This implies that the value tourists received from viewing the wildlife was more than the economic cost of hosting them. The surplus, which constitutes economic rent, is attributable to the wildlife (elephants) and, if retained (captured) these rents could be used to advance conservation. At least a portion of the tourists' willingness to pay for conservation could thus come out of these surpluses, and may be defined as direct non-consumptive use value. It is important to note, however, that the estimated non-use values, as summarised in both tables 5 and 7, are mostly only hypothetical values. Until institutional mechanisms are created through which such hypothetical values can flow and be materialised to the advantage of both people (through CBNRM or otherwise) and elephants, and to the nation as a whole, they remain hypothetical.

Economists (e.g. Bulte *et al.,* 2006; Van Kooten & Bulte, 2000; Kahn, 1998; Barbier *et al.,* 1990) seem to share the view that the use of markets through a well-designed institutional arrangement is a much better way of managing a precious resource over the long term, than an outright ban. This is so because markets offer more management options and flexibility than command and control mechanisms. Barbier *et al.* summarise this thought as follows (Barbier *et al.,* 1990, 147):

> The future of the African elephant is dependent upon the taking of immediate action. The ivory trade ban must be considered an interim measure, not a solution. Sustainable populations of the African elephant, as with so many other endangered species, will depend upon the development of reforms which constructively utilize the trade, rather than attempts to combat it. Institutional reforms to this end must be addressed now.

The development of market options has to be considered also from the perspective that official development aid, especially predominant in East Africa, is not sustainable in the long run and cannot sustain or improve conservation (Van Kooten & Bulte, 2000; Norton-Griffith, 2007). A further stimulus for the development of markets is provided by the emergence of the Far Eastern markets as significant role-players within the global ivory trade. This implies that the political-economic gridlock concerning the ban on trade in ivory cannot be maintained indefinitely. Leakage – both the legal and illegal trade in ivory – is likely to occur since sanctions and bans are imperfect measures in the long run. It is much more prudent to manage proactively and to introduce the use of markets and incentives measures in a controlled environment rather than to be confronted with the effects of leakage. Since the economic system is a self-organising system (Krugman, 1996) that requires adaptive management, markets and incentive measures are much more efficient and effective in achieving such desired behavioural change, if constituted and institutionalised appropriately, than are traditional command-and-control measures. In this context the use of market-based and command-and-control measures can occur in conjunction with each other for a period of transition, allowing markets to operate within a controlled environment and, progressively, to mature.

The time for such institutional change is ripe now. Almost two decades since the African elephant's listing as an endangered species, its numbers have increased by 50 per cent. Concurrently, much experience has been gained in incorporating CBNRM into the conservation framework and thereby distributing conservation benefits broadly, and this could include the sustainable direct use or extraction of elephants (Damm, 2002). Such direct use will reduce at least the growth in the number of elephants, but, as has been observed in Botswana, the numbers are likely to be relatively small. It should be noted that the sustainable use of elephants is, at least theoretically, not in conflict with the non-use values, but could instead be an important complement.

In parallel to the development of CBNRM and other institutional arrangements over the past two decades, much has been learnt since the late 1980s and early 1990s on how to establish and operate markets for ecosystem goods and services (Pagiola & Platais, 2007). Such a market would allow for the transfer of money, especially from Europe and the USA, to capture some of the non-use values of elephants. In so doing, the economic value of elephants can be optimised by capturing all the values (direct consumptive, direct non-consumptive, indirect, and non-use values) and, additionally, by releasing finances to both conserve the elephants, and increase their range to include human-occupied areas (Van Aarde & Jackson, 2007; Van Aarde *et al.*, 2006). This option would inject new streams of income into rural communities, all across South Africa, especially to those living in areas adjacent to elephant-containing ecosystems, many of whom have a formal land claim on currently protected land. This offers an opportunity to link the formal (first) economy of South Africa with the informal (second) one, and to inject finances into the second economy by embracing the two as partners and fellow custodians of the natural environment and national heritage. This option is becoming increasingly viable due to current and probable future socio-demographic changes, as South Africa undergoes a rapid increase in urbanisation and possibly even de-population of the rural areas.

CONCLUSION

Some values of the African elephant are clearly expressed in the market, such as tourist expenditures on elephant viewing, or the direct costs of trophy hunting; the direct use benefits from elephants include ivory, although banned, and other animal products. However, non-use values are generally not captured as income or observed, and are hence difficult, but not impossible, to determine. The willingness to pay to conserve elephants for future generations on the part of many people who may never even see an elephant in their lifetimes, is

Box 3: Key research questions

What is the economic value of elephants in South Africa?

- What is the most appropriate, desirable, and feasible institutional arrangement and market mechanism to realise the suite of economic values of elephants?
- How could elephant markets, realising the direct, indirect, and non-use values of elephants, benefit local populations adjacent to elephant containing ecosystems?
- What are the likely impact of the emerging ivory market in the Far Eastern countries on South Africa and the impact thereof on the elephant management options for South Africa?
- How can markets be constructed to assist in reducing the risk and uncertainty in managing elephants and elephant containing ecosystems to the advantage of both elephants and people?

generally only partially captured through donations and thus largely remains unexpressed. An interpretation of economic value thus goes beyond exchange values as measured through market-based transactions.

Although there are no studies on the TEV of elephants in South Africa, there is a rich knowledge base thanks to work done in Botswana, Zimbabwe, and Namibia. Based on these studies, there is evidence of (1) an increase in the proportional contribution of non-consumptive values to the TEV of elephants, but (2) a decline in the overall economic value derived from elephants after the CITES ban on trading in elephant products. There is mixed evidence of the extent of elephant damage to local communities' crops and infrastructure from studies done in Botswana and Namibia. In some cases it was less than the damage by livestock, but in other cases substantial losses were incurred. In Kenya it was estimated that benchmark damage costs to the Maasai amounted to US$2 470 km^{-2}.year. In South Africa it is more than likely that costs are substantially lower due to our formal elephant management system in fenced-in conservation areas. A list of pertinent research questions with specific reference to South Africa is listed in box 3.

The success of institutions to compensate local communities, on the one hand, for their loss in income of elephant and elephant products and, on the other, for damage costs, is also mixed. There is evidence of some success in distributing the economic value of conservation through CBNRM schemes in Namibia, but much less in Zimbabwe. The proper function of institutional success is a prerequisite for the effective internalisation of damages.

Based on evidence of international willingness to pay for the conservation of elephants, and the recent development concerning markets for ecosystem goods and services, ways have to be found to internalise this expressed willingness to pay to advance elephant conservation. Traditional policy options are limited in their scope as regards achieving this objective, but significant evidence exists that there is potentially sufficient international support to develop market-based alternatives. These high expressed non-use values for elephants are based on three factors, namely the fact that elephants exist – in other words that they *have to be preserved* for future generations; the ecological role they play within ecosystems; and the fact that people want to have the option to enjoy benefits from them in future. The preliminary meta-analysis presented in this chapter suggests that the non-use values from Europe and the US are three to four times higher than tourism values, 25 times higher then the benchmark compensation payments required to land owners, and almost 30 times higher than a high-end estimation of the total ivory market. This is a trend supported by De Boer *et al.* (2007). There is therefore abundant scope for the creation of markets and institutional strengthening.

ACKNOWLEDGEMENTS

The authors would like to thank James Aronson for comments provided on earlier drafts. Any remaining errors are, however, the sole responsibility of the authors.

REFERENCES

Barbier, E., J. Burgess, T. Swanson & D. Pearce 1990. *Elephants, economics and ivory*. Earthscan, London.

Barnes, J., J. Cannon & K. Morrison 2001. Economic returns to selected land uses in Ngamiland, Botswana. Unpublished report, Conservation International, Washington, DC, USA.

Barnes, J.I. 1990. Economics of different options of elephant management. *Proceedings, Kalahari Conservation Society Symposium: The Future of Botswana's Elephants*. Gaborone, Botswana, 10 November 1990, 60–66.

Barnes, J.I. 1992. The economic use value of elephant in Botswana in 1992. Unpublished paper, Department of Wildlife and National Parks, Gaborone, Botswana.

Barnes, J.I. 1996a. Changes in the economic use value of elephant in Botswana: the effect of international trade prohibition. *Ecological Economics* 18, 215–230.

Barnes, J.I. 1996b. Economic characteristics of the demand for wildlife viewing tourism in Botswana. *Development Southern Africa* 13(3), 377–397.

Barnes, J.I. 1998. Wildlife economics: a study of direct use values in Botswana's wildlife sector. Ph.D. thesis, University College, University of London, London.

Barnes, J.I. 2006. Economic analysis of human-elephant conflict in the Okavango delta Ramsar site, Botswana. Unpublished report, Natural Resources and People, Gaborone, Botswana.

Barnes, J.I., C. Schier & G. van Rooy 1999. Tourists' willingness to pay for wildlife viewing and wildlife conservation in Namibia. *South African Journal of Wildlife Research* 29(4), 101–111.

Barnes, J.I., J. MacGregor & L.C. Weaver. 2002. Economic efficiency and incentives for change within Namibia's community wildlife use initiatives. *World Development* 30(4), 667–681.

Blanc, J.J., R.F.W. Barnes, G.C. Craig, H.T. Dublin, C.R. Thouless, I. Douglas-Hamilton & J.A. Hart 2007. *African elephant status report: an update from the African Elephant Database.* IUCN, Gland.

Blignaut, J., J. Aronson, P. Woodworth, S. Archer, N. Desai & A. Clewell 2007. Restoring natural capital: a reflection on ethics. In: J. Aronson, S. Milton & J. Blignaut (eds) *Restoring natural capital: the science, business, and practice.* Island Press, Washington, DC, 9–16.

Bond, I. 1994. The importance of sport-hunted African elephants to CAMPFIRE in Zimbabwe. *Traffic Bulletin* (WWF/IUCN) 14(3), 117–119.

Bond, I. 1999. CAMPFIRE as a vehicle for sustainable rural development in the semi-arid communal lands of Zimbabwe: incentives for institutional change. Ph.D. thesis, Department of Agricultural Economics and Extension, University of Zimbabwe.

Brown, Jr., G. & W. Henry 1989. *The economic value of elephants.* Discussion Paper 89-12. London Environmental Economics Centre.

Bulte, E.H., R.B. Boone, R. Stringer & P.K. Thornton. 2006. *Wildlife conservation in Amboseli, Kenya. Paying for nonuse values.* Roles of Agriculture Project Environment Series, December. Agriculture and Development Economics Division, Food and Agriculture Organisation of the United Nations.

Care for the Wild International (CWI) 2007. US exposed as leading ivory market. *Care for the Wild International,* 5 June.

Child, G. & B. Child 1986. *Some costs and benefits of controlling elephant populations in Zimbabwe.* Proc. 7th and 8th Sessions of the AFC Working Party on Wildlife Management and National Parks, Arusha, Tanzania, 1983, and Bamako, Mali, 1986, FAO, Rome, Italy, 74–77.

Child, G. & J. White 1988. The marketing of elephants and field-dressed elephant products in Zimbabwe. *Pachyderm* 10, 6–11.

CITES 2007. Cites decision promotes illegal ivory trade. *Official Bulletin of Species Management Specialists Inc.* 14th conference of the parties to CITES, The Hague, Netherlands, 3–15 June 2007.

Cobb, S. (ed.) 1989. The ivory trade and the future of the African elephant. Volume 1: Summary and conclusions; Volume 2: Technical reports. Unpublished report of the Ivory Trade Review Group, Oxford, UK.

Damm, G. 2002. *The conservation game: saving Africa's biodiversity.* Rivonia: Safari Club International African chapter.

De Boer, F., J.D. Stigter & C.P. Ntumi 2007. Optimising investments from elephant tourist revenues in the Maputo Elephant Reserve, Mozambique. *Journal for Nature Conservation.* In press.

Ekins, P., S. Simon, L. Deutsch, C. Folke & R.S. de Groot 2003. A framework for the practical application of the concepts of critical natural capital and strong sustainability. *Ecological Economics,* 44, 165–185.

Farley, J., & E.J.B Gaddis 2007. Restoring natural capital: an ecological economics assessment. In: J.S. Aronson, S. Milton & J. Blignaut (eds) *Restoring natural capital: the science, business, and practice.* Island Press, Washington, DC, 17–27.

Geach, B. 1997. The Addo Elephant National Park as a model of sustainable land use through ecotourism. MSc thesis, University of Port Elizabeth.

Hökby, S. & T. Soderqvist 2003. Elasticities of demand and willingness to pay for environmental services in Sweden. *Environmental and Resource Economics* 26, 361–383.

Hunter, N., E. Martin & T. Milliken 2004. Determining the number of elephants required to supply current unregulated ivory markets in Africa and Asia. *Pachyderm* 26, January–June, 116–128.

Jones, B. & J. Barnes 2007. WWF human wildlife conflict study: Namibian case study. Unpublished report, Macroeconomics Programme and Global Species Programme, WWF, Gland, Switzerland.

Kahn, J. 1998. *The economic approach to environmental and natural resources.* Dryden.

Krugman, P. 1996. *The self-organising economy.* Blackwell, Oxford.

Leader-Williams, N. 1994. The costs of conserving elephants. *Pachyderm* 18, 30–34.

Libanda, B. & J. Blignaut 2007. Tourism's local benefits for Namibia's community based natural resource management areas. *International Journal for Ecological Economics and Statistics.* In press.

McCauley, D.J. 2006. Selling out on nature. *Nature* 443(7), 27–28.

Muchapondwa. E. 2003. The economics of community-based wildlife conservation in Zimbabwe. Ph.D. thesis, Department of Economics, Göteborg University, Sweden.

Muchapondwa. E., F. Carlsson & G. Kölhin 2003. Can local communities in Zimbabwe be trusted with wildlife management? Application of CVM on the elephant in Mudzi Rural District. Unpublished paper, Department of Economics, Göteborg University, Sweden.

Norton-Griffith. M. 2007. How many wildebeest do you need? *World Economics* 8(2), 41–64.

Novelli, M., J.I. Barnes & M. Humavindu 2006. The other side of the tourism coin: consumptive tourism in Namibia. *Journal of Ecotourism* 5, 62–79.

Pagiola, S. & G. Platais 2007. *Payments for environmental services: from theory to practice.* World Bank, Washington.

Smith, A. 1997. *The wealth of nations.* Penguin, London.

Sutton, W. 1998. *The costs of living with elephants in Namibia.* Proceedings of the Workshop on Cooperative Regional Wildlife Management in Southern Africa, University of California, Davis, California, USA, August 13–14 1998, 57–71.

Sutton, W.R. 2001. The economics of elephant management in Namibia. Ph.D. thesis, University of California, Davis, California.

Sutton, W.R., D.M. Larson & L.S. Jarvis 2004. *A new approach for assessing the costs of living with wildlife in developing countries.* Research Discussion Paper No. 69, Directorate of Environmental Affairs, Ministry of Environment and Tourism, Windhoek, Namibia.

Turner, R.K., D.W. Pearce & I. Bateman 1994. *Environmental Economics. An Elementary Introduction.* Harvester Wheatsheaf, New York.

Turpie, J.K. 2003. The existence value of biodiversity in South Africa: how interest, experience, knowledge, income and perceived level of threat influence local willingness to pay. *Ecological Economics* 46, 199–216.

Turpie, J.K. & A. Joubert 2001. Estimating potential impacts of a change in river quality on the tourism value Kruger National Park: an application of travel cost, contingent and conjoint valuation methods. *Water SA* 27(3), 387–398.

Turpie, J., J. Barnes, J. Arntzen, B. Nherera, G-M. Lange & B. Buzwani 2006. *Economic value of the Okavango Delta, Botswana, and implications for management.* Department of Environmental Affairs, Gaborone, Botswana.

ULG Northumbrian 2001. Economic analysis of commercial consumptive use of wildlife in Botswana. Unpublished report, Botswana Wildlife Management Association (BWMA), Maun, Botswana.

Van Aarde, R. & T. Jackson 2007. Megaparks for metapopulations: Addressing the causes of locally high elephant numbers in southern Africa. *Biological Conservation* 134, 289–297.

Van Aarde, R., T. Jackson & S. Ferreira 2006. Conservation science and elephant management in southern Africa. *South African Journal of Science* 102, 385–388.

Van Kooten, G. 1995. Elephant economics in the rough: an ivory trade model. Draft paper, Department of Economics, University of Victoria.

Van Kooten, G. & E. Bulte 2000. *The economics of nature.* Blackwell, Oxford.

Vredin, M. 1995. *Values of the African elephant in relation to conservation and exploitation.* Umeå Economic Studies No. 390, Department of Economics, Umeå University.

Vredin, M. 1997. *The African elephant: existence value and determinants of willingness to pay.* Umeå Economic Studies No. 441, Department of Economics, Umeå University.

Wilderness Conservancy n.d. Available at: http://www.wildernessconservancy. org/projects /ongoing.html.

WWF nd. History of the African elephant. Available at: http://www.wwf.org.uk/ filelibrary/pdf /africanelephant_tl_0804.pdf .

NATIONAL AND INTERNATIONAL LAW

Lead author: Lisa Hopkinson
Authors: Marius van Staden and Jeremy Ridl

PREFACE

THIS CHAPTER provides a synopsis of the law relevant to elephant management in South Africa. The authors provide an assessment of the law as a subset of the broader enquiry undertaken in the Assessment of Elephant Management in South Africa ('the Assessment'), and in so doing, highlight shortcomings that impact on the efficacy of elephant management practices and strategies.

The Assessment is intended to inform the Authorising Body (policy makers) by way of the provision of high level expert advice in order to develop policy and law to regulate the management of elephants in all of its facets in South Africa. This chapter assesses the current status of elephant-related law in order to assist management and limit the risks associated with policy formulation or promulgation of legislation and regulations.

In making policy decisions, the Authorising Body is often presented with differing interpretations of the law that appear to present options or alternative approaches. This chapter is intended to help policy makers act in accordance with the law, or where the law is seen to be lacking, they are given a sound legal basis for departing from conventional approaches or are able to consider legislative intervention. The authors accordingly base their conclusions on judicial interpretations of the law, state the law as it is generally accepted to be, and indicate where compliance is mandatory. The opinions of the authors have been clearly distinguished from statements of existing law.

THE AUTHORS' RESPONSE TO THE BRIEF

In giving effect to the requirements of the Assessment, the authors have adopted the following approach:

Methodology

An accurate statement of the law is provided. This is based on conventional legal principles and is as far as possible free of the authors' personal opinions or analyses, except where this is required by the context. Where reference is made to legal texts, the wording of the relevant statute or court judgement is used as far as practicable. Paraphrasing that may lose the import of the statements made is avoided unless the syntax otherwise requires.

The authors provide a summary of their conclusions and analysis of the law (strengths and weaknesses) in so far as it relates to elephant management and wildlife management generally. The conclusions of the authors are presented in such a way as to ensure that these are distinguishable from the legal texts themselves.

Outcomes

This chapter provides a description of the legal framework within which elephant management must be practised, highlights the difficulties experienced by managers (and the public) in interpreting and applying the law, and identifies potential legal interventions by appropriate authorities.

The definition of wild animals as *res nullius* presents problems that are not adequately dealt with by legislation or recent judicial interpretation of the law. It is submitted that it is this fundamental legal definition that is at the heart of conflicts over wild animals. Accordingly, an argument for the recognition of a 'new common law' is presented on the basis that South Africa's common law in relation to the management of wildlife is in conflict with the Constitution of the Republic of South Africa (as embodied in the Constitution of the Republic of South Africa Act No. 108 of 1996) ('the Constitution'), the recognition of environmental rights as fundamental human rights, and the judicial interpretation of South Africa's new environmental legislation, and is no longer justifiable in South Africa's open and democratic society. It is concluded that for the development and redefinition of South Africa's common law as required by section 9(3) of the Constitution, the changing attitudes of society to wild animals must be recognised and accommodated.

The concept of wild animals as *res publicae* (in public ownership) and as *res omnium communes* (those which by natural law are common to all but belong to no one) except where these are in private ownership is advanced as the more consistent understanding by society of their legal relationship with wild animals.

SOURCES OF SOUTH AFRICAN LAW

The sources of South African law which regulate the ownership, control, protection and utilisation of elephants and elephant products comprise South African common law, national and provincial statutory law, policy, norms and standards, and customary law as well as constitutional law. This legal framework is assessed as to the adequacy with which it protects rights of animal owners and possessors, regulates the risks and responsibilities associated therewith, and protects the rights and livelihoods of people living within the elephant range. The influence of relevant international law is also considered.

SOUTH AFRICAN COMMON LAW

Law deriving from historical sources, augmented by and developed through case law, and to a lesser extent, customary law, constitutes South African common law. Roman Law, Roman Dutch Law and English Law are historical sources of South Africa's common law. (*The Law of South Africa ('LAWSA')* 25(1) par 278.)

Classification of wild animals under common law

In South African law, animals are divided into two main categories, namely domestic animals and wild animals (*LAWSA* 1(2) par 454). This classification is important as it affects the rights of property in animals as well as liability for their behaviour (*LAWSA* 1(2) par 454). Wild animals (*ferae bestiae*) are classified as those animals that belong to a species that exists in a wild state anywhere in the world (*LAWSA* 1(2) par 456).

There are no specific criteria for ascertaining whether or not an animal qualifies to be wild, although various species regarded as 'game' have been recognised in South African case law as belonging to the class of wild animal species. Examples are wild ostriches (*De Villiers v Van Zyl 1880* F 77; *R v Bekker* (1904) 18 EC 128), wildebeest (*Richter v Du Plooy* 1921 OPD 117) and lions (*R v Sefula* 1924 TPD 609 610). Van der Merwe (1989, 218) suggests that the question whether a particular animal is domestic or wild depends on the view held by the community in which it occurs. It is generally accepted by the South African community that elephants occurring in a wild state in South Africa are wild animals. This forms the foundation for establishing property rights in these animals as well as liability for their behaviour.

Acquisition and loss of ownership of wild animals under common law

Acquiring ownership in wild animals

Wild animals enjoying a state of natural freedom are considered to be *res nullius* (i.e. belonging to no one). Because they belong to no one, they can be captured by any person and their capture does not amount to theft. If certain requirements are met, their capture may amount to *occupatio*, a method by which ownership in a wild animal can be acquired (*LAWSA* 1(2) par 461. *R v Bekker* (1904) 18 EDC 128; *R v Maritz* (1908) 25 SC 787).

The requirements for *occupatio* have been stated by our courts as being the following: (a) the wild animal must be ownerless (*res nullius*); (b) physical control must be exercised over the animal; and (c) the captor must have the intention to be the owner of the animal (*LAWSA* 1 (2) par 461).

What measure of physical control will be sufficient for a person to become the owner of a wild animal is a question of fact, although it would seem that a fairly strong degree of physical control is required (*S v Mnomiya* 1970 1 SA 66 (N); *Langley v Miller* 1848 3 M 584 (whales). In some cases physical capture of the animal has been required by our courts (*R v Mafohla* 1958 2 SA 373 (SR); *Reck v Mills* 1990 1 SA 751 (A)). Sonnekus in 1989 *TSAR* 727 states that total deprivation of freedom of movement of the pursued animal is apparently required.

Once a wild animal is captured, it remains the property of its captor as long as the latter retains sufficient control over it. Exactly what degree of ongoing control is sufficient to retain ownership of a wild animal is a question of fact which depends upon the circumstances of each case. In one case the court held that a fence of five and a half feet surrounding an area of 250–300 morgen was sufficient to control a herd of 100 blesbok (*Lamont v Heyns and Another* 1938 TPD 22), whereas in another case the court held that an ordinary fence enclosing a farm of 800 morgen was not considered sufficient to retain control over a herd of 57 wildebeest (*Richter v Du Plooy* 1921 OPD 117).

Losing ownership in wild animals

As soon as control over a wild animal is lost, it reverts to its state of natural freedom, ceases to be owned and becomes *res nullius* again, and is capable of being acquired by a new owner. In *Richter v Du Plooy* (*supra* at page 119) the court held that as soon as a wild animal emerges from its place of detention, it becomes *res nullius* and is capable of being acquired by *occupatio* by the first person who has the acquisitive instinct and the means to gratify it. Roman Dutch Law authors adopt a less stringent approach and have indicated that a wild animal previously owned is only regarded as having regained its natural freedom if it is no longer in sight, or still in sight but difficult to pursue (*LAWSA* 1 (2) at par 461).

Taking possession of wild animals on another's land

It appears to make no difference according to common law as to where a *res nullius* wild animal is captured and South African courts have held (in extremely old decisions), that a hunter becomes the owner of a *res nullius* wild animal irrespective of whether the hunter captures it on his or her land, or on another person's land or on land belonging to the State. This appears also to be the case even if a person expressly forbids a hunter to hunt on his or her land or prohibits entry upon his or her land for this purpose (*LAWSA* 1 (2) at par 461).

Taking possession of wild animals contrary to any laws

The question as to whether or not *res nullius* wild animals which are captured contrary to game laws and other statutory provisions can become the property of their captor, has also received the attention of our courts, and conflicting views on this exist. The Natal Supreme Court in *Dunn v Bower* (1926 NPD 516) held that a hunter does not become the owner of a wild animal which he/she captures contrary to statutory provisions. In contrast to this, the Cape Supreme Court in *S v Frost, S v Noah* (1974 3 SA 466 (C)) held that a captor becomes the owner of wild animals captured illegally and in contravention of game laws, fishing ordinances or other statutory provisions, unless the relevant legislation unequivocally provides otherwise. South African legal commentators seem to prefer the decision in *S v Frost, S v Noah* (*supra*).

Statutory law amending the common law

Intervention in favour of private game farmers

The common law principles regulating the acquisition and loss of ownership of wild animals on private game farms were radically modified by the recommendations of the South African Law Commission following its investigation into the acquisition and loss of ownership of wild animals in South Africa in 1988 (*SA Law Commission Working Paper No. 27*, Project 69 1989).

The investigation followed calls by various bodies and persons for more effective protection of game farmers. Poaching of wild animals was on the increase
and it was submitted to the Minister of Justice that the rights of game farmers
were not adequately protected when animals escaped, were stolen, poached
or lured away with the intent to steal. It was argued that these rights should be
protected to the same extent that ownership of agricultural stock is under the
Stock Theft Act 57 of 1959. The result was the promulgation of the Game Theft
Act 105 of 1991, which came into operation on 5 July 1991.

The Game Theft Act regulates ownership of 'game' which is defined in
section 1 of the Act as meaning 'all game kept or held for commercial or hunting
purposes, and includes the meat, skin, carcass or any portion of the carcass
of that game'. As will be observed below, the different meanings that may be
ascribed to the word 'game' as defined in the Act present difficulties.

The Act deals with ownership in two ways: section 2(1)(a) provides that
ownership is not lost if game escapes from land that is sufficiently enclosed
(as defined in section 2(2)(a) of the Act); and section 2 (1)(b) provides that
ownership is not acquired by any person of game that escapes from sufficiently
enclosed land, or is hunted, caught or possessed on the land of another without
the consent of the owner of that land.

Land is deemed to be sufficiently enclosed if the Premier of the Province in
which the land is situated, or his assignee, has issued a certificate stating that
the land concerned is sufficiently enclosed to confine to that land the species
of game mentioned in the certificate. A certificate is valid for three years and
can be renewed. The criteria for the issue of a certificate and the standards that
must be met before a certificate can be issued in respect of any land are not
prescribed although these would most often be contained in policy documents
adopted by the provinces. Once a certificate of adequate enclosure has been
issued in respect of land, the sufficiency of the enclosure of that land for the
purposes of the future application of the Act in any particular circumstance is
not determined as matter of fact, but as matter of law by virtue of the deeming
provision of section 2(2)(a) of the Act. In other words, the land is deemed to
be sufficiently enclosed for the purposes of subsections 2(1)(a) and (b) of the
Act simply by virtue of the landowner holding a valid certificate issued under
subsection 2(2)(a).

The Act provides for the disclosure of 'species' of game for which the
enclosure will be considered sufficient. There is no requirement for the
compilation of an inventory to be recorded in the certificate, and therefore the
owner will retain the evidentiary burden of proving ownership of the escaped
animal. This may be through some form of identification or mark or otherwise,

and that the animal is of the species identified in the certificate. Once this is established, the owner of such game is entitled to take all reasonable steps necessary to retrieve the animal. The requirement of the common law that sufficient physical control over the animal be proven as a fact gives way to an administrative determination of this issue by the issue of a certificate confirming the adequacy of the means of enclosure of the animal.

Thus, in complete contradistinction to the South African common law position that wild animals generally become *res nullius* upon their escape from the control of their previous captor, the Game Theft Act provides that ownership of a particular category of wild animal defined in section 1 of the Act is not lost upon its escape from the control of its owner provided the deeming provision of section 2(2)(a) is met. The common law position adopted by our courts in respect of these matters is thus overridden by this statutory intervention.

Reservations have however been expressed with regard to the scope of the main provisions of the Act (*LAWSA* 1(2) at par 462). Firstly, the Act applies only to 'game', and since this is not defined in the Act, is open to different interpretations. The ordinary dictionary meaning of this word generally refers to 'wild mammals or birds hunted for sport or food' (*Concise Oxford English Dictionary* 11th Edition). If 'game' is given this meaning, animals, including giraffe and baboons, for instance, that are generally not hunted for either sport or food, will fall outside of the ambit of the Act (*LAWSA* 1(2) at par 462).

Secondly, 'game' as defined by the Act refers only to animals kept for commercial or hunting purposes. Thus game kept in South Africa's system of protected areas (as defined in section 9 of the National Environmental Management: Protected Areas Act No. 57 of 2003) primarily for conservation purposes, is excluded from this definition. Arguably, the business activities of park managers in charging entrance fees to wildlife parks and the revenue generated by these protected areas from tourism activities and the sale of wildlife, constitutes commercial purposes. However, a counter to this proposition is the fact that in many instances, funds generated from these commercial activities are not intended for commercial gain but are generally reinvested in protected area conservation programmes. A further consideration is the fact that it would appear that it was the intention of legislators that the Act was to apply to game farming, as it was in this sector that problems were being experienced, and not to public conservation activities.

Thirdly, because of the size of many protected areas as well as considerations of practicality and cost, it is difficult to erect and maintain fences to the standards required for the issue of a certificate of adequate enclosure, and may be ecologically undesirable to do so.

It would therefore seem that the Game Theft Act has practical application in and is of benefit to the game farming industry and not to wildlife conservation.

Intervention in favour of the State

South Africa's conservation and biodiversity legislation does not deal with the acquisition, retention and loss of ownership in wild animals occurring in or escaping from protected areas. While statutory intervention has provided a measure of protection of the ownership rights of owners of wildlife in the private commercial game farming industry, the same has not occurred in the public wildlife conservation sector. In the absence of the common law being redeveloped and redefined as proposed in this chapter, ownership of wild animals occurring in or escaping from protected areas is dealt with exclusively by the common law, in terms of which they are owned by the State for so long as they are in the physical control of the State, and when such physical control is lost, they revert to the status of *res nullius* (belonging to no one) and are lost to the State.

Private property rights in respect of escaped elephants

The rights of protected area managers to recover elephants that have escaped from protected areas onto private land in circumstances where such elephants become *res nullius* are subject to the property rights of private landowners, the capacity of such landowners to appropriate and to exercise rights in and to such elephants, and the right of such landowners to prevent anyone else from exercising any rights in respect of such escaped elephants. These aspects are dealt with below.

Property rights of private landowners

Land is the primary component of immovable property. Land consists of the soil, its geophysical components such as minerals at and below the surface, and everything attached to the soil by natural means such as trees, plants, and growing crops. Other component parts comprising immovable property are all artificial annexures of a permanent nature, such as buildings and installations, as well as permanent and necessary attachments to such annexures (*LAWSA* 14(1) at par 3).

Immovable property is recognised as a 'thing' and ownership rights in immovable property are considered to be the most comprehensive real right which can be held in respect of a 'thing' under South African law. These rights confer upon the holder, in principle, complete and absolute control over the thing, and therefore the absolute and full control over the sum total of all possible rights over and capacities in respect of such thing, except insofar as this may be limited by common law or statute (*LAWSA* 14(1) at par 4).

The right of ownership of immovable property includes the right to posses, use, enjoy and alienate the property (*LAWSA* 14(1) at par 7). This includes the right to use and enjoy all natural resources (for example water) occurring on the land, provided this is not in conflict with any law (*LAWSA* 14(1) at par 7). All *res nullius* wild animals occurring on private land are natural resources occurring on the land, to which the private landowner has the right of use and enjoyment, without intervention or interference from others, for as long as they occur on his land, and provided that this does not conflict with any law. Accordingly, an escaped elephant that becomes *res nullius* is subject to this private property right.

Rights of private landowners in respect of escaped elephants

Having acquired this right, the landowner is entitled, among other things, and without being subjected to any interference from others, to allow the escaped elephant to continue roaming on his property in a state of natural freedom, to exercise physical control over the escaped elephant with the intention to become the owner of it through *occupatio*, or to exercise any other rights to which he may be entitled. This would include the right to apply for a permit under the relevant provincial or national conservation legislation to capture, hunt, keep, donate, sell or translocate the elephant. The financial rewards of any lawful use of an elephant accrue to the then owner of the elephant. This benefit is in the form of a windfall to the landowner and may be at the expense of conservation authorities. The future of an elephant so acquired may thereafter be determined by contract concluded on commercially expedient terms.

Rights of private landowner in respect of third parties

Having acquired real rights in and to the escaped *res nullius* elephant upon its mere presence on a landowner's land (as a natural resource occurring thereon) or by securing a sufficient degree of physical control over the elephant with the intent to become owner, such landowner can take whatever steps are available to protect such rights in and to the escaped elephant. This includes the prevention of any other person from entry onto the property should such other person wish to exercise any rights to the escaped elephant, and such person's ejectment if such entry is unlawful.

Any unlawful taking of the elephant by another person gives rise to an action for damages in the common law of trespass, the amount of such damages being the value of the elephant, fairly determined (SA Law Commission Working Paper No. 27, Project 69, April 1989 at page 20, par 2.35; LAWSA 1(2) par 461).

The Trespass Act 6 of 1959 also applies to the unlawful entry by a person onto the property of another and may result in such person being criminally prosecuted for trespass. The penalty is a fine not exceeding R2 000 or imprisonment for a period not exceeding two years or both.

In addition to this, if the landowner exercises sufficient physical control over the escaped elephant to constitute the landowner the owner of the elephant, such landowner, as owner of the elephant, has all of the usual common law remedies available to an owner of any movable property.

Implications for the State and for protected areas

Applying the principles set out above to elephants that become *res nullius* upon their escape from protected areas, such elephants will be lost to conservation upon their escape and neither the State nor protected area managers will have any rights of recourse to recover such loss.

Private landowner liability for damage caused by escaped elephants

The issue as to who is responsible for any damage caused by these elephants in these instances is will largely depend on the facts of each case. The law which regulates this is discussed in detail below.

Suffice to say, a protected area manager would only be liable for damage caused by an escaped elephant if he/she fails to discharge the common law duty of care that arises in the circumstances. A protected area manager would be obliged, in discharge of such duty of care, to pursue an escaped elephant in order to prevent or limit the damage caused, for the twofold purpose of preventing harm to others, and to mitigate the potential claim that such manager may face. If a private landowner were to refuse to allow such a manager access to his or her property, the following possible legal implications arise. By refusing access, the landowner may contribute to the damage that results and will be held accountable proportionately for the damage that may occur to his or her property, and to the property of others. Alternatively, the landowner may, by the exercise of sufficient physical control over the elephant with the intention to own it, become the owner, and would thereby voluntarily assume the risk and liability associated with the elephant. This would give rise to an estoppel in any action for damages against the protected area manager occurring after the assumption of ownership.

Common law liability for damage caused by wild animals

The principal common law actions whereby compensation can be claimed for damage caused by wild animals are: (a) the *actio de pastu*; (b) the *edictum de feris*; and (c) the *actio lex acquilia*.

The *actio de pastu*

This is a Roman law action for damages caused to land by grazing. Liability is strict and is based on the mere ownership of the wild animal without fault having to be proved. The *actio de pastu* is applicable in all cases where damage is caused to grass, crops, shrubs, trees, and the like. The actual patrimonial (pecuniary) loss which has been suffered can be claimed, which means that if crops are grazed, the value of the future crop can be claimed as damages. The action can be instituted by the owner, the holder of a servitude, a usufructuary or a lessee.

Defences which can be raised to the action are: *vis major* (an act of God), for instance where a gale blows open a gate or a flood destroys fences with the result that wild animals escape and graze on another person's land; and fault on the part of the injured party, for instance where as a reasonable person the injured party should have foreseen that the neighbour's animals might be able to escape onto his or her land because of inadequate or damaged fences and failed to take the reasonable steps available to him/her to guard against this (*LAWSA* 1(2) at par 470 to 474).

The *edictum de feris*

This edict has its origins in Roman Law as well as Roman Dutch Law and has received support from our courts (*LAWSA* 1(2) at par 475 to 480). Liability under the edict in Roman Law and in Roman-Dutch Law is strict and renders the person who keeps wild animals in the vicinity of a public place liable if the animal causes injury. There is however authority in South African case law for the view that liability under the edict is not strict but is rather based on fault or a presumption of fault (*LAWSA* 1(2) at par 476). The edict lays emphasis on the dangerous propensities of animals and the action is aimed at the recovery of damages caused by such wild animals.

Defences that can be raised to the *edictum de feris* could be that of the plaintiff's unlawful presence on the premises, *vis major* and fault on the part of the injured party.

The *actio lex acquilia*

This action lies in cases where damage is caused as a result of fault on the part of the owner or controller of an animal, either wild or domestic, and liability attaches where the owner or controller does not take reasonable care to prevent the animal from causing damage or injury to others (*LAWSA* 1(2) at par 481).

It is not necessary under this action for the defendant to be the owner of the animal. The basis of the action is the personal negligence of the defendant and it need therefore only be proved that a close enough relationship exists between the defendant and the animal that the defendant can be presumed to have a duty to prevent the animal from doing harm (*LAWSA* 1(2) at par 482).

To succeed, the plaintiff must prove either intention or fault on the part of the defendant. Various factors are taken into account in establishing negligence, and these include knowledge on the part of the defendant of the harmful characteristics of the animal, the class to which the animal belongs, the individual characteristics of the particular animal, the manner in which the damage was caused, the nature of the damage caused, the use to which the animal had been put and the place where it did the damage.

This action has in the past successfully been relied on to claim damage suffered as a result of *inter alia* (1) the instrumentality of a wild animal; (2) failure to secure a wild or vicious animal properly; or (3) damage caused by wild animals straying onto a public road (*LAWSA* 1(2) at par 484). Each of these causes of action is discussed below.

Damage caused through the instrumentality of a wild animal

There are many possible ways in which a delict may be committed through the instrumentality of a wild animal under one's control and there are many steps that one can take to limit one's exposure to a claim of this nature. What will be considered by our courts to be reasonable steps will depend on the facts and circumstances of each case. For example a manager of a hotel who rented out horses for riding was found to be negligent when the horse suddenly broke from the line, threw its rider and dragged the child along the ground. The negligent act relied upon was that the manager provided an unsuitable horse for the seven year old girl and that his employees failed to supervise the riding adequately. The principle applied in such circumstances is that an owner or controller of a wild animal who is aware of the harmful characteristics of an animal or the characteristics of a particular animal, has a duty to guard against harm to others. If such person fails to take measures to inform visitors to the property that the animal found thereon is wild, vicious or dangerous, and fails to take reasonable and effective measures to guard against harm to such visitors by such animal, he/she will be liable for negligence and could face a claim for damages.

Damage caused through the failure to secure wild animals properly

Damage is frequently also caused because of a failure to secure wild or vicious animals properly. Damages have been awarded by our courts in cases of harm caused by inadequately secured wildebeest and the like (*LAWSA* 1(2) at par 484). The question as to whether or not, and in what circumstances, a person has a duty to secure wild animals in order to prevent them from leaving one's property and causing harm to a neighbour, has been the subject of some debate and has received the attention of our courts.

In *Sambo v Union Government* (1936 TPD 182), Greenberg J held that where a person introduces a dangerous wild animal onto his or her property, such person is obliged to prevent such wild animals from leaving such person's property and causing damage or harm elsewhere. The *ratio decidendi* of the court in this matter would apply to the management of elephants.

In contrast to this, in *Mbhele v Natal Parks, Game and Fish Preservation Board* (1980 (4) SA 303 (D)), the facts of which were that the plaintiff had been attacked and seriously injured by a hippopotamus in an area near one of the defendant's game reserves, the court held that where a landowner simply lets nature take its course, he/she is under no duty to act to prevent wild animals on his or her property from escaping and causing damage to others.

The court found that the defendant did nothing more than simply let nature take its course, that hippopotamus in the reserve were protected and that although their numbers had increased since the establishment of the reserve it was not overstocked. Further, that the hippo were not introduced by human agency, that there was nothing artificial about their existence or their numbers, and that the defendant's control of the reserve went no further than to allow nature to take its course.

By applying this principle to the management of elephants it would mean that where elephants occur naturally on private or public land, have not been introduced by human agency, and the owner, lawful occupier or manager of such land does nothing to influence or interfere with the presence of such elephants on his land, and lets nature take its course, such landowner, lawful occupier or manager is under no duty to prevent such elephants from escaping from the property and causing damage to others.

However, in the light of the extensive management practices adopted for elephants in protected areas and private land, the use of a wide range of artificial mechanisms to control movement, population and disease, the management of habitats and the provision of supplementary food, it is difficult to conceive of circumstances in which elephant management would amount to no more than 'nature taking its course'. Although the circumstances are likely to determine the outcome of a particular case, it is submitted that the *ratio decidendi* of *Mbhele* would not be applicable to elephants managed as they presently are in South Africa.

Once such a legal duty is established, the manager of the protected area could be held liable for any damage caused by elephants escaping or straying from the protected area and causing damage or harm elsewhere. Liability arises if the protected area manager is found to have been negligent in discharging this duty. Negligence arises if a diligent *paterfamilias* (reasonable man) would have foreseen the reasonable possibility of harm, would have taken reasonable steps to guard against it, and fails to take reasonable steps to guard against the possibility of such harm being caused (*Kruger v Coetzee* 1966 (2) SA 428 (A)).

Reasonable steps which can be taken to discharge this duty would depend on the particular circumstances and may include the erection and ongoing maintenance of adequate fences; the use of other means to prevent elephants from escaping; mechanisms set up by the responsible person to enable third parties to easily report elephant escapes or any matters relating to problems with fencing which facilitate elephant escapes; steps taken to educate neighbours to enable them to protect themselves against the risks associated with elephant escapes; as well as mechanisms set up by the responsible person to enable the responsible person to respond immediately to reported escapes and related threats.

Measures to be adopted need to be reasonable and realistic in the circumstances. Account must be taken of the likely occurrence of elephant escapes, the degree of harm potentially caused to others, and the frequency of such escapes if they have occurred in the past. The danger posed to the lives and livelihoods of people living within the elephant range, as well as considerations of practicality and cost, must be balanced against the likelihood of the risk materialising (see *Mbhele* supra).

Under the *actio lex acquilia*, it must be shown on a balance of probabilities that the elephant that caused the damage or harm in fact escaped from or strayed from the property owned, lawfully occupied or lawfully managed by the defendant. If a claim is successful, the damages to be awarded will be limited to the actual damage proved to have been suffered.

Damage caused by wild animals straying onto a public road

Where an owner or controller of an animal through negligent action allows a collision to take place on a public road between the animal and a vehicle on the road, the owner or controller may be liable for damage. Negligence must be established on the facts of each case. There appears to be no hard and fast rule indicating what conduct constitutes negligence.

Defences

Defences to actions under the *lex acquilia* for damage caused by wild animals are the usual defences that can be raised against any delictual action and include necessity, provocation, self-defence, as well as fault on the part of the claimant.

The fact that a wild elephant may be classified as *res nullius* immediately prior to causing harm to others, is not *per se* a defence to a claim for damages made under the *actio lex acquilia*. Liability for damage under this action is not reliant on ownership but is based on whether or not the person from whose land the elephant escaped had a duty of care to prevent such animal from escaping and causing harm to neighbours or any other person, and if so, was negligent in discharging or omitting to discharge this duty.

Apportionment of damages

If a claimant is found to have contributed to a negligent act, in other words, some measure of negligence can be attributed to the injured party, the claimant's damages will be apportioned in accordance with the Apportionment of Damages Act 34 of 1956.

Challenges posed for the State by the common law

The State faces many challenges with the common law legal regime within which it is required to protect its rights to wild animals occurring in and escaping from protected areas under its management and control.

Elephants in protected areas at risk of appropriation by others

Since a strong degree of ongoing physical control over wild animals is required at common law for any person to establish and retain ownership rights in such wild animals, where elephants occurring in a protected area under the management and control of the State are not sufficiently restrained to that protected area through adequate fencing or other measures to constitute the State the owner of such elephants, such elephants are considered in common law to be *res nullius* and are therefore at risk of being appropriated by others.

In consequence of this, where conservation legislation does not specifically provide that the ownership of wild animals occurring in protected areas shall not vest in any person who, contrary to the provisions of such legislation or without the written consent of the protected area manager, hunts, catches or takes possession of such wild animals, ownership of such wild animals (and therefore also elephants) taken unlawfully or without such consent would vest in their captor. In other words, the captor (for example a poacher) would become the owner of the captured animal.

This must however be considered against the various statutory provisions that provide for criminal sanction, jail sentences as well as fines to be imposed on a captor in these circumstances.

State investment in elephants lost upon their escape from protected areas

As the common law dictates that as soon as a wild animal emerges from its place of detention, it becomes *res nullius* and is capable of being acquired by *occupatio* by the first person who has the acquisitive instinct and the means to gratify it, one can accept that as soon as an elephant escapes from a protected area it also becomes *res nullius* and is also capable of being acquired by others by *occupatio*. Animals so acquired are lost to the State, giving rise to a fortuitous gain by the person that acquires ownership thereof. The future of the escaped elephant then lies in the discretion of the new owner, subject to such controls as may be imposed over the common law property rights of the new owner by any relevant permitting laws, and may even become the subject of a contract concluded on commercially expedient terms.

A further consequence of this is that, in the absence of contractual arrangements concluded for the establishment and management of Transboundary TFCAs or the removal of fences between protected areas and private land providing otherwise, wild animals roaming out of protected areas and across international boundaries or onto private land, that are not actually or deemed to be owned by the State or any other person, become *res nullius* when they stray from the protected areas to private or foreign land. These animals fall outside of the control of the State, will form part of the property rights of private landowners or be subjected to the laws of neighbouring countries, and would therefore be capable of falling within the ownership rights and capacities of neighbouring governments or neighbouring landowners, as the case may be. These animals will be at risk of being lost to the State.

Further to this, when land that is formally protected and managed as a protected area or as part of a protected area under national or provincial conservation legislation, is awarded to private individuals or communities under the land restitution process currently under way in South Africa, and such land is de-proclaimed as protected area land, the wild animals occurring on such land, if not removed by the State prior to de-proclamation, will become *res nullius*. As such, they will form part of the private property rights of the new landowners and will be lost to the State.

Acquiring ownership in elephants contrary to conservation laws

Where legislation contains measures aimed at completely prohibiting the taking of possession of wild animals in protected areas in all circumstances, or the taking possession of wild animals in protected areas at certain times, unless such legislation unequivocally provides that ownership in an elephant captured in contravention of such legislation will not vest in the captor, the captor may nevertheless, become the owner of the elephant so captured or hunted, if it is deemed to be *res nullius*. This must however be considered against the statutory provisions that provide for criminal sanction, jail sentences as well as fines to be imposed on a captor in these instances.

Managing human-wildlife conflict

As can be observed from the case law dealt with above (*Sambo v Union Government* supra; *Mbhele v Natal Parks, Game and Fish Preservation Board* supra), wherever human-wildlife conflict has come for consideration before the courts, the decisions have invariably gone against the complainants and claims for damages suffered have been dismissed. This is largely because of the application by the courts of a common law that has not been redeveloped and redefined as required by the Constitution and does not reflect the change in values of the society in which it is applied.

There is growing dissatisfaction being expressed by and on behalf of communities living adjacent to protected areas at the lack of consistent, clear and unambiguous policy and guidelines being provided by the State in respect of matters relating to State's responsibility (or lack thereof) for the management of wild animal escapes from State owned or controlled protected areas, the manner in which these escapes are to be managed, as well as the general unwillingness on the part of the State to accept responsibility for damage arising out of such escapes.

Conservation experience has shown that failure on the part of governments world wide to adequately address these issues has resulted in confrontation between government and the communities most directly impacted by this, the undermining of government conservation efforts, and often even the exploitation of State conservation efforts for the benefit of only a select few.

SOUTH AFRICA'S NATIONAL LEGISLATION

Introduction

As mentioned previously, South Africa's conservation and biodiversity legislation does not deal with the acquisition, retention and loss of ownership in wild animals occurring in or escaping from protected areas. This legislation does however deal with many other aspects related to the management of elephants, and a synopsis of this legislation is provided below.

The National Environmental Management: Protected Areas Act 57 of 2003 (NEMPAA)

Objective of the NEMPAA

The NEMPAA came into effect on 1 November 2004 and at this time only regulated protected area matters of concurrent national and provincial legislative competence. Protected area matters of exclusive national legislative competence, such as the establishment, management and regulation of national parks, as well as the continued existence, powers and functions of South African National Parks, was inserted into the NEMPAA at a later stage by the National Environmental Management: Protected Areas Amendment Act 31 of 2004 which came into operation on 1 November 2005. The objective of the NEMPAA as set out in section 2 is to provide for the declaration and management of a national system of protected areas in South Africa within the framework of national legislation as part of a strategy to manage and conserve its biodiversity. In so doing it seeks to ensure the protection of the entire range of biodiversity, comprising natural landscapes and seascapes, the entire range of natural ecosystems and all biodiversity found in protected areas. It is intended to coordinate the declaration and management of protected areas and all biodiversity found in these protected areas.

System of protected areas

The NEMPAA categorises the different kinds of protected areas in South Africa. These include special nature reserves, national parks, nature reserves (including wilderness areas), protected environments, world heritage sites, marine protected areas, various specially protected forest areas, forest nature reserves and forest wilderness areas as well as mountain catchment areas (section 9).

Declaration of protected areas

Chapter 3 of the NEMPAA deals with the purpose of protected areas, states the various criteria which must be met before an area can be declared a protected area under the Act, and prescribes a range of procedures, including consultation and public participation procedures, which must be followed before any kinds of areas can be declared to be or to form part of a protected area as provided for in the NEMPAA.

Management of protected areas

Chapter 4 of the NEMPAA deals with the management of protected areas and applies generally to the management of special nature reserves, nature reserves and protected environments only. The provisions of this Chapter therefore do not find application in respect of the other categories of protected areas identified in section 9 (such as marine protected areas, specially protected forest areas and the like) unless specifically provided otherwise in this Chapter.

This Chapter provides for the assignment of responsibility for the management of the defined protected areas to which it relates to stipulated management authorities set out in the NEMPAA (section 38). It also deals with the preparation of management plans for these protected areas and provides for the objects and criteria for such management plans. It specifically provides that a management plan for a protected area should at the very least contain the terms and conditions of any applicable biodiversity management plan for the protected area to which it relates, a coordinated policy framework, planning measures, controls and performance criteria, a programme for its implementation and its costing, procedures for public participation, where appropriate, the implementation of community-based natural resource management, as well as a zoning of the area indicating what activities may take place in different sections of the protected area (section 41(2)).

SANParks as management authority of national parks

SANParks (a national organ of State and independent statutory body capable of suing or being sued in its own name and appointed to manage South Africa's system of national parks in terms of section 92 of the Act) is appointed the management authority for all existing national parks (section 92(1)(a).), and detailed provisions regulating its functions and powers (section 86-88), the composition and appointment of its governing board (section 57-66), operating procedures (section 67-71), financial regulation (section 74-77) and general administration (section 72-73), are contained in the NEMPAA.

SANParks relies primarily on the provisions of the NEMPAA as well as the regulations promulgated under the NEMPAA for the administration and management of the national parks assigned to it for management under the NEMPAA. Regulations for the proper administration of special nature reserves, national parks and world heritage sites, were issued in terms of section 86(1) of the National Environmental Management: Protected Areas Act No. 57 of 2003, in *Government Gazette* No. 28181 dated 28 October 2005, Notice No. R 1060.

Provincial spheres of government as management authorities of provincial protected areas

South Africa's provincial protected areas are currently managed by provincial departments responsible for environmental matters for each province, and in some cases by independent provincial statutory bodies established for this purpose (often referred to as provincial conservation agencies), in terms of the relevant conservation legislation of each province, with only certain overriding general provisions contained in the NEMPAA and being stated to have specific application to provincial protected areas, being applicable to the declaration and management of such areas.

Ownership of wild animals occurring in protected areas

The NEMPAA is silent on the question of ownership of wild animals occurring in protected areas. While section 3 of the NEMPAA requires the State, through the various organs of State implementing legislation applicable to protected areas, to act as trustee of protected areas in South Africa, this applies only to the management of the protected areas and the land set aside by the State for these protected areas.

Managing elephants under the NEMPAA

Detailed plans for the management of elephants in each protected area will generally be provided for in the management plan of each protected area. These plans have the force of the law and a management authority is obliged to manage a protected area in accordance with the management plan approved for the area by the Minister (section 92(1)(b)(i)). A public consultation process needs to be followed before any management plan can be submitted to the Minister for approval (section 39(3)).

Protecting elephants in protected areas

Some of the provisions contained in the NEMPAA which protect elephants are those aimed *inter alia* at preventing any persons, without the written authority of the management authority of the area, from: intentionally disturbing or feeding any species (regulation 4); hunting, capturing or killing a specimen (regulation 45(2)(a)(i); possessing or exercising physical control over any specimen (regulation 45(2)(a)(iv); conveying, moving or otherwise translocating any species (regulation 45(2)(a)(vi).

Any person who contravenes a regulation is guilty of an offence (regulation 61(1)).

The law enforcement, offence and penalty provisions contained in the NEMPAA for the protection of wild animals, and therefore also wild elephants, pose certain challenges and are currently being subjected to a process through which they will in due course be improved through various specific legislative interventions.

So for example, while the NEMPAA provides that a person convicted of an offence under the NEMPAA is liable on conviction to a fine or to imprisonment for a period not exceeding five years or to both such fine and such imprisonment (section 89 and regulation 64), having provided no specific fine amounts to be imposed in respect of specific offences, one becomes obliged to apply the provisions of the Adjustment of Fines Act 101 of 1991 read with the provisions of section 92 (1) (a) and (b) of the Magistrates' Court Act 32 of 1944), which has the effect that, if an offence committed under the NEMPAA is heard by a district magistrate's court, the sentence or fine to be imposed by such court cannot exceed three years or ZAR60 000 respectively. Where the matter is heard by a court which is a regional magistrate's court, the sentence or fine to be imposed cannot exceed five years (could have been 15 years had the NEMPAA not limited this to 5 years) or ZAR100 000 (could have been ZAR300 000 had it not been for the limitation of five years provided for in the NEMPAA).

Considering the extent of fines and penalties that can be imposed by the State in respect of environmental crimes under the National Environmental Management Act 107 of 1998 (NEMA), in some instances amounting to fines of up to ZAR5 million and imprisonment for periods not exceeding 10 years (section 24F of NEMA), these provisions appear to be inadequate.

In addition to this, none of the provisions contained in NEMA providing for offenders to be subjected to damages awards (section 34(1) and (2)), to fines based on the monetary value of the advantage gained or likely to be gained by a criminal in consequence of an offence committed under NEMA (section 34(3), and to cost orders based on the costs incurred by the State in respect of investigations and prosecutions of offences (section 34(4), vicarious criminal liability for employers in respect of the criminal acts of their employees, managers or agents, and vice versa, (section 34(5) and (6)), liability for body corporates and other legal entities for the acts of their directors, partners, members of boards, members of executive committees or members of other managing bodies, and vice versa (section 34(7) and (8)), have been carried over into the NEMPAA or in any way made applicable to any offences committed under the NEMPAA.

Applying this to a practical example, where five large tusk male elephants with a hunt value of ZAR100 000 each are stolen from a special nature reserve, national park or world heritage site, or where five black rhino worth ZAR350 000 each are so stolen, the sentences and fines that can be imposed in these instances could never exceed a sentence of five years or a fine of ZAR100 000, and no further damages or costs awards can be made against offenders, nor can any persons be held vicariously liable for the acts of others in their employ or under their control.

Whilst no criminal remedies for damages or vicarious liability exist under the NEMPAA, civil liability for such damages as well as vicarious liability under a civil claim for damages, are available to the State.

Addressing human-wildlife conflict in the case of wild animals escaping from protected areas

The NEMPAA does not address any issues relating to the escape or straying of wild animals from such areas and their propensity to cause damage or harm to others. As such the NEMPAA provides no guidance as to how human-wildlife conflict is to be addressed when wild elephants escape from protected areas and cause injury or harm to others.

The National Environmental Management: Biodiversity Act 10 of 2004 (NEMBA)

Objective of NEMBA

The NEMBA came into operation on 1 September 2004 and provides for the management and conservation of South Africa's biodiversity within the framework of the National Environmental Management Act (FN. No. 107 of 1998) (NEMA); the protection of species and ecosystems that warrant national protection; the sustainable use of indigenous biological resources; the fair and equitable sharing of benefits arising from bioprospecting involving indigenous biological resources; the establishment and functions of the South African National Biodiversity Institute; and for matters connected therewith.

Integrated biodiversity planning, monitoring and research

Chapter 3 (sections 38-50) provides for integrated and coordinated biodiversity planning, the monitoring of the conservation status of various components of South Africa's biodiversity, and for biodiversity research.

A national biodiversity framework for South Africa

Section 38 requires that the Cabinet Member responsible for national environmental management in South Africa (the Minister) must prepare and adopt a national biodiversity framework for South Africa within three years of the date on which NEMBA takes effect and it is expected that this framework will provide for an integrated, coordinated and uniform approach to biodiversity management by organs of State in all spheres of government, non-governmental organisations, the private sector, local communities, other stakeholders, and the public.

Creating and protecting bioregions

Section 40 provides for the Minister or a member of the Executive Council of a province who is responsible for the conservation of biodiversity in the province (the MEC) with the concurrence of the Minister, to determine a geographic region as a bioregion for the purpose of NEMBA if that region contains whole or several nested ecosystems and is characterised by its landforms, vegetation cover, human culture and history, and publish a plan for the management of biodiversity and the components of biodiversity in such region. Bioregional plans for such bioregions can be prepared and applied and the Minister may enter into an agreement with a neighbouring country to secure the effective implementation of any such plans.

Management plans for specific ecosystems and listed threatened or protected species

Section 43 provides that any person, organisation or organ of State desiring to contribute to biodiversity management may submit to the Minister for his or her approval, a draft management plan for *inter alia* an ecosystem listed in terms of section 52 of NEMBA or an indigenous species listed in section 56 of the NEMBA. The African wild elephant is a section 56 listed species.

Section 45 requires that the biodiversity management plan must be aimed at ensuring the long-term survival in nature of the species or ecosystem to which the plan relates, must indicate who will be responsible to implement this, and must be consistent with *inter alia*, the NEMBA, all national environmental management principles, the national biodiversity framework, any applicable bioregional framework, any environmental implementation plans and management plans referred to in Chapter 3 of NEMA, any municipal integrated development plan, any other plans prepared in terms of national or provincial legislation, and any relevant international agreement binding on the Republic of South Africa.

Public consultation

A consultative process as described in sections 99 and 100 has to be followed before the national biodiversity framework or a bioregional or a biodiversity management plan can be adopted or approved by the Minister. Similar provisions apply to the adoption or approval of a bioregional or biodiversity management plan by the MEC of a Province who is responsible for the conservation of biodiversity in that province.

Protection of threatened or protected species

Chapter 4 provides for the protection of species that are threatened or in need of protection to ensure their survival in the wild. Section 56 provides for the listing by notice in the national *Government Gazette* of species that are critically endangered, endangered, vulnerable and protected, and are in need of national protection. Such lists have been published in *Government Gazette* No. 29657 dated 23 February 2007, Notice No. ZAR151, (the TOPS Species Lists), and the African wild elephant is identified in the list of TOPS Species as being a 'protected species' which in terms of section 56(1)(d) of NEMBA means a species which is of such high conservation value or national importance that it requires national protection.

Section 57 prohibits the carrying out of any restricted activity involving a specimen of such listed species without a permit issued in terms of Chapter 7. Section 57 also provides that the Minister may by notice in the *Government Gazette* completely prohibit the carrying out of any activity which is of a nature that may negatively impact on the survival of such listed threatened and protected species.

The definition section contains a list of 'restricted activities' and these include *inter alia* activities aimed at hunting, catching, capturing, killing, importing, exporting, having in possession or exercising physical control over, breeding, conveying, moving or otherwise translocating, selling or otherwise trading in, buying or in any way acquiring or disposing of, any specimen of a TOPS Species.

Section 59 obliges the Minister to monitor compliance with section 57 insofar as this relates to trade in TOPS Species in South Africa. Section 60 provides for the establishment of a Scientific Authority which is obliged *inter alia* to assist the Minister with regulating and restricting the trading in TOPS Species in South Africa as well as monitoring both legal and illegal trade in such species.

Permit requirements and risk assessments

Chapter 7 deals with permit requirements that have to be complied with before a permit can be issued authorising the conduct of a restricted activity involving a specimen of a listed threatened or protected species, and section 89 goes as far as to enable the issuing authority, before issuing a permit, to require that the applicant furnish to it in writing, at the applicant's expense, an independent risk assessment or such expert evidence as the issuing authority may determine.

Issuing authority and regulations to be promulgated by the Minister

The issuing authority for any permits required to authorise the conduct of a restricted activity involving a listed threatened or protected species is the Minister or an organ of State in the national, provincial or local sphere of government designated by the Minister by regulation in terms of section 97 to be an issuing authority for such permits. Section 97 then provides for the Minister to make regulations for the designation of organs of State to be the issuing authorities for permits as well as for other matters such as for the carrying out of restricted activities involving listed threatened or protected species; the assessment of risks and potential impacts on biodiversity of restricted activities involving listed threatened or protected species; the conditions subject to which issuing authorities may issue permits; the procedures to be followed and the fees to be paid; factors that must be taken into account when considering applications, etc.

Norms and standards

Section 9 also provides for the Minister to, by notice in the *Government Gazette*, issue norms and standards for the achievement of any of the objectives of the NEMBA and required for the management and conservation of South Africa's biological biodiversity and its components or the restriction of activities which impact on biodiversity and its components. Such norms and standards may apply nationwide or in a specific area only or to a specific category of biodiversity only.

Such norms and standards have been tabled for the management of elephants in South Africa, and are dealt with in more detail below.

Law enforcement, offences and penalties

It is an offence in terms of section 101(a) for any person to conduct a restricted activity in respect of an African wild elephant without a permit issued in terms of Chapter 7. As with the offence and penalty provisions contained in the NEMPAA, the offence and penalty provisions contained in the NEMBA pose challenges and require improvement.

So for example, a person who hunts, captures, kills, imports, exports, translocates, conveys, moves or sells or trades in African wild elephants without the necessary permit will at most face imprisonment not exceeding five years or a fine not exceeding ZAR100 000, with no further damages or costs awards being capable of being made against offenders and without any provision being made for vicarious liability of persons for the acts of others in their employ or under their control.

Whilst no criminal remedies for damages or vicarious liability exist under the NEMBA, civil liability for such damages as well as vicarious liability under a civil claim for damages, are available to the State.

Addressing human-wildlife conflict

The NEMBA does not address any issues relating to human-wildlife conflict. However, the TOPS Regulations (dealt with below) do provide for a process to deal with damage-causing animals.

The National Environmental Management: Biodiversity Act No. 10 of 2004 – Threatened and Protected Species Regulations as published in Government Notice No. R 152 published in Government Gazette No. 29657 dated 23 February 2007 (implementation date – 1 February 2008) (the TOPS Regulations)

Objectives of the TOPS Regulations

The TOPS Regulations were essentially promulgated to further regulate the permit system set out in Chapter 7 of the NEMBA insofar as that system applies to restricted activities involving TOPS Species; to provide for the registration of captive breeding operations, commercial exhibition facilities, game farms, sanctuaries, rehabilitation facilities and the like; to provide for the regulation of hunting of TOPS Species; to completely prohibit the carrying out of certain activities in respect of certain of the TOPS Species; to provide for the protection of wild populations of TOPS Species; and to provide for the composition and operations of the Scientific Authority (Regulation 2).

The Regulations are intended to serve concurrently with any provincial legislation that regulates similar matters relevant to TOPS Species, wherever they occur in South Africa. Should any conflict occur as between the Regulations and any provincial legislation, such conflict will have to be resolved as required by the Constitution (section 146(1)). National legislation will prevail over provincial legislation if it is *inter alia* necessary for the protection of the environment (section 146(2)(c)(vi) of the Constitution). This is however subject to the proviso that a law made in terms of an Act of Parliament can only prevail if the law has been approved by the National Council of Provinces as provided for in section 146(6) and 146(7) of the Constitution.

Impact of the TOPS Regulations on the keeping and management of elephants

It is intended that the TOPS Regulations come into effect on 1 February 2008. From this date, no person shall be entitled to engage in or carry out a restricted activity in respect of a TOPS Species (which includes the African wild elephant) unless such person is in possession of a permit issued under the TOPS Regulations. As such, no person shall be entitled to hunt, catch, capture, kill, import, export, be in possession of or exercise physical control over, breed, convey, move or otherwise translocate, sell or otherwise trade in, buy or in any way acquire or dispose of an African wild elephant, unless such person is in possession of the required permit.

All protected area managers, owners of game farms, owners of commercial exhibition facilities (defined in NEMBA as meaning a facility, including zoological gardens and travelling exhibitions, that keeps elephants for display purposes), owners of sanctuaries (defined in NEMBA as meaning a facility in which a permanent captive home is provided in a controlled environment for elephants that would be unable to sustain themselves if released), and owners of rehabilitation facilities (defined in NEMBA as meaning a facility equipped for the temporary keeping of elephants for treatment or recovery purposes, for the rearing of young orphaned elephants, for the keeping of elephants for quarantine purposes, or for the keeping of elephants for relocation purposes, with the overall intent to ultimately release the elephants), at which elephants are kept, will have to be in possession of the various permits required by the TOPS Regulations for such persons to be in possession of and to conduct any restricted activity in respect of any elephants found in such areas, farms or facilities. Failure to be in possession of a valid permit constitutes a criminal offence (Regulation 73(1)(a)). A person convicted of an offence is liable to a fine of R100 000 or three times the commercial value of the specimen in respect of which the offence was committed, whichever is the greater, or to imprisonment for a period not exceeding five years or both (Regulation 74). It is submitted that the offence and penalty provisions, as is the case with the NEMPAA, require improvement.

Permits and compulsory registration of certain facilities

The TOPS Regulations provide for different kinds of permits to be issued for different periods (Regulation 22) and contain a host of provisions relating to who may apply for the different kinds of permits (Regulation 5), application procedures (Regulation 6), period of validity of permits (Regulation 22), factors to be taken into account when considering permit applications (Regulations 10, 12 and 13), the compulsory registration of *inter alia* commercial exhibition facilities, sanctuaries and rehabilitation centres (Regulation 27), the voluntary registration of game farms (Regulation 28), as well as compulsory conditions that are to be applied to registered commercial exhibition facilities, rehabilitation facilities and sanctuaries (Regulation 35).

Protected area managers are able to apply for a 48 month standing permit to conduct identified restricted activities in respect of elephants under their management and control that are necessary for their management in accordance with the management plan for the area (Regulation 5(2)(c)). In the absence of successfully applying for a 48 month standing permit, a protected area manager would have to apply for a separate permit for each and every restricted activity to be conducted in respect of any elephants occurring in such protected areas.

The owners of commercial exhibition facilities, sanctuaries and rehabilitation facilities will not be entitled to conduct their business unless and until such facilities are registered with the responsible issuing authority. Registration permits are valid for 36 months and can be renewed. Once the facility is registered, the owner of the facility will automatically qualify for a 36 month standing permit authorising the conduct of such restricted activities involving elephants held at such facilities, as may be necessary, in the case of a commercial exhibition facility, for the purpose for which the commercial exhibition facility is registered, and in the case of registered sanctuaries and rehabilitation facilities, for their treatment or care. Certain transitional provisions (Regulation 71) provide that the owner of such facility has a period of 3 months from the date on which the TOPS Regulations come into effect to apply for registration of the facility. If the application is declined, the owner has a further period of 9 months after the refusal to comply with all requirements and to reapply for registration. If again declined, the facility will have to be closed down. During the application process, the owner will have to apply for and be in possession of a separate permit for each and every restricted activity to be conducted by such owner in respect of any elephants held at such facilities.

It is not compulsory for a game farm owner to register his game farm under the TOPS Regulations. If an owner of a game farm however elects not to register his facility, such game farmer will have to apply to the responsible issuing authority for a separate permit under the TOPS Regulations for each and every restricted activity that such game farmer may wish to conduct on his game farm in respect of elephants held on such farm. However, if an owner of a game farm registers his game farm under the TOPS Regulations, such game farm owner will be able to apply for a 36 month standing permit (regulation 22(2)(a)(ii)) which will entitle him to conduct the various restricted activities required for the management of his game farm without having to apply for separate permits for each and every such activity.

Control over hunting and trade in elephants

Since no person is able to hunt, import, export, sell or otherwise trade in, buy or in any way acquire or dispose of a TOPS Species or any product of such Species without being in possession of a permit issued under the TOPS Regulations, the Minister is able to exercise a control and monitoring function over *inter alia* hunting and trading in elephants and elephant products in South Africa.

Further monitoring and control over hunting activities are exercised by requiring *inter alia* that the holder of a hunting permit is obliged to have all permit documents in his possession at the time of the hunt, and is further obliged to furnish a return of the hunt to the issuing authority within 21 days of the hunt specifying *inter alia* the permit number, date of issue, species, sex and number of animals hunted, location where the hunt took place (Regulation 21). In addition to this, hunting by persons who are not resident in South Africa and who pay or reward professional hunters for or in connection with the hunting of a TOPS Species, is prohibited, unless such person is accompanied by a professional hunter (Regulation 21(1)(c)). A professional hunter is a person licenced in terms of provincial legislation as a professional hunter. In this way the quality, qualifications and experience of persons overseeing hunts by overseas clients are controlled.

Certain hunting methods are also prohibited, such as hunting by poison, traps, snares, automatic rifles, darting (except for veterinary purposes), shotgun, air gun or bow and arrow (regulation 26(1)(a)). (By implication, the 'green hunting' of elephants is also prohibited.) In addition to this, the use of floodlights or spotlights, motorised vehicles or aircraft for hunting is also prohibited unless this is required to track an elephant over long ranges or to cull elephants (Regulation 26(5)). All of this is stipulated in the interests of promoting acceptable methods of hunting in South Africa.

Protection of wild elephant populations

Protection of wild elephant populations (defined in the definition section of the TOPS Regulations as meaning a group or collection of wild specimens that are living and growing in natural conditions with or without human intervention) occurring in South Africa, is provided by prohibiting the translocation of elephants into extensive wildlife systems (defined in the definition section of the TOPS Regulations as meaning a system that is large enough and suitable for the management of self-sustaining wildlife populations in a natural environment which requires minimal human intervention in the form of the provision of water, the supplementation of food (except in times of drought), the control of parasites or the provision of health care) which fall outside of the natural distribution range of those elephants (Regulation 23(a)). Translocations of elephants into extensive wildlife systems which are protected areas are also prohibited (Regulation 23(a)).

Regulating possession of and trade in elephant ivory

Any person who is in possession of any elephant ivory must, within three months of the commencement of the TOPS Regulations, apply in writing to the issuing authority in the relevant province, to have such elephant ivory permitted, and if applicable, marked and registered on the national database for elephant ivory (Regulation 70(1)). Only ivory which is 20 centimetres or more in length, or more than 1 kilogram in weight, whether carved or not, needs to be marked and registered as prescribed. All marking needs to be made with a punch-die, or if not practicable, with indelible ink, and needs to comprise a certain formula made up of the country-of-origin two letter ISO code and the last two digits of the particular year followed by a forward slash, the serial number of the particular year followed by a forward slash, and the weight of the ivory in kilograms (regulation 70(3)). Marking is to be done at the expense of the applicant. Any person in possession of elephant ivory without the required permit will be guilty of a criminal offence (Regulation 73(1)).

This mechanism enables the Minister to exercise a control and monitoring function over all possession, sale or trade in all elephant ivory in South Africa.

Addressing human-wildlife conflict

Regulation 14 sets out provisions relating to damage-causing animals. Subregulation (1) requires the provincial department responsible for the conservation of biodiversity in a province, to determine whether a listed threatened or protected species can be deemed to be a damage-causing animal. In the case of a damage-causing animal originating from a protected area, subregulation (2) requires that the following control options be considered by the provincial department referred to in subregulation (1) or the management authority of a protected area: capture and relocation by the provincial department referred to in subregulation (1) or the management authority of the protected area; control by the provincial department referred to in subregulation (1) or the management authority of a protected area by culling or by using methods prescribed in subregulations (4), (5) and (6); or control by a person, other than a hunting client, designated in writing, by the provincial department referred to in subregulation (1) or the management authority of the protected area to capture and to relocate or to control by means of methods prescribed in subregulation (4), (5) and (6).

Subregulation (1) does not prevent a landowner from killing a damage-causing animal in self-defence where human life is threatened. If a damage-causing animal is killed in an emergency situation subregulation (3) requires: the landowner to inform the relevant issuing authority of the incident within 24 hours after it has taken place; and the issuing authority to evaluate the evidence, and if justified, to condone the action in writing or if necessary, take appropriate steps to institute criminal proceedings.

Subregulation (4) allows the holder of a permit referred to in regulation 5(2)(a) and (c) to hunt a damage-causing animal by the following means, as specified on his or her permit: poison, which has in terms of applicable legislation been registered for the purpose of poisoning the species involved and as specified by the issuing authority; bait and traps, excluding gin traps, where the damage-causing animal is in the immediate vicinity of the carcass of domestic stock or wildlife which it has or apparently has killed, or is about to cause damage to domestic stock or wildlife; the use of dogs, for the purpose of flushing the damage-causing animal or tracking a wounded animal; darting, for the subsequent translocation of the damage-causing animal, and the use of a firearm suitable for hunting purposes. In terms of subsection (5) the holder of a permit referred to in Regulations 5(2)(a) and (c) may hunt a damage-causing individual by luring it by means of sounds and smell, and in terms of subregulation (6) may hunt a damage-causing animal by using a motorised vehicle and flood or spotlights.

Regulation 14 is at most a fragmented attempt to address only a very limited aspect of the complex and legally challenging issue of wildlife-human conflict that arises when wild animals escape from State-owned or controlled protected areas, and cause damage or harm to others. The Regulation makes no attempt to address the more critical issues relevant to ownership rights outlined above, nor does it provide a mechanism for dealing with the financial implications of damage caused by animals. In essence, the regulation is a restatement of the common law with the imposition of control over the exercise of common law rights in property by way of permitting.

The Animal Diseases Act 35 of 1984

While elephants are not generally themselves carriers of disease, their breakouts can facilitate the escape of other wild animals that have the propensity to carry with them or to contract various animal diseases. This Act has therefore been included in this chapter for completeness.

Objective of the Animal Diseases Act

This Act came into operation on 1 October 1986 and its purpose is to provide for the control of animal diseases and parasites, for measures to promote animal health, and for matters connected therewith. The Minister of Agriculture is vested with responsibility for the administration and implementation of this Act.

Duties imposed on owners and managers of land and animals

The Act imposes a primary duty on any owner or manager of land on which there are animals, as well as the owner in respect of animals, *inter alia* to take all reasonable steps necessary to prevent the infection of the animals with any animal disease or parasite and the spreading thereof from the relevant land or animals; to apply prescribed treatments whenever animals become or can reasonably be suspected of having become infected with an animal disease or parasite; and to report any suspected incident of an animal becoming or suspected to have become infected with an animal disease or parasite (section 11). Such steps would include erecting and maintaining adequate fences where required and necessary.

The owner of land is defined in section one and includes a wide range of persons in physical or legal control over land. The owner of an animal is defined as meaning the person in whom the ownership in respect of such animal is vested, including the person having the management, custody or control of such animals. Ownership will be determined as a matter of law, either by virtue of the common law or the Game Theft Act dealt with above. Management, custody or control will be determined as a matter of fact.

Control measures prescribed by the Minister

The Minister is entitled for any controlled purpose to prescribe general control measures, or particular control measures in respect of particular animal diseases and parasites. A controlled purpose is defined in the Act as meaning the prevention of the bringing into the Republic, or the prevention or combating of or control over an outbreak or the spreading or the eradication of any animal disease or parasite. Control prescribed in section 9 includes measures regarding the powers and duties of owners and managers of land, and owners of animals, in respect of infectious or contaminated things or animals, and with regard to controlled veterinary acts or any other examinations or acts in connection with such animals or things.

Fencing and cost recovery issues

The Minister may by written notice served on the owner or manager of land, declare that he assumes from a specified date, control over the land, including all fences and structures on the land for a specified purpose and for a specified period, and may during this time perform any act on the land which the owner or manager of the land is required in terms of the Act to perform, and to recover any expenditure connected therewith from the owner or manager (Section 14.).

An officer of the Department of Agriculture may also, for any controlled purpose, erect fences along the boundaries of any land, and maintain such fences as may be necessary for such controlled purpose (Section 18(1)). Where the director is of the opinion that the erection of a fence will be of advantage to an owner or manager of relevant land, the director may recover any portion of the relevant costs determined by him/her from the owner or manager (Section 18(5)).

Compensation for animals destroyed

The owner of any animal which has been destroyed or otherwise disposed of pursuant to a control measure, may submit an application for compensation for the loss of the animal and will be compensated the fair market value of the animal (section 19).

Role of the Department of Agriculture

Primary responsibility to prevent the spread of disease from land on which wild animals occur lies with the owner of the land or the manager of the land on which such animals occur, and with the owner of the animals. While the Department of Agriculture is of necessity required to provide general direction to, support for and administrative control over issues related to the occurrence and spread of animal disease across the country, and in many instances to take steps to avoid, prevent or to control the occurrence or spread of disease, primary responsibility remains with the landowner or animal owner. The costs of interventions by the State where owners have not discharged their duties adequately are generally recoverable from such persons by the State.

Liability for damage caused by the spread of disease

Liability of owners and managers of land as well as owners of animals for the spread of disease and the consequences arising out of this is dealt with in accordance with the usual principles of delict dealt with above. However, section 27 limits liability in respect of persons, including the State, for anything done or omitted to be done in good faith in the exercise of a power or the performance of a duty under or by virtue of the Act, or in the rendering of any service in terms of the Act.

The Animals Protection Act 71 of 1962

The purpose of this Act is to consolidate and amend the law relating to the prevention of cruelty of animals. An animal includes any wild animal, bird or reptile which is in captivity or under the control of any person. Control for this purpose would mean *de facto* control or deemed control under any law. The Act would therefore apply to all animals except those that are not in captivity or under the control of any person. Any person who conducts any of the acts of cruelty identified in the Act in respect of an animal in captivity or under his or her control is guilty of an offence. An act of cruelty performed on an elephant which is in a free roaming state would however not fall within the ambit of the Act.

The Performing Animals Protection Act 24 of 1935

The purpose of this Act is to prohibit the exhibition or training for exhibition of any animal in the lawful ownership or custody of a person unless such person is the holder of a licence. The Act therefore applies to the use of elephants in zoos, circuses and other forms of exhibition, but would not apply to elephants used as beasts of burden, including for elephant-back safari purposes.

The Society for the Prevention of Cruelty to Animals Act 169 of 1993

The purpose of this Act is to establish the Society for the Prevention of Cruelty to Animals. The Society is the nominated authority responsible for the enforcement of the Animals Protection Act and the Performing Animals Act.

SOUTH AFRICA'S PROVINCIAL LEGISLATION

The five provinces in which most elephants occur, namely Limpopo, Mpumalanga, North West Province, KwaZulu-Natal and the Eastern Cape, each have their own legislation dealing with the management, control, and hunting of wild animals.

Provincial legislation deals with wildlife in various ways but primarily by species listing, the allocation of levels of protection to such species and the permitting of uses of such species. The Limpopo Environmental Management Act, 7 of 2003 lists elephants as a 'specially protected wild animal'. The Mpumalanga Nature Conservation Act, No. 10 of 1998 likewise regards an elephant as 'specially protected game'. The Cape Nature and Environmental Conservation Ordinance, No. 19 of 1974 (applicable to the Eastern Cape Province), however, regards elephants as 'protected wild animals', a lesser protected status. The Bophuthatswana Nature Conservation Act, No. 3 of 1973 (applicable to the North West Province) regards elephants as 'specially protected game'. The Nature Conservation Ordinance, No. 15 of 1974 (applicable to the KwaZulu-Natal Province), treats elephants as 'specially protected game'.

The provincial ordinances referred to do not deal with the question of ownership of wild animals and the common law is therefore not altered by such legislation. Nor do they deal consistently with human-wildlife conflict.

POLICY, NORMS AND STANDARDS UNDER SOUTH AFRICAN LAW

Policies, norms and standards are further legal instruments that assist with the interpretation and administration of Acts of Parliament and the regulations promulgated thereunder. Section 146(1) of the Constitution requires uniformity in the application of national legislation which is achieved by establishing norms and standards, frameworks and national policies. These are so called 'soft law' documents that do not have legal or binding effect and are usually determined to assist officials charged with the implementation of the law.

The development of policy is often a precursor to legislation and involves a consultative process in which all stakeholders, including the public, are invited to participate. The Consultative National Environmental Policy Process (or CONNEPP) led to the *White Paper on an Environmental Policy for South Africa,* which was the foundation for the promulgation of NEMA.[1] Similarly, the Integrated Coastal Management Policy Process became a *Green Paper* and then a *White Paper for Sustainable Coastal Development* and is presently the Integrated Coastal Management Bill (http://www.mcm-deat.gov.za/indexpage_DOCS/ICM%20Bill%20Draft%2010_.pdf), approved by Cabinet but awaiting promulgation as an Act. A White Paper is Cabinet-approved national policy and guides the interpretation of laws within its purview. Policy is not law, and is not enforceable as such.

In contrast to this, norms and standards may be developed to provide technical and practical guidance for the implementation of legislation, if provision is made in the relevant legislation for their adoption. Norms and standards must be applied to give effect to the legislation and are not enforceable in their own right. Furthermore, they do not stand as policy directives for the interpretation of the statute under which they are developed. The draft norms and standards for the sustainable utilisation of large predators published by the Minister of Environmental Affairs and Tourism in terms of section 9(1) of the National Environmental Management Biodiversity Act, 2004 is an example of the invocation of this power. Draft norms and standards for the regulation of the hunting industry and for the management of elephants in South Africa (dealt with below) are examples of the same.

In KwaZulu-Natal, Ezemvelo KZN Wildlife has announced its intention to make recommendations to the MEC Agriculture and Environmental Affairs on a policy for the keeping of wild animals in captivity, the development of norms and standards for this, as well as the development of norms and standards for the keeping of primates generally, and specifically vervet monkeys.

It should be noted that property rights in animals are protected by section 25 of the Constitution which provides that no one may be deprived of property arbitrarily. Section 36 of the Constitution provides for the limitation of rights in the bill of rights but only in terms of law of general application and then only to the extent that the limitation is reasonable and justifiable in an open and democratic society, and after taking into account all relevant factors. The so called soft laws, being policies, norms and standards, referred to above, are not laws of general application and therefore cannot be applied in such a way as to cause either a deprivation of property rights or a limitation of any other right. On the other hand, the principal legislation under which the policies, norms or standards are developed may be used to limit rights in animals provided the circumstances justify such a limitation.

The National Environmental Management: Biodiversity Act No. 10 of 2004 – Draft Norms and Standards for the Management of Elephants in South Africa

Background

Norms and standards for the management of elephants in South Africa were published by the Department of Environmental Affairs and Tourism in February 2008 in *Government Gazette* No. 30833 dated 29 February 2008, under general notice no. 251, and came into operation on this date.

The norms and standards were prepared in terms of section 9 of the NEMBA which provides for the Minister to, by notice in the *Government Gazette*, issue norms and standards for the achievement of any of the objectives of the NEMBA and required for the management and conservation of South Africa's biological biodiversity and its components or the restriction of activities which impact on biodiversity and its components. The norms and standards apply to both wild and captive elephants.

Purpose and application

Paragraph 2 states the purpose of the document is to set national norms and standards to ensure that:

   a.   elephants are managed in the Republic in a way that –
       i.   ensures the long-term survival of elephants within the ecosystem in which they occur or may occur in the future;

    ii. promotes broader biodiversity and social goals that are ecologically, socially and economically sustainable;

    iii. does not disrupt the ecological integrity of the ecosystems in which elephants occur; and

    iv. enables the achievement of specific management objectives of protected areas, registered game farms, private and communal land;

    v. ensures the sustainable use of hair, skin, meat and ivory products;

    vi. is ethical and humane; and

    vii. recognises their sentient nature, highly organised social structure and ability to communicate;

b.   the management of elephants is regulated –

    i. in a way that –

      aa. is uniform across the Republic;

      bb. takes into account the Republic's international obligations in terms of international agreements on biodiversity management binding on the Republic; and

    ii. in accordance with national policies on biodiversity management and sustainable development.

The norms and standards are informed by the principles contained in paragraph 3 and apply to the management of elephants wherever they occur within the Republic.

Guiding principles

The principles set out in paragraph 3 require any person executing a function or exercising a power or carrying on an activity that relates, directly or indirectly, to an elephant to do so with regard to the following further principles, which are largely ecological rather than legal in their nature:

1.   elephants are intelligent, have strong family bonds and operate within highly socialised groups and unnecessary disruption of these groups by human intervention should be minimised;

2.   while it is necessary to recognise the charismatic and iconic status of elephants and the strong local and international support for their protection, proper regard must be given to the impacts of elephants on biodiversity or people living in proximity to elephants;

3.   elephants are recognised engineers of habitat change and their presence or absence has a critical effect on the way in which ecosystems function;

4.   the movement of elephants throughout their historical range has been disrupted by the activities of people over the last two centuries;

5.   careful conservation management has led to the significant growth of elephant populations and human intervention may be necessary to ensure that any future growth occurs in a manner that does not result in the loss of biodiversity, ecosystem function and resilience or human life, or the compromise of key management objectives for protected areas, registered game farms or private or communal land;

6.   elephants often exist in close proximity to people, with the result that the elephants potentially pose a threat to the well-being of people and management measures must endeavour to limit these threats;

7.   measures to manage elephants must be informed by the best available scientific information and, where the available scientific information is insufficient, adaptive management forms the cornerstone of the management of elephants and adaptive decision making tools must be adopted;

8.   management interventions must, wherever practicable, be based on scientific knowledge or management experience regarding elephant populations and must –

   a.   take into account the social structure of elephants;

   b.   be based on measures to avoid stress and disturbance to elephants;

9.   where lethal measures are necessary to manage an elephant or group of elephants or to manage the size of elephant populations, these should be undertaken with caution and after all other alternatives have been considered;

10. while efforts should be made to ensure that elephants continue to play an important role in an already well established nature-based tourism sector this should not occur in an inappropriate, inhumane or unethical form or manner;

11. in the context of objective-based management of complex ecological systems elephants should not be accorded preference over other elements of biodiversity;

12. every effort must be made to safeguard elephants from abuse and neglect; and

13. elephant population in the wild should be managed in the context of objective-based management of the complex ecosystem in which they occur.

The norms and standards provide guidance for the application of the TOPS Regulations in respect of elephants and no restricted activity may be undertaken in respect of elephants without having due regard for the applicable provisions of the said Regulations.

Addressing human–wildlife conflict

Paragraph 25 provides a process to be followed when elephants escape from protected areas or from adequately enclosed areas and is to some extent aligned with the provisions of Regulation 14 of the TOPS Regulations dealing with similar issues, although the provisions requiring the written approval of the owner or manager or other person in control of the property onto which an escaped elephant has escaped, to hunt or to destroy the escaped elephant, are in direct contradiction to the provisions of Regulation 7(2) of the TOPS Regulations and place an unnecessary burden on the State.

General comment

The norms and standards touch on major legal risk and liability issues for both private individuals and the State. Their implementation imposes significant financial obligations on those involved in the keeping and management of elephants.

DEVELOPING SOUTH AFRICA'S COMMON LAW AND STATUTORY LAW

The new constitutional dispensation adopted by South Africa had an immediate and profound effect on most areas of law, particularly in the area of human rights. Racial discrimination and oppression as matters of law were swept away in an instant. It was recognised that rights in and to the environment, natural resources and land were not equitably distributed. Similarly, account was taken of the environmental injustice of burdening poorer communities with the negative aspects of industrial development in the form of pollution of the air and water. The response of the legislature was to set about the promulgation of some of the most comprehensive and progressive environmental laws in the world. This legislation dealt primarily with natural resources that support a fundamental quality of life. More recently, the legislature has grappled with issues of biodiversity and protected areas management in the legislation, as will have been observed above. It will have been noted that the common law has not followed this progressive trend in its treatment of wild animals. In the sections that follow, the role of customary law and international law in the development of South Africa's statutory and common law will be explored.

Customary law

Customary law needs to be considered as part of the body of law which may influence issues and attitudes relating to the ownership, possession, control, or management of wild animals in South Africa. South Africa's Constitution formally acknowledges Roman-Dutch law and customary law as the major components of the State's legal system. Customary law comprises the various laws observed by communities indigenous to the country. This is sometimes referred to as 'indigenous law' but in this chapter, the term 'customary law' is used because it has a wider currency in Africa and because it is used in the Constitution. In the present context, 'customary law' denotes only laws that have historical roots in the societies of pre-colonial South Africa. The more general meaning of 'custom' as referring to practices of religious communities, commercial institutions and the like, does not form part of customary law for the purposes of this Assessment.

A custom will be found to constitute law if it has existed for a long time, has been uniformly observed by the community concerned, is reasonable, and certain. In deciding when to apply customary law, it has generally been a matter of judicial discretion, with the result that judges have tended to decide each case on its merits. Although a casuistic approach such as this may achieve justice in individual cases, it does so at the cost of legal certainty. The vague application of customary law because of the absence of a uniform source of reference on which to draw has the potential to undermine the individual's right to certainty in the administration of justice. The South African Law Commission has sought to address this by proposing a Customary Law Act to regularise the application of customary law in civil and criminal litigation. (See: The South African Law Commission Project 90: *The Harmonisation of the Common Law and the Indigenous Law: Report on the Conflicts of Law* (September 1999)).

It is not clear when customary law is applicable, for the rules on application are fragmentary, vague, badly drafted, and out of date. At present, the principal rule is one of 'recognition'. This principle is contained in the Law of Evidence

Amendment Act 45 of 1988 (which is concerned with the evidence necessary to prove both customary and foreign systems of law). This rule gives no guidance to courts wishing to discover when customary law is applicable.

Section 1(1) of the Law of Evidence Amendment Act provides that:

> Any court may take judicial notice of the law of a foreign state and of indigenous law in so far as such law can be ascertained readily and with sufficient certainty: Provided that indigenous law shall not be opposed to the principles of public policy or natural justice: Provided further that it shall not be lawful for any court to declare that the custom of *lobola* or *bogadi* or other similar custom is repugnant to such principles ...

The matter has now been clarified by the Constitutional Court in *Bhe and Others v Magistrate, Khayelitsha and Others; Shibi v Sithole and Others; SA Human Rights Commission and Another v President of the RSA and Another* 2005 (1) BCLR 1 (CC) (at page 15). Langa DCJ (as he then was) puts customary law into its proper context when he states:

> It follows from this that customary law must be interpreted by the courts, as first and foremost answering to the contents of the Constitution. It is protected by and subject to the Constitution in its own right. It is for this reason that an approach that condemns rules or provisions of customary law merely on the basis that they are different to those of the common law or legislation, such as the Intestate Succession Act, would be incorrect. At the level of constitutional validity, the question in this case is not whether a rule or provision of customary law offers similar remedies to the Intestate Succession Act. The issue is whether such rules or provisions are consistent with the Constitution.

He points out further that this status of customary law has been acknowledged and endorsed by the Constitutional Court in quoting from *Alexkor Ltd and Another v Richtersveld Community and Others,* 2003 (12) BCLR 1301 (CC) in which the following was stated:

> While in the past indigenous law was seen through the common-law lens, it must now be seen as an integral part of our law. Like all law it depends for its ultimate force and validity on the Constitution. Its validity must now be determined by reference not to common-law, but to the Constitution.

With regard to wild animals, the problems lies less with the application of customary law *per se*, than with the identification of uniform principles and practices that may properly be regarded as forming part of South African customary law. As will have been observed in the section on the common law, in matters involving conflict over wild animals, the issues have been resolved almost exclusively in terms of the (Roman-Dutch) common law. Sources of African customary wildlife law are limited and vest largely with fast disappearing oral repositories. To the extent that there is an identifiable body of customary law, this has not been treated as authoritative, and has given way to the more conventional application of statute and common law.

There is authority in South African customary law for the application of a public trust doctrine to the use and ownership of wild animals, as opposed to the application of the conventional common law principles of *res nullius* which have to date been applied by our courts.

It has been said that the Zulu people 'have a tradition of understanding nature. Their conservation awareness goes back to the foundations of their society. Because they lived close to nature, they lived in harmony with and a balance was maintained between man and his environment' (Steele, 1988, 111.) Magqubu Ntombela, co-founder with Ian Player of the Wilderness Leadership School, widely accepted as an authoritative oral repository of Zulu custom, explains the relationship between the Zulu people and their environment thus:

> KwaZulu was once a land full of wild animals like the elephant, rhino, kudu and crocodiles. We lived with and knew these animals. I was born amongst them. This animal is highly respected by our people. ... We did not kill the animals without permission from our traditional king, King Dinizulu. He did not allow people to kill the animals and any person caught was severely punished. ... I think that it is a very good thing that we should stick to the old traditional ways of living so as to protect the future for our children, so that our children will understand what a wild animal is. ... I understand the plants and the animals, birds and insects. I can tell when the rain is coming. All this knowledge is in my blood. ...We once had a way of living in the world and knowing what was happening on the land. We were in tune with all that lived and sang. (Ntombela, 1988, 288–291.)

This thinking is consistent with the development of the science of ecology and a better understanding of the linkages between the different components of the environment. Shaw observed: 'Human survival is inextricably linked to the continued performance of the myriad of energy flow and cycling processes of the earth's ecosystem. Wild animals are an essential component of this complex system' (Shaw, 1984, 223).

In South Africa, Glavovic has also argued for the recognition of wildlife law as a 'discrete body of law' in which a public trust doctrine in respect of wildlife was recognised. He postulated this:

> There are several ways in which wildlife law may be expanded in public law, one of which is by the adoption and extension of the concept of a public trust doctrine. Assuming, arguendo, that wildlife is a public resource and nothing more, it is a resource which should be protected and administered in the public interest. The state could legislatively assume ownership of wildlife as a public trust, to be held on behalf of the nation, with the effect that the state as trustee will have not only the right but also the obligation to deal with the resource, which is the corpus of the trust, in the best long term interests of present and future citizens as the beneficiaries (Glavovic, 1988, 519).

The public trust doctrine is not new to African ideology. Land generally is regarded as being part of the public domain, held in trust for the tribe or community, wherein bare ownership vested in the Chief and beneficial ownership in the individual (See Bennett, 1985, 173.) Elias (quoted in Bennett, n 25) cites a Nigerian chief in a statement to the West African Lands Committee in 1912 in which it is said: 'I conceive that land belongs to a vast family of which many are dead, few are living and countless numbers are unborn.' If one accepts that in conventional ecological wisdom, animals are an integral part of a functioning whole, inextricably linked to the land, the extension of the public trust doctrine to wild animals is logical.

It follows, therefore, that customary law must similarly recognise wild animals as being part of the public domain unless privately owned. Although the application of customary law is sometimes obligatory, the courts have gone to great lengths to preserve the integrity of their own systems by the use of 'avoidance devices' to justify the application of the common law ahead of customary law. In this way, the courts have avoided the problem of elaborating new terms and concepts and accommodating them within the system (see Bennett, 1985, 183).

It is submitted that in the light of the Constitutional imperative that customary law receive proper recognition, this practice is no longer permissible. The judiciary generally must now begin to recognise and apply customary law where this is required, and in which the public ownership of wild animals is endorsed.

The Constitution

Section 24 of the Bill of Rights in the Constitution 1996 provides as follows:
Environment – Everyone has the right –

a.   to an environment that is not harmful to their health or well-being; and
b.   to have the environment protected, for the benefit of present and future generations, through reasonable legislative and other measures that –
  i.   prevent pollution and ecological degradation;
  ii. promote conservation; and
  iii. secure ecologically sustainable development and use of natural resources while promoting justifiable economic and social development.

The distinction is made in subsection (a) between health and well-being, arguably to provide for both the physical and spiritual components of human existence. The right is expressed in the negative implying that it is the right not to be physically or emotionally harmed that is created, rather than the positive right to access to health care under section 27. The use of the word 'everyone' implies first that the right is available to humans, and second that it is available to all. The issue of *locus standi* (legal standing) has long been a vexed issue in environmental litigation. The reference in section 24 to everyone, read with the provisions of section 38, makes it clear that the right to approach the court extends to individuals, groups and classes of person, to persons acting in the interests of others and to persons acting in the public interest. This is further clarified in section 32 of the National Environmental Management Act 107 of 1998 dealt with below. Constitutional rights generally exist 'vertically' (i.e. between a person and the State), but the language of section 24 suggests that it also has 'horizontal' application (i.e. between individuals). This is also consistent with the nature of the environmental right that takes on a public law or group character, and is available to everyone.

While the right described in subsection 24(a) is clearly a fundamental ('first generation') right, subsection 24(b) is socio-economic (or 'second generation') in character, and imposes on the State the obligation to secure the rights of the individual to have the environment protected through the reasonable legislative and other measures described. It is important to note the parallel obligation imposed on the State while protecting the environment to 'promote justifiable economic and social development'.

Judicial recognition has been given to the justiciability of environmental rights in unequivocal terms in the landmark decision of the Supreme Court of Appeals in *Director: Mineral Development, Gauteng Region and Sasol Mining v Save the Vaal Environment and others* 1999 (2) SA 709 SCA at 719 when it was held:

> Our Constitution, by including environmental rights as fundamental justiciable human rights, by necessary implication requires that environmental considerations be accorded appropriate recognition in the administrative process in our country.

This approach has been followed with approval in subsequent matters and most recently by the Constitutional Court in *Fuel Retailers Association of SA (Pty) Ltd v Director-General, Environmental Management, Mpumalanga, and Others* CCT 67/06.

The term 'environment' is not defined in the Constitution, and must therefore be given its widest meaning unless otherwise statutorily constrained. The National Environmental Management Act 107 of 1998 ('NEMA') is one of the legislative measures taken by the State in discharge of the constitutional imperative imposed by subsection 24(b).

Ngcobo J in *Fuel Retailers* (supra) at [67] puts NEMA in the following context:

> NEMA principles 'apply ... to the actions of all organs of state that may significantly affect the environment'. They provide not only the general framework within which environmental management and implementation decisions must be formulated, but they also provide guidelines that should guide state organs in the exercise of their functions that may affect the environment. Perhaps more importantly, these principles provide guidance for the interpretation and implementation not only of NEMA but any other legislation that is concerned with the protection and management of the environment.

Albeit in a dissenting judgment, Sachs J in the same matter reaffirms the influence of NEMA when he states at [113] 'Running right through the preamble and guiding principles of NEMA is the overarching theme of environmental protection and its relation to social and economic development.'

NEMA defines the environment in section 1(xi) as:

> the surroundings within which humans exist and that are made up of –
>
> i.   the land, water and atmosphere of the earth;
> ii. micro-organisms, plant and animal life;
> iii. any part or combination of (i) and (ii) and the inter-relationships among and between them; and
> iv. the physical, chemical, aesthetic and cultural properties and conditions of the foregoing that influence human health and well-being.

Clearly wild animals are included in this definition. What is not immediately apparent is the right associated with wild animals as a component of a constitutionally entrenched environmental right. In order to create a jurisprudential link, it is necessary to interpret NEMA as part of the imperative imposed by subsection 24(b) of the Constitution and then to admit the following logic:

- Section 2(4)(o) of NEMA determines that the environment is held in public trust for the people, that the beneficial use of environmental resources must serve the public interest and that the environment must be protected as the people's common heritage.
- Wild animals by definition are inextricably linked to and are part of the environment.
- *A fortiori*, wild animals form part of and must be protected as the people's common heritage.
- All laws, including the common law and customary law where it relates to wild animals, must be interpreted subject to this principle as a component of the constitutional imperative imposed by section 24.

As such, elephants form part of and must be protected as the people's common heritage, must be held in public trust for the people, and their beneficial use must serve the public interest. Any classification of elephants in protected areas under State control as *res nullius* is clearly inconsistent with this as well as with section 24(b) of the Constitution and does not promote conservation or the protection of our environment for the benefit of present and future generations. This aspect is dealt with in more detail below.

International law

The shift in the 1970s to a 'one world' perspective of global responsibility has produced a sense of interdependency and awareness of global commons and international heritage which is clearly demonstrated by the number and pattern of international treaties that have been concluded in recent years.

As to the nature of public international law, it is said that *ubi societas ibi ius* – where there is a society there is law. There is continuing debate as to whether this aphorism holds true for the community of nations, whether public international law is 'law' properly so called. However law may be defined, its primary function must be to regulate the conduct of the members of a society for their common good. The question that arises is whether there is indeed an international society.

In a sense the international community is a political community without a sovereign. There is no central government or effective judiciary or police force. In strict juristic theory it is perhaps more correct to classify international law as a branch of ethics rather than of law. However, in practice questions of international law are generally regarded as being of legal rather than purely moral character. The existence of international law stems from general assent and recognition by member states of the international political community, notwithstanding the absence of a sovereign or effective police force capable of imposing sanctions to ensure adherence to its rules.

Compliance without compulsion is generally a matter of self-interest. Most international law rules are respected and adhered to by the majority of nations notwithstanding the apparent weakness of effective organised coercion. However, states may employ counter-measures as a form of sanction against internationally recognised legal wrongs, such as economic sanctions, reprisals, use of force, and even war, all of which would otherwise be regarded as unlawful. Violations are rare. States observe international law because it is politically expedient for them to do so.

The rules of international law are divided into three main categories or law-creating processes: treaties, international customary law, and the general principles of law recognised by civilised nations. Judicial decisions of the World Court, the writings of respected jurists, and the resolutions of the General Assembly of the United Nations are other sources of international law. With technological advancement and population increase, more and more activities

Box 1: The more important multilateral environmental agreements that affect wildlife are the following:

1946	International Convention for the Regulation of Whaling (Washington)
1971	Convention on Wetlands of International Importance Especially as Waterfowl Habitat (Ramsar)
1972	Declaration of the United Nations Conference on the Human Environment (Stockholm)
1972	Convention for the Protection of the World Cultural and Natural Heritage (Paris)
1973	Convention of International Trade in Endangered Species of Wild Fauna and Flora (Washington) – CITES
1973	International Convention for the Prevention of Pollution by Ships (London) – MARPO
1979	Convention on the Conservation of Migratory Species of Wild Animals (Bonn)
1982	United Nations Convention on the Law of the Sea – UNCLOS
1985	Convention on the Protection of the Ozone Layer (Vienna)
1987	Montreal Protocol on Substances that Deplete the Ozone Layer
1989	Convention on the Control of Transboundary Movements of Hazardous Wastes and Their Disposal (Basel)
1992	United Nations Convention on Biological Diversity
1992	United Nations Framework Convention on Climate Change
1994	United Nations Convention to Combat Desertification

Regional and sub-regional environmental instruments include:

1968	African Convention on the Conservation of Nature and Natural Resources
1981	Convention for Cooperation in the Protection and Development of the Marine and Coastal Environment of the West and Central African Region
1985	Convention for the Protection, Management and Development of the Marine and Coastal Environment of the Eastern African Region
1987	Agreement on the Action Plan for the Environmentally Sound Management of the Common Zambezi River System
1991	Convention on the Ban of the Import into Africa and the Control of Transboundary Movement and Management of Hazardous Wastes Within Africa (Bamako)
1994	Lusaka Agreement on Cooperative Enforcement Operations Directed at Illegal Trade in Wild Fauna and Flora
1995	Protocol on Shared Watercourse Systems in the Southern African Development Community
1999	Protocol on Wildlife Conservation and Law Enforcement
2003	Revised Protocol on Shared Watercourse Systems in the Southern African Development Community
2003	African Convention on Conservation of Nature and Natural Resources 1968 ('Algiers') (Maputo)

require international cooperation. In recent years treaties have proliferated, and more treaties are now concluded in the course of a year than were concluded in the first two decades of the twentieth century.

A major problem in international law is the question of enforcement, because there is no international police force or administrative machinery to implement the decisions of the International Court of Justice. However, international trade, politics and public policy ensure that international agreements more often than not become translated into parties' national systems, and treaties are in practice generally well enforced. Treaties can be bilateral or multilateral. The more parties there are to a treaty – in large multilateral agreements as many as 130 states may be bound – the weaker and more ambiguous it is often likely to be because of the compromises made to achieve acceptability by all the states involved. Because wildlife treaties (for example) usually affect several states, they tend to suffer from this weakness; but they have generally proved to be reasonably effective for the purposes for which they were designed.

Arbitration cases as well as cases before the International Court of Justice (at The Hague) are rare. The Court has heard fewer than 50 contentious cases since its establishment in 1946 as the principal judicial organ of the United Nations. States are reluctant to take each other to the International Court, partly because it is seen as a politically unfriendly act to be avoided if possible and partly because it is often difficult to achieve a satisfactory remedy by this means.

Many conventions recording a wide range of international agreements on environmental matters have been adopted by the international community (see box 1). The content of these instruments, while of importance to wildlife management generally, does not contribute to the context of this chapter. What is important is the common thread that runs through these international law instruments. It is recognised that the environment generally is a global commons in which member states have a duty to protect the natural environment against harm from human conduct, encourage the conservation and sustainable use of biological resources, protect biodiversity, and to recognise sites of international conservation significance as means of protection. Environments are now accepted at both scientific philosophical perspectives to be holistic entities in which the individual components are interdependent. As a matter of logic and law, animals form part of this common heritage and must be treated as such in customary international law.

Section 232 of the Constitution confirms the common law position that customary international law is recognised as law in the Republic unless it is inconsistent with the Constitution or an Act of Parliament. Section 233 of the Constitution requires every court when interpreting any legislation that is consistent with international law to prefer any reasonable interpretation of the legislation that is consistent with international law over any alternative interpretation that is inconsistent with international law.

International agreements are binding on the Republic when they are approved by both the National Assembly and the National Council of provinces, unless they are technical, administrative or executive in nature and do not require either ratification or accession. In such cases, they must be tabled in the National Assembly within a reasonable time. Section 231(4) of the Constitution provides for the enactment of international agreements as law in our national legislation. Examples of the use of this provision are the World Heritage Convention Act 49 of 1999 and the National Environmental Management: Biodiversity Act 10 of 2004.

HIERARCHY OF SOUTH AFRICAN LAW

The Constitution

The Constitution is the supreme law of the Republic. Law or conduct that is inconsistent with the Constitution is invalid. The obligations imposed by it must be fulfilled.

Legislative powers exist at three levels: of the national sphere of government, by Parliament, by way of national statutes, and subsidiary legislation in the form of regulations by the relevant minister; of the provincial sphere of government, by the provincial legislature in the form of provincial statutes; and at the local level by Municipal Councils in the form of bylaws.

Functional areas of concurrent national and provincial legislative competence are set out in Schedule 4 of the Constitution and those of exclusive provincial legislative competence are set out in Schedule 5. In the present context, the matters of concurrent competence which are of importance are the administration of indigenous forests, agriculture, animal control and diseases, environment, indigenous law and customary law (subject to Chapter 12 of the Constitution), and nature conservation (excluding national parks, national botanical gardens and marine resources).

Section 41(1) of the Constitution sets out principles of co-operative government and intergovernmental relations. It requires all spheres of government and all organs of State within each sphere to respect the constitutional status, institutions, powers, and functions of government in the other spheres; not to assume any power or function except those conferred on them in terms of the Constitution; to exercise their powers and perform their functions in a manner that does not encroach on the geographical, functional or institutional integrity of government in another sphere; and to co-operate with one another in mutual trust and good faith.

Potential conflicts between the national and provincial legislation are dealt with in section 146 and 147 of the Constitution. National legislation will prevail over provincial legislation if a matter cannot be effectively legislated by the provinces individually; if the legislation is required to ensure uniformity across the nation; if the legislation is necessary for the maintenance of national security, the maintenance of economic unity, the protection of the common market in respect of the mobility of goods, services, capital, and labour, the promotion of economic activities across provincial boundaries, the promotion of equal opportunity or equal access to government services, or the protection of the environment; or if the legislation is necessary to prevent unreasonable action by a province.

If conflicts cannot be resolved by a court, national legislation is deemed to prevail and the conflicting provincial legislation will be inoperative while such conflict remains. In the interpretation of all legislation, a court considering the matter must prefer interpretations that avoid conflicts.

In the context of elephant management, differing provincial wildlife laws create the potential for conflicts that may require resolution under the provisions of the Constitution or determination by applying national legislation in order to create the uniformity required by section 146(2). An example of this would be the application of the TOPS Regulations over the provisions of conflicting provincial Acts or Ordinances in the issue of permits to allow the movement of animals between provinces without the need for multiple permit applications. Uniformity in regard to the criteria set for the issue of permits and the conditions affixed thereto could be addressed in this way.

Administrative overlapping

In the administration of wildlife laws, the creation of separate national and provincial authorities to manage and control the utilisation of wildlife resources creates overlapping areas of responsibility geographically, functionally and institutionally. In principle, this offends against the provisions of section 41(1)(g) of the Constitution. Most of the provinces are operating exclusively under the provisions of Ordinances that pre-date the Constitution by decades.

With the commencement of the TOPS Regulations on 1 February 2008, there will be national application of regulations that will overlay provincial legislation and administration. It is proposed that the provincial authorities will be the designated implementing agents and will be charged with the duty of applying separate and potentially conflicting laws. This has the potential to create conflicts that may compromise the administration of justice, particularly with the issue of permits. Moreover, there are no indications that the provincial authorities will be financially compensated for the additional administrative burden placed on them, and they will therefore have 'unfunded mandates' to discharge.

Capacity and skills

Wildlife administrators nationally and provincially face shortages in capacity and skills to fulfil their mandates and this is exacerbated by the generally low financial priority given to environmental portfolios at the national and provincial levels. This compromises the ability of the State to discharge its constitutional obligation to ensure the right of everyone to just administrative action as provided for in section 33 of the Constitution. Section 33(3) requires national legislation to be enacted that gives effect to this right, and more specifically in subsection (c), to legislation that 'promotes an efficient administration'. This efficiency occurs at two levels: by the promulgation of appropriate legislation that directly promotes efficient administration (e.g. the Promotion of Administrative Justice Act 3 of 2000), and more generally that all legislation should promote the efficient administration of the matters within its ambit. It is submitted by us that present and proposed legislation is not conducive to efficient administration given the resources available to the relevant authorities charged with its implementation. In the assessment of South Africa's wildlife laws, account will have to be taken of the efficiency with which it is administered and enforced.

CONCLUSIONS ON THE STATE OF SOUTH AFRICAN LAW

In describing wildlife law as a discrete branch of the law into which the Assessment must be located, a complex, inconsistent, disparate, conflicting and inefficient system of legal rules, policies and administration is disclosed.

It is observed that one of the foundations of our legal system, the common law, is not compatible with our Constitution, is rooted in socio-economic conditions of ancient Eurocentric culture that no longer has relevance in a modern South Africa, but nevertheless, plays a dominant role in the legal relationship between wild animals and society. Customary law, which should have equal status with the common law and should be applied where circumstances dictate in preference to common law, is usually avoided in favour of the more conventional and comfortable common law. This is generally because customary law suffers from a paucity of accurate records and authorities, and its application carries the inherent risk of a lack of true validity, and may be applied inconsistently. A plethora of national and provincial environmental legislation is described, in which overlapping and sometimes conflicting administrative functions are created. Wildlife management occurs separately at the national and provincial levels, and there appears to be no uniformity between national and provincial legislation or as between the different pieces of provincial legislation, nor is there uniformity between the national and provincial rules, regulations and policies applicable to wild animals. For effective elephant management policy and law, these shortcomings will have to be addressed.

All of this is an indication of a need for consolidation and coherence of wildlife law. It has become a trend to use NEMA as the framework within which separate statutes, regulations, norms and standards are promulgated, all of which have the potential to exacerbate an already overly bureaucratic administration. To make matters worse, it would seem that inadequate human and financial resources are generally applied to conservation management.

In developing a workable legal framework within which elephant management may be practised, cognisance should be taken of the evolving nature of law generally, and particularly in South Africa, in which the evolutionary process of wildlife management and the development of wildlife laws have been significantly accelerated by our Constitution.

A BETTER UNDERSTANDING OF SOUTH AFRICA'S COMMON LAW

The role of the law as an instrument of change is a matter of some debate. Is, or should, the law merely be reflective of societal values or should it perform a normative function by providing rules to guide society in a particular direction? In the case of the evolution of South African wildlife law, it should probably serve both functions.

As has been observed, the common law is not reflective of societal values, and customary law has not attained a sufficient status or recognition to be influential in the administration of justice where this has concerned wild animals. There may therefore be a need for legislation to be an agent of change and to reinforce current attitudes to wildlife. Legislative intervention that clarifies the constitutional imperative, restates the common law and recognises customary law, both international and South African, may be a necessary precursor to the establishment of an appropriate legal framework within which elephant management may be effectively practised. In the development of such legislation, a major challenge facing the legislature will be to accommodate the ethical shift from anthropocentric to ecocentric (biocentric) values in our attitude to animals. (While he was not the first author to note this shift, Jan Glazewski describes the trend succinctly in his work Environmental Law in South Africa, 2005, 6–8.)

In the result, there will of needs be a revision of the common law concepts of ownership, not by way of any change to the common law, but by recognising that South Africa's common law is no longer rooted inextricably in Roman and Roman-Dutch law principles and has a character of its own that is representative of traditional values of our culture, diverse as it is.

Wild animals as part of the public estate

It would seem that there is a strong legal argument to be made in favour of moving away from the traditional application of the common law principles of treating wild animals as *res nullius* in a number of circumstances and rather moving towards principles that treat all wild animals which are not in lawful private ownership as being public goods and part of the public estate.

Wild animals as *res publicae*

Things (such as wild animals) may be classified either as out of commerce (*res extra commercium*), things which cannot be privately owned, or in commerce (*res in commercio*), things which can be privately owned or can be objects of other real rights (for example land over which a person holds a registered servitude). Things out of commerce may be divided into common things (*res omnium communes*), public things (*res publicae*), and in some instances things belonging to corporate bodies that serve a communal function (*res universitatis*). *Res nullius* (things belonging to no one), are *res in commercio*, and are susceptible to private ownership (Silberburg & Schoeman, 2006, 24).

Res publicae on the other hand are owned by the State and are intended for the general benefit and usage of the public. *Res publicae* are available to the general public, but unlike *res omnium communes*, which are things that are common to all, they belong to the State, not in the same way as private individuals own property, but for the public benefit. *Res publicae* include harbours, public rivers, public roads and public buildings, the sea and the seashore and national parks. The right in common of all to use public assets (be they *res publicae* or *res omnium communes*) may be subject to statutory restrictions and controls, for example, the control of access by the public to the seashore by motor vehicles. A logical extension of this reasoning is that the State is obliged to defend the public ownership of wild animals against *occupatio* by private individuals for as long as they are public assets, either as *res publicae or res omnium communes*, through reasonable legislative and other measures.

It is submitted that the classification of elephants in protected areas under State control as *res nullius*, is inconsistent with section 24(b) of the Constitution, in that they form part of the environment that must be protected for the benefit of present and future generations. In terms of Section 39(2) of the Constitution, our courts are obliged to develop the common law to promote the spirit, purport and objects of the Bill of Rights, including the rights set out in section 24(b). It follows that the State, as trustee of the environment for future generations, is obliged to conserve wild animals that are part of the public estate, and is more specifically, in terms of Section 17(c) read with Section 3(a) of the NEMPAA, obliged to conserve all wild animals occurring in protected areas which form part of this estate.

International experience

Internationally, wildlife is regarded generally either as part of the rights of ownership over land or as State property (see generally, Cirelli, 2002, par 4.1). In Morocco, as in South Africa, wildlife is classed as *res nullius*, whereas in Uganda, ownership of wild animals is vested in the State on behalf of and for the benefit of the people. In the law of Tajikistan, animals are subject to State ownership and are 'common property of all citizens' and a similar position obtains in China where wildlife is the property of the State (Cirelli, 2002, par 4.1.1). In countries where wildlife belongs to the State, this is either generally or because it occurs on State land (Cirelli, 2002, par 4.1.1).

Namibia expunged the *res nullius* category from its wildlife law by adopting Article 99 of its Constitution which states that all natural resources belong to the State unless otherwise owned by law. While a similar approach may be appropriate for South Africa, it may not be an immediate solution because of the difficulties associated with making any amendments to the Constitution.

South Africa's wildlife common law

In the final analysis, it is suggested that wild animals as part of the public estate are *res publicae*, and not *res nullius*, but may move between a classification as *res extra commercium* while publicly owned, and *res in commercio* when in private ownership. However, private ownership would not be based on an original mode of acquisition (i.e. *occupatio* of a *res nullius*) but by derivative acquisition (i.e. the transfer of rights from one person as owner, to another).

It may be necessary to consider wild animals thus: Wild animals in protected areas constitute *res publicae* and would be owned by the State, which could with justification, and subject to the usual rules relating to the disposition of State assets, transfer ownership to private individuals or corporate bodies. Upon such animals escaping or straying from protected areas, the State would be entitled, as the owner of the animal, to take all steps reasonable and necessary to retrieve the animal. Wildlife which is not *res publicae* or in private ownership, but which occurs on private or public land where there is no intention or physical ability to own such animals would be *res omnium communes* and be common to all. The latter animals could be acquired by the State and the public alike, subject to appropriate controls, in much the same way as water as a public resource may be acquired in accordance with the National Water Act 36 of 1998. Existing legislation in terms of which control over animal ownership and use is exercised would apply to such acquisition. Animals in private ownership, having been legitimately taken from *res publicae* or *res omnium communes*, would become *res in commercio* and would be owned by a person as *res alicuius* (belonging to someone), either individually as *res singulorum* (belonging to an individual) or by corporate bodies as *res universitatis* (belonging to corporate bodies).

In the result no wild animal is unowned. It is either in private ownership and is protected as private property under the Constitution, the common law or customary law, as may be appropriate, or in public ownership by the State for public benefit, and as trustee of the common estate, and protected in accordance with the consitutional imperative imposed on the State to do so through reasonable legislative and other means.

CONCLUDING REMARKS

By recognising wild animals as a category of property more properly reflective of societal needs, namely that they form part of the public estate where they are not privately owned, the determination of the rights and obligations associated therewith becomes more relevant to prevailing circumstances. In so doing, most of the inadequacies of the law identified in this chapter, where it deals with the financial loss to the State of animals from protected areas, the liability for damage-causing escapee animals, difficulties with the crossing of provincial borders and the movement of animals between private and public land, are largely resolved. Finally, the variable treatment of wild animals in an unowned state from an animal welfare perspective will be given more clarity.

ENDNOTE

1. The Bill was introduced by the Minister on 8 May and will now go through the process.

REFERENCES

Bennett, T.W. 1985. Terminology and land tenure in customary law: An exercise in linguistic theory. In: *Land ownership – changing concepts Acta Juridica* 1986, 173.

Cirelli, M.T. 2002. *Legal trends in wildlife management.* Legislative Study 74, Food and Agriculture Organisation of the United Nations.

Du Plessis, L.M. 30 April 2001. Statute law and interpretation. In: *The Law of South Africa,* Volume 25 (first reissue volume). Lexis Nexis Butterworths, Durban, par 278–365.

Erlich, P.R. & A.H. Erlich. 1982. *Extinction: The causes and consequences of the disappearance of species.* Victor Gollancz Ltd, London.

Fuggle, R.F. & M.A. Rabie (eds) 1992. *Environmental management in South Africa.* Juta & Co, Cape Town.

Glavovic, P.D. 1988. An introduction to wildlife law. *South African Law Journal* 105, 519–527.

Glavovic, P.D. 1995. *Wilderness and the law.* Law Books Press, Durban.

Glazewski J. 2005. *Environmental law in South Africa.* Second edition. Lexis Nexis Butterworths, Durban.

Lund, T.A. 1980. *American wildlife law.* University of California Press, London, Berkley and Los Angeles.

Martin, V. (ed.) 1988. *For the conservation of the Earth.* Fulcrum Inc., Golden, Colorado.

Myers, N. 1979. *The sinking ark: A new look at the problem of disappearing species.* Pergamon Press, Oxford.

Ntombela, M. 1988. The Zulu tradition. In: V. Martin (ed.) *For the conservation of the Earth.* Fulcrum Inc., Golden, Colorado, 288–291.

Peterson, G.L. & A. Randall (eds) 1984. *Valuation of wildlife resource benefits.* Westview Press, Boulder and London.

Scheepers, J.H.L. 2001. Land. In: *The Law of South Africa,* volume 14 (first reissue volume). Lexis Nexis Butterworths, Durban, par 1–110.

Shaw, W.W. 1984. Problems in wildlife valuation in natural resource management. In: G.L. Peterson & A. Randall (eds). *Valuation of wildlife resource benefits.* Westview Press, Boulder and London, 223.

Silberberg, H. & J. Schoeman 31 December 2006. *The law of property.* Fifth edition. Lexis Nexis Butterworths, Durban.

South African Law Commission Project 90. *The harmonisation of the common law and the indigenous law,* September 1999.

South African Law Commission Working Paper 27, Project 69. *Acquisition and loss of ownership of game,* April 1989.

Steele, N. 1988. KwaZulu – conservation in a Third World environment. In: V. Martin (ed.). *For the conservation of the Earth.* Fulcrum Inc., Golden, Colorado, 113–118.

Van der Merwe, C.G. 1989. *Sakereg.* Butterworths, Durban.

Van der Merwe, C.G. 31 October 2001. Things. In: *The Law of South Africa,* volume 27 (first reissue volume). Lexis Nexis Butterworths, Durban, par 211–227.

Van der Merwe, C.G. & M. Blackbeard. January 2003. Animals. In: *The Law of South Africa,* volume 1. Second edition. Lexis Nexis Butterworths, Durban, par 454–541.

TOWARDS INTEGRATED DECISION MAKING FOR ELEPHANT MANAGEMENT

Lead author: Harry C Biggs
Author: Rob Slotow
Contributing authors: Robert J Scholes, Jane Carruthers, Rudi van Aarde,
Graham HI Kerley, Wayne Twine, Douw G Grobler, Henk Berthshinger,
CC (Rina) Grant, HPP (Hennie) Lötter, James Blignaut, Lisa Hopkinson,
and Mike Peel

> In answer to the question 'Is containment of a population eruption desirable?' Graeme Caughley replied 'This is not a scientific question. I can boast of no qualifications that would make my opinion any more valuable than those of my two immediate neighbours, a garage mechanic on the one hand and an Air Vice-Marshall on the other.' (*Caughley, 1981*)

INTENTION AND APPROACH

THIS CHAPTER draws on material from previous chapters and builds linkages among them. We supply some theoretical background that may help explain the consequences of various approaches to the 'elephant problem' as currently framed, a 'problem' which has arisen in conjunction with the growth of human settlements and activities across the landscape. We construct and discuss an integrative framework, and then summarise and synthesise the main points from the contents of Chapters 1–11 into this framework.

Using the above analysis, we then suggest how decision makers might most usefully approach and formulate elephant issues. We present a range of options for particular circumstances, at the level of societal influences, strategy and practical implementation, and the integration of these three. Finally we list what we see after the assessment as important gaps, and conclude.

MAKING COMPLEX ISSUES TRACTABLE

One underlying reason why the 'elephant problem' appears so intractable is that it is complex (Chapter 1). This affects decision making. Kinnaman & Bleich (2004) describe a range of responses, from toleration through to full collaborative behaviour, where there are different combinations of agreement and certainty (figure 1). The elephant issue clearly falls into the zone of complexity. Therefore it should not come as a surprise that reductionist 'command-and-control' policies (Chapter 1) have not succeeded. Even if they had been correct in assessing the biodiversity outcomes as simple and predictable (and there is serious doubt that this is the case (Chapter 3)), there is no doubt that the associated social responses (Chapter 4; Chapter 9), and hence the problem as a whole, are complex. Some even feel it is a 'wicked problem' (Conklin, 2006), insoluble because of ever-shifting goalposts.

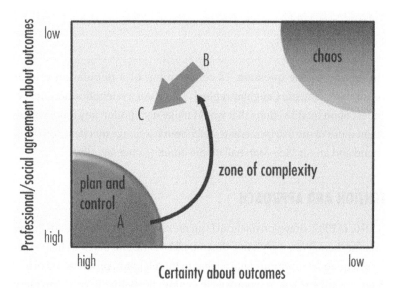

Figure 1: Feasible response zones for different levels of agreement (y-axis) and certainty around outcomes (x-axis). Elephant management in South Africa began (mistakenly) at A, but failed because it overstated the certainty on both axes. It is possible that this assessment initiative can help move the situation from the current position B to a hypothetical position C (modified after Kinnaman & Bleich, 2004)

Forming collaborative partnerships is central to the resolution of such issues. Figure 1 suggests that the predominantly unilateral management of elephant in the past operated in the command-and-control domain, and was therefore unlikely to lead to lasting solutions of any kind (Chapter 1). Furthermore, the different parties involved in the search for a solution must have sufficiently overlapping understanding of a problem (Abel *et al.*, 1998) or enough of a shared rationale, to succeed. Holling (2001) asserts that 'there is a requisite level of simplicity behind the complexity that, if identified, can lead to an understanding that is rigorously developed but can be communicated lucidly'. This chapter, indeed this assessment, attempts to crystallise out such requisite simplicity, that might then permit agreement from most stakeholders, and assist understanding, communication and action.

One of the challenges to effective management is co-ordinating not only the linkages within a level (such as say, the province) but also the vertical or inter-level linkages in a way that serves the overall purpose, and that works for almost everyone at the different levels (figure 2).

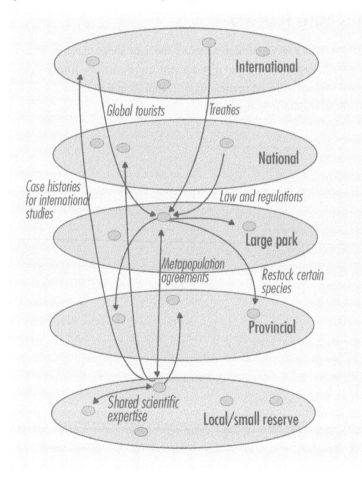

Figure 2: Social networking across and between levels, required for successful natural resource management. Conceptually illustrated for a large and a small park, with selected concrete examples of links in italics (modified after Olsson *et al.*, 2003)

This chapter emphasises management of elephants in single protected areas, where most day-to-day decisions are made, but it must be remembered that this is nested in a wider decision making and management context, as shown in this figure. People at the various levels, and those operating among levels, are all searching for clearer guidance for decision making concerning elephants. Keeping these kinds of linkages in mind usually helps decision makers arrive at more useful, robust and inclusive decisions.

AN INTEGRATIVE FRAMEWORK

Figure 3 presents a way of linking together the wide range of issues dealt with in this Assessment. According to this schema, there are three primary clusters of interest that are believed to meaningfully represent bundles of issues in the 'real world' of elephant management:

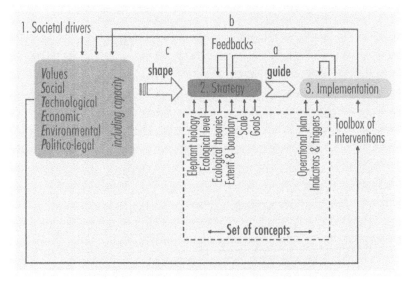

Figure 3: How societal drivers shape strategy, which guides implementation, with several feedback loops. Numbering and lettering corresponds with description in text

1.   The *societal drivers* of attitudes to elephant issues can be analysed in terms of a 'V-STEEP' (Values, Social, Technological, Economic, Environmental and Politico-legal) framework of Rogers (2005), an extension of the SEEP framework of Campbell & Olson, 1991. Although individual issues allocated to the various subdivisions could arguably belong in more than one of the six (e.g. 'animal rights' might feature

under 'values' or under 'social'), this is not seen as a problem, as long as the approach helps us comprehensively elicit the full range of important drivers. We have generally not written such overlapping issues under more than one of the headings. We shall refer to these as broader societal drivers, or where the context makes it clear, simply drivers.

2.   The *strategic or explicit philosophical approach* towards elephant management. A strategic paradigm usually underlies the actual strategy, though sometimes strategy development is absent as an explicit step in peoples' thinking. In such cases, one can sometimes infer a plausible strategy and underlying paradigm from the tacit assumptions made. The strategy, whether explicit or implicit, is shaped by all the drivers in (1), as well as many concepts from the 'set of concepts' depicted in figure 3. The terms 'goals' and 'objectives' are used synonymously in this chapter, and in such a way as to incorporate their full range of meaning.

3.   The *ultimate implementation or deployment* 'on the ground'. This is guided by strategy and also informed by certain concepts (operational plan; indicators/triggers for action). The 'toolbox of available interventions' (culling, contraception, translocation, etc.) is in turn renewed particularly by technological innovations, but also by other changes in the drivers.

Drivers help shape the particular strategy, if not explicitly then subconsciously in peoples' minds, or *de facto*. Similarly, strategy should form the guiding basis for implementation. Furthermore, there exist three important feedbacks:

a.   *Implementation to strategy*: implementational realities often affect the way the strategy can be derived. For instance, if contraception is possible and being considered, any ecological threshold levels in the strategy need to take into account the longer lag period till population reduction can be achieved, as opposed to, say, the immediate population reduction effect after culling.

b.   *Implementation to drivers*: experiences of consequences, including successes and failures, can feed directly back to technological innovations as improved or new technological ideas, or modified societal values, as with experiences with Scoline during culling operations, which was forced by societal pressure out of the allowable toolbox (Chapter 8).

c. *Strategy to drivers*: Similarly, learning accumulated from experiences in the use of strategy feeds back to modify drivers. For instance, infeasibility of certain key ideas that were deemed necessary in a strategy may lead to a reappraisal of ecological theory.

There are 'internal feedbacks' in each step. For instance, the main drivers co-evolve and influence each other along the way. Readers may like to add other feedbacks that are important in their particular situations. For example, we have added a direct line of influence from the technology driver to the toolbox of interventions.

Each feedback is an important step in adaptive learning which decision makers should be encouraged to use. This will allow the spirit of the Norms and Standards (Department of Environmental Affairs and Tourism, 2008) to be upheld.

FACTORS INFLUENCING MANAGEMENT DECISION MAKING FOR ELEPHANTS

The discussion below has elements of summary, analysis and synthesis, often using a chronological development sequence.

Societal drivers

Values

Values are deeply-held beliefs, sometimes explicitly espoused, but often unstated. They hold the underlying key to understanding where our elephant management approaches come from, and in which direction they are likely to be heading. Values interact and co-evolve with all five of the other drivers but perhaps represent the most fundamental level of human aspiration that ultimately determines, or at the least significantly influences, elephant management (Chapter 9). In eras showing unequal power among different stakeholder groups, the dominant values driving the system tend obviously to be those of the powerful, while widely differing values may be held by others, and these may or may not be documented or even well understood.

By the end of the nineteenth century we observe (Chapter 1) exploitation values (that had been supporting the by then nearly exhausted ivory trade) and recreational values (underlying sport hunting) as dominant, with conservation values just emerging (driving preservation of elephant, seen by society as threatened). During the second half of the twentieth century, as elephant numbers increased in conservation areas in southern Africa (supported by growing conservation values), managers drew mainly on a belief that consequences of abundant elephants were unacceptable in the ecosystem as a whole (Chapter 1). They therefore strove to reduce elephant densities (and at the same time expand elephant range for conservation through translocation) using the many tools that technology was developing. These tools included culling (Chapter 8) as a major option. The belief around the unacceptability of elephant impacts was justified through the criterion of 'exceeding the carrying capacity' (Chapter 1). This justification appeared, after explanation, to be widely accepted by society, though with little input from them in the process other than insisting that, when culling was used, it was done humanely. A minority voice of animal rightists condemned killing in principle. As a largely separate development to the rights-based ones, eventually the simple use of the notion of 'carrying capacity' was also challenged by an increasing number of scientists, this being driven by a growing recognition of complexity (Chapter 1). By the 1990s, the influence of democratisation had opened the playing field for a much wider range of societal values to be drawn into the debate. One outcome (but not one emanating from the previously unfranchised majority of South Africans, several of whose representatives were talking about possible benefits from use of elephants and their products) was that elephant culling was placed under moratorium in South Africa (Chapter 1), pending further discussion of which this assessment forms part.

Given wide agreement on the elephant conservation value, the simple dichotomous moral dilemma posed by contrasting 'culling to protect the ecosystem and other species' with 'not killing elephant' (Chapter 9) may be in the process of growing into a multi-way moral dilemma. The outcomes are now seen as more complex than simply 'elephants vs. other organisms'. For the first time, the rural poor in southern Africa (victims especially of elephant crop raids) have a voice that is being heard, against elephant conservation (Chapter 4), though many recent examples exist where benefits of elephant utilisation (for tourism and especially trophy hunting) lead these communities on balance to want to promote or maintain elephant numbers (Chapter 10). Aesthetic values (the simple preference for landscapes with tall trees, elephants being blamed for their loss) may now be taken more seriously (Chapter 9) because such emotions are no longer considered necessarily weaker than so-called 'concrete' objectives like preventing biodiversity loss. Wilderness values (non-disturbance of pristine landscapes by humans) have been present for almost a century, but their proponents have not effectively influenced interventions on elephants (Chapter 9) apart from suggesting that mechanised transport (as used in culling) or 'non-natural' interference (as in contraception) should not be allowed in wilderness areas. With such a five- or six-way 'tug-of-war', the suggestion of moral pluralism (allowing all points of view some practical outlet) as a 'solution' appears attractive. But who will then allow what and where? Perhaps management strategies could be agreed on locally, bearing in mind that international pressures also have a bearing. One thing the decision maker needs to know is that this moral dilemma is likely to persist, unless one value becomes dominant over others (Chapter 9).

Without going into detail, it is important to note that the unfolding of elephant-related values in East Africa has differed considerably from the southern African narrative, and has led to a very different trajectory taken in elephant management there (Chapter 1). While this has involved a differing balance of unfolding values, several other formative drivers have also been very different in East Africa.

Social

Social drivers reflect revealed values manifesting as individual or group expressions, or 'social movements' or reactions.

Human-elephant conflict, especially in Africa north of South Africa, has proved to be a perennial issue, those worst affected being the rural poor (Chapter 4). Generally, and especially in South Africa, levels of human-elephant conflict are low (Chapter 4), but they can be locally severe, and clearly devastating in the case of occasional resultant human deaths (Chapter 4). There are many helpful remedies (such as fencing and other barriers, conditioning, and killing of habitual offenders) but the conflict continues as elephants learn to avoid or overcome these deterrents (Chapter 4). Increased interaction between humans and elephants, as happens when people encroach on elephant habitat, or elephant habitat expands (the main mechanism in South Africa) may mean higher levels of conflict (Chapter 4). Decision makers need to acknowledge the reality of the resultant negative sentiment towards elephants and conservation in general (Chapter 4). Economic opportunities based on elephants may counterbalance these negative effects and sentiments in many communities, but this requires careful institutional arrangements to ensure that the benefits do indeed reach the affected parties (Chapter 10).

Grassroots conservation responses in favour of elephants have arisen repeatedly, and this civic society 'movement' can be expected to continue, especially from middle classes relatively safe from elephant depredations and damage (Chapter 4). Another frequent, almost universal response is the reported awe (Chapter 9) with which humans view elephants, even from societies very exposed to threats by elephants, and much symbolism in these societies reflects elephants (Chapter 1).

Technological

Technological developments have had spectacular impacts, especially in the last few decades, on possibilities for elephant management. Between the 1950s and 1990s, managers tended to readily embrace whichever technical option they could use, and to an extent these tools appeared to 'lead' elephant management (Chapter 1).

From the early 1900s onwards, roads began making an enormous difference to elephant management, for instance in terms of access for control of poaching. Coupled with the later development of elephant-proof fences (Chapter 7), a defined area could be protected in a way that allowed realisation of a command-and-control management style (Chapter 1).

Effective culling methods (Chapter 8), chemical immobilisation for translocation (Chapter 5), and eventually contraception and sterilisation (Chapter 6), all led to major management uses or possibilities, often aided by airborne support (helicopters in particular), along with electronic tools such as GIS/GPS, that also facilitated effective counting of elephant. When elephant densities were the primary criterion for decision making, such technology was paramount. A relatively unexplored area, except for experiments in so-called 'disturbance culling' (Chapter 8), is that of behavioural modification. This new stream of scientific work (Chapter 7) promises possibilities of promoting avoidance of certain areas by wild herbivores.

Fencing (including all forms of barriers, repellent plants, and even protection of individual trees) is dealt with in Chapter 7, which also touches on the effects of artificial water provision on elephant distribution and numbers – the latter clearly influential under arid conditions. Fences in our landscape are invariably an integrated expression of various influences. They often have more to do with veterinary legislation, direct demarcation or protection of property than with ecologically influencing elephant populations (Chapter 7). Electric fencing has made a major difference to controlling movement of elephants, and indeed in many circumstances, to maintaining wildlife reserves amidst other land uses (Chapter 7). Some believe that fence maintenance is straightforward, and should be diligently practised by authorities to limit occurrence and effects of breakages, while others point out how difficult maintenance can be in certain topographies, and under certain social circumstances (Chapter 7). However, fencing can have an obvious major disruptive effect on ecological processes, affecting dispersal of many species including elephants. Fences are also useful research tools (for instance, for excluding elephants to study effects) (Chapter 7). When fences are removed for whatever reason, a lag period can be expected before elephants colonise adjacent areas (Chapter 7).

Economic

In broad terms, elephants and humans compete for similar habitats. The result is that in the modern era, humans have generally marginalised elephants, reducing their range (Chapter 1). Re-establishing connected corridors between current regions of elephant distribution has been suggested (Chapter 2), but will be practically difficult in the densely populated and developed regions of South Africa where these are needed, though more possible in less-developed countries. Translocating elephants into small areas from which it is impossible to allow range expansion should be viewed with great circumspection, as this creates a whole host of ecological and management challenges.

TEV (including use and non-use values) has not been calculated (Chapter 10) in South Africa, but some such exercises have been carried out in neighbouring countries. Results are strongly influenced by the social values predominant at the time, and the general context. In other words, drivers, other than economic ones, modulate economic outcomes as exemplified by CITES bans (Chapter 10).

Elephants can have a negative economic value (e.g. landscape degradation or crop-raiding by elephants). Positive economic values are not necessarily additive (e.g. spatial separation between tourism and trophy hunting) (Chapter 10). Findings of the studies in Namibia, Botswana and Zimbabwe show that economic benefits from elephant often outweigh negative effects, though acceptable and effective rules and arrangements are crucial to the realisation of these benefits, and have not always been possible to make (Chapter 10).

Demand for ivory from the Far East peaked in the 1970s (Chapter 10). The subsequent CITES bans on elephant trade effectively caused a shift in profitability and operations for legal and illegal markets (Chapter 10). Since these bans, estimates of willingness-to-pay for elephant survival have been measured in some northern hemisphere countries, and can cover negative effects of elephant damage and still show a surplus (Chapter 10). This opens possibilities for, for instance, payment of compensation for lost revenue (elephant damage; foregone cost of ivory that can no longer be sold) and effective conservation (Chapter 10). Ultimately such market solutions may supersede regulatory incentives such as bans (Chapter 10) though both 'stick' and 'carrot' approaches are usually needed.

Environmental

The concern that elephants could be degrading landscapes mostly arose post-1960 following growth of elephant populations in South Africa, and their confinement to certain areas (Chapter 1). It is clear that losses of other species can occur due to habitat modification by elephants, but the occurrence and extent (and particularly societal perception of the acceptability thereof) of this varies under different circumstances (Chapter 3).

Ecological views also shifted markedly in the late twentieth century. Notions of simple causality, stability, and 'balance of nature' gave way to complexity and to views allowing ecosystems to vary over space and time, thus yielding other interpretations of the undesirability of such changes brought about by elephants (Chapter 1). Although ecosystems may be subjected to multiple drivers, it is usually only two or three that are the major determinants of system behaviour (Holling, 2001). In semi-arid savanna systems, where most southern African elephants occur, rainfall, fire and herbivory (including by elephants) are key factors. How they play out is mostly determined by underlying geology, soils and landscape structure. Current philosophy of ecosystem management emphasises the dangers of modifying, or attempting to control, single drivers because ecological systems generally require the action of a full suite of varying drivers to maintain heterogeneity and system resilience (Levin, 1999). This has led to the idea of managing to allow for high, medium and low elephant (and other driver) impacts at different places and times as the strategy most likely to guarantee a wide range of biodiversity (Chapter 1); some feel this is most likely to be achieved by varying water provision across the landscape. The appreciation of complexity and change placed a premium on a rapid rate of ongoing learning (Chapter 1), thought to be best achieved by adaptive management with clear initial goals and an anticipation of surprise. Pushing and probing the system to gain knowledge, and abstractly modelling system behaviour to promote understanding and generate predictions or scenarios, form the basis of such an approach (Chapter 1).

The global biodiversity crisis (Cracraft & Grifo, 1999) is relevant in that the results of inappropriate elephant management could be seen as further worsening the worldwide decline of biodiversity through habitat homogenisation or degradation (Chapter 3). Conversely, without elephants effecting seed distribution (certain plants are distributed by elephants) or creating necessary disturbance (producing heterogeneity), species could also be lost (Chapter 3) in terms of Levin's (1999) hypothesis of heterogeneity as the basis for biodiversity. A common theory proposed (also with respect to likely elephant effects) is the intermediate disturbance hypothesis. It posits that intermediate disturbance produces higher biodiversity than low or high disturbance, although Mackey & Currie (2001) show that evidence for this is by no means universal. However, different species are found at different disturbance levels, so that over a bigger area a patchwork of low, medium and high disturbance should give the greatest overall diversity (Chapter 3). Disturbance is itself a complex phenomenon, characterised by severity, frequency and extent (Chapter 3).

Global environmental change not only includes land use change (already discussed under economic drivers) but also rising levels of CO_2 and resultant climate change which could in combination lead to possible increase in tree cover in grassland areas.

Politico-legal

The major statutory protected areas in South Africa were proclaimed in the twentieth century, reflecting a growth of the conservation belief in society among the white population who had a near-monopoly of power during this period; one of the goals was to save the elephant species in South Africa.

After the Union of South Africa was created in 1910, land occupation and ownership was increasingly segregated by race. At that time there were very few elephants in South Africa. The few in the Addo and Knysna areas were nearly exterminated by government efforts to protect the white agricultural community, commercial agriculture and elephants being largely incompatible. Over time, the growing density of rural black communities also meant that further parts of South Africa became unsuitable for elephants. A broadly similar narrative applies to South Africa's neighbours Namibia and Zimbabwe. Land restitution in South Africa (Chapter 11) includes many claims inside statutory and private protected areas. It is anticipated that as this process unfolds, it will have consequences for elephant management, some hard to anticipate. For example, there could plausibly be greater demand for the lucrative and easy-to-start trophy-hunting option; or equally plausibly, pressure to cull elephants to limit their damage to crops; or to protect elephants in support of ecotourism.

After the culling era, governments and agencies have tended to skirt the elephant management issue (Chapter 4), creating the impression of inaction through lack of political will. Recent developments in South Africa, which include this assessment, represent a move towards a more explicit and accountable policy formulation. In the adaptive approach, making mistakes as a result of actions taken is seen as an important source of learning. Mistakes should thus be embraced rather than avoided or feared, as the no-action option can lead to even greater problems (Maguire & Albright, 2005).

Several global conventions are related to elephant management (Chapter 11). The Convention on Biodiversity obliges signatories (including South Africa) to achieve biodiversity conservation targets and to move towards benefit-sharing with local communities in areas of resource utilisation, while the Convention on Trade In Endangered Species restrains international trade in elephant products.

A spate of post-apartheid legislation (Chapter 11) has significantly altered the politico-legal landscape. The South African Constitution established a goal of a healthy environment and the notion of participatory governance (Chapter 11). The National Environmental Management suite of Acts enforces many biodiversity obligations, including norms and standards on elephant management and on threatened species (Chapter 11). Animal disease legislation (Chapter 11; Chapter 7) is particularly influential in the realm of fencing (where elephants are often responsible for breakages and hence indirectly for disease outbreaks). However, there are serious gaps in the legal frameworks in South Africa regarding ownership and responsibility for wildlife (Chapter 11). By contrast, some recent legislation (such as for threatened and protected species) seems over-cumbersome, especially in an environment where ensuring compliance is likely to be difficult (Chapter 11). All these influence decision making regarding elephants.

Strategy

Strategy refers to the intentions and broad roles relating to elephant management. We discuss it below under the set of influencing ideas referred to in the conceptual framework.

Relevant features of elephant biology

Key aspects of elephant biology, such as growth and reproduction, endocrinology, social behaviour, musth, and communication are well documented (Chapter 2; Chapter 6). Movement behaviour (Chapter 2) and diet (Chapter 3), as well as the effect of variation in habitat (heterogeneity) on them, are well studied.

Natural mortality of elephants from droughts or predation occurs especially in very young calves and then again just after weaning (Chapter 2). Elephants are not particularly vulnerable to disease (anthrax and elephant myocarditis may cause sporadic deaths) (Chapter 7). Kruger has a good 50-year time series of elephant population size, but not of the sex and age structure, whereas Addo has an almost complete record of both (Chapter 2). Any relatively young population will show rapid growth and is likely to overshoot its key resource, causing a later correction (invariably involving deaths), and possibly associated with habitat change (Chapter 2 and Chapter 8). If specific age-classes are missing from a population, this can be disruptive to behaviour (Chapter 8). Use of immunocontraception and vasectomy will take a long time to reduce populations (Chapter 6). Large males are important gene contributors (Chapter 2).

Elephant are megaherbivores, consuming vast quantities of food per animal, and are known as 'wasteful feeders' (Chapter 3). They are regarded as a keystone species, meaning that their presence is important for other species and for the functioning of the ecosystem (Chapter 3; Chapter 4). Elephants can play both a competitive and a facilitatory role relative to other species (Chapter 3). They are important in nutrient cycling and seed dispersal, and elicit plant defence and growth responses (Chapter 3). Elephants and fire are regarded as drivers of alternate states in ecosystems (Chapter 3). It is difficult to disentangle the relative roles of elephant, fire, drought, disease, and other browsers in tree population patterns (Chapter 3). Limited palaeo-ecological results over thousands of years suggest tree densities have fluctuated in Kruger, but with no long-term trend (Gillson & Duffin, 2007), implying that pre-ivory trade impacts of elephants on vegetation were not uniformly higher. Elephants are known under certain circumstances to cause local extinctions of other species (Chapter 3) and known to also have significant effects on structure of vegetation (Chapter 3). Adult males are larger and kill or damage larger trees, also disproportionately pushing them over (Chapter 3). Elephant effects vary spatially, and piosphere effects (meaning the appearance of bare ground around waterpoints) are partly attributable to elephants (Chapter 3). When elephants are removed from a system, equally drastic changes may occur (Chapter 3).

There are important interactions between particular interventions and elephants. For instance fencing off, especially of water (Chapter 7), changes ranging behaviour. 'Overabundance' effects are often ascribed to the fact that elephants were fenced into parks (Chapter 3), but subsequent dropping of fences does not necessarily result in a quick reduction in elephant numbers (Chapter 7). The relative effects in small fenced-off parks is uncertain, with some studies indicating that increased homogeneity results from impacts, and others increased heterogeneity over the landscape (Chapter 3). Elephant, mostly males, break fences, providing a conduit for disease transmission (Chapter 7).

Regarding manipulation of reproduction, oestrogen implant experiments were stopped due to unacceptable side-effects (Chapter 6). pZP immuno-contraception, on the other hand, has a seven-year study at one site, and has been implemented at another four, with few detectable side-effects at individual or population levels (Chapter 6), though aging effects obviously occur in the population (Chapter 6). Some work has been done on stopping male musth with GnRH vaccine. Contraception can stop population growth within two years (Chapter 6), but does not reduce population size until the older elephants reach the end of their life spans (Chapter 6).

Translocation of only young animals resulted in them forming secretive mobs, and aggression towards fences, humans and rhinos (Chapter 5). Habituated elephants, or elephants from wilderness areas, tend to retain their behavioural characteristics after translocation (Chapter 5). Females in small populations may display abnormal aggression (Chapter 8).

Ongoing culling of problem animals in peripheral 'sink areas' may erode trophy quality and have genetic implications (Chapter 8). Large-scale culling can lead to high rates of population growth once the culling is stopped (Chapter 8). In arid areas, limiting the distribution of surface water can influence elephant distribution and limit populations, the latter partly through increasing juvenile mortality; while in well-watered landscapes, elephants still concentrate more along riparian areas than elsewhere (Chapter 3; Chapter 7).

Artificially small populations easily develop genetic and behavioural problems (Chapter 3).

Level of ecological organisation

This refers to the target level of management, from individuals to populations to species to ecosystem, and ultimately, biodiversity in its broadest sense. The level being addressed influences decision making.

During the mid-1900s wildlife conservation tended to focus on the protection of individual species, as exemplified by the Addo Elephant and Mountain Zebra National Parks, and Tembe Elephant Park. These parks were initially run according to the overriding management needs of particular species (Chapter 8). It was soon realised that these focal species were part of larger animal and plant communities, and the focus was widened accordingly, for example to include threatened plant communities in both Addo and Tembe

(Chapter 3). The swing of emphasis to ecosystem management (Meffe *et al.*, 2002) that followed was endorsed by the IUCN in the late 1980s (McNeely, 1993), and many parks, such as KNP, can be seen to be effectively following those tenets today. In many respects, agencies are still grappling with the complexities of conserving biodiversity (Noss, 1990); the full definition includes diversity of structure, composition and function at the genetic, species, population and ecosystem levels. Ecosystem management may indeed cover these needs (Hunter, 1991). However, mission statements and objectives are being fine-tuned to accommodate biodiversity, and the actual target formulation and resultant monitoring programmes in particular are reflecting these demands. Generally, defining function (i.e. process) is the most difficult aspect of biodiversity, and reliance is still placed on the more feasibly measured structural (meaning pattern, across scales) and compositional aspects (meaning genes, species, communities, and ecoregions).

Ecological theories

Ecological theories have changed over the years (Chapter 1) and have been very influential in setting strategy. Early ideas suggested that the main interactions in ecosystems were relatively simple cause-and-effect relations, leading to, for instance, an orderly succession of vegetation following disturbance. Equilibrium or 'balance-of-nature' concepts meshed well with the notion of maximising productivity, an idea arising from the strong influence of agriculture. In more recent decades there has been a shift to viewing ecosystems as complex and less predictable, with non-linear responses, and (often delayed) feedbacks that make the notion of causality difficult to pin down (Chapter 1). In this view variation over space and time is considered crucial to ecosystem health and resilience. The differing ideas described above have appeared in a loose progression, the more recent ones partly replacing the older ones. The ideas co-exist to some extent, in that one aspect or facet is explained by one mode of thinking, and another component by another set of theories.

Issues of scale, extent and boundary conditions

This refers to the size of the management area, and the length, shape and permeability of its boundary, as well as the perceived relationship with a broader area beyond. It is often stated that small parks need more intensive and 'less natural' management than large parks, although this can sometimes be used as a reason for perpetuating invasive practices in small parks.

'Fortress conservation' (Brockington, 2002) refers to carrying out rigid conservation inside a tight and usually defended boundary, with little concern for what is occurring beyond. Movements that 'looked outwards' – for instance the KNP 'beyond the fence' approach used especially in river management – became more common once limitations of the 'fortress' approach became clear. Once land use outside a protected area became sufficiently similar to that inside, people automatically started referring to terms such as the 'greater Kruger National Park'. Such developments have allowed elephant range to expand, even without translocation. Under the National Environmental Management: Biodiversity Act bioregions will become formal parts of the South African geography, with some large multi-owner elephant-containing regions already existing (such as the UNESCO-designated Kruger-to-Canyons biosphere; www.kruger2canyons.com/biosphere.htm). These ideas now extend across international boundaries through the establishment of Transfrontier Conservation Parks and Transfrontier Conservation Areas, the latter including the wider area around the Parks.

A key strategic issue is the extent to which both planning and implementation approach spatial and temporal scaling, and how importantly they rate the consequences as intrinsic to their philosophy. The scale at which elephant impacts on management take place now occupies a central position in the way these are visualised (Chapter 1). Decision makers may commission planning at one scale and land up having unexpected effects or consequences at very different scales (Chapter 8). This also happens in the time dimension, where what seems to be an outcome that is acceptable in the present and the immediate future, turns out to be unacceptable over longer time scales (Chapter 8) yet the investments have been made. Because of their longevity (Chapter 2) individual elephants can carry their experiences through several successive tenures of say, protected area managers. Finally, scale issues apply to the human system as well, and problems can arise through not paying attention to them or because of mismatches in the whole interacting elephant-habitat-human system between the key biophysical scales and the scales at which management is attempted (Chapter 8).

Goals

Management goals are usually a direct consequence of the dominant ideas and legislation of the time, as influenced by societal drivers. Recognisable categories of goals may include preservation, conservation, benefit-maximisation, and more recently objectives arising from a 'desired future state' as set under the

National Environmental Management: Protected Areas Act. This 'desired state' (actually a set of varying conditions desired for the future), if it exists at all in goals, can be implicit, explicit but general, or explicit and articulated in detail.

Management styles

Management styles tend to follow mainly one of the following (Chapter 1):

- Indiscernible (absent, inconsistent or ineffectual)
- 'Laissez-faire' ('leave it to nature')
- Command-and-control, where man's superiority, in terms of clear actions over a mechanistic nature, is assumed
- Management by intervention, cognisant of a more dynamic nature, but still with a 'central balance' or 'optimal point' and corresponding intervention, sometimes locally or for an isolated reason. These interventions could be widespread and far-reaching (e.g. culling in Kruger), and at other times may have constituted opportunistic local initiatives. This transitional style could be seen as an early bridge towards the later styles, but still very much rooted in command-and-control assumptions.
- Passive adaptive management. Recognition of a greater nature and ongoing change often beyond our control, and adapting accordingly to maintain particular goals.
- Active adaptive management. Further recognition that unless managers push and probe the system, we will not learn fast enough to adapt and manage successfully.

These management paradigms are ultimately only convenient (and hopefully largely appropriate) pigeonholes to help describe what in practice can be more nuanced hybrid styles. Different styles might be used at different spatial scales, or for different aspects of a system, depending on objectives, state of knowledge, and degree of effective control.

Implementation

On-the-ground management translates strategy (explicit or implicit) into action. We have chosen two focus areas (operational plan, and indicators and triggers) as key themes around which to gain insights into these outcomes. The pool of ideas from which interventions are chosen can be considered a 'practical toolbox', which is discussed later in table 3.

Operational plan

Operational plans come in a wide range of forms, and reflect the way in which the 'on-the-ground action' is conceptualised and deployed. In a closely coupled system, the structure of the operational plan takes its lead from the strategy in a clear and logical way. At other times and places, very practical plans have existed in isolation, or almost in isolation, of an explicit strategy, indicating poorly backed-up but directed action-on-the-ground. Conversely, sometimes a developed strategy exists with a poor operational plan. There are also cases where there has simply been no defined operational plan at all, either implying that no action was being taken, or that no justification or guidance was needed to take any actions that were decided on, presumably then in a very opportunistic or arbitrary way.

In the latter half of the twentieth century it became common for conservation agencies to have several plans for key species at the operational level. Most reserves with elephants had elephant management plans (required prior to introductions), often with large sections on culling. These plans often existed in relative isolation from plans for other species and sometimes even from system drivers, but served a particular role, for instance, in conservation of a threatened species, or management of a problematic species, elephants qualifying for both. Objectives tended, accordingly, to be isolated, such as introduction of elephants for genetic purposes; or the management of breakouts. Over time the wider interconnections among the growing number of ideas in ecology, and the wider range of concerns about multi-species, community and systems issues, led to clearer articulation of such operational plans in response to this wider battery of drivers.

Strategic plans should be in place and agencies ready to adequately operationalise them. Reflecting on IUCN conservation effectiveness evaluations (Hockings *et al.,* 2000), satisfactory biodiversity outcomes can result from carefully planned objectives that are in turn based on clear visioning. Many process steps (such as standard operating procedures) are needed to operationalise management in a routinised way on the ground. Even if there are such documented procedures available in certain parks, they may be functionally isolated from the strategy, adjustments being needed at both ends to harmonise the two. Scenario planning, because of uncertainty about which outcomes may unfold as one manages, can also be very useful, even at this operationalisation level.

Implementation lags of many kinds, i.e. periods that elapse before action occurs, bedevil management (Pfeffer & Sutton, 2000). Offsetting such lags includes shortening the period between knowing and doing, effectively translating good policies into action, avoiding a culture of non-compliance, and building the human capacity to enforce and monitor. If further lags occur between action and outcome, these can make it even more difficult to understand drivers and to adapt appropriately.

Indicators and triggers

Indicators are the elements that are measured to enable decisions to be made about implementing management, and triggers are the final signals that elicit action. There is a tendency for indicators to become more complex over time and then be re-simplified, or to converge into all-encompassing general indicators of system well-being. Administrators and scientists should expect such changes. The National Environmental Management: Protected Areas Act places emphasis on adaptive management. This section thus incorporates monitoring concepts required for understanding elephant management, which are not included in previous chapters (see particularly box 1).

Under a carrying-capacity paradigm, elephant density (the number or biomass per unit area) is a pivotal indicator, particularly when introducing elephants to small reserves. The state of other species (in practice, particularly those obviously impacted on by elephants, such as plants) has proved a regular indicator, most often expressed as structure (height classes) and composition (species) of plant communities. The complexity of covering an even wider range of species and features potentially impacted on by elephants has led to the use of surrogates, or related indicators that can be more easily measured than the actual species that are of direct concern. An example might be the vegetation height and cover profile and number of downed trees as indices of how much suitable habitat there is in elephant range for other smaller animal species.

Increasing interest in the wider context outside of pure biodiversity indicators has led to social and economic indicators coming into use – for instance, the number of crop raids into neighbouring farmland, or the financial benefit accruing to a village from elephant hunting.

In structured adaptive management, indicators are only chosen after concrete objectives are identified around actual aspirations or concerns, and in some versions of adaptive management only after a mental model of cause-and-effect suggests a conceptual threshold where a system is likely to pass into an undesirable state.

Some triggers are opportunistic. For instance, elephants may be introduced to a small population to widen a genetic base, the trigger having been a concern about the genetic bottleneck. Similarly, killing or translocation (the latter usually unsuccessful in removing the tendency in the individual) of the animal may be triggered by a damage-causing elephant.

Most agencies with monitoring programmes tend to choose too wide a set of variables, trying to 'mean everything to everybody', and usually later find they need more focus simply to do a reasonable job under realistic constraints. Reference is sometimes later made to 'core' and 'peripheral' indicators (Palmer Development Group, 2004). Some parks with focused objectives strive to only measure those few parameters that directly reflect issues of immediate or serious long-term concern, and drop the rest so as to concentrate resources. The wider the set of objectives (such as in the case of Kruger), the wider the suite of thresholds and set of monitoring variables tend to be. Learning can generate efficiency, and in its first five-yearly revision Kruger has honed down its suite of thresholds, believing this can still meet all its needs. However, it requires more detailed monitoring on the ground in appropriate categories to properly service the thresholds retained (J. Kruger, pers. comm.). This entire process requires ongoing iterations of evaluation and adjustment. There are obvious risks in investing too little or too much in monitoring, and the judgement calls can be difficult.

A final issue relates to the link from trigger to selection of action. Upon triggering, operational procedures can be 'hard-wired' into a very clearly defined set of practical steps. Care however needs to be taken to not codify these too firmly and rob managers of the space to manage adaptively, one of the failures of an over-emphasis on decision-support systems (Hayman, 2004). There is a fine balance between keeping on a strategic course for long enough to learn, and allowing implementers to intelligently choose the practical options most appropriate to the local context, at each decision call.

FORMULATING ELEPHANT ISSUES

The intention of this section is to help guide decision making concerning elephants. This is done through a series of key questions or steps. It sets out to be practical, but does not intend to be prescriptive.

Box 1: Thresholds, targets and process-based management triggers

Triggers are the final signs or flags which elicit action; they are the 'endpoints' of particular indicators. For instance an ecological threshold is derived from the mental model of where the system is likely to 'fall over the edge of the cliff', or in ecological jargon, to change state fairly quickly into an undesirable alternative. Some time before this (but not too long before, as it is believed that systems need to be allowed to vary, to stay resilient) is a point called the threshold of potential concern. This is the 'amber light' indicating that the system is moving fast enough in an undesirable direction towards the ecological threshold, and action must now be formally considered. To operationalise such a threshold of concern (which may be, for example, that species richness is being lost at more than a specified rate) an indicator, such as a broadly representative list of reptiles, birds, and insects, could be monitored. The trigger would be the exceeding of that threshold when the rate of loss is higher than the specified rate. A closely related but more widely used concept in regional biodiversity planning is that of targets, which normally are set as an intention to secure (or maintain at least) x% of a certain vegetation type under conservation. If this is achieved, these veld types are considered adequately conserved, or said another way, safe from the risk of passing into an unacceptable state. As soon as there is evidence that the targets are already at, or clearly heading to, a point outside the 'desired state', action is triggered. If targets or thresholds are set at different scales, they can be nested under each other. For instance, elephant-related thresholds of potential concern in Kruger (each articulated at sub-park scale) take into account what regional targets (of the South African National Biodiversity Institute) are of the same vegetation type in the region as a whole. This implies that thresholds of concern might allow greater change (even perhaps with inter-generational consequences) in vegetation types which are well represented and safe outside the park, while the desired state of vegetation types which do not occur (or are poorly protected) outside the park is defined more tightly. The fact that administrative boundaries do not always coincide with ecological boundaries adds challenge to implementation.

There are many similar constructs to targets and thresholds, such as limits of acceptable change, a concept never widely used in South Africa. For Ezemvelo-KZN Wildlife, a central idea is that the relatively small or intermediate size of their reserves means that natural ecological processes are significantly altered. Under this assumption, they have developed a philosophy of process-based management which then allows human-made interventions to make good the shortfall. For instance, they cull antelope in certain parks which cannot house large predators in a predator simulation programme; and they remove rhino from broad areas around the edges of an intermediate-sized park, to simulate source-sink dynamics (meaning the rhino breed up in the central core area and then disperse concentrically, but fences curtail this, justifying this intervention). Each of these process themes has a target or trigger to guide it, such as an expected number of a particular species that would have been taken by predators. Ignoring the underlying assumption for a moment, process-based management can be considered very advanced in that it tries to tackle ecosystem function and not mainly composition (species) and structure (such as tree height patterns) on which most other agencies concentrate. Most agencies claim that function is extremely difficult to understand, monitor and manipulate, and currently focus on composition and structure, hoping that this reflects healthy ecosystem processes as well.

Where do you fit into the decision-making process?

Decision making occurs at different levels. Most of the material presented here is pitched at the protected area level. If you are a private owner of a smaller reserve (or making recommendations to such a person), you may find that the context of the landscape and society around the reserve has an overriding effect on your decision. Alternatively, in a large park, you may indeed be able to take a very individualistic stance concerning elephant management. In both cases, following through the framework will help you determine an appropriate approach. You may decide to gloss over certain of the headings in an effort to 'get to the point' further along the framework, but resist the temptation to do so too cursorily, as your situation is materially affected by all the headings, though you may feel some of their outcomes are for all practical purposes 'givens' for your particular situation. It is good to recognise these clearly and highlight them, and sometimes even to challenge what appears to be already 'fixed'. Go to the trouble of drawing the linkages between these and your final decision, as this will make it more justifiable, better-rounded and more durable.

If you are responsible for influencing or taking decisions at a level above the protected area level (for instance, metapopulation management in a wider region; or national policy; or even international or transboundary policy), then you will need to take a very broad view and will probably be working with multiple values and paradigms, and several decision-making levels. Again, the framework should assist in comprehensively identifying and balancing the issues. To help place where your decision fits in:

1. Determine the level and scope of your own decision-making position relative to the three 'compartments' (drivers, strategic and implementation) (figure 3). Ensure that the parts of that overall process you are not dealing with are somehow adequately covered and feed into your own decision-making process. If they are not adequately covered, decide explicitly how you will deal with that shortcoming.

2. Consider how you will contribute towards the need for bridging between other stakeholders on the same level, and particularly among levels (see figure 2).

Building a goal-orientated adaptive approach

An adaptive approach is mandated by the norms and standards for elephant management (Department of Environmental Affairs and Tourism, 2008), certainly at the property or reserve level. This is in line with management philosophy (Chapter 1) as it has developed, and there is no reason to believe that other levels of decision making would not benefit equally from this approach. A structured adaptive approach can be built using the following steps:

1. Define your particular context, such as the boundaries (also abstract ones, like social impacts) of the system you are dealing with for the decision at hand. This or the next step will require the listing of relevant stakeholders. It is usually very helpful to limit your focus through explicitly listing and exploring the special and unique attributes of your particular system.

2. Generate a balanced understanding of societal drivers, in this way helping to take care of the relationship between broader society and your management decision. Relative importance and balance will differ in different situations, but ignoring any driver category or assigning even too low or high a profile in the overall portfolio, may lead to less effective results. Public facilitation to elicit this understanding may be

prescribed or desirable. Review this understanding from time to time, say every five years.

3. Ensure transparent (at least to all the relevant stakeholders in the particular context) and clear setting of objectives consistent with the values recognised. This represents the upper or conceptual part of the 'desired state'.

4. Generate targets or thresholds that represent an initial stab at whether the objectives are indeed being met. Have a mechanism for revising these in the short-term as feedbacks come in and people learn. These targets represent the operational endpoints of the 'desired state'.

5. While doing all this, be very cognisant of your capacity to actually implement the plan. In figure 3 this is indicated as a cross-cutter for consideration perhaps while dealing with broader societal drivers. The way that, for instance, a park plan is designed and deployed, may place too high an institutional burden on the agency, and may preclude effective operationalisation of the strategy and plan. At the same time, this should not be used as an excuse for ineffective planning or implementation.

It is realistic to anticipate drivers shifting in future. Scenario planning (the use of plausible narratives about possible, as opposed to predicted, futures) is a very useful technique for testing the robustness of your plan, and can be commissioned by a conservation agency, government department, or NGO at any appropriate level, even a village level. Scenarios should be developed participatively. Particular elephant management scenarios may be built on existing and already available wider (global or regional) scenarios, such as those released by the Millennium Ecosystem Assessment. The basic reason for this is to heighten resilience through preparedness for any surprises, rather than thinking about these for the first time after they have happened. Interesting scenarios to develop around elephants would include issues such as an ascendancy of animal rights values, a domination by utilitarian values, or the effect of big swings in exchange rates or oil prices on ecotourism and hunting.

Promoting learning

Adaptive management (especially active adaptive management) encourages practical ongoing learning, and the casting of management as a series of sensible experiments from which valuable experiential and scientific learning is possible. The following points constructively promote learning within and among disciplines and stakeholders:

1. If relevant to your level of decision, ensure a healthy science-management link, a difficult task in many agencies. Case history experience almost dictates that the science component should include reputable researchers and experts external to the agency. Long-term nurturing of this science-management link will be essential. The explicit formulation of targets, thresholds or triggers based on a shared vision and objectives constitutes first prize. Partnerships between external scientists and a particular agency must carry mutual respect, and may have to develop through phases initially requiring less trust, such as looser collaboration.

2. At the same time as ensuring tighter science-management links, it is important to retain a measure of identity and independence for both groups. When completely merged, there may arise a strong and inflexible 'groupthink' that may work against longer-term success, and raise questions about bias.

3. Biological scientists and conservation managers, particularly in the elephant debate, should be required to confront and appreciate broader societal values, and to not view their 'authoritative' results in isolation.

4. Design the management so that (at least within the limitations of your context) maximum learning is possible. This often involves a measure of responsible experimentation, and where possible, comparison of tools. This requires good documentation of the reasons actions were taken, and of the outcomes.

5. Link yourself to other sources of learning that are relevant. As essential as local learning is, you will not be able to learn fast enough without external inputs.

Dealing with change and diversity

From the history discussed in this assessment, it is clear that the pace of change concerning inputs to (not necessarily decisions about) elephant management has accelerated in recent decades. While this may level off, it is unlikely that the broad range of drivers and factors described will suddenly become streamlined into a smaller set. Whether this happens or not, we are currently faced with ongoing change and the reality of diversity of opinions. The following general guidelines will assist us:

1. Be cognisant that today's approaches and paradigms are themselves fallible, and be open to the reality that others will emerge and the possibility that older ones may re-emerge in slightly different form.

2. Be open to the possibility of moral pluralism, in that contradictory values may have to be accommodated to some or even a great extent.

3. Expect that as values change and management systems evolve, old and new narratives may contain contradictions to each other. These may need to be managed to move forward, keeping an eye on the longer-term vision

4. Ensure that all the important feedbacks in adaptive processes are taking place. Many elephant management initiatives to date are, in practice, almost devoid of adaptive feedbacks such as shown in figure 3. For adaptive practitioners looking for a listing of generic feedbacks in adaptive cycles, see Biggs *et al.* (2003).

5. Adaptive systems should be designed to not become paralysed by differences in scientific opinion. Such differences may require contrasting recommendations arising from both points of view, in the full spirit of a well-motivated and thought-through adaptive experiment.

GUIDELINES FOR SPECIFIC DECISIONS

Decision making is a complex science and art in its own right, and this chapter can do no more than synthesise the best current knowledge coming through the assessment and apply it in the following guidelines. This will be done domain by domain, remembering that the interlinkages and feedbacks among these are as important as those within domains.

Societal drivers domain

Table 1 helps ensure that the full range of broader societal drivers is used appropriately in decision making. It contains information to help ascertain when and in what depth particular drivers should be examined, and broadly how this can proceed and be interpreted.

Strategic domain

Table 2 summarises the major guidelines for consideration when a decision maker is concentrating on strategic factors as outlined in the framework. It is intended as an overview of key concepts under each heading, and their broad relevance in decisions.

Implementation domain

The decision maker needs to decide on the character of the operational plan, for instance, whether or not it deals with elephants in relative isolation, the extent to which it allows local flexibility in decision making supporting the goal, and whether standard operating procedures are included. Importantly, and as discussed earlier in this chapter, coupling to the strategy needs to be made explicit. Finally, in any adaptive system there need to be at least some triggers/targets/thresholds and decisions on the indicators being measured that provide this information. These need to be sensibly chosen, in the light of all the discussion in this chapter, bearing in mind the absolute need to monitor adaptively, but also remembering likely limitations of capacity to do so. It may be that capacity simply has to be expanded to at least provide basic feedback, else no learning takes place, except by inference from elsewhere. Justification based on such evidence from elsewhere may not be acceptable in terms of the norms and standards (Department of Environmental Affairs and Tourism, 2008), as each park needs to justify at least the basics of its own case.

Finally, but nested inside all of the decisions taken till this point, comes the final decision whether or not to intervene, and if so, then the actual choice of intervention or interventions. Regardless of the type of trigger framework used, once a monitored indicator passes a threshold, a management action will invariably be elicited in an attempt to shift the system to meet the objectives of the plan. Table 3 compares key attributes of interventions to assist decision making at this level. When compatible, multiple interventions can be selected – that is, they are not necessarily mutually exclusive.

Feedbacks, and integrating all three domains of decision making

Decision makers who have been through the process above need to think through the linkages among the different elements. For instance, it does not help to have excellent implementational plans not grounded in good strategy, as little as it does to ignore practical feedbacks from implementers on the ground.

Societal driver category	When are these needed?	How are these elicited?	What weighting should these have?
Values	When framing the broader issue. If never framed, or no authoritative recent value determination can be cited, ensure this happens and states values explicitly, or risk failure (in any functional democracy). Do this genuinely but do not over-engage. Needs only one or two 2-hour sessions, often connected to related issues at a longer meeting. Repeat say every 5 years to check for value drift	By adequate and representative socially sensitive facilitation which produces a legitimate result. Depending on context, this may be a small set of stakeholders, but inclusivity of those is crucial. Outcome is usually sufficient consensus, but some cases may be messy. Agencies may assert own values but recognise that these should influence and be influenced by other values	Explicit identification of values (which can vary widely in different contexts) is almost an absolute prerequisite. Insufficient explicitness will result in ongoing misunderstandings and 'muddling on'. Some values preclude others, 'losers' should feel process was reasonable, and be given another chance in say 5 years' time
Social	Whenever people are significantly involved as stakeholders around an elephant issue, e.g. human-elephant conflict; or right to significant concern, e.g. citizens awe-inspired by elephants	Usually at meetings to specifically discuss grievances or at joint-management fora; or public meetings around park management plans; or with specific interest groups	Depends on context. Social issues may dominate decision driver framework or may be tiny (e.g. private well fenced-off reserve with few expectations or possibilities of societal interaction)
Technological	These are realities of the decision-making landscape, the error is to not recognise or ignore their effect or potential effect in influencing decisions	Usually made known by technical salespeople or researchers with altruistic or vested interests. Options, pros and cons usually need to be summarised and possibly evaluated for decision makers	Wildcards which can overawe decision makers to the exclusion of other factors, but can also create a whole set of new positive outcomes. Best measured against values and defined objectives. May require initial experimentation
Economic	Whenever benefits and costs need to be explicitly evaluated. (May be less necessary in luxury projects with overriding values)	Ultimately, requires total economic value studies in a holistic framework by economists sensitive to complexity. These studies are expensive and rare, so decision-makers may use the best available information from studies in a similar enough context	Crucially important when changing land-use options and/or livelihoods of poor rural people are the main issues. If institutions can be put in place, may critically enhance African elephant conservation through even non-use values

Societal driver category	When are these needed?	How are these elicited?	What weighting should these have?
Environmental	If degradation or species loss is believed to be an issue (perceived to be particularly so in smaller properties)	Through scientific evidence relating to roles and effects of elephants. Elephants have 'good' and 'bad' effects, and caution is recommended in blanket use of e.g. the Precautionary Principle (Cooney, 2004)	If 'true' degradation is present or reliably predicted, place a high value on such environmental drivers. If not, be sure to evaluate elephant effects relative to values and resultant objectives – environmental issues still important but tailored to these
Politico-legal	Whenever political compliance is sought. Desire for legal compliance should be universal, though this assessment points to important gaps in some key issues of the legal framework	By being aware of current policies, trends, laws, norms and standards. These can also be influenced by you or a representative, theoretically with ongoing chances for updating over time	Generally regarded as minimum standard of legal compliance, in reality much flouting and confusion as legal landscape changes. Policy compliance can be optional depending on (e.g. private) context

Table 1 : Key guidelines regarding use of societal drivers in decision making

Key factor	Key concepts	Broad relevance
Elephant biology	Ecological role Reproductive biology Population growth Behaviour Genetics	Biodiversity and ecosystem management Manipulation of reproduction Expectations of growth or control Intervention impacts, e.g. translocation/culling/fencing Small populations/translocation
Level of ecological organisation	Continuum from species-community-ecosystem-biodiversity	Species concerns more straightforward; ecosystem management takes 'big-picture' view relying on healthy processes; biodiversity targets full spectrum of organisms and features
Ecological theories	Rough continuum of ideas from succession-production-equilibrium-resilience-scale and variability	Succession and production aims at highest yields. Equilibrium tries to maintain static balance. Resilience allows or encourages variation also at various differing scales
Extent and boundary	Rough continuum from fortress-look outward-greater ecosystem-bioregion-TFCA	Decide on most appropriate domain for your strategy and develop understanding for that context
Scale	Deals with the spatial and temporal scales of both planning efforts and consequences of these. Also social scale	Decision makers must ensure clarity of thinking about scale to avoid mismatches. Be explicit about both spatial and temporal scale of implementation as well as consequences
Goals	Focus on one or more of these: Preservation Protection Conservation People-and-parks Desired state-objectives	Preservation keeps unchanged. Protection prevents asset erosion. Conservation accepts change inside a hard barrier; People-and-parks focuses on interactions with neighbours; NEMA-mandated parks demands objective-driven planning accountable to stakeholders
Management style	Undefined Laissez-faire Command-and-control Passive AM Active AM	Unclear Watch but don't interfere Intervene: and reestablish fixed state Accept bigger dynamic and adapt Ditto but perturb to understand

Key: NEMA – National Environmental Management Act; TFCA – Transfrontier conservation area

Table 2: Guideline summary of key factors for strategic decision making for elephant management

Intervention	Comparative values	Incompatible values	To meet what objectives	Finances (ballpark)	Short-term stress	Long-term stress	Short-term population effects	Long-term population effects	Effect of stopping	Uncertainties and risks
Do nothing	Naturalness Wilderness Tourism Welfare	Conservation?	Natural processes	None initially	None	Contentious, none to serious stress if fenced in	'Normal growth'	Stabilisation or population crash	May predispose dramatic later intervention	What limits populations? Is there density dependence? Fail to meet objectives. Higher HEC?
Culling	Conservation Sustain. use	Wilderness Naturalness Animal rights	Biodiversity (comp.; structure) Sustain. use Learning	R6 600/ele cost; profit R7 000/ele (excl. ivory)	High disturbance	High disturbance with behavioural effects	Immediate reduction Faster growth	Eruptive growth Social disruption Avoid resource depletion	Rapid growth Social problems	Long-term stress effects. Hidden processing costs. Self-perpetuating need Public aversion Behavioural effects
Translocation	Conservation Tourism	Naturalness	Biodiversity (comp.; structure) Sustain. use	R5 000/ele	Short term disturbance	Cannot translocate twice (N&S)	Immediate population reduction but limited impact	Eruptive growth Social disruption	Rapid growth Social problems	No acceptable estimations No market
Problem animal control	Sustainable, use HEC	Animal rights	HEC	High to return animal, low to shoot	Short term disturbance	High stress	Minimal (depends on population size)	None	None	Underlying cause remains Neighbour relations
Trophy hunting	Sustainable, use HEC Tourism	Animal rights/ welfare	Biodiversity Pop structure HEC	High return	Short term disturbance	Almost none	Minimal	Possible depletion of trophy classes	None	Public aversion

Intervention	Comparative values	Incompatible values	To meet what objectives	Finances (ballpark)	Short-term stress	Long-term stress	Short-term population effects	Long-term population effects	Effect of stopping	Uncertainties and risks
Immuno-contraception	Conservation Welfare (debatable)	Sustainable, use naturalness	Biodiversity (comp.)	R900/ele	Short term disturbance	Possibly none	None for 2 years, then reduced growth rate	Reduced growth rate; eventually reduced pop. size. Incr. natural mortality. Incr. density dependence. May avoid resource depletion.	Slower growth	Are there similar long-term stress and/or behavioural consequences? Neighbour relations i.t.o. foregone opportunity to cull or hunt
GnRH treatment	Conservation Welfare	Sustainable, use naturalness	Biodiversity (comp.) HEC	Unknown	Short term disturbance	Unknown (none after 7 years)	Minimal	Minimal except in very small populations	Unknown	See 'Immuno-contraception' Chapter 6
Fencing	HEC Disease control	Naturalness Wilderness Sustainable, use (ouside)	HEC Habitat protection Protect resources	Capital & maintenance Aesthetic cost	Minimal local disturbance; high stress for migrants caught against fence	Unknown No direct	None	May lead to need to intervene	Attraction to elephants when fence removed; or may leave legacy. Increased HEC	Creating local hotspots of high impact. Poor maintenance. Side-effects on certain spp. May exclude human communities from parks

Intervention	Comparative values	Incompatible values	To meet what objectives	Finances (ballpark)	Short-term stress	Long-term stress	Short-term population effects	Long-term population effects	Effect of stopping	Uncertainties and risks
Water provision	Sustain. use Tourism Overcome upstream influences Humaneness in droughts	Naturalness Wilderness Conservation	Buffering Productivity Tourism	Capital & maintenance Aesthetic cost	None	None	Higher growth rate	Higher density, overshoot resource base, delays density dependence	Short term mortalities; changes in spp comp and numbers	Increased risk to biodiversity. Legal cost
Waterpoint removal	Naturalness Conservation	Tourism Welfare	Biodiversity (struc. and comp.)	Dams and gravel pits v. expensive	High	None? may return to look for water	Incr. calf mortality, esp. in arid systems	Decr. growth rate	See 'Waterpoint provision' in Chapter 7	May have limited effect in mesic systems. Possible tourism backlash
Range expansion	Naturalness Wilderness Conservation Tourism Welfare	Sustain. use	Biodiversity (comp, process, structure) Conservation are expansion	May form regional economic node and/or loss of local resources; high cost	None	Reduced	Little	Decr density. Lower prob of over-shooting resource base in original areas but eruptive in new areas	Development pressure on corridors	Forced HEC and/or land/use change. Changes in sentiment. Only while extra range in available

Key: HEC – human–elephant conflict; N&S – Norms and Standards; pop. = population; comp. = composition; sustain. = sustainable

Table 3: Guidelines to relative assessment and selection of different management interventions. Note: except where the context dictates otherwise, effects are deemed to be those applying to animals that stay behind post-intervention

Trace influences and feedbacks by keeping figure 3 in mind, so that your overall decision will more likely be balanced and effective, in both the short and longer term. You may need to add peculiarities or particular details to figure 3, in line with your specific context. To achieve this in practice you may want to:

1. Check whether you are satisfied that all the influences feeding forward into the decision blocks of interest to you have been taken into account. For instance, has the prospect of a future eruptive population been factored into a decision to cull, or has lost revenue for joint owners (claimants) been considered in a decision to contracept? In the first example, a legacy value is being evoked, and in the second a sustainable use value.
2. Check whether the decisions you are taking, or the information in certain blocks, is in fact feeding back into the areas where people can learn, and thus hopefully produce better decisions in future. There are often natural lags in societal responses, and you may need to be persistent in helping promote or 'market' such feedbacks so that they actually do eventually improve learning. Do not be disappointed if the first, second or even later attempts appear to 'fall on deaf ears': this is normal – there is usually a premium on inertia in society, and adaptive management very often leads to change. As an example of feedback from the implantation to the strategy box, if the proposed monitoring for a particular indicator is too complex, too expensive, or found to not reflect the underlying element of concern, this should necessitate a revision of that particular threshold and associated procedures, or should result in it being removed or replaced. As another example of feedback from implementation to drivers, we might find that culling by helicopter stresses elephants in the long term, and that there may be a change to ground culling in response to an animal welfare value.

WORKED EXAMPLES

The process described in this chapter may seem fine in principle, but raises the question 'how exactly does this work in practice?' This assessment avoids quoting prescriptive figures as direct guidelines, such as numbers of plants damaged, or (less commonly used nowadays) densities of elephants, because objectives differ widely, as do the different landscapes and situations in which decisions must be taken. In addition, a particular institution will monitor their adaptive goals in a particular way, and may not be able to employ a method used by another property or agency. In other words, one size does not fit all, a principle recognised in the Norms and Standards for elephant management (DEAT, 2008). However, we provide four worked examples from widely varying situations to show the application in practice. These contain an illustrative set of actual figures of the targets set to achieve the goals of those situations. They are not intended to be directly for use elsewhere, at least not without very careful thought and possible modification.

The four examples are selected to illustrate a range of ecosystems and management objectives. Kruger represents a large savanna national park without many sensitive endemic species, with established ecosystem-level biodiversity goals, but also goals for tourism and the maintenance of wilderness areas. Tembe Elephant Park is a small to mid-sized provincial protected area with elephants as a major tourist attraction, but that also contains patches of narrowly endemic plant communities actively used by elephants. Madikwe represents a mid-sized provincial protected area with clear job creation and financial objectives, based on maximum development of nature-based tourism. Balule Nature Reserve within the Associated Private Nature Reserves, west of central Kruger Park, represents a small to mid-sized privately owned protected area, based on tourism and recreational objectives.

In all cases there are stakeholder-based processes (varying from widely participative in the case of Kruger, to mainly internal and implicit in the case of Balule) which guide the choice of targets and indicators to be used. These target or threshold values drive elephant management by triggering response actions, hopefully in a fully adaptive process characterised by the feedbacks described in this assessment (see figure 6 in the Summary for Policymakers).

Three typical thresholds used in Kruger have been selected from their longer list, and are depicted in figure 4a. One deals with loss of the least-common plants in each landscape, one takes care of major shifts in herbivore dominance (possibly grazer-browser shifts due to changing vegetation), and the third measures the loss of large trees. Unless the given thresholds are exceeded (or are likely to be exceeded in the near future) the ecosystems are left alone. This allows flux or dynamism in the system, an approach that now de-emphasises the Precautionary Principle (Cooney, 2004) which could previously, albeit wrongly, be argued in support of opposing ends – elephants or large trees. Management of ecosystem drivers may have to be instituted many years before unacceptable thresholds are crossed, making these important long-term management decisions in the park. An important amendment to the Kruger thresholds has been the consideration of the Kruger biodiversity targets in the context of the regional biodiversity targets established by the South African National Biodiversity Institute (SANBI). There is some residual debate as to whether, in an *a priori* sense, to actively manage towards low and high impact zones (including elephant impacts), or to allow these zones to emerge through management actions resulting only from predicted exceeding of thresholds set differentially after taking the SANBI information into account. The more likely latter route (the one depicted in the figure) is consistent with broader threshold-based management, while the former would require a procedural modification supporting highly active adaptive experimentation and possibly quicker learning. At the time of writing, only the large tree threshold has been exceeded in parts of the park. The impact of elephants on large trees is receiving much attention in terms of modelling, in order to understand the trajectories better.

The loss of Sand Forest canopy in Tembe (figure 4b) is already close to the threshold, requiring urgent management. In Madikwe (figure 4c), management concerns are driven mainly by the desire to minimise incidents involving elephants that place tourists at risk. Exceeding this threshold also precipitates immediate action.

In Balule (figure 4d), a sophisticated equilibrium approach is used to bring down herbivore biomass (currently dominated by elephants, but including a range of other mammals) whenever the threshold level is approached or exceeded. The calculations involve useful energy flows into a system minus a certain fraction that is reduced by overheads (the so-called 'environmental loading' (EL)). This is subtracted from the metabolisable energy of the total amount of measured forage, taking into account seasonal variation and proportion of forage actually available to animals. For equations and details, see Peel (2005). The approach is based on the philosophy of managing for a productive system rather than commodities within a system, by managing the context (Allen *et al.*, 2003), and may be especially appropriate to systems which have been re-scaled by humans through erection of fences and supply of extra water points.

Currently, park management plans are pending approval in a new process under the Protected Areas Act. For more information on each of the four examples, consult the plan (once available), relevant management authority, or person entitled to speak on their behalf in this regard.[1]

A

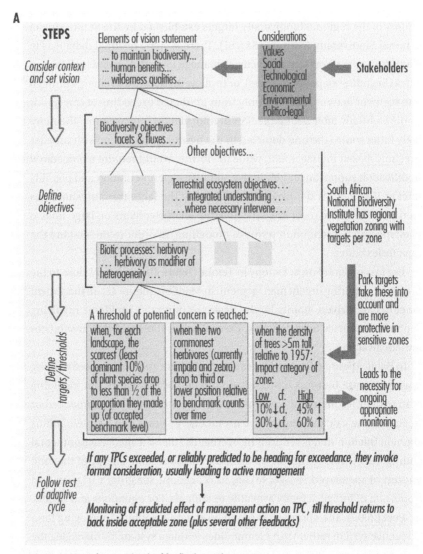

See Figure SPM.6 for more details of feedbacks in adaptive process

Figure 4: Graphic schemas of worked examples of elephant-related decision making in four different cases, showing derivation and type of objectives through to actual numeric examples of thresholds or targets. The localities are (A) Kruger National Park, (B) Tembe Elephant Park, (C) Madikwe Game Reserve, and (D) Balule Nature Reserve (BNR). TPC = Threshold of potential concern. In (D) EL = environmental loading, described in text

B

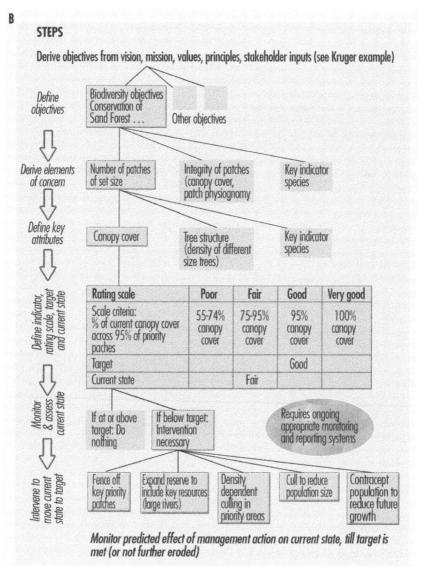

STEPS

Derive objectives from vision, mission, values, principles, stakeholder inputs (see Kruger example)

Define objectives

Biodiversity objectives
Conservation of
Sand Forest . . .   Other objectives

Derive elements of concern

Number of patches of set size

Integrity of patches (canopy cover, patch physiognomy

Key indicator species

Define key attributes

Canopy cover

Tree structure (density of different size trees)

Key indicator species

Define indicator, rating scale, target and current state

Rating scale	Poor	Fair	Good	Very good
Scale criteria: % of current canopy cover across 95% of priority patches	55-74% canopy cover	75-95% canopy cover	95% canopy cover	100% canopy cover
Target			Good	
Current state		Fair		

Monitor & assess current state

If at or above target: Do nothing

If below target: Intervention necessary

Requires ongoing appropriate monitoring and reporting systems

Intervene to move current state to target

Fence off key priority patches

Expand reserve to include key resources (large rivers)

Density dependent culling in priority areas

Cull to reduce population size

Contracept population to reduce future growth

Monitor predicted effect of management action on current state, till target is met (or not further eroded)

Follow rest of adaptive cycle (see Figure 6 in the Summary for Policymakers for more details)

C

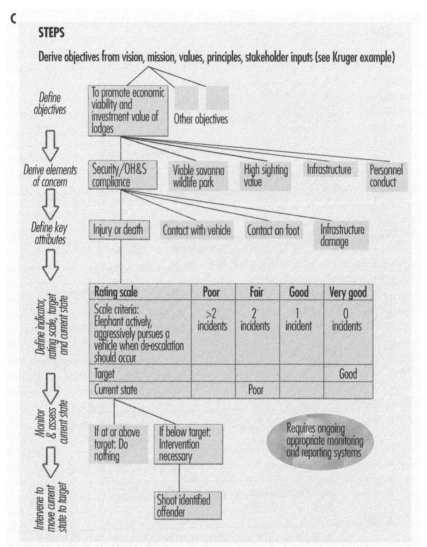

STEPS

Derive objectives from vision, mission, values, principles, stakeholder inputs (see Kruger example)

Define objectives
To promote economic viability and investment value of lodges   Other objectives

Derive elements of concern
Security/OH&S compliance   Viable savanna wildlife park   High sighting value   Infrastructure   Personnel conduct

Define key attributes
Injury or death   Contact with vehicle   Contact on foot   Infrastructure damage

Define indicator, rating scale, target and current state

Rating scale	Poor	Fair	Good	Very good
Scale criteria: Elephant actively, aggressively pursues a vehicle when de-escalation should occur	>2 incidents	2 incidents	1 incident	0 incidents
Target				Good
Current state		Poor		

Monitor & assess current state
If at or above target: Do nothing   If below target: Intervention necessary   Requires ongoing appropriate monitoring and reporting systems

Intervene to move current state to target
Shoot identified offender

OH&S = occupational health and safety
Follow rest of adaptive cycle (see Figure 6 in the Summary for Policymakers for more details)

D
STEPS

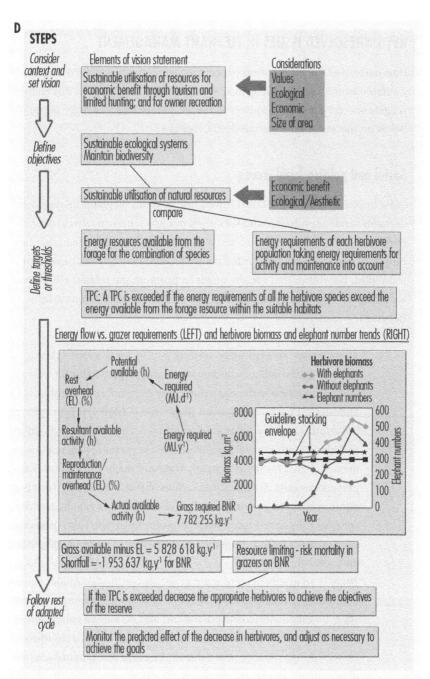

Follow rest of adaptive cycle (see Figure 6 in the Summary for Policymakers for more details)

KEY UNRESOLVED ISSUES IN ELEPHANT MANAGEMENT

Comparison of the information concerning elephant decision making made available by the assessment against the framework suggested in this chapter reveals several gaps and unresolved issues. References to Kruger are made below on the basis of the case study in Chapter 1.

Social and politico-legal issues

Most important and urgent unresolved issues about elephant management decision making revolve around social and politico-legal concerns in adaptive management rather than environmental ones. This does not mean that new biophysical or technological information around elephant management is unhelpful, but rather that society needs to promote the related ethical and social aspects and in that way 'catch up' and enable overall decisions to be more balanced.

1. There exists an extremely serious gap in legal terms relating to ownership of wildlife, including elephants, responsibility for wildlife, and compliance with regulations, affecting not only elephants in and around parks, but species in general (Chapter 11). In spite of the comprehensive revamp of so much South African legislation, including environmental legislation, these key areas appear not to have been screened for revision in any way, and major uncertainties and perversities exist. An urgent and serious rework is necessary if our best efforts are not to be unexpectedly thwarted by this gap. By contrast, some well-intended recent legislation for threatened and protected species poses serious impediments (Chapter 11) to achieving these and other objectives.

2. There is an urgent need to more explicitly clarify stakeholder values (Cumming & Jones, 2005) in relation to assessing society's 'desired state' for elephants in ecosystems and among human communities. South Africa has made advances in this regard inasmuch as the new Protected Areas Act requires values to be elicited as part of each park plan, so that all parks with elephants (that have so far submitted plans) have at least confronted this explicitly for the first time.

3. We need a better understanding of the consequences of espoused versus revealed values. This is apparent from the case history of seemingly paradoxical behaviour in the history of Kruger

management, where for example, managers spoke of 'minimum interference' but often made significant interventions, and talked about the overriding ecological imperative but then let tourism interests override in practice. These are understandable and very human responses (and very generic elsewhere), and this critique is only possible because persons in the Kruger system had the transparency to document intentions, decisions and outcomes, and to try to understand why certain stakeholders felt this was paradoxical. With the implementation of the Protected Areas Act such transparency will become the order of the day, and we should not be surprised to find such alleged discrepancies more generally. The concept of 'mental models' (Abel *et al.*, 1998) is an important aid in assessing this, rather than resorting to additional auditing. Clarity of setting objectives, and a clear and shared understanding of contrasting issues, is paramount.

4. The elephant-ecosystem-human interaction is more complex than is allowed for by the simpler models. Management approaches often still seem to be more about perpetuating or conserving the status quo (as captured by the commonly cited ranger mandate of 'maintaining territorial integrity' that is certainly an important issue in its own right), and less about learning how to change to adapt in a bigger system which itself is changing. These contrasting layers of thought need to be internalised and somehow practised together. This requires ongoing changes in attitude (Brock & Salerno, 1998) if elephant management is not to run into the same problems as in the past and lock itself into another impasse.

5. Change management: changes in policy often proved gradual, with overlapping (contradictory) statements during transitional phases which introduced the new idea but left enough of the old to help laggards adapt, or to temper early adopters. The relative strength of early adopters versus laggards needs to be managed for successful transitions, not necessarily always in favour of the fastest change. Such inertia in Kruger appears to have 'held back' the elephant management policy by about a decade relative to other major policies, though this may also have been due in part to the very wide range of stakeholders and ethical viewpoints.

6. Differing attitudes towards external scientists and ideas are evident in agencies, and the same splits may even be seen within the staff of one agency. In Kruger there has been a major drive to engage

in partnerships with outside scientists during the last decade. This happened to a far lesser extent regarding elephant ecologists, a situation that has started changing since the Great Elephant Indaba (SANParks, undated) and Luiperdskloof meeting (Grant, 2005). Elephant decision makers can materially assist the evolution of such science-academic partnerships, a fuller subject in its own right. The challenges establishing such a relationship should not be used as an excuse for inaction.

7.   The 'moral dilemma' over elephant management has evolved into a very complex circumstance, largely due to the continued addition of new perspectives as more stakeholders participate. We may therefore now have the opportunity to simplify the 'moral dilemma' to its key attributes, and need to consider whether this is achievable. Alternatively, we need to better understand that the persistent moral dilemma in particular may cast the problem as a 'wicked problem' (Conklin, 2006), one with no definitive solution in which society 'muddles along' as the problem evolves.

Monitoring for adaptive management

Thresholds or targets and the indicators and monitoring programmes that support elephant management are in an ongoing evolution to meet the demands of adaptive management, and significant financial and moral support is needed from broader society if this is to succeed. This is part of a more general issue, but one in which elephants, wherever present, are key elements. This aspect of servicing monitoring (that has both biophysical and social science components) is a key bottleneck.

1.   A variety of slightly differing approaches (e.g. thresholds of concern, process-based management targets, etc.) are in use by different agencies, and these are often seen as in competition with each other. This assessment suggests that in fact these are more similar than different, and in any case provide important alternative learning paths that should co-evolve. To promote this, more sharing of knowledge and results, and some harmonisation of vocabulary or jargon, is essential. This is often more challenging across international boundaries.

2.   The fact that elephants provide such a variety of effects at different scales and under different circumstances means that not only research, but in particular adaptive monitoring will be essential

if this complexity is ever to be unravelled fast enough to manage elephants effectively. In line with the general adaptive approach prescribed under the Protected Areas Act, this requires effective ongoing resourcing that appears not to be in place, especially not for monitoring. Such a successful thrust is likely, within a decade or two, to move understanding of elephant effects from the contentious state it is currently in, to one of limited consensus (based on expert opinion) and later to one of wider consensus based on factual evidence. A clear exposition of research gaps in terms of elephant effects on ecosystems is given in Chapter 3.

Effect of interventions

Apart from studies on side-effects of contraception (more of which are required), there has been little interest in understanding the effect of interventions. As far as the ethical basis of interventions and their consequences is concerned, only culling has received some attention. This assessment highlights these as significant practical gaps.

1. The legacy consequences of current and past intervention (including culling) need to be clarified and resolved, particularly regarding any residual long-term stress and population effects on current populations of elephants.
2. Ongoing studies are needed on potential side-effects of several promising novel interventions, such as immuno-contraception, and the creation of relatively large fenced exclosures that are permeable to species other than elephants.
3. Ethical consequences of the full range of interventions need as much study and discussion as culling receives.

Economic gaps

There is a dearth of economic information on elephants in South Africa.

1. Economic studies, appropriately done, have major implications for elephant decisions, especially in the crucial area of livelihoods of the rural poor, and generally in land-use and resource allocation decisions. Investment in wildlife often proves to be economically efficient, and achieves upliftment and conservation.

2. National and subcontinental issues need to be weighed up against wider, and in practice often contrasting, demands such as those emanating from central African elephant conservation requirements. Central and southern African elephant management trajectories have differed markedly (Chapter 1). The bans and donor subsidies that are instrumental in maintaining the East African system are unlikely to be sustainable in the longer run, particularly in the face of growing Asian economic influences (Chapter 10). Therefore it may be a better long-term strategy to start introducing markets into the central African system, than to expect southern Africans to continue limiting their choices in sympathy to central African needs. Such an altered mechanism will require altering existing CITES agreements (Chapter 11).

Bioregionalism

Greater bioregional emphasis is a reality for most decision makers, and the implications of this for elephant management (land-use change, fences, corridors, etc.) need to be both studied and learnt about in practice.

Technological innovations

By their very nature, these can be unpredictable and can make a big impact on possibilities for elephant management.

The possibility of intrinsic behavioural modification of free-ranging elephants has never been examined, but promising results at practical scales, based on well-understood scientific rationales, are now being reported from other wild herbivores (Chapter 7). In principle, being intelligent animals, such approaches should be applicable to elephants.

CONCLUSION

This chapter has attempted to bring together the diverse ideas around elephant management, using an integrative framework. It has suggested a set of approaches that may materially assist decision makers working in this tricky area. Contention may persist, but the Assessment offers structured defensible processes to follow in reaching decisions. This Assessment can be expected to promote a partly shared rationale, or at the very least some empathy, among parties with differing viewpoints. It carries the longer-term possibility of the problem solutions converging to some extent, as depicted by the arrow in figure 1.

Our synthesis confirms that the adaptive approaches suggested in DEAT Norms and Standards (2008) appear to offer a sensible way forward. It must be remembered that the discipline of adaptive management itself is evolving fast, showing improvements over earlier versions. This development needs to be supported. If monitoring under this mantra of adaptive management can be funded, and carried out to confirm the ideas generated under differing and clearly set objectives for management, we should generate a stream of learning that enables us to justify an ever-improving basis for managing these intriguing animals. Ongoing change in approaches can be expected because of the evolution of broader societal drivers, especially as values shift. Greater clarity is urgently required around elucidating current and evolving values, as these turn out to be pivotal in deciding on how elephants should be managed. Indeed, society may turn out to be working towards a new and very different 'social contract' with elephants. Moral pluralism is currently advocated because of widely varying values and needs in different circumstances. Major and potentially very deleterious gaps have been discovered in legislation relating to species (including elephants) in and around parks, and this shortfall will need to receive urgent attention.

Economic studies offer particular hope for understanding and influencing land-use change as related to elephants, in particular as this relates to the welfare of poor rural communities.

Social attitudes and constructs amongst stakeholders concerned with elephant management should be influenced in ways that allow greater sharing of information and values, but also allow for the promotion of moral pluralism. Polarisation of the kind that characterised the 'culling versus anti-culling debate' led to stalled options, and to unsatisfactory progress in adaptive learning and ecosystem management.

ACKNOWLEDGEMENTS

We are grateful to Kevin Rogers, Stefanie Freitag-Ronaldson, Peter Novellie, and the many reviewers and editors in the Assessment who helped make this a more effective synthesis and a clearer chapter.

ENDNOTE

1. Contacts for more information on each of the four examples in figure 4:
    Kruger: draft revised plan at http://www.sanparks.org or via Dr Stefanie
    Freitag-Ronaldson at stefanief@sanparks.org
    Tembe: Wayne Matthews at waynem@icon.co.za
    Madikwe: Pieter Nel at hpnel@mweb.co.za
    Balule: Mario Cesare at olireserve@worldonline.co.za or Dr Mike Peel at
    mikep@arc.agric.za

REFERENCES

Abel, N., H. Ross & P. Walker. 1998. Mental models in rangeland research, communication and management. *Rangelands Journal* 20(1), 77–91.

Allen, T.F.H., J.A. Tainter & T.W. Hoekstra. 2003. *Supply-side sustainability.* Columbia University Press, New York.

Biggs, H.C. & K.M. Rogers. 2003. An adaptive system to link science, monitoring and management in practice. In: J. du Toit, K.M. Rogers & H.C. Biggs (eds). *The Kruger experience: ecology and management of savanna heterogeneity.* Island Press, Covelo, 59–80.

Brock, L.R. & M.A. Salerno. 1998. *The secret to getting through life's difficult changes.* Bridge Builder Media, Washington DC.

Brockington D. 2002. *Fortress conservation: the preservation of the Mkomazi Game Reserve, Tanzania.* Indiana University Press, Bloomington.

Campbell, D.J. & J.M. Olson. 1991. Framework for environment and development: the kite. *Occasional Paper of Michigan State University* 10, 30.

Caughley, G. 1981. Overpopulation. In: P.A. Jewell, S. Holt and D. Hart (eds). *Problems in management of locally abundant wild mammals.* Academic Press, New York, 7–19.

Conklin, J. 2006. *Dialogue mapping: building shared understanding of wicked problems.* Wiley, New Jersey.

Cooney, R. 2004. *The precautionary principle in biodiversity conservation and natural resource management: an issues paper for policy-makers, researchers and practitioners.* IUCN, Gland, Switzerland and Cambridge, UK.

Cracraft, J. & F.T. Grifo (eds). 1999. *The living planet in crisis. Biodiversity science and policy.* Columbia University Press, New York.

Cumming, D. & B. Jones. 2005. Elephants in Southern Africa: management issues and options. *WWF–SARPO Occasional Paper* 11. WWF–SARPO, Harare.

DEAT 2008. Draft National Norms and Standards for the management of elephants in South Africa. Department of Environment and Tourism. *Government Gazette*, 8 May 2008.

Gillson, L. & K.I. Duffin. 2007. Thresholds of potential concern as benchmarks in the management of African savannahs. *Philosophical Transactions of the Royal Society of London Series B-Biological Sciences* 362, 309–319.

Grant, C.C. 2005. Elephant effects on biodiversity: an assessment of current knowledge and understanding as a basis for elephant management in Sanparks: a compilation of contributions by the scientific community for SANParks. *Scientific Report 3/2005*. Skukuza, Scientific Services.

Hayman, P.T. 2004. *Decision support systems in Australian dryland farming: a promising past, a disappointing present and uncertain future.* Proceedings of the 4th International Crop Science Congress. 'New directions for a diverse planet', 26 Sept–1 Oct 2004, Brisbane, Australia. Published on CD-ROM. Web site: http://www.cropscience.org.au.

Hockings, M., S. Stolton & N. Dudley. 2000. *Evaluating effectiveness: a framework for assessing management of protected areas.* IUCN, Cardiff

Holling, C.S. 2001. Understanding the complexity of economic, ecological and social systems. *Ecosystems* 4, 390–405.

Hunter, M.L. Jr. 1991. Coping with ignorance: the coarse filter strategy for maintaining biodiversity. In: K.A. Kohm (ed.). *Balancing on the brink of extinction: the Endangered Species Act and lessons for the future.* Island Press, Covelo.

Kinnaman, M.L. & M.R. Bleich. 2004. Collaboration: aligning resources to create and sustain partnerships. *Journal of Professional Nursing* 20(5), 310–322.

Levin, S. 1999. *Fragile dominion: complexity and the commons.* Helix Books, Boston, Massachusetts.

Mackey, R.L & D.J.Currie. 2001. The diversity-disturbance relationship: is it generally strong and peaked? *Ecology* 82(12), 3479–3492.

Maguire, L.A. & E.A. Albright. 2005. Can behavioural decision theory explain risk-averse fire management decisions? *Forest Ecology and Management* 211, 47–48.

McNeely, J.A. 1993. (ed.) *Parks for life: Report of the Fourth World Congress on National Parks and Protected Areas*, 10–21 February 1992. IUCN Protected Areas Programme; WWF. Gland.

Meffe, G.K., L.A. Nielsen, R.L. Knight, & D.L. Schenborn. 2002. *Ecosystem management. Adaptive community-based conservation.* Island Press, Washington.

Noss, R.F. 1990. Indicators for monitoring biodiversity: a hierarchical approach. *Conservation Biology* 4, 355–364.

Olsson, P., L. Schultz, C. Folke & T. Hahn. 2003. Social networks for ecosystem management: a case study of Kristianstads Vattenrike, Sweden. Available at: http://ma.caudillweb.com/documents/bridging/papers/olsson.per.pdf.

Palmer Development Group. 2004. *Development of a core set of environmental performance indicators.* Final report and set of indicators. Department of Environmental Affairs and Tourism, Pretoria. Available at: http://soer.deat.gov.za.

Peel, M.J.S. 2005. Towards a predictive understanding of savanna vegetation dynamics in the eastern Lowveld of South Africa: with implications for effective management. Ph.D. thesis. University of KwaZulu-Natal.

Pfeffer, J & R.I. Sutton. 2000. *The knowing-doing gap: how smart companies turn knowledge into action.* Harvard Business School, Massachusetts.

Rogers, K.H. 2005. *Biodiversity custodianship in SANParks: a protected area management planning framework.* Report to South African National Parks, Pretoria.

SANParks. Undated. *The Great Elephant Indaba. Finding an African solution to an African problem.* Minutes of a meeting held at Berg-en-Dal. 19–21 October 2004. SANParks, Pretoria.

GLOSSARY

Active adaptive management: see Adaptive management

Adaptive management: integrates research, planning, management, and monitoring in repeated cycles of learning how to better define and achieve objectives. It is built on the assumption that natural systems are complex, our knowledge is imperfect but we can learn from purposeful, documented objectives and actions (Rogers, 2005). Active adaptive management is characterised by testing and investigations into how a system functions. In strategic adaptive management future objectives are set; these objectives are expected to change with increasing knowledge of the system in question.

Allometry: relative changes in proportions, of morphological body parts or physiological measurements. Changes can occur during the evolution of the species or in the growth of the individual.

Animal rights (see also **Animal welfare**): the viewpoint that nonhuman animals are entitled to certain basic rights and should not be used by humans or regarded as their property.

Animal welfare (see also **Animal rights**): the viewpoint that humans may use nonhuman animals for food, clothing, entertainment, and in scientific research so long as unnecessary suffering is avoided.

Anthropocentric (the antonym is **biocentric**): value orientation which focuses on human uses and benefits from nature as defined by the Rio Declaration on Environment and Development in 1992.

Assemblage: see Species assemblage

Biocentric or **ecocentric** (see also **anthropocentric**): value orientation which considers society as part of nature and emphasises the non-use values of biodiversity. Includes both the traditional African world-view and more western models.

Biodiversity (a contraction of **biological diversity**): the full range of natural variety and variability within and among living organisms, and encompassing multiple levels of organisation, including genes, species, communities, and ecosystems. Noss (1990) developed a definition of biodiversity which includes composition, structure, and function: 'composition has to do with the identity and variety of elements in a collection, and includes species lists and measures of species diversity and genetic diversity. Structure is the physical organization or pattern of a system, from habitat complexity as measured within communities to the pattern of patches and other elements at a landscape scale. Function involves ecological and evolutionary processes, including gene flow, disturbances, and nutrient cycling.'

Bottom-up control (see also **Top-down control**): the ecological scenario in which abiotic resources and primary productivity control the dynamics and processes within a community or ecosystem.

Browse (see also **Graze**): verb – to eat leaves, buds, twigs, shoots of trees and shrubs; noun – woody vegetation.

Browsing-lawn: a short-grass patch created in response to grazing pressure.

Bull: adult male elephant.

Bush encroachment: an increase in the relative dominance of woody plants in a savanna or grassland.

Calf: young elephant, generally less than 4 years old.

Calving interval: the time that elapses between births.

Carrying capacity: the maximum population of a species that can be sustained in a specific area.

CITES: the Convention on International Trade in Endangered Species of Wild Fauna and Flora is an international agreement between governments signed in 1975. Its aim is to ensure that international trade in specimens of wild animals and plants does not threaten their survival (see www.cites.org).

Command-and-control: a management approach in which protected area managers attempt to stabilise, maintain, and engineer the ecosystems they manage. Also known as 'management by intervention'.

Complexity: deals with partly and often poorly predictable patterns in systems. This uncertainty is pervasive in ecosystems and social-ecological systems. Complexity involves the study of linkages between system components/processes, and the feedbacks which these generate, which in turn cause trajectories into differing system states separated by so-called thresholds, invariably characterised by lags and emergence of interactions across scales.

Confidence limits: an upper and lower statistical value which reflect the probable range in which the true value lies. When confidence limits are apart, estimates are imprecise.

Congeners: refers to species belonging to the same genus.

Conspecific: refers to individuals of the same species.

Consumptive use: the reduction in the quantity or quality of a good available for other users due to consumption (MA, 2005).

Coppice: re-growth of damaged woody vegetation.

Cow: adult female elephant.

Culling (Definitions from the National norms and standards for the management of elephants in South Africa, DEAT 2008):

a. in relation to an elephant in a protected area or on a registered game farm, an operation executed by an official of, or other person designated by, the responsible person to kill a specific number of elephants within the area in order to manage elephants in the area in accordance with the management plan of the area; or

b. in relation to an elephant that has escaped from a protected area and has become a damage causing animal, an operation executed by an official of, or a person designated by the issuing authority to kill the elephant.

Damage causing animal (National norms and standards for the management of elephants in South Africa, DEAT 2008):
refers to an individual elephant that

a. has caused and threatens to cause losses to stock or to other wild specimens;

b. has caused and threatens to cause excessive damage to cultivated trees or crops or natural flora or other property;

c. presents an imminent threat to human life; or

d. alone or in conjunction with other elephants is materially depleting agricultural grazing.

Decision maker: a person whose decisions, and the actions that follow from them, can influence a condition, process, or issue under consideration (MA, 2005).

Density dependence: an effect on either the birth or death rate in a population that is sensitive to the number of animals in the population per unit area, for instance, a birth rate that declines as the population size increases, or a death rate that goes up as the population size increases.

Ecological climax: a theory that the end point of the process of succession is a relatively stable community of predictable species that is in equilibrium with environmental conditions.

Ecosystem: a dynamic complex of plant, animal, and microorganism communities and their non-living environment interacting as a functional unit (MA, 2005).

Ecosystem process: refers to a physical, chemical or biological action or events (or series thereof) that link organisms to one another and their environment.

Environmental indicator: a parameter which signifies the condition of the environment or the impact of a perturbation or disturbance on a system.

Eruptive growth: the rapid growth of a consumer population in a system, until some peak density is reached, followed by a period of rapid decline due to the large discrepancy between available resources and consumer density.

Extinction: the irreversible condition of when a species or genus is no longer in existence anywhere in the world (as compared to extirpation).

Extirpation: the loss of a local population although the species still lives elsewhere (as compared to extinction).

Facilitation: an interaction between two species in which one or both benefit but neither is harmed. In a mutualistic interaction both species benefit. In commensalism one species benefits and the other is unaffected.

Family unit: related adult females and their immature offspring.

Fitness: the reproductive success of individuals of a particular genotype. This is a relative measure, calculated relative to the other genes or organisms that are present in the population.

Forage: noun – the plant material eaten by grazing animals; verb – the act of searching for food.

Gestation period: period between conception and birth of a calf. Gestation lasts 22 months in African elephants, which accounts for approximately 50 per cent of the intercalving period. This means that for a period of up to two years after calving cows do not show an oestrous cycle.

Gonadotropin releasing hormone (GnRH): a hormone produced by the hypothalamus: in the brain. GnRH binds to the pituitary gland and stimulates it to produce luteinising hormone (LH) and follicle stimulating hormone (FSH) and therefore controls the functioning of the ovaries.

Graze (see also **Browse**): to eat grass, forbs, etc. (i.e. herbaceous).

Heterogeneity: a measure of the diversity and variability of parts or processes within a system. Spatial heterogeneity refers to the diversity of parts, usually called 'patches' (habitat/vegetation class), within a defined area. Temporal heterogeneity refers to the diversity between different parts within an area over time. Pattern is a key component of heterogeneity.

Hierarchical patch dynamics: a paradigm for viewing ecological systems which includes spatial and temporal dynamics and the explicit linkage between scale and heterogeneity. Ecosystems are seen as consisting of a hierarchy of 'nested' patches of resources which occur in mosaics; these patches change in time and space. Both environmental stochasticity and biotic feedback interactions can cause instability and contribute to the dynamics observed at various scales (Wu & Loucks, 1995).

Hindgut fermenter: non-ruminant herbivore, such as elephant, in which breakdown of cellulose occurs in the caecum and large intestine.

Home range: the home range of an elephant represents the area it traverses in its normal activities of food gathering, mating, and caring for young. Home ranges can be measured at various time scales (e.g. monthly, seasonally, annually), and provide a measure of elephant spatial use in relation to various biotic and abiotic factors. Rainfall apparently plays an important role in determining home range size and location.

Human–elephant conflict(HEC): situations where elephants and humans come into conflict, e.g. crop-raiding, attacking livestock or humans.

Hypothalamus: a small and important organ in the centre of the brain that interprets signals from the environment and controls body temperature, breathing, heart rate and reproduction. Emotional signals are translated into hormonal and other changes and brain signals.

Immunocontraception: a method of using an elephant's immune response to reduce fertility by controlling or preventing conception and pregnancy.

In situ: in the original place.

Inherent value: refers to a value that exists as an intrinsic characteristic of a thing simply due to its existence and independent of its usefulness to humans.

Inter-calf interval: interval between births by a given female; in elephants usually ranges from 3 to 9 yrs depending on environmental conditions.

Inter-musth: periods between bouts of musth, shown primarily by younger (25–35 years) males, which seem to go in and out of musth more than older males.

Jacobson's organ: area located in the roof of the mouth that is sensitive to olfactory cues, especially those associated with urine.

Juvenile: sub-adolescent individual; in elephants this is often divided into young juvenile (2–5 years old) and old juvenile (5–10 years old).

Keystone species: A species that has major ecological effects on its habitat and, therefore, on other species living in the same area.

!Kung: a southern African people living in the Kalahari Desert in Namibia, Botswana and in Angola who traditionally followed a hunting and gathering lifestyle.

Landscape: an area of land that contains a mosaic of ecosystems, including human-dominated ecosystems (MA, 2005).

Landscape Functionality Index: a simple indirect measure of the change or degradation in landscape function. The index is derived following the methodology of landscape function analysis (LFA) (Ludwig *et al.* 1997).

Life history: the complete suite of traits that an organism has; it may change through the organism's life.

Matriarch: mature female who acts as leader of a family unit; typically the eldest and most experienced individual in the group.

Megaherbivore: a terrestrial herbivore that attains an adult body mass in excess of 1 000 kg. This group includes elephants, rhinoceroses and hippopotamuses.

Metapopulation (see also **Source-sink**): a group of spatially separated populations of the same species which interact through dispersal. A metapopulation is considered more stable than one single population.

Mortality: referring to the death-rate, or loss, in a population; includes factors such as disease, accidents, starvation and predation.

Musth: period of heightened sexual activity in mature elephant males characterised by urine-dribbling, strong odour, increased aggression, swollen temporal glands, temporal gland secretion, and elevated testosterone levels.

Oestrus: a phase of the reproductive cycle which occurs in sexually mature elephant females; associated with ovulation and the time that conception is most likely to occur. Only during this period do elephant females permit copulation by males; they usually show a preference for older males, especially those in musth. Oestrus cycles in elephants are approximately 12–17 weeks, and last 2–6 days.

Overpopulation: when an organism's numbers exceed the carrying capacity or the stocking density of the area.

Pachyderm: historical name for large, thick-skinned hoofed mammal (e.g. elephant, rhino). This is not a taxonomic grouping.

Piosphere: area around a water source that shows ecological effects (i.e. damage), owing to proximity to that water source.

Policy maker: a person with power to influence or determine policies and practices at an international, national, regional, or local level (MA, 2005).

Population (biological): a group of individuals of the same species occupying a defined area. All the elephants in a region, including sub-populations of females and their offspring, plus the adult males; they may all have some contact with each other, especially during the wet season when large aggregations may form.

Population (human): a group of people in a given area.

Population density: average number of elephants per unit area in a region; usually given as elephants.km^{-2}.

Precautionary principle: this is an approach to uncertainty, it provides for action to avoid serious or irreversible environmental harm in advance of scientific certainty of such harm (Cooney, 2004).

Proboscidean: a member of the order Proboscidea. Refers to elephants and elephant relatives with a long, flexible snout, such as a trunk.

Recent: the present period, or the Holocene (the past 12 000 years).

Recruitment: increase in a population, usually as the result of births exceeding deaths; may also be augmented by immigration.

Refugia: locations where conditions for organism survival are maintained, and where species can persist.

Res alicuius: belonging to someone.

Res nullius: belonging to no one.

Res omnium communes: common to all but belonging to no one.

Res publicae: in public ownership.

Res singulorum: belonging to an individual.

Res universitatis: belonging to corporate bodies

Ringbarking: the complete removal of a strip of bark around the outer circumference of a tree, usually leading to the death of the tree.

Rogue: vernacular term for a particularly aggressive and dangerous elephant, most often a bull. The animal may be sick or injured and may also be in musth.

Ruminant: a mammalian herbivore with a four-chambered stomach (called a rumen) and even-toed hooves. Food is partially digested in the rumen and regurgitated for additional chewing, thus a ruminant is referred to as an 'animal that chews its cud'.

Scientific Round Table (or **SRT**): a panel consisting of 18 internationally recognised elephant scientists, convened in January 2006 by South African minister of environmental affairs and tourism, Marthinus van Schalkwyk, to advise on policies regarding elephant management.

Scoline: a compound formerly used during the culling of elephants. Scoline was used to render the target elephant immobile before a brain shot was administered.

Senescence: deteriorating physical condition, owing to old age.

Sexual dimorphism: physical or behavioural differences between the sexes.

Source-sink (see also **Metapopulation**): a model used to describe how variation in habitat quality may affect the population growth. Source patches are high-quality habitat in which a population will increase. Sink patches are low-quality habitat that would not be able to support a population. When an excess of individuals occurs in the source area, individuals move to the sink, allowing the sink population to persist.

Species assemblage: a set of species co-occurring in a particular area.

Species Survival Commission: an IUCN commission tasked with gathering information regarding the current status of plant and animal species. The Species Survival Commission consists of specialist groups, each of which focuses on a specific taxon.

Stocking density: the number of animals per unit area of managed land.

Strategic adaptive management: see Adaptive management.

Sub-population: families that occupy distinct dry-season home ranges but may mix freely in the wet season.

Succession: (see Ecological climax) the process by which organisms replace each other over time at the site.

Succinylcholine chloride: see **Scoline**

Sustainable use: the use of natural resources in such a way that: 1) populations of the species are biologically viable for the long term, 2) declines in biodiversity are avoided. Use such that the potential to meet the needs and aspirations of future generations is not compromised.

Taxonomy: branch of science dealing with the classification of organisms.

Temporal gland: a gland located midway between the eye and ear, which resembles salivary tissue and secretes temporal gland secretion. It swells significantly in males during musth.

Temporal gland secretion: there are two types of secretions:1) a watery and short-term secretion in males, females and young which signifies social excitement or stress; or 2) a more viscous and durable secretion which occurs in males only and which signifies musth.

Thresholds of Potential Concern: the upper or lower limit for an environmental indicator which triggers for decision-making when it is reached. Two actions can occur: (1) a management intervention to moderate the cause of the exceedance or (2) a recalibration of the threshold.

Top-down control (see also **Bottom-up control**): the ecological scenario in which ecosystem processes or dynamics are determined by predators or herbivores.

Translocation: the process of capture, transportation, and release of animals into a new area.

Ungulate: a hoofed animal.

Urine-dribbling: leakage of urine from the sheathed penis, shown by males in musth. During full musth bulls may lose more than 300 l of fluid per day in this manner.

Vagrant: see Rogue.

Weaning: cessation of nursing; usually starting at 1–2 years and usually completed by 4–5 years in elephants; occasionally continues until about 8 years.

Wild elephant (definitions from the National norms and standards for the management of elephants in South Africa, DEAT 2008): an elephant that –

a. is not a captive elephant or is in temporary captivity, pending release into a limited or an extensive wildlife system; or

b. is in a limited or an extensive wildlife system.

REFERENCES

Cooney, R. 2004. *The Precautionary Principle in biodiversity conservation and natural resource management: An issues paper for policy-makers, researchers and practitioners.* IUCN, Gland, Switzerland and Cambridge, UK.

DEAT 2008. National norms and standards for the management of elephants in South Africa. Department of Environment and Tourism. *Government Gazette*, 8 May 2008.

Kahl, M.P. & C. Santiapillai. Elephant voices glossary. Accessed at: http://www.elephantvoices.org.

Ludwig, J., Tongway, D., Freudenberger, D., Noble, J. & Hodgkinson, K., 1997. Landscape Ecology: Function and Management. CSIRO Publishing, Collingwood, Victoria, Australia.

Millennium Ecosystem Assessment (MA) 2005. *Ecosystems and human well-being: multiscale assessments: Findings of the Sub-Global Assessments Working Group.* Island Press, Washington.

Noss, R. 1990. Indicators for monitoring biodiversity: a hierarchical approach. *Conservation Biology* 4(4), 355–364.

Owen-Smith, R.N. 1992. *Megaherbivores: the influence of very large body size on ecology.* Cambridge University Press, Cambridge.

Rogers, K.H. 2005. Biodiversity custodianship in SANParks: a protected area management planning framework. Report to South African National Parks, Pretoria.

Wu, J. & O.L. Loucks 1995. From balance of nature to hierarchical patch dynamics: a paradigm shift in ecology. *The Quarterly Review of Biology* 70, 4, 439–466.

INDEX

This index lists terms and subjects mentioned in the text; figures and tables are not indexed. Animals are listed under their common and scientific names, while for plants only scientific names are used. Abbreviations used: CBNRM – community-based natural resource management; KNP – Kruger National Park; TFCA – Transfrontier Conservation Area. (Compiled by Marthina Mössmer)

Printed and bound by CPI Group (UK) Ltd, Croydon, CR0 4YY

16/04/2025

14658448-0002